MATERIALS SCIENCE AND TECHNOLOGIES

POLYMER FILMS

PROPERTIES, PERFORMANCE AND APPLICATIONS

MATERIALS SCIENCE AND TECHNOLOGIES

Additional books in this series can be found on Nova's website under the Series tab.

Additional E-books in this series can be found on Nova's website under the E-books tab.

MATERIALS SCIENCE AND TECHNOLOGIES

POLYMER FILMS

PROPERTIES, PERFORMANCE AND APPLICATIONS

STEFAN A. ROMANO
AND
GEORGE P. SOMNERS
EDITORS

Nova Science Publishers, Inc.
New York

For permission to use material from this book please contact us:
Telephone 631-231-7269; Fax 631-231-8175
Web Site: http://www.novapublishers.com

Library of Congress Cataloging-in-Publication Data

Polymer films : properties, performance, and applications / editors, Stefan
A. Romano and George P. Somners.
p. cm.
Includes bibliographical references and index.
ISBN 978-1-61209-201-0 (hardcover)
1. Plastic films. 2. Polymers. 3. Polymerization. 4. Coatings. I.
Romano, Stefan A. II. Somners, George P.
TA418.9.T45P645 2010
668.4'95--dc22
2010047113

Published by Nova Science Publishers, Inc. † New York

CONTENTS

PREFACE

Polymer film coatings protect metals from the effects of aggressive media. They are used extensively as semi-permeable membranes, electrode materials in chemical batteries and in protective materials. This new book presents current research in the study of polymer films, including electropolymerization production of polymer films and coatings; polymer surface modification by electromagnetic radiation processing; modified perfluorinated polymer films and stability of self-assembled polyelectrolytes.

Chapter 1 - This chapter examines the possibilities of producing functional polymer films, coatings and semi-permeable membranes by the method of electrochemical initiation of polymerization (EIP) of monomers. It was demonstrated that the EIP method had a number of advantages in comparison with other generally-used approaches: high rate of polymer formation on the electroconducting substrate; single-stage character of the respective technology; the possibility of controlling the composition, structure and porosity of films by varying the monomer nature and concentration, solution pH, and electrolysis conditions; the simplified method of producing composite membranes and membrane units; and the process of automation. The peculiarities of the mechanism and kinetics of the electrochemical synthesis of polymer films was discussed. The dependence of the polymer composition on the monomer complexing ability was revealed. It was shown that acrylic acid and acrylamide formed the complex with the monomer ratio 3:1, which results in the synthesis of a copolymer product with the same monomer cycles ratio in the chain.

The physical-chemical and structural properties of films and their dependence on the EIP conditions were investigated, and the possibilities of controlling the synthesized polymers properties in order to obtain materials with specified properties were demonstrated. The method of *in situ* visualization of the electropolymerization process using video microscopy was suggested. The possibility of estimating the kinetics of film formation on the change of the growth rate of a polymer layer was established.

The surface morphology and microgeometry and the structure of electrosynthesized polymer films were investigated by means of AFM, SEM, TEM and optical video microscopy with differential interference contrast methods. It was established that polymers contained amorphous and crystalline phases. The conditions of formation of the porous structure of polymer semi-permeable membranes were examined. It was shown that, due to the monomer molecules orientation along the electric field force line, the channels inside the film were oriented perpendicularly to its surface. The membranes pore sizes were estimated on the microscopic data, films electrical resistance and test particles with known radii. It was established that the membrane was asymmetric and includes a thin barrier layer formed on the membrane side adjacent to the cathode and a loose matrix layer. The model of formation of the supramolecular structure of polymer membranes was suggested. It was shown that the film growth is accompanied by the synthesis of high- and low-molecular polymer fractions. The removal of soluble fractions from a coating results in the formation of the film porous structure. The antiseptic properties of membranes on the basis of acrylamide and formaldehyde were established, and the possibilities of these membranes' biological degradation were demonstrated.

The post-polymerization effect was discovered, the respective process mechanism and kinetics were determined, and the role of the zinc catalytic effect on the polymer film formation was demonstrated. The induction period of post-polymerization was established, and its quantitative estimation was performed.

Chapter 2 - Polymers have been applied successfully in many different fields such as membrane and thin-film technologies, biomaterials and biocompatibility, sensors, adhesion, protective coatings, microelectronic devices, composites, etc. Success or failure in a particular application depends many times on special surface properties of the material with regard to chemical composition, hydrophilicity, roughness, crystallinity, conductivity, lubricity, and cross-linking. Polymers very often do not possess the surface properties needed for these applications. However, surface modification techniques, which can transform these inexpensive materials into highly valuable finished products, have become an important part of the plastics and many other industries. In recent years, many advances have been made in developing surface treatments to alter the chemical and physical properties of polymer surfaces without affecting their excellent bulk properties. In particular, the past two decades have seen the rapid development of new strategies for the design of polymer functional surfaces. As one of the main techniques developed to achieve efficient surface modification of polymeric materials, electromagnetic radiation-induced surface modification has been widely applied as simple, useful and versatile approach to improve the surface properties of polymers. This chapter surveys the recent advances in electromagnetic radiation–induced surface modification of polymers, predominantly focusing on the research work carried out in the last decade and covering the modification and functionalization of polymer surfaces using: (1) Ultraviolet (UV); (2) Vacuum-UV (VUV); (3) Laser and (4) Synchrotron radiation sources. Each radiation source combined with traditional or new experimental approaches has the ability to tailor specific applications.

The academic interest in this area is driven for two reasons: the first one is the advance in the understanding and manipulation of nano and microstructures designs at surfaces together with the chemistry involved at the molecular level. The second is associated with the unique potential in use light to transform and pattern surfaces. The extensive base of knowledge of photochemistry accumulated in the last half century, in particular the advances in surface photochemistry, offers exciting potential for modification of surfaces with control of both the density and nature of surface chemical groups. Electromagnetic radiation permits a high selective control of the changes produced at the surfaces due to quantum allow transition rules. Simultaneously the use of ultrafast laser sources, where ablation takes place because of non-resonant process by multi-phonon absorption, can be used even in materials normally transparent to the laser wavelength. Additionally, electromagnetic radiation treatments offer solutions in terms of industrial applications, which need for materials to compete in the increasing low-cost consumer market. The materials for these applications should be cheap, easy to make, and, last but not least, be environment-friendly, combustible, or better yet biodegradable or photodegradable. These characteristics, which are not usually found in traditional wet and dry surface modification techniques of polymer surfaces, offer great potential for expansion of modification methods based in electromagnetic radiation

Chapter 3 - One of the most important applications of electrodialysis using ion exchange membranes is the recovery of strong acids from effluents of hydrometallurgy or electroplating industries. Rinse streams of these processes are generally concentrated equally in multivalent metal ions. All these different industrial applications of electrodialysis, in order to be feasible, require the use of some specific cation exchange membranes which are preferentially selective to monovalent cations. It was recognized that depositing a thin polyamine layer on the membrane surface which becomes positively charged in contact with a mixed solution of acid and metallic multivalent salts, thus creates an electrical repulsion barrier towards the highly charged and large hydrated ions is a well-known technique for enhancement of proton selectivity. This chapter describes the current state of the art in the authors' understanding of surface modification routes involving plasma process, electrochemical and chemical methods to obtain the proton permselective membranes. Authors also report their efficient and one-step chemical process through the diazonium chemistry which can be considered within the field and provide some directions for future endeavours. Evidently, proton selectivity of the modified membranes as well as the important parameters with respect to the electrodialytic efficiency such as electrical resistance, ion exchange capacity, and long-term stability will be discussed herein.

Chapter 4 - This short communication provides a review on Cytop™, a commercial perfluorinated polymer, as a barrier- and dielectric layer. It also describes ideas for future work on optimizing Cytop™ for aforementioned applications. Solution-deposited films of Cytop™ have shown great promise as dielectric films and/or encapsulation barrier layers for organic field-effect transistors (OFETs) and/or other organic electronic devices, e.g. organic light-emitting diodes (OLEDs) and organic photovoltaic (OPV) devices. OLEDs encapsulated with Cytop™ exhibited up to five times longer continuous operation under identical environmental and driving conditions compared with devices that were not encapsulated with Cytop™. A general method has also been developed to make thin and smooth films of Cytop™ encapsulation barrier layers against water and oxygen permeation. Based on a phenomenological analysis, it was found that disturbances in flow due to Rayleigh-Benard-Marangoni instability during drying of spin-cast perfluorinated polymer films caused high surface roughness and the formation of "pinholes". Casting films from solutions with higher polymer concentration and from solvents with higher viscosity suppressed the instability and significantly reduced the roughness. Suppression of the instability resulted in improved barrier properties as indicated by the calcium thin film optical transmission test. Spin-coated films of Cytop exhibit good gate insulating properties and also provide excellent OFET operational stability. However, these devices operate at large voltages since a dielectric thickness of ~ 450-600 nm is required for stable device operation. A cross-linked Cytop (C-Cytop) formulation deposited by spin-coating enabled the fabrication of OFETs with excellent performance using C-Cytop dielectric films as thin as 50 nm. The use of anti-plasticizing additives, possibly in combination with cross-linkers, may further improve the performance of dielectric films and/or encapsulation barrier layers for OFETs and/or OLEDs and OPVs.

Chapter 5 – The authors discussed the influence of post-treatment rinsing after the formation of selfassembled polyelectrolyte films made with the different polymers pairs: PAH/PSS, P4VP/PMA, PMA/QPVP. The stability of the film was studied using optical fixedangle laser reflectometry to measure the release of polymeric material, AFM experiments to reveal the change of morphology and thickness, ellipsometry, wetting characteristics, confocal laser scanning microscopy. They found that the polymer films were stable upon rinsing when the pH and the ionic strength were the sames in the solution as those used in the buildup. The films released most of the polymeric material when rinsed with pH values where the polymer charge decreases, but a layer remained that corresponded to a monolayer directly bound with the silica surface. Films containing at least four bilayers were stable upon rinsing at pH values where the interaction charge increases, but the stability of thinner films depended on the type of the last polymer deposited. They were stable in the case of polymer diffused on the film as an outermost deposit, but they released a large part of their material in the case of polymer adsorbed on the substrate and on opposite polymer groups pointing out of the macromolecular complexes. The stability results were determined using a simple model of the step-by-step assembly of the polymer film described formerly.

Chapter 6 - This work aims to investigate the initial build-up stages of polyelectrolyte multilayers formed by the successive deposition of the weakly charged Poly (Allyl amine, HCl), PAH and the strong acid Poly (Styrene Sulfonate), PSS, on silica at pH=9. Complementary tools were used: laser reflectometry to determine the step-by-step deposited weight of each polymer, Atomic Force Microscopy (AFM) to characterize the topography and measure the thickness of the film and adsorption on silica colloids to measure successive adsorption isotherms, the effect of maintaining a constant pH, the charge balance and the counterions co-adsorption in the successive steps. The experiments show that the deposited weight of PAH is the same at each step (around 0.5 $mg.m^{-2}$), whereas that of PSS increases more or less linearly along with the step number. Starting at the second polymer deposit (PSS), the structure of the film is very heterogeneous, with a thin polymer layer and bumps that are attributed to the presence of PSS/PAH hydrophobic complexes. The amount of bumps and their height were determined at each of the ten first polymer deposits. The film thickness between the bumps was evaluated by scratching small areas with an AFM tip. The authors' results support a former description of film growth, based on the electrostatic interactions between the two polymers and the substrate. They also reveal different the roles of the two polymers in the growth of the films: PAH is responsible for the progressive bi-dimensional coverage of the surface area and PSS for the increase in the thickness of the film.

Chapter 7 - This chapter addresses the growth mechanism of the self assembled film of higher thickness. Different weak polymer pairs are investigated as PAH/PSS, PLL/HA, PAH/(PGA_x-PSS_{1-x}), using the fixed-angle laser reflectometry, and the Quartz Crystal Microbalance to follow in situ the deposition of self-assembled polyelectrolyte films in a large domain of thickness. It is observed that a mixture of both components leads to a hybrid growth law. At the beginning of the construction, the thickness of the film increases exponentially with the number of deposited bilayers. Once a given thickness is reached (from 150 to 300 nm depending on the polymer pair), one observes the crossover to a linear growth regime. The experiments lead to refinement of the buildup mechanism model, introduced previously for the exponentially growing films. The mechanism no longer needs all the "in" and "out" diffusing polyanions or polycations to be involved in the buildup process to explain the linear growth regime but merely relies on the interaction between the polyelectrolytes with an upper zone of the film. This zone is constituted of polyanion/polycation complexes which are"loosely bound" and rich in the polyelectrolyte deposited during the former deposition step. The morphology and the topography of the higher thickness film were analyzed by Atomic Force Microscopy (AFM) and Focused Ion Beam (FIB) technique. The film thickness was determined from the FIB, the ellipsometric data and the reflectometric output interpreted in terms of refractive index and thickness way above the usual range which covers the earlier deposition steps of the polymers (i.e., 30-50 layers, depending on the film composition).

Chapter 8 - Conducting polymers (CPs) are organic polymers that exhibit electronic, magnetic and optical properties of metals and semiconductors while retaining the attractive mechanical properties and processing advantages of polymers. These features, along with chemical sensitivity, room temperature operation, and tunable charge transport properties, have launched conducting polymers as a major class of chemical transducers, creating powerful thin/thick film sensors for over two decades. Additionally, their synthetic routes are compatible with several functionalization schemes, including covalent bonding to the monomer, entrapment during synthesis, and surface chemistry, which can impart a high degree of selectivity. CPs can be used as the selective layer in sensors (sensor receptor) or as the transducer itself. Physical transducers varying from optical, mass sensitive and electrochemical are all discussed in this chapter. In comparison with most of the commercially available sensors, based usually on metal oxides and operated at high temperatures, the sensors made of CPs have many improved characteristics. They have high sensitivities and short response time and more interestingly they work at room temperature. As a result, more and more attentions have been paid to the sensors fabricated from CPs.

This chapter covers main applications of conducting polymers in chemical sensors and biosensors. The properties, synthesis and processing of CPs into films are reviewed and discussed in the first part. Following this, the concept of chemical sensors and biosensors along with a commercial perspective are presented. There is always a desire to further optimize CPs when targeting a specific application by increasing their affinity to particular analytes. In this respect, the modification strategies of CPs with synthetic or natural receptors to meet specific application are introduced. Follows, the applications of CPs to sensors classified by transducing mode (optical, mass, and electrochemical) and application mode (pH, humidity, gas/odor and ions selective). Additionally, CPs have attracted much interest in the development of biosensors. The developing role of CPs as health care biosensors, immunosensors, DNA biosensors, biosensors for environmental monitoring of hazardous chemicals and biosensors for agriculture and food industry are extensively reviewed. The authors will basically consider the emerging trends in CPs based sensors and biosensors during the last 5 years. Finally, a section is presented on challenges and future directions.

In: Polymer Films: Properties, Performance and Applications ISBN 978-1-61209-201-0
Editors: Stefan A. Romano and George P. Somners

Chapter 1

Electropolymerization as the Method of Producing Functional Polymer Films and Coatings

L. G. Kolzunova
Institute of Chemistry, Far Eastern Branch of Russian Academy of Science, Vladivostok, Russia

Abstract

This chapter examines the possibilities of producing functional polymer films, coatings and semi-permeable membranes by the method of electrochemical initiation of polymerization (EIP) of monomers. It was demonstrated that the EIP method had a number of advantages in comparison with other generally-used approaches: high rate of polymer formation on the electroconducting substrate; single-stage character of the respective technology; the possibility of controlling the composition, structure and porosity of films by varying the monomer nature and concentration, solution pH, and electrolysis conditions; the simplified method of producing composite membranes and membrane units; and the process of automation. The peculiarities of the mechanism and kinetics of the electrochemical synthesis of polymer films was discussed. The dependence of the polymer composition on the monomer complexing ability was revealed. It was shown that acrylic acid and acrylamide formed the complex with the monomer ratio 3:1, which results in the synthesis of a copolymer product with the same monomer cycles ratio in the chain.

The physical-chemical and structural properties of films and their dependence on the EIP conditions were investigated, and the possibilities of controlling the synthesized polymers properties in order to obtain materials with specified properties were demonstrated. The method of *in situ* visualization of the electropolymerization process using video microscopy was suggested. The possibility of estimating the kinetics of film formation on the change of the growth rate of a polymer layer was established.

The surface morphology and microgeometry and the structure of electrosynthesized polymer films were investigated by means of AFM, SEM, TEM and optical video microscopy with differential interference contrast methods. It was established that polymers contained amorphous and crystalline phases. The conditions of formation of the porous structure of polymer semi-permeable membranes were examined. It was shown that, due to the monomer molecules orientation along the electric field force line, the channels inside the film were oriented perpendicularly to its surface. The membranes pore sizes were estimated on the microscopic data, films electrical resistance and test particles with known radii. It was established that the membrane was asymmetric and includes a thin barrier layer formed on the membrane side adjacent to the cathode and a loose matrix layer. The model of formation of the supramolecular structure of polymer membranes was suggested. It was shown that the film growth is accompanied by the synthesis of high- and low-molecular polymer fractions. The removal of soluble fractions from a coating results in the formation of the film porous structure. The antiseptic properties of membranes on the basis of acrylamide and formaldehyde were established, and the possibilities of these membranes' biological degradation were demonstrated.

The post-polymerization effect was discovered, the respective process mechanism and kinetics were determined, and the role of the zinc catalytic effect on the polymer film formation was demonstrated. The induction period of post-polymerization was established, and its quantitative estimation was performed.

Keywords: Electropolymerization; Acrylic monomers; Mechanism and kinetics of film formation; Polymer film structure; Porosity, pore shape and radius; Semi-permeable membranes.

1. INTRODUCTION

The modern society demand of polymer films of various purposes is huge and increases continuously. The polymer coatings protect metals from the effects of aggressive media; impart necessary electric insulating, antiseptic, anti-friction, decorative and other specific properties to different types of articles. The polymer films are extensively used as semi-permeable membranes, electrode materials in chemical batteries, protective materials etc. A wide range of practical problems, which can be solved using the functional films and coatings, requires versatile methods for their production. The majority of the existing technologies used for the above purposes are complicated and have multiple stages. They include the initial synthesis of a polymer binder, its conditioning as applied to the method of deposition on a substrate and/or processing into a film, and the actual formation of the polymer layer. However, in some technologies, it is possible to combine all the mentioned processes in a single cycle. The technologies of the latter type include: a) high-energy irradiation by a flow of fast electrons of the substances capable to poly-addition on the articles-substrates; b) application of ultraviolet irradiation; c) application of electric current in gaseous and condensed phases. In these cases, it becomes possible to obtain polymers of unconventional composition, pre-determined structure, quality and thickness, which enables one to substantially extend the range of polymers themselves as the fields of their application and improve economic, technological and ecological parameters of the manufacturing process.

Among the state-of-the-art technologies of the above type, one should mention electrochemical methods of polymer films and coatings formation: electrochemical initiation of monomer polymerization and electrochemical polycondensation. These methods have a number of advantages as compared to chemical ones. The main advantage is the fact that the polymerization and formation of a polymer film (coating on metal surface) occurs within one stage. In this case, the product comprises the film localized on the surface of the conducting substrate. Thus, the possibility is created to exclude from the technological setup the stages of preliminary synthesis of the polymer substance, its further dissolution or dispersion and final fixation, by combining these stages into a single process. Another important advantage consists in good dissipative properties of electropolymerization, which enables one to deposit uniform polymer layers on complex-shape articles with high rate. The third positive factor is the possibility of the process of automation. The fourth factor consists in rather strict stoichiometry of the process, which allows obtaining a product of high purity. Finally, the formation kinetics and polymer film properties can be controlled in the process of synthesis.

Both electropolymerization topics are under extensive development [1-46]. In the studies related to electrochemical synthesis of conducting polymers, the primary attention is paid to advanced problems of material science – production of novel materials for electronics capable to combine unique mechanical properties with electrical and optical characteristics of semiconductors and metals. The obtained results enable us to predict the application of such polymers in producing working electrodes in conventional chemical energy sources to replace deficit non-ferrous metals by more widely spread and less expensive materials with stable parameters. The prospective fields of application of novel polymer materials include electrochromic, electrosensor and biosensor devices, integral circuits and electrochemical displays of electro-optical installations, manufacture of selective electrodes and gauges [47, 48]. The appearance of works in this field enables one to reevaluate prospects of the monomer electropolymerization method and is an evidence of the fact that, up to present, the method possibilities have not yet been fully revealed.

The method of electrochemical initiation of polymerization (EIP) of monomers is used for the synthesis of electrochemically inactive polymers. The method enables one to control the reaction rate through changing the strength of the current passing through the system, while adjusting the electrolysis potential makes it possible to control the polymerization mechanism and, as a result, obtain whether homopolymers or copolymers with pre-determined ratio of monomer units in the polymer chain. Thus, modification of the electrolysis conditions results in variation of the polymer composition and molecular weight and, therefore, the possibility of synthesis of coatings with new useful properties, which are impossible to obtain by other methods. The EIP special features – strict control of the stages of polymerization excitation and stopping and macromolecules orientation in the electrode direction – opens prospects to synthesis of "live" and crystalline polymers, high-molecular compounds with narrow molecular weight distribution, and coatings with unique physical-chemical characteristics.

In the EIP process, the excitation of monomer/polymer transformations is realized by active particles – the electrolysis intermediate products – and occurs under mild conditions: current densities or potentials not exceeding the redox-potentials of initial monomers, room temperature and atmospheric pressure. Such conditions not only simplify the technological process, but also eliminates the chances of degradation of electrochemically and/or thermally unstable substances. Coatings and films can be obtained using the EIP method whether only on cathodes or on both electrodes at reversing the current.

One should mention that polymer layers (simultaneously with electropolymerization or on consecutive stages) can be electrochemically and/or chemically modified through the formation of surface complexes or immobilization of organic substances or metal powders into the polymer matrix, which enables one to obtain coatings with various functional properties and composite materials with the polymer matrix-immobilized compounds. The method also allows producing film membranes for baromembrane filtration and electrodialysis and ion-exchange materials. The above mentioned advantages of the method resulted in the fact that the processes of electrochemical initiation of polymerization were under intensive development in Russia during 1970-1990s. The basics of the mechanism and kinetics of films formation using the EIP method were published in [49-61]. However, since 1993, the results of research works have not been systematized. At the same time, there appeared the papers developing the theoretical notions on the regularities of electrochemical synthesis of polymer layers electroconducting substrates, EIP kinetics, and the films structural features. The latter direction of studies is of special interest, since the properties of polymer layers are determined by their structure on both molecular and supramolecular levels. The structural parameters, such as volume orientation of polymer chains, the crystalline/ amorphous phase ratio in the polymer, the degree of the polymer layer asymmetry, determine the transport and kinetics of redox-transformations of depolarizers as inside the polymer film and at the metal/polymer layer/solution interface. The above parameters are responsible for the lyophilicity and permeability of polymer layers, control protective and other special properties and, finally, establish the range of possible applications of EIP-synthesized films and coatings. As a result of the above studies, the fields of efficient practical use of the method were extended, in particular, on the account of the possibility of the synthesis of membranes for baromembrane processes, composite materials and different types of sensors. Since at the present time the method is under intensive development, a review of the recent publications is of urgent importance. The author of the present chapter made a substantial contribution to the EIP method development. So this chapter comprises the results of the authors' studies on electrochemical formation of non-electroconducting polymer films on the surfaces of metals and alloys by means of cathode electropolymerization of monomers published in Russian scientific literature which is not available internationally. In addition, the chapter contains novel data so the chapter can be interesting for researchers.

The analysis was focused on:

- development of monomer compositions for electropolymerization;
- development of novel approaches to the advanced study of the mechanism and kinetics of polymer layers formation on electroconducting surfaces;
- studies of the structure and morphology of electropolymerized polymer films;
- search of new fields of practical use of polymer films obtained by the EIP method.

2. Fundamentals of Polymer Films Formation on Electroconducting Surfaces Using the Monomer EIP Method

Since theoretical fundamentals of EIP are comprehensively summarized in [49, 51, 55, 59, 60, 61], the present chapter will list the most important notions of the process mechanism and kinetics. Polymerization is considered to be electrochemically initiated, if the monomer/polymer transformations are induced by active particles formed at current transmission through the monomer-containing system. Unlike galvanic processes, only the initial electrode reaction of active particles (intermediates) formation follows the Faraday law. The electron transfer from electrode to depolarizer (electroreduction) or from an electroactive substance to electrode (electrooxidation) occurs at the electrode/solution interface. The chain initiation occurs in the close vicinity of the electrode surface due to high reactivity of intermediates.

The electron transfer between an electrode and a molecule can be realized directly (direct initiation) and through substances – electron carriers (indirect initiation). The *direct initiation* is possible, if the energy of the lowest vacant orbital of a monomer-electron acceptor is the lowest (cathode process) or the highest (anode process) among all the composition components. The *indirect initiation* is realized, if the electron carriers (ions, ion-radicals, neutral particles or complexes) have lower electroreduction (electrooxidation) potential than other solution components. The electron affinity and components ionization potential of the composition under study are considered as the criteria of direct/indirect division of the EIP processes [49, 51, 55].

The direct initiation is characteristic for acrylamide (AA), acrylonitrile (AN), styrene (ST), methylmethacrylate (MMA), 4-vinylpyridine, other vinyl monomers and their derivatives, phenylisocyanate in aprotonic solvents providing high electrolyte dissociation degree (DMFA, DMSO, N,N'-dimethylacetamide, acetonitrile, pyridine, liquid ammonia) [49, 51]. The solvent nature could determine the polymerization kinetics. For example, the EIP of trioxane in 1,2-dichloroethane starts immediately after a short impulse of the current with a linear dependence on the monomer concentration, while in dichloromethane, the polymerization start is preceded by a long induction period.

The difference in nature particles formed on cathode and anode can induce ionic, radical or ion-radical polymerization [103]. When a fast combination of ion-radicals is hindered, for example, due to steric factors, while the monomer concentration around the electrode surface is high, it is possible that the polymer chain grows simultaneously from one end in accordance with the ionic mechanism and from another end – on the free radical mechanism. Such a combined mechanism is realized at copolymerization of AN and ST with MMA in DMFA [49].

The cathode polymerization mechanism depends on the solvent nature, medium pH, electrode material, and the working potential. In a protogenic solvent (for example, in water), reduction of the monomer protonated molecule produces free radicals $MH^{\bullet}$ capable whether to initiate the radical polymerization or to dimerize, the reaction rate is proportional to the current density [49, 51]. These processes will be examined below in more detail.

The monomer electrooxidation results in the formation of free radicals, cation-radicals, actions capable to induce polymerization in accordance with cationic, radical or mixed mechanism. At direct initiation, there occurs the electron transfer from the monomer molecule to anode with formation of the cation-radical $M^{\bullet+}$ which can be oxidized to dication. As a rule, the monomers with electrodonor substituents (alkoxy-groups, phenyl, vinyl and 1,1-dialkyl) are used in the cation polymerization [61]. The compounds like ST, benzene, p-dialkylbenzene, phenylsalicylate, 4-aminobyphenyl form insoluble polymer layers on the electrode at direct anode initiation [49, 62, 63]. The necessary condition of anode film formation consists of the application of insoluble electrodes and polar solutions (H_2SO_4, HF, acetonitrile, liquid sulfur dioxide, hexafluoroantimonic acid) [59].

The mechanism of the anode polymerization is determined by the nature of the background electrolyte. For example, the EIP of benzene in the presence of Bu_4NBF_4 proceeds through the formation of the cation-radical due to direct benzene oxidation in accordance with a single-electron mechanism, followed by its binding to benzene and proton removal. The complete transfer of polymerization to the surface and film formation on metal requires the benzene oxidation at a potential above 2 V (Ag/AgCl electrode) and temperature -75°C. Alternatively, using Bu_4NClO_4 as an electrolyte the benzene oxidation potential is lower and comparable with that of perchlorate ion. The latter could be related to the oxidation benzene by perchlorate radicals with the formation of a benzene phenyl radical which undergoes oxidation easier than benzene.

Direct cathode (anode) EIP of monomers does not provide the required polymerization rate and is accompanied by elevated power consumption. The impossibility of producing polymers of high molecular weight using the ionic polymerization results in low quality of coatings, which can not be markedly improved even at the application of copolymerization [20, 49, 60]. On the other hand, *indirect initiation* enables one to significantly reduce the potential of active particles formation [22-24, 59-61, 65, 66] and energy consumption, by eliminating the monomer and polymer destruction, solvent and electroconducting salt decomposition. Moreover, such mild conditions prevent gas emission on electrodes, resulting in convective shift of polymerization into solution, blistering and perforation of growing films, their peeling from metal, which is very important to obtain quality coatings.

The sources of active particles at indirect cathode initiation include the H^+, Me^+ ions, ammonium and phosphonium salts, carbonyl compounds, peroxides (H_2O_2, organic hydroperoxides, perborates, peroxosulfates of alkali metals and ammonium) [49, 59, 67]. In the acidic or neutral medium of non-degasified solutions, initiation by the products of electroreduction of monomer peroxides is possible or, alternatively, oxygen itself is reduced to H_2O_2 and, further, to $OH^\bullet$ radicals [65]. The reduction rate of the bond –O–O– increases in the presence of ions of variable valency metals (Fe, Cu, Al, In, V, Mo etc.) [49, 59, 68, 69]. Such redox systems metal ion/peroxide can be used for the initiation of polymerization of AA, AN, acrylic acid (AAc), vinylacetate (VA), methacrylate, other vinyl compounds [59, 68, 70].

A solvent capable to form in the reduction process the products reacting with monomer (for example, 4-vinylpyridine-pyridine-$NaBPh_4$) can serve as an electrochemical initiator of the anionic polymerization. In the aprotonic medium, the polymerization initiators may include active particles formed at cathode reduction of tetraalkylammonium salts or *p*-nitroaniline. The anion-radicals and dianions of dianthracene and dimethylfulvalene induce polymerization of ST and 1-triethylsilylbutadiene-1,3, those of benzophenone and 2,2′-bipyridyl – anionic polymerization of AN. It is possible to indirectly initiate electropolymerization of ST, AN, MMA by active particles, which are the products of different degrees of electroreduciton of 9,10-anthrachinon, 9-fluorenon and their derivatives [51, 60]. In the latter cases, the monomer EIP is significantly affected by the nature of the background electrolyte cation, which results from changes of the reactivity of active centers due to their participation in association processes with the formation of ion $R^{\bullet -}Me^+$ (Na^+, K^+) or triplets (Li^+) [51], whose reactivity is lower that for free anion-radicals. The macromolecule growth rate and polymer yield decrease with an increase of the counter-ion radius ($Et_4N^+ > K^+ > Na^+ > Li^+$), i.e. with an increase of the dissociation constant of cations with anion-radicals [60]. The polymerization initiation by solvated electrons, products of background electrolyte cations electroreduciton, alkali metals and zinc salts is described in [1, 5-7, 53]. Hydrogen emitting on an electrode comprises the simplest cathode initiator for monomers in aqueous media [49].

The monomer ability to form stable coordination complexes with halogenides of Zn, Al and other metals was used in EIP of AA, AN, N-vinylcarbazole, 2-vinylnaphtalene, butylacrylate, methylenebisacrylamide (MBAA) [51, 22-24, 65]. As was demonstrated in [61], the specific effect of halogenide ions was related to their coordination by the monomer amide group. Redistribution of electron density due to coordination results in strengthening the polarization of the bond -C=C-, thus promoting its break and active particles addition. In this group, zinc chloride is the most efficient polymerization initiator. The hydrolyzed zinc ion forms a complex with the monomer, the carbonyl group electron density shifts to the zinc cation, which makes the bond electron-deficit and facilitates the cathode reaction [65]:

$$ZnOH^{+}M + \bar{e} \rightarrow ZnOH^{+}M^{\bullet-}; ZnOH^{+}M^{\bullet-} + H^{+} \rightarrow ZnOH^{+}M^{\bullet}; ZnOH^{+}M^{\bullet} + \\ + n\,M \rightarrow \text{polymer}$$

The use of Li, Al and Sn chloride or $ZnSO_4$, which are also capable to form complexes with monomers, is less efficient than $ZnCl_2$ [61].

In the indirect anode EIP. the following substances oxidation is used: carbonyl acids, background electrolyte anions (SO_4^{2-}, $S_2O_8^{2-}$, ClO_4^-, NO_3^-), solvent [49, 51, 59, 71, 62, 63], aromatic hydrocarbons (9,10-diphenylanthracene, 9-phenylanthracene, 9,10-dimethylanthracene, triphenylene, 1,3,6,8-tetraphenylpyrene and perylene) [49, 51, 60, 72, 73, 74].

The initiation efficiency increases in the presence of the variable valency metals. The mechanism of cationic electropolymerization of vinyl esters is determined by the nature of background electrolyte.

For example, in the presence of $(C_4H_9)_4NClO_4$, $NaClO_4$ and $NaNO_3$ initiation is induced by the cation-radicals formed at monomer oxidation, while at using $(C_6H_5)_4BNa$, the oxidation product $(C_6H_5)_4B^-$ serves as an initiator. Tetrabutylammonium iodide is oxidized with anode emissions of iodine which initiates polymerization of isobutylvinyl ester in 1,2-dichloroethane on [49, 51].

The cationic polymerization can also be initiated by the particles formed at electrooxidation of monomer donor-acceptor complexes (2-vinylnaphtalene, 9-vinylanthracene, N-vinylphalimide) with $ZnCl_2$ [49, 51, 55, 59-61]. The electropolymerization on the cation mechanism is not limited by monomers with double bonds. Heterocyclic compounds (lactones, lactams, cyclic esters, sulfides, imines, epoxy-compounds, trioxane 1,3-dioxolane), during the electrolysis in non-aqueous solvents, also form polymers through cycle opening. The rate of the ionic polymerization indirect initiation is determined by the electrode potential monomer concentration and electrochemical activator.

3. Peculiarities of EIP Mechanism and Kinetics

The EIP kinetics is determined, first of all, by the current density and electrode potential, since the formation of active particles occurs as a result of electrode reactions. Further, the active particles can interact with monomer (polymerization initiation), undergo recombination or disproportionation or reduction (oxidation). The theoretical analysis of the kinetics of the active particles' electrochemical generation is presented in [61]. However, in most cases, the kinetic equations are formal, derived with a large number of simplifications and do not reflect the real EIP picture.

The important role in the EIP kinetics and mechanism is the monomer nature and concentration. According to the kinetic equations, the polymerization rate must have the first order in respect to the monomer concentration [49, 51, 55, 59-61]. Sometimes the reaction order can be higher if a monomer participates in the reactions of polymerization initiation, transfer or chain stopping [59]. The EIP kinetics and mechanism become more complicated at conducting copolymerization. Depending on the monomer nature, homopolymers as copolymers enriched by some specific monomer can be formed [49, 51, 60, 64]. Such a course of the process is concerned, in particular, with different adsorption capacities of monomers, which are determined by both the monomer molecule structure and the electrode material [59]. The temperature effect on the EIP rate is also ambiguous [49, 51, 55, 59-61].

A substantial effect on the EIP rate and mechanism is provided by the solvent, which, as early as on the initiation stage, determines the nature of active centers [49, 59, 60, 77-81]. In the protogenic medium, only the radical polymerization is possible, while for the aprotonic medium the ionic one should also be taken into account. The solvent restricts the potential range, in which the electrochemical processes preventing polymerization do not occur. The solvent electron affinity and ionization potential determine the degree of its sorption on the electrode, participation in the chain transfer and stopping reactions, as well as its own ability to polymerize (tetrahydrofuran, THF) or copolymerize with a monomer (caprolactam – THF). The solvent can also act as the polymerization inhibitor and initiator [49, 51].

The aqueous monomer compositions, whose application enables one to avoid some ecological problems [81], constitute a special interest for EIP study. Besides, water, as a polar liquid, provides the electrolyte ionization and high solution electroconductivity, which allows reduction of the bath voltage. On the other hand, since water has higher polarity than organic solvents, as well as higher surface tension, it facilitates the adsorption of less polar-organic molecules on metal. The latter factor is a necessary condition of deposition of a high-adhesion polymer layer on an electrode. The disadvantages of this solvent application include a limited set of soluble monomers and narrow range of working potentials, beyond which there occurs water decomposition with the emission of hydrogen or oxygen. The gas emission facilitates the transfer of active particles into the bulk solution, results in the coating damage (perforation, blistering, and peeling) and prevents obtaining a monolith film. Another disadvantage consists of the fact that water-soluble monomers form polymers swelling or solving in water, so that the obtained films cannot be applied as protective coatings. However, the swelling films are interesting as optical, electron- and ion-exchange materials, membranes for baromembrane processes and pervaporation, and temporary coatings [82]. Such films can be obtained in aqueous solutions of AA, AAc, VA, MMA, *p*-styrene-sodium sulfonate, methacrylic and itaconic acid, maleic anhydride and other monomers [22-25, 44].

The higher the solvation of the electrolyte ion, the lower is the effect of electroconducting additives on the polymer chain growth and stopping. Since the solvation decreases with the increase of the ion radius, this latter parameter characterizes the degree of the salt effect on EIP [51, 60, 83, 84].

The electrode material effect on the EIP rate and mechanism is determined by the electrode processes over-potential and the adsorption of monomer and reaction products on the electrode. For the cathodes with higher over-potential and lower hydrogen adsorbability [Sn, Pb, Zn, Cd, Ag, Cu, Ni, Fe], the polymerization rate is higher, since in this case, a lower part of current is spent on the side reaction of hydrogen emission. The electrode material with higher polymer adsorbability is more suitable for EIP, since here the polymer molecular weight and adhesion are higher. The metal-coating oxides inhibit the active particles generation and polymer chain growth. Usually, the following materials are used as cathodes: graphite, Pt, Al, Cu, Fe, Hg, Zn, Cd, Ni, Ti, other metals and alloys and metal ceramics [49, 85]. Special attention should be paid to the anode material selection due to its electrocorrosion at the initiation potential [49, 86, 87] and high sensitivity of the polymerization reaction to traces of admixtures.

4. General Regularities of Film Formation on Electrode Surface

The peculiarity of the monomer EIP consists in the heterogeneity of the elementary reaction of the polymer chain initiation. That is why the process of the polymer coating formation depends on multiple factors related to the solved substances composition and properties, electrode surface conditions (charge, microgeometry, purity etc.). The contribution of each of the factors is individual for different monomer compositions.

Orientation effect of the electric field on the positions of monomer molecules in respect to the electrode and, as a result, the polymer structure, was established at electropolymerization of thiophene, aniline, pyridine [60, 88], phenylisocyanate [59], benzoic [89] and phthalic [90] acids. This effect must take place only at the polymer deposition thicknesses of a few polymer layers, since for thicker films (>10 nm), the deposition upper layers are formed at the conditions when the force field of the electrode surface is shielded by the polymer [59, 60].

Monomer adsorption is of crucial importance for electropolymerization localized on the electrode surface [59, 79, 91, 92]. The monomer's strong chemosorption provides high adhesion of the polymer coating to the substrate [91]. High occupation density of the electrode surface by monomer prevents recombination and creates favorable conditions of the formation of dense polymer networks, which serve as a base to obtain coatings of good physical-mechanical and corrosion-resistance properties.

The organic molecules containing electro-acceptor substituents – CN, CNS, CNO, C=O, CHO, NH2 – are characterized with high adsorption activity on metals with partially occupied orbitals of the *d*-type. The adsorption of unsaturated compounds of the vinyl type is subject to general regularities of simple aliphatic compounds with a small number of carbon atoms in the chain. The abnormally high adsorbability is observed for the compounds with multiple

C=C and C≡C bonds (θ = 0.4-0.8) [4]. A polymolecular character of adsorption was registered for a wide class of monomers in organic solvents [26].

The elementary reaction of *active centers formation* occurs in the monomer's adsorbed layer. Furthermore, the active centers can undergo dimerization, hydrodimerization, and deactivation or interact with the monomer, thus giving rise to initiation and growth of polymer chains. In the latter case, two variants are possible [59]:

- the chain growth occurs in the monomer condensed two-dimensional layer in parallel to the electrode plane (adsorption-monolayer mechanism);
- the active center desorbs from the electrode surface resulting in the polymerization transfer to the solution.

Film formation models. Under the conditions of the macromolecules high solubility, the polymer concentration in solution increases, and there emerges the concentration gradient directed to the electrode plane. Since polymers have higher surface activities than other solution components, macromolecules are transferred from the bulk phase to the surface. In the case of the polymer's low solubility, the disperse phase, which is thermodynamically unstable and coagulates in the near-electrode area, is formed.

The total EIP rate is determined by that of the slowest stage. The latter could be pre-adsorption, formation of active centers or initiation, growth or stopping of polymer chains. In the case of the adsorption-monolayer mechanism, at relatively high monomer concentration, the occupation of the electrode surface by monomer and establishment of the adsorption equilibrium occur rather fast. The opinions on the rates of other reactions vary. Some of the authors consider the limiting stage to be the electron transfer, others – the reaction of the polymer chain growth [49, 55, 59, 93].

In accordance with another model [49, 51, 55, 59, 60, 61, 94], the formation of the polymer layer on the electrode passes through the stages of monomer sorption on metal, initiation of polymerization sites, two-dimensional growth of randomly deposited germs of a new phase and their spreading over the electrode until merging in a solid coating. The mathematic description of such a model and the analysis of the surface processes kinetics are presented in [49, 51, 55, 59, 60, 61]. According to [95, 96], the real film growth occurs at high germs density, and the two-dimensional progressive germs formation proceeds simultaneously with their three-dimensional growth.

At the EIP initial stage, the coating formation is affected by the current distribution over the electrode [97]. Since the current density by the electrode edge is always higher than in the center, the former area is enriched by the oligomeric gel-fraction at the diffusion-controlled EIP process. That is why the film grows from the edges to the center, while the film itself is heterogeneous over the thickness. When the degree of the metal coating by the deposit approaches 1, one can observe the polymer layer thickness growth which is subject to diffusion limitations resulting from the transfer of reacting substances to the electrode through pores and cavities formed by fluctuations of polymer macromolecules [51, 60, 98].

The layer thickness on the electrode surface depends on the polymer solubility in the applied solvent. If the polymer solubility rate is higher or equal to the chain growth rate, the deposit on the electrode does not form, while the polymer is accumulated in solution. However, if the rate of the macromolecules diffusion into the solution is less than the polymer solution rate, it is possible to observe the formation of a thin viscous film on the electrode due to oversaturation of the near-electrode layer. Such a film does not produce diffusion hindering for the reacting substances transfer to the electrode surface. Finally, if the polymer solution rate is lower than the chain growth rate, a deposit is formed on the electrode and its thickness increases over time. Here, the substances diffusion to the substrate serves as the process limiting stage.

The film thickness depends on the strength of the coating binding with the electrode: the higher the adhesion, the higher the film yield in respect to current [49]. Since the application of monomers with low surface tension and solvents with high values of this parameter (water) facilitates the monomer adsorption on metal, the adhesion of coatings deposited from aqueous solutions of vinyl monomers is higher than for coating produced from organic media.

According to [51, 60], the radical polymerization is more efficient at producing coatings than the ionic one, since the formation of radicals and their interaction with monomer is concentrated in the double electrical layer. The reactivity of ion-radicals is lower, while the lifetime is higher than for radicals, then ion-radicals can transfer to the solution. The process develops more actively if an electrode and ion-radicals have the same sign of the charge. As a result, the ionic polymerization occurs predominantly in solution.

5. Monomer Compositions for EIP

The monomers capable to form polymer coatings on the *cathode* include styrene (ST) and its derivatives [51], isocyanates, methylmethacrylate (MMA) [7, 8, 9, 51, 60, 100] and its copolymers with ST or esters of acrylic (AA) and methacrylic (MAA) acids (MAK) [51, 101], methylvinylketone [101], onium salts, vinylpyridine complexes of ruthenium, osmium and iron [99], acrylonitrile (AN) and methacrylonitrile (MAN) [15, 16, 17, 26, 31, 33-35, 49, 51, 59, 102-107], acrylic (AAc) an terephthalic acids, phthalic and maleic anhydrides [26, 49, 101], silicon-containing vinyl monomers [61], acroleine []101 and caprolactam [49], diphenylamine and *m*-toluidine [27], oligoesteracrylayes (OEA) [55], vinylacetate (VA) [101], *p*-chloro-*α*-methylstyrene [108, 109], acetylene [18], p-xylilene-bis-(triphenylphosponium bromide) [71], viologen [14, 19, 20, 110], [(vinyloxyetoxy) methyl]oxyrane (vinylox, VO) [101], other vinyl monomers [55, 59, 6, 28-30].

In spite of a substantial number of monomers capable to form polymer layers on metals in the EIP process, the properties of the obtained coatings do not meet the practical requirements. The most interesting compositions, in view of the formation of quality polymer films on the cathode, are those on the basis of *acrylamide* (AA) and its derivatives. These compositions enabled one to obtain electro-insulating, decorative, protective, anti-friction, anesthetic, thrombus-resistant polymer coatings, membranes for baromembrane filtration and dialysis, selective electrodes, and biosensors on metals in aqueous and organic media [22-25, 44, 49, 55, 61, 64, 65, 68, 72, 86, 87, 111-117].

Direct reduction of AA with the formation of a polymer on metal surface is possible for Hg or amalgamated Ag electrodes. However, the process goes on with an insignificant current-related yield (~0.25%) [55]. The deposition rate and coating limiting mass increase at using the AA derivatives: N,N′-methylenebisacrylamide (N,N′-MBAA), N,N′-ethylenebisacrylamide, N-methylolacrylamide (MAA), and peroxide initiators (H_2O_2, H_2BO_4,

$S_2O_8^{2-}$, ROOH) in acidic polar solvents (pH<3; HCl, H_2SO_4, $ClCH_2COOH$) [61, 86, 87]. In these cases, the polymerization rate in water is higher than in alcohols, dioxane, DMFA. In the aqueous medium, poly-AA coatings are formed at a high rate on Ni, Cu, Pt cathode in the presence of Zn, Al, Sn chlorides [49, 61, 65, 68, 72, 111, 115, 116]. The addition of DMFA into an aqueous composition reduces the polymerization rate and poly-AA molecular weight [61]. In the course of the potentiostatic electrolysis, the coating mass quickly attains the limiting value due to electrode insulation [49]. Under galvanostatic conditions, the coatings ohmic resistance changes insignificantly with their high swelling ability and porosity [61]. In both cases, after the solid film formation, the EIP rate becomes completely dependent on the monomer diffusion to the electrode through the coating. Nitrates, perchlorates and sulfates inhibit the AA electropolymerization in aqueous media [61, 65, 99, 115].

Copolymerization of AA with N,N′-MBAA or polyvinyl alcohol allows increasing the mechanical strength, thickness and coating mass [61, 116, 117]. EIP of AA with AAc in aqueous solutions results in formation of glass-like films with the swelling ability ~40-200%. The addition of formaldehyde (F) into the (AA-AAc) composition increases the strength of the amide bond through substituting the amide group hydrogen atom by an organic radical, which enables one to reduce the swelling ability down to 10 % [101]. The coatings comprise glass-like, colorless, transparent films with high adherence. The film thickness can be controlled by the electrolysis time and current density. Due to the presence of COOH and NH_2 groups, the copolymer (AAc-AA-F) could manifest ion-exchange properties [64, 101].

6. Mechanism and Kinetics of Film Formation at Electrocopolymerization of Acrylic Acid, Acrylamide and Formaldehyde

Due to a prolonged induction period (15-40 min) and comparatively low rate of film formation, the composition (AA-AAc-F)-$(NH_4)_2S_2O_8$ comprises the most convenient model system for studying the mechanism and different stages of EIP kinetics. As was shown in [59], EIP of AAc and AA is preceded by the formation of complexes.

6.1. A Role of Complexes in Formation of the Polymer Film [64]

In the electrochemical initiation of polymerization (EIP) of monomers, the electrode reactions can be complicated by preceding and subsequent chemical transformations, whose contributions are determined by the reactivity of the original compounds and the intermediate products. This must be taken into account when a mixture of monomers undergoes EIP in which both copolymers and a mixture of homopolymers can be formed. In the majority of cases, the possibility of copolymerization of monomers is qualitatively estimated by the difference in $\varphi_{1/2}$ of the polarographic reduction waves of compounds tested under strictly identical conditions. For copolymerizing monomers, the value $\Delta\varphi_{1/2}$ usually does not exceed 0.2 V [54], however, in some cases, monomers with $\Delta\varphi_{1/2} > 0.2$ V are able to copolymerize. EIP of acrylic acid (AAc) with acrylamide (AA) in an aqueous solution of ammonium persulfate (APS) can serve as an example. This results in the formation of polymer coatings on cathodes [68]. The physicochemical properties of these coatings, which do not change for several years, differ from those of homopolymers of AAc and AA. This allows one to draw the conclusion that AAc and AA form a copolymer product during EIP, although the $\Delta\varphi_{1/2}$ value of the original monomers is ~0.3 V (Figure 1).

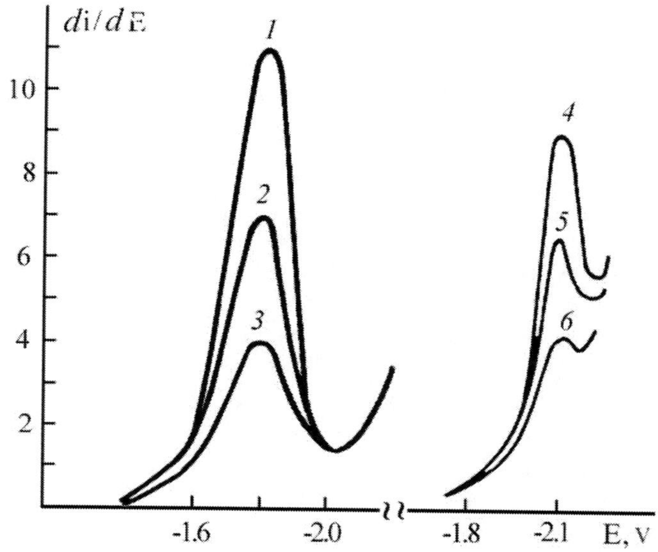

Figure 1. Differential oscillographic polarograms of acrylic acid (M): 8.3 10^{-4} (*1*); 4.16 10^{-4} (*2*); 2.08 10^{-4} (*3*) and acrylamide (*M*): 2.53 10^{-4} (*4*); 1.69 10^{-4} (*5*); 0.84 10^{-4} (*6*) (background is 0.1 *N* $LiClO_4$).

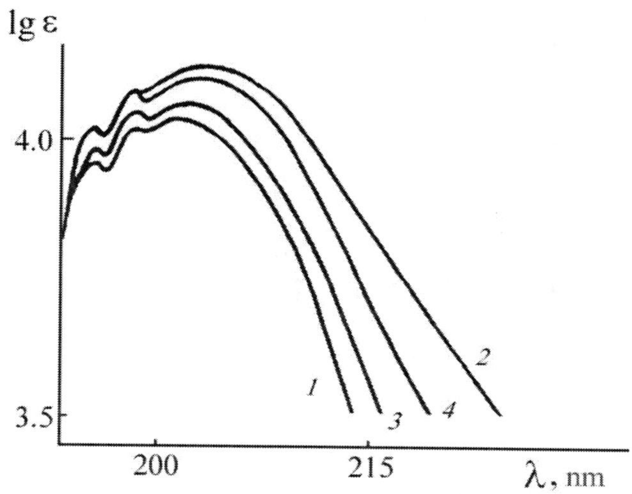

Figure 2. UV spectra of acrylic acid (*1*), acrylamide (*2*) and their mixtures in molar ratios 3:1 (*3*) and 1:1 (*4*) (concentration of monomers is 8.3 10^{-3} *M*, *l* = 0.01 cm).

A comparison of the UV spectra of a mixture of monomers with those of the individual compounds (Figure 2) reveals some deviation of the experimental values of optical density and the extinction coefficient from those calculated with the assumption that the components do not interact. This deviation, which reaches 9-11 % in the region of the maximum for the mixture AAc:AA=1:1, is possible if intermolecular interaction occurs between the compounds studied. This conclusion is supported by polarographic studies. For individual AAc and AA, the height of the peaks of their differential polarograms is proportional to the concentration of these compounds in the solution (Figure 1). However, this regularity is violated when both AAc and AA are present in the solution (Figure 3). For example, when AA is added to the solution of AAc, some decrease in the height of the differential peak of AAc is observed. This phenomenon cannot be explained only by the high surface activity of AA [54], because, first, increasing the AA concentration in the AAc and AA model system shifts the potential of the peak of the AAc wave a little to more negative values. Second, the addition of AAc, which does not possess enhanced sorption ability, to the AA solution causes a decrease in the height of the peak on the differential polarogram of AA (Figure 3) and completely suppresses it at the molar ratio of AAc and AA equal to 3:1 (Figure 4, curve *1*). It should be mentioned that the concentration boundary of suppression of the polarographic wave of AA is very sensitive to the molar ratio of the original monomers.

Even an insignificant excess in the AA concentration over the indicated ratio [AAc]:[AA] = 3:1 leads to the appearance of a distinct differential wave of AA (Figure 3, curve *3*; [AAc]:[AA] = 3:1.01). On the other hand, dilution of the monomer mixture [AAc]:[AA] =3:1 is accompanied by the appearance of the polarographic peak of free AA, whose height increases with dilution (Figure 4). This data allows us to draw the conclusion that acrylic monomers form a complex containing a threefold excess of AAc relative to AA in an aqueous solution. The reduction potential of the complex is similar to that of AAc.

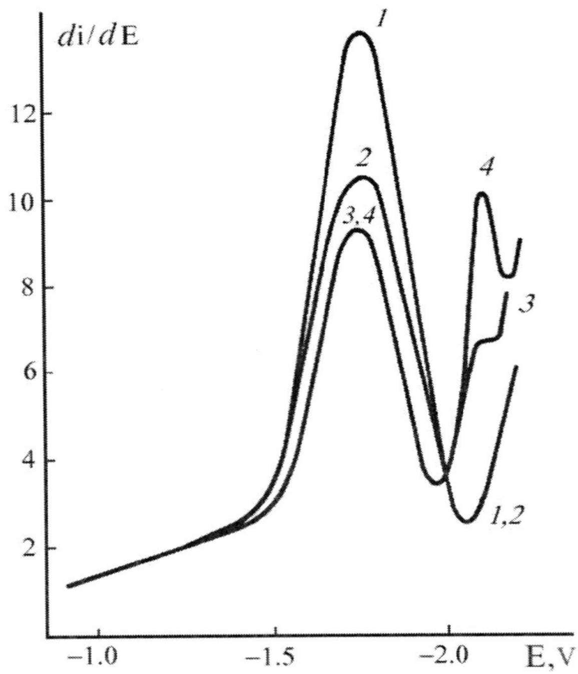

Figure 3. Differential oscillographic polarograms of the mixture of acrylic acid and acrylamide at concentrations (*M*) respectively: 8.3. $10^{-4}$and 0.56-10^{-4} (*1*), 1.68 10^{-4} (*2*); 2 8 10^{-4} (*3*); 4.2 10^{-4} (*4*) (background is 0.1 *N* $LiClO_4$).

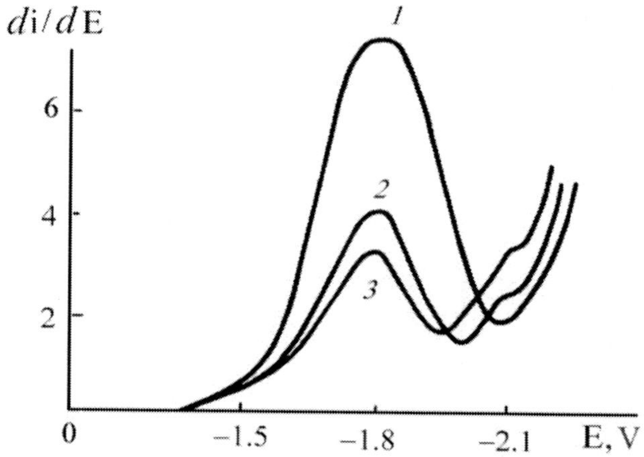

Figure 4. Differential oscillographic polarograms of a mixture of acrylic acid and acrylamide (8.3 10^{-4} and 2.76 10^{-4} *M* (*1*)) on twofold (*2*) and fourfold (*3*) dilution (background 0.1 *N* $LiClO_4$).

The decrease in the diffusional current (i_d), which corresponds to the peak on the differential polarogram of AAc, as the AA concentration in the solution increases, also favors AAc-AA complex formation. This indicates that the diffusional coefficient (*D*) of AAc bound in the complex decreases. At [AAc]:[AA] = 3:1, the height of the AAc wave is ~77 % of that in the absence of AA. Using the Il'kovich equation, in which $i_d \sim kD^{1/2}$, where *k* is a proportionality constant, we obtain $D^{1/2}_{compl} \sim 0.77\ D^{1/2}_{AAc}$. Taking into account the known value $D_{AAc} \sim 5\ 10^{-6}$ cm s^{-1}, we calculate $D_{compl} \sim (0.77)^2 D_{AAc} \sim 3\text{-}10^{-6}$ cm s^{-1}.

Complex formation is also confirmed by the concentration dependence of the AA peak height on the differential polarogram in the solution at a steady-state concentration of AAc. As follows from the data in Figure 5, the height of the polarographic peak of free AA sharply increases when the ratio of [AAc] to [AA] changes from 3:1 to 3:2 and 1:1. This stepwise pattern of the curve can attest to the existence of several stages of dissociation of the complex:

$$[3\ AAc\bullet AA] \leftrightarrow 3\ AAc + AA \quad (1)$$

$$[3\ AAc\bullet 2\ AA] \leftrightarrow [3\ AAc\bullet AA] + AA \quad (2)$$

$$[3AAc\bullet 3AA] \leftrightarrow [3\ AAc\bullet 2\ AA] + AA \quad (3)$$

Using Eqs. (1)-(3), taking the results presented in Figures 4 and 5 into consideration, and estimating the concentration of free AA from the height of the corresponding peak on the differential polarogram and the data of the calibration curve obtained for pure AA, the values of the dissociation constants of tentative complexes were calculated: $K_{[3\ AAc\bullet AA]} = 2\bullet\ 10^{-15}\ M^3\ l^{-3}$, $K_{[3\ AAc\bullet 2\ AA]} = 0.8\ 10^{-3}$ M, $K_{[3AAc\bullet 3AA]} = 0.72\ 10^{-2}$ M. The dissociation constant for the [3AAc•AA] complex was determined starting from the model mixture [AAc]:[AA] = 3:1.

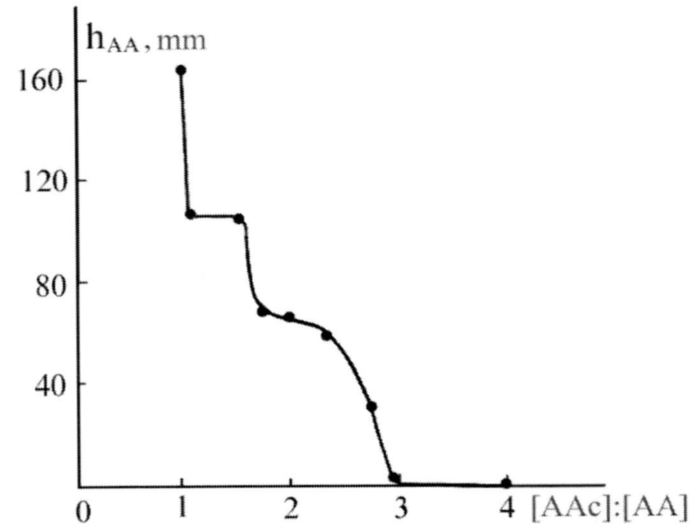

Figure 5. Dependence of the height of the peak of acrylamide on the differential polarogram on the ratio [AC]:[AA]; [AC] = 8.3 10^{-3} *M.*

The differential polarogram of this mixture does not display the reduction peak of AA. The account is made for dilution of this mixture down to the concentration at which the peak of free AA appears and increases with dilution. The constants for the other complexes were calculated, taking into account the change in the height of the reduction peak of free AA as a function of its content in the solution relative to AAc.

A comparison of the dissociation constants does not allow us to conclude clearly which of the complexes is most stable. However, the data in Figure 3 (curve *3*) and Figure 4 (curve *1*) indicates that the [3 AAc•AA] complex is the most stable. This was confirmed by polarographic monitoring of the course of EIP and by a physicochemical analysis of the polymeric film synthesized.

Since $\varphi_{1/2}$ of AAc and $\varphi_{1/2}$ of the complex are similar and occur at less negative potentials than $\varphi_{1/2}$ of AA, it is most probable that these are the species that participate most in the polymerization. Actually, the polarography of monomer solutions subjected to EIP shows that the height of the peak of free AA is almost unchanged, but that of the peak with $\varphi_{1/2}$ in the range of the reduction of AAc and the complex compound decreases. However, despite the observed constancy of the concentration of free AA in the solution during electropolymerization, the polymer formed during EIP contains not only AAc, but AA as well. This follows from the adsorption band in the 1540-1560 cm^{-1} range observed in the IR spectra of the polymeric film. This band is characteristic of secondary amides and is due to the deformation oscillations of the NH-group (Figure 6, *a*). This is important when a complex compound containing both of the original monomers is consumed in the formation of the polymeric film. Elemental analysis shows that the nitrogen content in the copolymer film is ~5 %, which corresponds to a threefold excess of AAc over AA. This suggests that the copolymerization is preceded by the formation of the [3AAc•AA] complex in the solution, which then takes part in propagation of the polymeric chain. The existence of complexes of other compositions is also possible, but their fraction in EIP is insignificant due to their high instability constants.

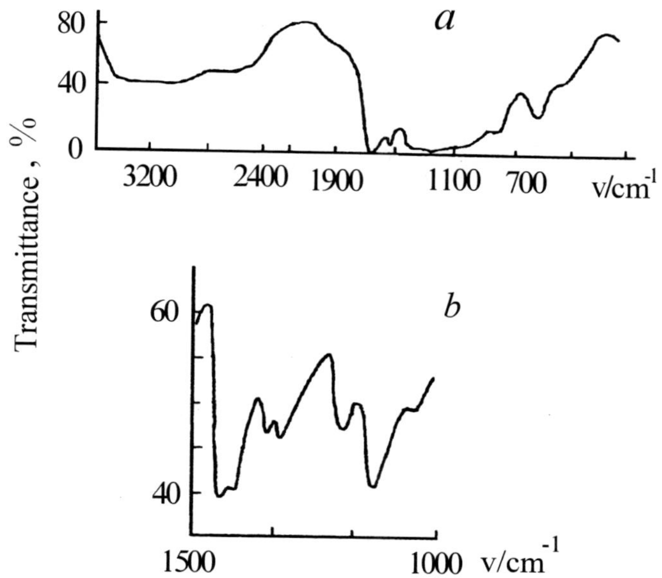

Figure 6. IR spectra of polymers obtained from AC and AA (*a*) and AC, AA, and F (*b*).

Complex formation also decisively affects the composition of the polymers formed during EIP of more complicated monomer compositions containing components in addition to AA and AAc that participates in the film formation. In particular, the formation of complexes manifests itself during EIP of the AA-AAc-formaldehyde (F)-APS mixture. Since the coating formed on the cathode surface is a cross-linked copolymer containing ionogenic groups, the composition of the copolymer AAc-AA-F can be established by the determination of its exchange capacity with respect to carboxyl groups, which was found to be 10 mg-eq g^{-1}. Since formaldehyde interacts only with AA, but not with AAc, it becomes obvious that the molar ratio (A) of AA to F in the copolymer must be equal to or greater than one. Then, taking into account that cross-linking is accompanied by polycondensation with detachment of one molecule of water from each of the two elementary units of the copolymer [118], one can determine the ratio (*N*) between the number of AAc and AA molecules in the copolymer from the formula:

$$N = \frac{MM_{AA} + x \cdot MM_F - 0.5 \cdot MM_{H_2O}}{1 - E \cdot MM_{AAc}} \tag{4}$$

where MM_{AAc}, MM_{AA}, MM_F, and MM_{H2O} are the molecular weights of AAc, AA, F, and water, respectively; x is the number of F molecules in an elementary unit of the copolymer; E is the exchange capacity of the polymer.

The calculations show that $N = 3$ at $x = 0.75$. The value $x<1$ suggests either incomplete hydroxymethylation of the amide groups in the copolymer, which is unlikely due to the excess of formaldehyde, or that cross-linking occurs simultaneously according to two mechanisms:

1) by polycondensation of preliminarily hydroxymethylated amide groups [118]:

$$n\left[\text{-RCONH}_2\right] + n\,\text{CH}_2\text{O} \longrightarrow \left[\text{-RCONHCH}_2\text{OH}\right]_n \tag{5}$$

$$2\left[\text{-RCONHCH}_2\text{OH}\right]_n \longrightarrow \left[\begin{array}{l}\text{-RCONHCH}_2\\ \text{-RCONHCH}_2\end{array}\!\!>\text{O}\right]_n + n\,\text{H}_2\text{O} \tag{6}$$

2) by direct interaction between F and AA [6]:

$$n\left[\text{-RCONH}_2\right] + n\,\text{CH}_2\text{O} + n\left[\text{RCONH}_2\right] \longrightarrow \left[\begin{array}{l}\text{-RCONH}\\ \text{-RCONH}\end{array}\!\!>\text{CH}_2\right]_n + n\,\text{H}_2\text{O} \tag{7}$$

The existence of such cross-links is confirmed by IR-spectroscopic data: a series of bands in the 1035, 1330, and 1440 cm^{-1} range corresponds to oscillations of the OH group of primary alcohols; the band in the 1540-1560 cm^{-1} range corresponds to oscillations of the NH group of monosubstituted primary amides; the band in the 1130 cm^{-1} range corresponds to oscillations of the group $-CH_2-O-CH_2-$ in ethers (Figure 6, *b).*

In the first case (Eq. (5)), one molecule of formaldehyde interacts with one AA molecule, and in the second case (Eq. (7)), it interacts with two AA molecules. Then, assuming equal probability of both of the cross-linking processes, the molar ratio between AA and F must be 1:0.75. This assumption is confirmed by the composition of the elementary unit of the copolymer [3AAc-AA-0.75F] found from experimental estimates of the exchange capacity. The ratio of molar concentrations of AAc and AA in the copolymer equals 3:1 as before, which also indicates that the [3AAc AA] complex takes part in EIP.

As it is known, during the cooperative cathode discharge of two substances that produce a chemical compound with a gain of Gibbs energy ΔG, the equilibrium potential of the component that is reduced at more negative potentials shifts to the anode region by the value $\Delta\varphi = \Delta G/nF$, approaching the equilibrium potential of the second substance [119]. Apparently, a similar effect takes place in the case considered, resulting in the electrochemical synthesis of copolymers from the mixture of monomers AAc and AA, whose reduction potentials differ considerably from one another. Owing to this effect and complex formation between AAc and AA, their electrocopolymerization occurs more easily than the competing homopolymerization of AA.

6.2. Dependence of the Deposition Rate and Polymer Film Quality on the Monomer and Initiator Concentrations and Electrolysis Conditions

For the implementation of the EIP method in order to obtain polymer coatings and films on metals, it is of crucial importance to know which processes are responsible for electropolymerization transfer to shift the film formation to the required direction. As was stated above, the predominant polymer deposition on the electrode is facilitated by direct initiation, high adsorption degree, and insolubility of the formed polymer in the liquid phase. Besides, the factors determining the coating deposition rate include the induction period and possibility of polymer fraction's migration into the solution.

The works [68, 120] present the results of the influence of the electrolysis conditions on the induction period length and the inter-phase distribution of the copolymer formed from aqueous solutions of *acrylamide, acrylic acid and formaldehyde*. Formation of a polymer deposit on the cathode is possible at direct EIP of monomers, but the process rate is very low. The addition of peroxide initiators (persulfates, hydrogen peroxide) enables one to accelerate EIP, reduce by an order of magnitude the working current density (CD) and the induction period from 54 hr at CD= 1 A/m^2 down to 0.3-2.0 hr at CD=0.1 A/m^2. The polymerization rate in the presence of peroxides increases with increasing electrolysis potential (current), initiator and monomer concentration [10]:

$$V_{pol} = k\{[S_2O_8^{2-}] \exp(-\alpha \, E \, F / RT)\}^{1/2} [M]$$

However, to attain a predominant polymer deposition on the electrode surface, the conditions providing the highest polymerization rate are inacceptable. At the same time, the EIP transfer to the metal surface is one of the most important criteria of the respective technology advantages. Therefore, one should select a set of optimal concentrations of the monomer and initiator, at which the rate of film formation on the metal surface is accompanied by a minimal polymerization in solution.

It was established for the composition under study that the optimal concentration of $(NH_4)_2S_2O_8$ was equal to 0.45-2.20 M, acrylamide – 0.056 M. Lower concentrations of these components increase the induction period and decrease the coating formation rate, while higher concentrations result in the polymerization transfer to solution. The addition of acrylic acid (AAc) and its concentration increase up to 1.10 M, resulting in the reduction of the induction period from 20 down to 0.3 hr. The bulk polymerization increases at $C_{AAc} > 1.10$ M. The change of the formaldehyde concentration from 2.6 M up to 10.4 M also reduces the induction period down to 0.5 hr, but results in increasing the polymerization in solution and worsening the coatings quality.

The following composition appears to be optimal as from the point of the film growth rate as in view of the polymer coating quality:

1.1 M AAc, 0.056 M AA, 2.6 M F, $1.75 \cdot 10^{-3}$ M $(NH_4)_2S_2O_8$ (*electrolyte 1*).

Such a solution is stable over time and post-polymerization does not occur more than 1 month after the current shutoff.

The copolymer formation on metal with fulfilling the conditions of sufficiently high rate, good film quality and minimal polymerization product removal requires an optimal current density. In the composition AA-AAc-F-$(NH_4)_2S_2O_8$, the dependence of the polymer coating mass on the cathode current density, at fixed electrolysis time, goes through a maximum at J=2 A/cm^2 (Figure 7). Low polymer yield at J<2 A/cm^2 results from low concentration of active centers and, therefore, low initiation rate. The deposit yield reduction at J>2 A/cm^2 is related to the reduction of anion-radicals $SO_4^{-\cdot}$ into inactive ions SO_4^{2-}, thus, in turn, reducing the initiating particles concentration, as to spending a part of the current power to the side process of hydrogen reduction, near-electrode layer mixing, anion-radicals removal into solution and bulk polymerization increase. The best quality coatings are formed at *J*=1-2 A/см^2.

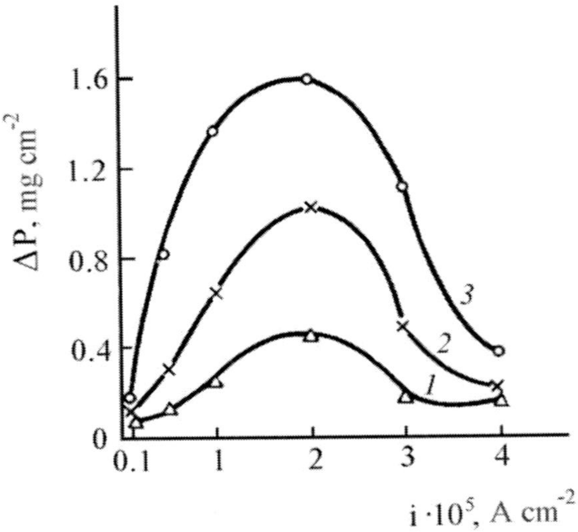

Figure 7. Effect of current density on polymer product deposition on cathode. Electrolyte 1. Deposition time after induction period (min): 20 (1), 40 (2), 60 (3).

Under the galvanostatic electrolysis conditions, the polymer film yield in the composition AA-AAc-F has a complex dependence on the polarization time (quantity of electricity) (Figure 8). On the respective curve, one can see a distinct induction period *ab*, within which the coating does not form, but the conditions for starting of its formation are created.

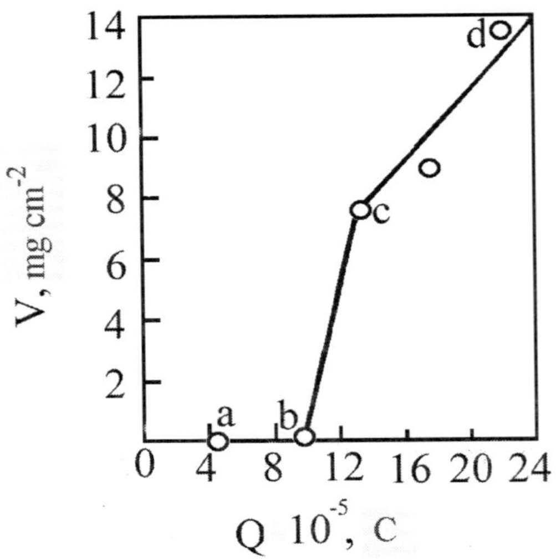

Figure 8. Dependence of polymer deposition rate on quantity of electricity Q. *ab* – induction period, *bc* – layer growth over surface, *cd* – layer growth in thickness. Electrolyte 1. *i*=0.1 A/m^2.

The duration of the induction period is proportional to the current density and monomer and initiator concentrations. The part *bc* corresponds to high film formation rate.

Microscopic and visual observations demonstrated that the initial stage was characterized by the emerging of the deposit formation sites. Then the growth centers spread over the whole metal surface, merging into a solid coating. After the electrode coating by the polymer, the layer thickness starts to increase (*cd*). Here, the observed decrease of the film formation rate results from the limiting action of the coating on the $S_2O_8^{2-}$ ions motion to the electrode and reduction of the radicals $SO_4^{-\cdot}$ concentration. The film thickness can be controlled by the electrolysis duration and current density.

The marked inhibiting effect on the film growth rate is provided by oxygen, as emitted on the anode as present in a non-degasified solution (the film growth rate is 3.5-fold lower, while the induction period is 4-10-fold larger than in a solution degasified by argon). That is why one should use the cells with separated cathode and anode spaces and electrolyte purging by an inert gas. In the oxygen-free solutions, the monomer EIP starts at more negative potentials and lower current densities (Figure 9), which indicates to easier, as compared to monomers, oxygen reduction that must be related to the formation of peroxide compounds inhibiting polymerization.

The monomer solution saturation by hydrogen results in an increase of the rate of polymerization on metal (Figure 10) and a decrease of the induction period from 2-3 hr down to 20-30 min. At the same time, hydrogen induces an intensified polymerization in the catholyte bulk, and the process is possible even without current switching on. The latter must be concerned with the fact that the atomic hydrogen is itself a good polymerization initiator for vinyl monomers. At the initiator addition, the polymer is distributed between metal (deposit yield is negligible) and the bulk of the solution. The combined initiating action of persulfate-ions and hydrogen accelerates the deposit formation on the metal surface.

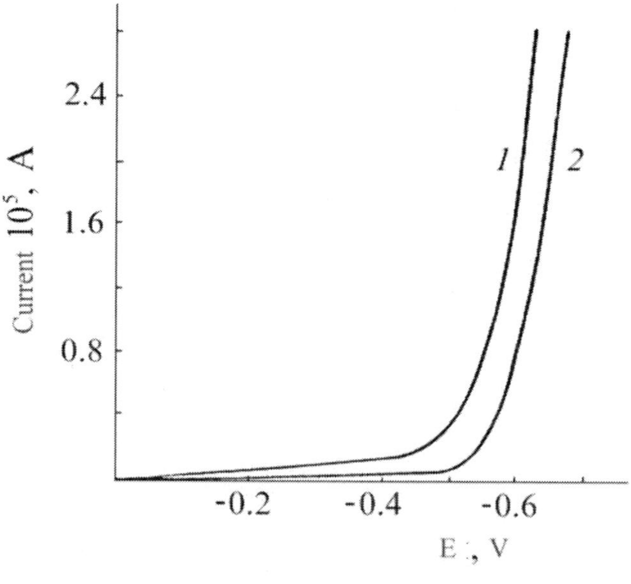

Figure 9. Polarization curves in degasified (1) and non-degasified (2) solutions. Electrolyte 1. Cathode – steel.

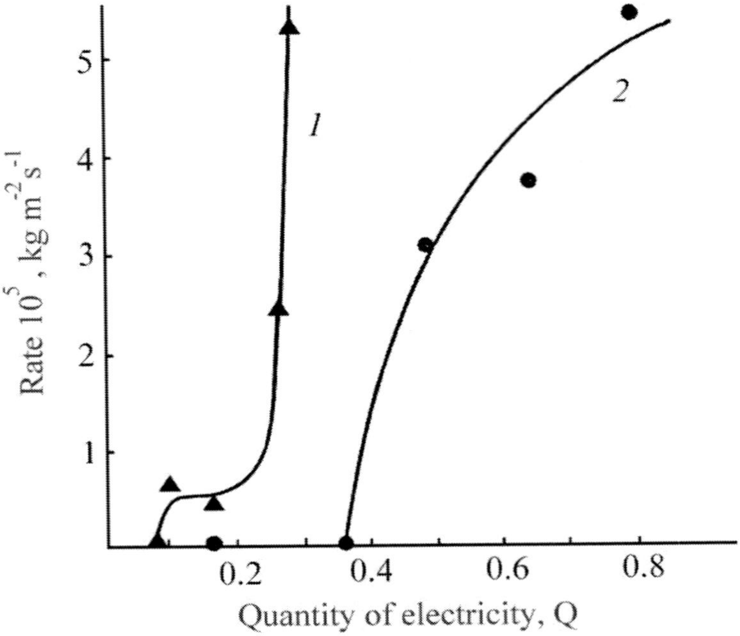

Figure 10. Dependence of polymer deposition rate on quantity of electricity in hydrogen-saturated (1) and degasified (2) solution. Electrolyte 1. $i = 0.1$ A/m^2..

The process of polymerization of the composition under study depends on the cathode material (Table 1). Since for a majority of metals the hydrogen overpotential is lower, the lower are the polymerization rate and deposited polymer amount, the copper electrode is the least suitable metal for the polymer film formation. In this case, EIP is additionally limited by the copper ions present in the near-electrode area (pH of the solution under study ~1–2), which act as strong polymerization inhibitors.

Table 1. Electrode material effect on electropolymerization. Electrolyte 1. $J = 0.1$ A/m^2

Metal	τ, h	Film mass (Δ m), g	Electrode area, cm^2	Δ m, g/h·cm^2	Superpotential, V
Pb	3.0	0.0690	10.0	0.0023000	1.56
Cd	3.5	0.0694	13.6	0.0015000	1.40
Zn	3.5	0.0176	13.6	0.0003700	1.24
Al	3.5	0.0016	13.6	0.0000340	1.00
Cu	18.0	0.0008	6.8	0.0000065	0.87
Fe	1.5	0.0745	8.8	0.0057000	0.70

The possibility of metal ions transfer into solution can not be neglected at such low current densities. For example, the highest polymerization rate was detected at using Fe cathodes, in spite of the lowest hydrogen overpotential on this metal [68]. The latter is concerned with the fact that the standard iron exchange current is equal to 0.5 A/m^2, while the cathode current density in a cell is equal to 0.1 A/m^2. That is why one inevitably observes the presence of Fe^{2+} ions, which facilitate the formation of an additional number of radicals near the electrode:

$$S_2O_8^{2-} + Fe^{2+} \longrightarrow Fe^{3+} + SO_4^{2-} + SO_4^{\cdot -}$$

This results in an increase of the rate of polymerization on the iron cathode surface. The cathode material determines not only the coating growth kinetics, but also the EIP induction period, which changes according to the order below:

iron < lead < cadmium < zinc < aluminum < copper.

The oxides on the metal surfaces also inhibit the polymer layers initiation and growth. In particular, on the oxidized Al, the induction period is twice as long than on that mechanically or chemically pre-treated for oxide removal.

6.3. Effect of Additives on Polymer Distribution between the Electrode Surface and Solution

Possible ways of EIP transfer to the electrode surface have been investigated [120]. It was established that the composition AA-AAc-F-$(NH_4)_2S_2O_8$ was a typical example of the systems in which the film formation on the electrode is related to the accumulation of oligomer products in the solution. This process is characterized by a prolonged induction period. The film deposition on metal starts after the expiration of some specific time period, when a sufficient amount of oligomers is accumulated near the cathode, and the insoluble fractions start to deposit due to the solution oversaturation. At high current densities, the radicals generation accelerates, thus increasing the formation rate and concentration of oligomer products in the near-electrode area. So one can reduce the polymerization induction period 3-4 times by conducting preliminary electrolysis at 1 A/m^2 and, further, reducing J down to 0.1 $A/м^2$ (this value is optimal for the polymer film formation).

It is evident that all the phenomena inhibiting the oligomer products removal into solution must facilitate polymer redistribution on the electrode surface. This can be achieved by two ways: a) by *reducing the solubility of polymerization products* due to polymer molecular weight increase or medium variation or b) by maximally *increasing the polymer concentration gradient* between the metal surface and solution bulk.

Indeed, the predominant polymer formation on the cathode is facilitated by the reduction of initiator and monomer concentrations and current density down to the value corresponding to the optimal EIP rate.

The above parameters reduction helps to reduce the amount of radicals and, therefore, improve the electrode-formation of products, which are less soluble and have higher molecular weight. Such conditions enable one to obtain a homogeneous and transparent film on the electrode at the minimal rate of the polymer accumulation in solution (Figure 11, curves 1, 2).

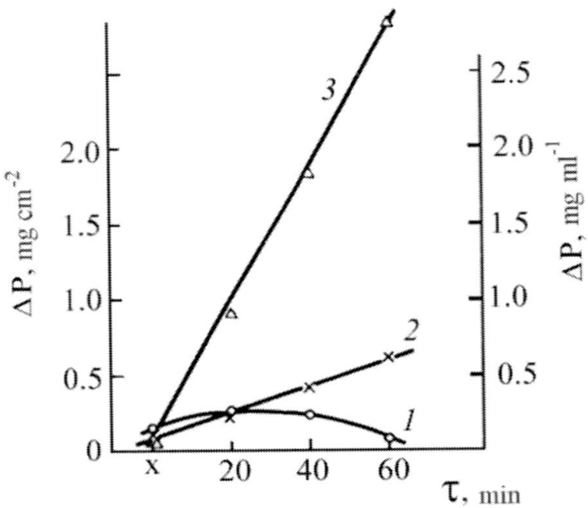

Figure 11. Polymer product formation rate in solution (1), on cathode (2) on cathode in the presence of 0.28 M Na_2SO_4 (3). Electrolyte 1. $i = 0.1$ A/m^2. x – start of film formation.

The polymerization products removal from the electrode surface can be reduced by: 1) increasing the solution viscosity, 2) decreasing the electromigration of oligomers $^{\bullet}(M\text{–}M)_nSO_4^-$ in the anode direction, 3) decreasing the solubility of the formed macroions.

The polymerization products solubility and their migration into solution can be attained by introducing a salt background to the monomer composition. Indeed, the addition of Na_2SO_4 into the electrolyte 1 (0.28–1.0 M) dramatically shifts the equilibrium to the cathode, thus reducing the amount of polymer in solution down to almost zero and shortening the induction period by two times (Figure 11, curves 1, 3).

The use of alcohols (methanol, ethanol, glycerol, polyethylene glycol) for the above purpose is inefficient, since they inhibit the monomer EIP by 10-30-fold reduction of the polymerization rate. In this case, the coatings quality also worsens. In particular, addition of ethylene glycol (3.5-11.0 M) makes the film hygroscopic.

The concentration gradient near the electrode (ΔC) is determined by the relation:

$$\Delta C = j\,\delta / D$$

where j – the substance deposition on the electrode proportional to J and the monomers concentration; δ – the diffusion layer thickness; D – the polymer diffusion coefficient.

Since the diffusion layer thickness $\delta = k\,\omega^{-1/2}\,V_\kappa^{1/6}\,D^{1/2}$ [121] and the diffusion coefficient D ~ 1 / MW, where V_k – the kinematic viscosity; ω – the solution mixing intensity; MW – the polymer molecular weight, it is evident that the near-electrode layer oversaturation must be facilitated by all the factors thus reducing the electrolyte motion near the metal surface, including the natural convection.

First of all, these factors comprise the medium viscosity increase and the use of moderate temperatures as well as the formation of products of high molecular weight. Indeed, the removal of the polymerization products from the electrode surface can be limited, in the case of the composition AA-AAc-F, by increasing the solution viscosity through adding of, for example, gelatin and gum arabic.

6.4. Use of Non-Stationary Electrolysis Conditions in Optimization of Coatings Formation on Metal

The use of non-stationary electrolysis conditions and different forms of the rectified current enables one to control the quantity of the formed active centers and, therefore, the polymer molecular weight, reducing the probability of side reactions occurring and equalizing the concentrations of reacting substances in a composition. Thus, the non-stationary electrolysis is of interest as one of the factors which enable one to modify the electropolymerization kinetics and coatings quality.

EIP was conducted under the conditions of combination of an asymmetric alternate current with a direct current (0.1 A/m^2). The ratio between the direct and alternate currents densities was maintained at the levels 1:0.5; 1:1; 1:3. The alternate current frequency was varied from 20 up to 200 Hz [120].

For all of the examined electrolysis conditions, the use of an asymmetric current results in the shortening of the induction period by 10-25%, the increase of the polymer coating deposition on the electrode and the reduction of the polymer concentration in solution (Figure 12).

Shortening of the induction period depends insignificantly on the alternate current amplitude. The largest effect of the polymer deposition rate increase on the electrode (2-fold) was attained at $i_{DC} : i_{AC}$ = 1:0.5 and 1:1 and the alternate current frequency 20 Hz. This data is in good agreement with the results of the direct current density effect on the film yield (Figure 7), since under the above conditions, total current density attains 0.15-0.20 A/m^2 during the cathode semi-periods that corresponds to the curves maximum in Figure 7. Further alternate current amplitude growth results in the decrease of the accelerating effect (Figure 12).

The positive effect of the asymmetric alternate current on the cathode polymer deposition can be caused by:

1. the periodic character of radical formation (at $i_{DC} : i_{AC}$ = 1:1 or 1:3), which results in the synthesis of higher molecular weight products, whose removal from the cathode is complicated;
2. more abrupt increase of the EIP rate with increasing the current J values and, therefore, making more polymer-oversaturated solution near the electrode;
3. the emerging of the electrode dissolution products (for example, Fe^{2+}), which are capable to catalyze polymerization, in the near-electrode area during the anode semi-periods (at $i_{DC} : i_{AC}$ = 1:3) [68];
4. the predominant chain growth on the electrode surface during the anode semi-periods (at $i_{DC} : i_{AC}$ = 1:3) due to increased adsorbability of negatively charged radicals $SO_4^{\bullet -}$ and $^{-}SO_4{-}M^{\bullet}$ on the positively charged surface.

The first explanation is the most probable, since others are not corroborated experimentally. Indeed, the second explanation can not be right, since the dependence of the EIP rate on the cathode i goes through maximum at i=0.2 A/m^2 (Figure 7). The last two hypotheses are also unreal, since the polymerization rate at $i_{DC}:i_{AC}$ = 1:3 is lower than at $i_{DC}:i_{AC}$ = 1:0.5 (Figure 12).

The periodic radical generation could result in the formation of products of higher molecular weights. So one may expect that the electrolysis using the rectangular pulsing current of a constant medium intensity, but of different off-duty ratios, would facilitate the polymer redistribution to the electrode. However, it turned out that at the impulse electrolysis under the selected conditions (Table 2; Q = 0.1 C/s·m^2), the polymer did not deposit on the cathode surface at all, while the polymerization in solution proceeded with virtually the same rate as at electrolysis under direct current.

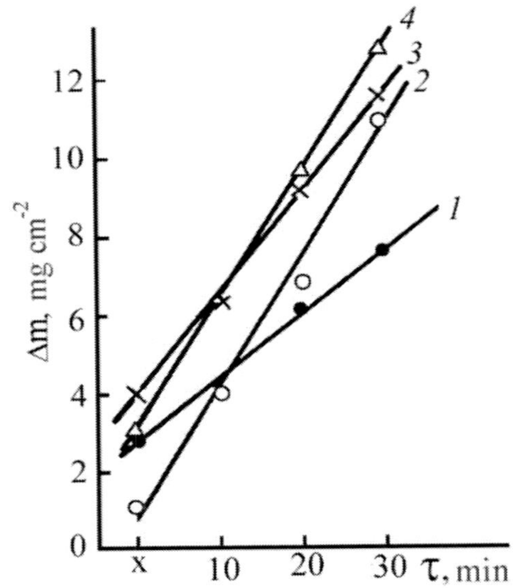

Figure 12. Effect of direct (1) and asymmetric alternate (2–4) current on the rate of polymer deposition on cathode. DC 0.1 A/m^2. $i_{DC} : i_{AC}$ = 1:1 (2), 1:3 (3) and 1:0.5 (4). The alternate current frequency 20 Hz. Electrolyte 1. x – start of coating formation.

Table 2. Impulse electrolysis conditions at monomer EIP. Electrolyte 1

Ratio impulse/pause	Impulse time, s	Pause time паузы, s	EIP time, s	Current frequency, Hz	Average current density, A/m^2
1:2	0.01	0.020	0.030	33.33	0.1
1:2	0.05	0.100	0.150	6.66	0.1
1:10	0.01	0.100	0.110	9.09	0.1
2:1	0.01	0.005	0.015	66.66	0.1
5:1	0.01	0.002	0.012	83.33	0.1

Such an effect is, most probably, related to the elimination of the near-electrode layer oversaturation by oligomers during the pause. The above EIP conditions can be favorable for the polymerization in solution.

In summary,, the application of non-stationary electrolysis conditions enables one to control the process of electrochemically initiated polymerization by transferring the polymer product output to the electrode surface or into the solution.

6.5. A Role of Hydroquinone in EIP Film Formation on Metal

Hydroquinone (HQ) is a typical inhibitor of the radical polymerization. At EIP in aqueous solutions, when the radical mechanism appears to be the only one possible, the addition of HQ into the solution must result in the slowing down or complete suppression of polymerization. However, the simultaneous presence of HQ and $SO_4^{\bullet-}$ radicals in an aqueous mixture of acrylic monomers electropolymerization was, in opposite, accelerated [122-123].

The accelerating effect of HQ on the polymer film formation on the cathode was observed in a specific concentration range and did not depend on the persulfate/hydroquinone volume ratio. The EIP inhibition started if the hydroquinone concentration (C_{HQ}) exceeded that of persulfate ($C_{S_2O_8^{2-}}$).

The abnormal accelerating hydroquinone effect on EIP can be explained as follows: In the absence of HQ, electropolymerization in the composition AA-AAc-F-$(NH_4)S_2O_8$ occurs under the effect of ion-radicals $SO_4^{\bullet-}$ formed at the reduction of sulfate-ions on the cathode. The addition of hydroquinone into the solution results in a chemical interaction between persulfate and HQ with the formation of an extra amount of ion-radicals $SO_4^{\bullet-}$ (see Scheme). Simultaneously, the formed semiquinone radicals could inhibit the polymerization by recombining with the monomer radicals $R^{\bullet}$ or $SO_4^{\bullet-}$. However, due to the electrochemical reduction of semiquinone in HQ, the amount of $SQ^{\bullet}$ radicals reduces, just like their inhibiting effect on EIP. Thus, the mechanism under examination comprises a typical case of EIP combination with an additional redox-initiation of polymerization, where persulfate acts as an oxidizing agent, while a reducing agent is represented by the cathode reaction of quinone (Q) and $SQ^{\bullet}$ transformation into HQ [123] (Scheme):

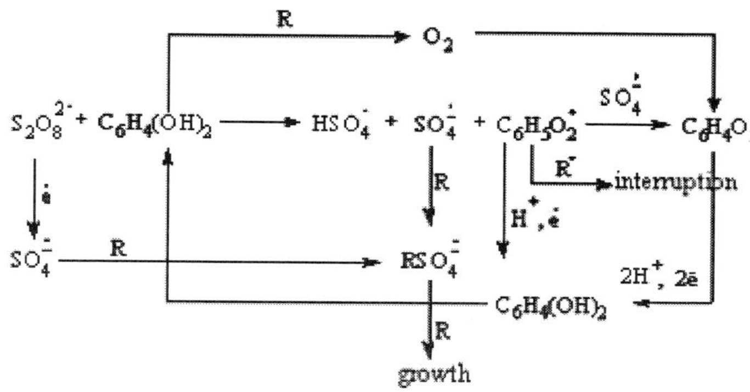

Scheme of interaction between persulfate and hydroquinone.

At C_{HQ} = 0-10^{-4} M (C_{max}) (Figure 13) the specified cathode current density (1 A/m^2) provides a complete regeneration of $SQ^{\bullet}$ into hydroquinone resulting in an increase of total amount of ion-radicals $SO_4^{\bullet-}$. In this case, the polymer film growth rate undergoes a 5-fold increase. At $C_r > C_{max}$, not all of the $SQ^{\bullet}$ radicals have time to be reduced. Their excess induces the stopping of polymer chains growth and deactivation of $SO_4^{\bullet -}$, thus decreasing the HQ initiating effect.

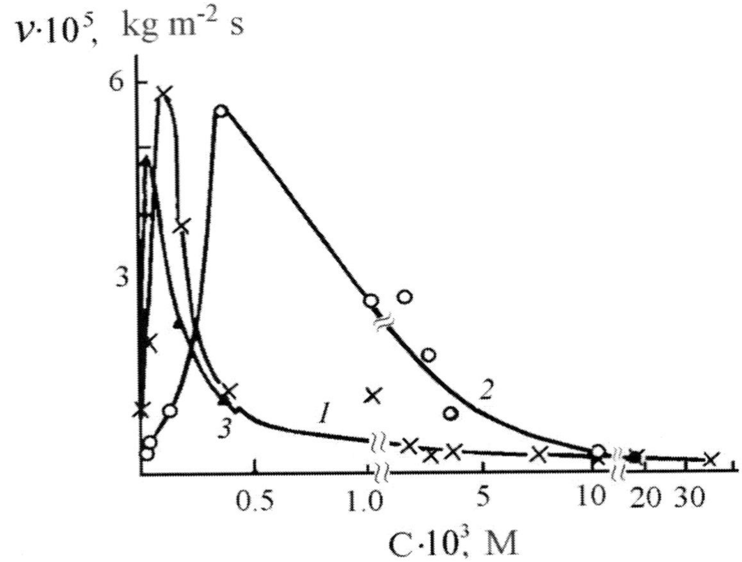

Figure 13. Effect of hydroquinone (1, 2) and quinone (3) concentrations on the polymer film growth rate on the cathode in degasified (1, 3) and non-degasified (2) solutions. Electrolyte 1. i=1 A/m^2.

At the equality of the persulfate and hydroquinone concentrations ($C_r = 1.8\times10^{-3}$ M), any number of ion-radicals $SO_4^{\bullet -}$, formed as a result of the chemical reaction, corresponding to the identical number of non-reduced semiquinone radicals $SQ^{\bullet}$. Finally, the additional initiating effect of the radicals $SO_4^{\bullet -}$ is fully compensated by chain stopping the radicals $SQ^{\bullet}$ are responsible for. As a result, the polymerization becomes the same, as in the absence of HQ. Further increase of C_{HQ} yields a classical inhibition of electropolymerization on the metal surface.

As seen from Figure 13 (curve 3), the addition of quinone into the electrolyte 1 also has an initiating effect, in this case, the polymer electrode-deposition rates are similar for Q and HQ. However, since in the Q-containing solution a part of the current is spent on the transformation of Q into HQ, here lesser amount of the radicals $SQ^{\bullet}$ can be reduced on the cathode. As a result, the completeness of the $SQ^{\bullet}$ reduction required for the maximal polymer film growth rate is provided at $C_Q < C_{max}$.

The addition of HQ to the composition under study also reduces the polymerization induction period τ_{ind} 15-20-fold (Figure 14, curve 1), while τ_{ind} is minimal at C_{max}. Further increase of C_{HQ} does not virtually change the length of τ_{ind} until the concentration equals the persulfate molar concentration. Thereafter, τ_{ind} again increases and tends to the value characteristic for the solution without additives. The above regularity can be explained by the fact that the polymer deposition on the electrode depends on its molecular weight MW and the polymer concentration C_p in solution [122, 123]. The higher are C_p and MW, the lower are the macromolecules mobility and solubility and the probability of oligomers removal into the solution. Therefore, $\tau_{ind} \sim 1/(MW\cdot C_p)$.

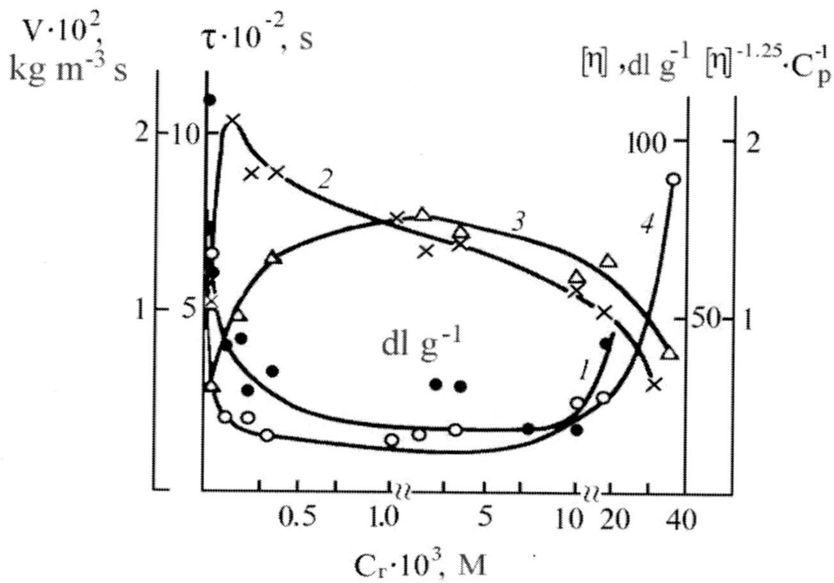

Figure 14. The effect of HQ concentration on the induction period of the polymer film formation (1), solution characteristic viscosity (2), polymer concentration in solution (3) and the value $[\eta]^{-1.25}\cdot C_p^{-1}$ (4). i=1 A/m². Electrolyte 1.

The average oligomer molecular weight can be estimated on the solution characteristic viscosity $[\eta]$: $MW^{\alpha} = [\eta] / K$ (α for the acrylic monomers ~0.8). Then $\tau_{ind} \sim [\eta]^{-1.25}\cdot C_p^{-1}$, which is corroborated by the data of Figure 14 (curves 1 and 4). The products with the highest molecular weights are formed in the solution at $C_{HQ} = C_{max}$ (Figure 14, curve 2). The latter provides the minimum of τ_{ind} and the highest rate of the polymer deposition on the electrode. Formation of the polymer lower molecular weight at $C_{HQ} < C_{max}$ is concerned with the lower EIP rate, while at $C_{HQ} > C_{max,}$ it can be related to chain stopping due to the emerging of the $SQ^{\bullet}$ radicals excess in the solution.

The accumulation of oligomer products in the solution goes on with the highest rate at $C_{HQ} = C_{S_2O_8{}^{2-}}$, when the maximum concentration of the radicals $SO_4^{\bullet -}$ is provided (Figure 14, curve 3). However, here the oligomers MW are small (Figure 14, curve 2). A further increase of C_{HQ} inhibits the polymerization due to the interaction of the HQ excess with the growing polymer chains.

According to the Scheme, the presence of oxygen in the solution must facilitate the HQ→Q transition, i.e. must reduce the amount of hydroquinone reacting with persulfate. Taking into account that the air-saturated solution contains ~3×10^{-4} M of oxygen [124], the amount of HQ participating in the reaction $C_6H_4(OH)_2 + {}^1/_2\ O_2 \rightarrow C_6H_4O_2 + H_2O$ must be equal to ~6×10^{-4} M. That is why in the degasified solutions, the position of the maximum (C'_{max}) on the curve of the film formation rate depends on C_{HQ} shifts to the range of higher concentrations of HQ (Figure 13). However, this shift comprises as little as 2.6×10^{-4} M. Therefore, in the process of EIP, the oxygen effect is not limited by its interaction with HQ. A part of the oxygen amount is spent on the polymerization inhibition, since in the degasified solution at $C_{HQ} = 0$ the coating growth rate is 3.5-fold lower, while the induction period is 4-

10-fold longer (Figure 13, curves 1, 2). As a result, at C_{max}, total film growth rate increases 17.5-fold in the presence of oxygen. There also increases the hydroquinone concentration, at which the polymerization rates on the cathode in or without the presence of hydroquinone become equal. The ratio $C_{S_2O_8^{2-}}/C_{HQ}$ comes to 1:3. The solution pH increase shifts the quinhydrone electrode to the range of higher cathode potentials, which complicates the Q and $SQ^{\bullet}$ reduction into HQ. That is why the initiating effect of HQ is maximal at pH=1–2, reduces with the pH growth and completely vanishes at pH > 3.5 (Figure 15, curve 2).

For each C_{HQ}, there exists a minimal current density J_{min} providing a complete regeneration of $SQ^{\bullet}$ into HQ. At $J \neq J_{min}$, the polymer chains length and, therefore, the rate of polymer growth on the electrode, will be lower. The reason for this at $J<J_{min}$ is the excess of $SQ^{\bullet}$ stopping the polymer chains, at $J>J_{min}$ – the excess of $SO_4^{\bullet -}$ resulting in simultaneous initiation of a large number of growing polymer molecules and decrease of the polymer MW due to the reactions of recombination and disproportionation (Figure 15, curve 1).

The changes of the film growth rate and induction period depend on the metal nature (steel, Zn, Al, Ni, Pb, Cd, Fe, Cu, Ti) in different ways. For example, at $C_{HQ}=C'_{max}$ and current density 1 A/m^2, the induction period on steel, Al, Pb decreases 36-, 12- and 2-fold, respectively.

The studies of these processes demonstrated that the EIP of acrylic monomers stabilized by hydroquinone, in the case of initiation by persulfate salts, can be performed without preliminary monomer purification from the inhibitor, since the HQ concentration conventionally used for such stabilization (0.001–1%) falls into the range providing a maximum positive effect. On the other hand, non-coincidence of the polymerization rate maximums on the metal surface and in the bulk solution, the influence of the electrolysis conditions on the EIP rate and the induction period enables one whether to obtain a film on the metal with maximal rate and minimal τ_{ind} at limited polymer accumulation in the solution or to shift the EIP process almost completely to the composition bulk phase.

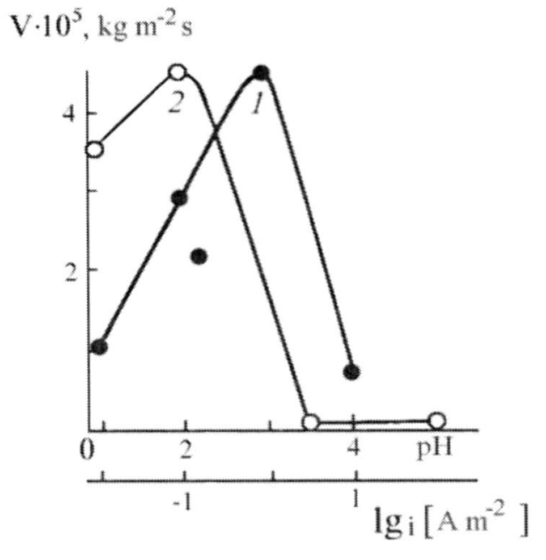

Figure 15. Effect of cathode current density (1) and solution pH (2) on the polymer film growth rate. Electrolyte 1. $C_{HQ} = 10^{-4}$ M.

7. Physical-Chemical Regularities of Polymer Films Formation At Electrocopolymerization of Acrylamide and Formaldehyde in the Aqueous Medium

Among the compositions allowing formation of quality polymer films on metals by the EIP method, the aqueous monomer compositions on the basis of acrylamide and its derivatives attract the most substantial interest. However, polyacrylamide coatings themselves are brittle and soluble in water [116]. One of the ways of improving these coatings quality consists of the copolymerization of acrylamide with other monomers. As was shown above, the polymer coatings from aqueous solutions of AA-AAc-F-$(NH_4)_2S_2O_8$, in spite of a number of advantages, have substantial disadvantages: high swelling ability, limiting their fields of application, and long induction period. The exclusion of acrylic acid from the copolymer composition (ionogenic –COOH groups) must reduce the water-absorbing capacity. On the other hand, in order to transfer polyacrylamide into the insoluble state, one can use its treatment by formaldehyde (45-50°C, pH=10-10.5, 2 hr), resulting in the formation of methylol derivatives:

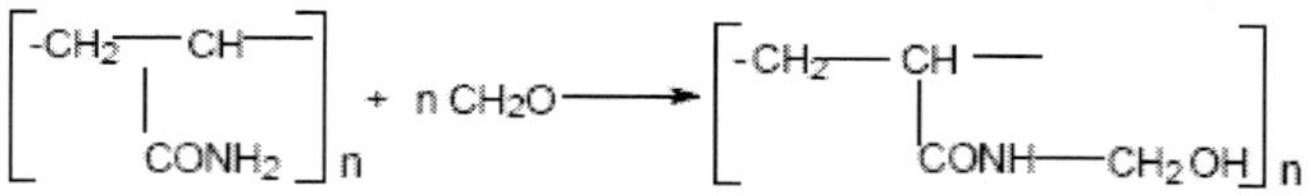

(8)

The studies demonstrated that the EIP method enables one to simplify the technology of the above compounds production by integrating the polymerization, methylolation and film formation on metal into a single process [95, 119]. In this case, one excludes heating, while the polymer layer formation period is shortened down to 0.5-5 min, unlike multi-hour chemical treatment. The process can be conducted in neutral and acidic media, since the alkalization required for the reaction (1) is achieved in the near-electrode layer due to electrochemical decomposition of water. The possibility of using acidic solutions is of special value, because in acidic media, methylamides are capable to undergo condensation with the formation of methylene or ester compounds of spatial structures, which are insoluble in water [125].

7.1. Initiation Mechanism of Polymerization of Acrylamide and Formaldehyde in the Aqueous Medium by the EIP Method

In the process of potentiostatic electrolysis (E = -0.4 ÷ -1.2 V) of the solution of AA and F (3-7 M) taken at equimolar ratios or formaldehyde excess, the cathode is coated by a transparent, strongly adhered to metal film comprising the 3D-cross-linked polyacrylamide. Such a film is insoluble in organic solvents, water, acids and alkalis and manifests the maximal swelling ability in petroleum ether and 15% NaOH. The use of zinc chloride as an initiator (0.01–0.3 M) accelerates the film formation process tens-fold, so that a homogeneous coating is obtained within 20 sec – 3-10 min of electrolysis.

The established factors – a) formation of the polyacrylamide film on the cathode as in the presence of zinc chloride, as without it; b) possibility of EIP in zinc-containing solutions at potentials lower than that of zinc reduction; c) substantial differences of the film formation rates and qualities at = (-0.5)–(-1.05) and E ≥ -1.1 V – allowed suggesting the existence of different mechanisms of initiating the acrylamide and formaldehyde copolymerization.

Chronovoltammetric studies confirmed the difference in natures of initiating particles formed in different potential ranges. It was established that at cathode potentials below (-1.1) V, the radicals formation occurred as a result of the hydrogen ions cathode reduction, while at more negative cathode potentials – at reduction of the acrylamide complex with zinc. Three distinct waves can be observed on the cathode polarization curves (Figure 16). The first wave ($E_{1/2}$ ~ -0.5 ÷ -0.6 V) increases in height with pH reduction and manifests even in the absence of zinc salts and formaldehyde in the composition. The latter allows assigning this wave to the reaction of the hydrogen ions reduction.

At the second-wave potentials, one can observe deposition of metal zinc on the cathode and abrupt acceleration of the polymer film formation. The combination of factors – the wave height growth with the increase of $ZnCl_2$ and formaldehyde concentrations, the semi-wave potential value (about -1.2 V) and its shift to the cathode area with the increasing AA content in the polymer composition – indicates to proceeding, in the above potentials range, of parallel reactions of reduction of formaldehyde and acrylamide complex with zinc [126]. The end product of these reactions comprises the particles initiating polymerization.

The third wave can be observed only in the case when a solid polymer film is formed, shields the electrode, thus resulting in a marked growth of the ohmic polarization, which induces the current drop. That is why its interpretation on the semi-wave potential value (~ -1.56 V ÷ -1.7 V) is ambiguous.

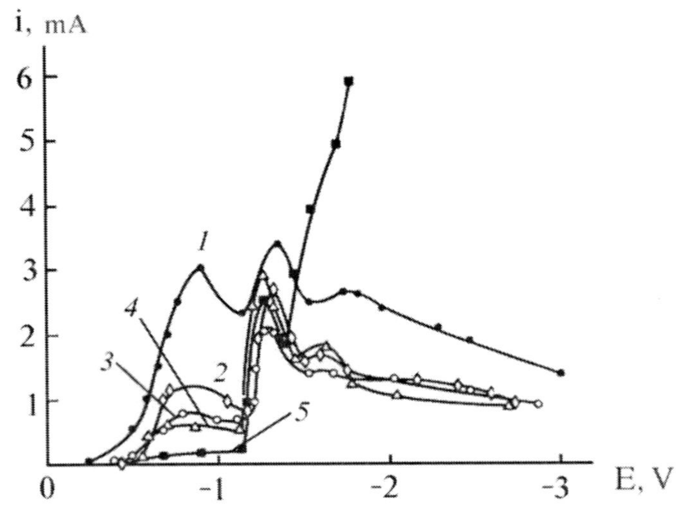

Figure 16. Cathode polarization curves obtained at different pH: 1 (1), 2 (2), 3 (3), 5 (4), 6 (5). Solution composition (M): acrylamide – 7, formaldehyde – 7, zinc chloride – 0.05.

However, since the wave height increases with the increasing F concentration and is absent in the solutions with zero F content, then, taking into account that $E_{1/2}$ of formaldehyde in alkali medium is equal to -1.59 B, and the formation of a solid polymer coating is accompanied with a strong alkalization of the near-electrode layer [126], one may suggest that the third wave on the polarization curves results from the catalytic reduction of formaldehyde limited by its diffusion through the film. All of the current-drop parts of the polarization curves are related to the formation of the polymer film shielding the surface. The film formation on the cathode starts simultaneously with the start of the hydrogen ions discharge, i.e. when the reduction of F and AA is still impossible (Figure 17, curve 1). The latter enables one to suggest that in the potential range -0.4 ÷ -1.0 V, EIP proceeds in accordance with the mechanism [127]:

$$AA + H^{+} + \bar{e} \longrightarrow AA^{\bullet} + H \quad (9)$$

$$AA^{\bullet} + AA \longrightarrow AA\text{–}AA^{\bullet} \quad (10)$$

The rate of the polymer formation V_m^H as a result of the above process is very small. Only after attaining the reduction potentials of the acrylamide-zinc complex and formaldehyde (E ≥ -1.1 V), there occurs a significant acceleration of the polymerization indicating the emerging of more efficiently acting initiating particles (Figure 17, ascent on curve 1). Simultaneously, the film transforms from a viscous to a solid state, its swelling ability becomes low, while its adhesion to metal increases. The increase of the cathode polarization higher than -1.4 V decreases the film growth rate (Figure 17, descent on curve 1), at this time due to the excess of initiating particles that results in the accumulation of low-molecular water-soluble polymer product fractions, whose washing out of the coating during the polymerization results in the loss of the deposit mass.

The linear dependence of the film formation rate (V_m) on the root of the zinc chloride concentration ($C^{1/2}_{ZnCl_2}$) (Figure 18) and the inhibiting effect of hydroquinone (Table 3) indicate that at potentials more negative than -1.1 V, polymerization occurs also in accordance with the radical mechanism. However, since the polymer formation on the cathode occurs in accordance with the mechanism (5.1.)-(5.2.) as well, then at C_{ZnCl_2} = 0 the rate $V_m = V_m^H > 0$.

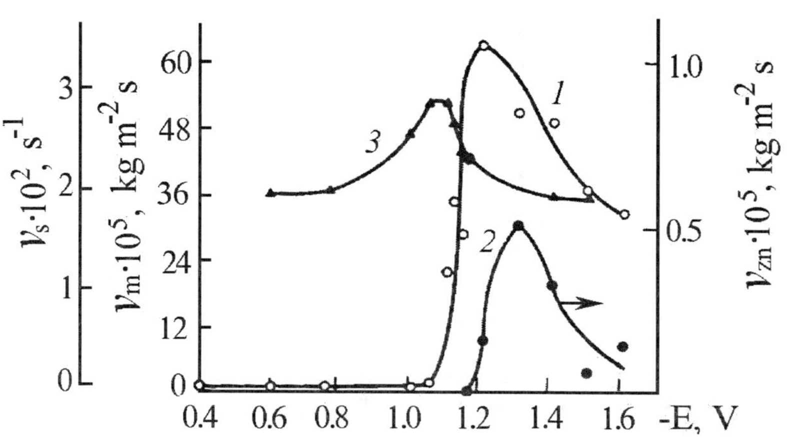

Figure 17. Effect of cathode potential on the rate of deposition of polymer (1), zinc (2) and the rate of electrode covering (3). Solution composition – as in 5.1.

Table 3. Effect of hydroquinone on polymer film formation rate. Solution composition (M): acrylamide – 7, formaldehyde – 7, zinc chloride – 0.05. E = -1.16 V. pH = 5

Concentration HQ, M	$V_m \cdot 10^4$, kg/m^2·s
0.0000	4.22
0.0005	3.62
0.0182	2.02

The $CH_2OH^{\bullet}$ radicals formed in the protogenic medium at formaldehyde electroreduction could be EIP initiators in the composition under study [126]. Indeed, the introduction of formaldehyde into the aqueous mixture of acrylamide and zinc chloride increases the EIP rate 1.5-fold at a potential -1.2 V. However, the crucial role in the polymerization initiation belongs to zinc-containing radicals, because under similar polymerization conditions, the addition of zinc chloride to the acrylamide-formaldehyde mixture increases the film formation rate 10-20-fold at virtually complete absence of polymerization in the solution.

The data on the composition of the zinc-containing acrylamide complex and EIP-initiating particles is rather contradictory.

According to [128], EIP is initiated by active particles formed at electroreduction of zinc compounds with molecular oxygen. The authors of [129] state that in this case, the role of theinitiator belongs to the anion $AAOH^-$ formed after the reduction of the zinc-acrylamide complex. Finally, the authors of [65] consider as active the radicals formed during reduction of the complex ion $[ZnOH^+AA]$.

Nevertheless, in accordance with our data, oxygen inhibits the acrylamide polymerization by 3-4-fold reduction of the film growth rate, which makes almost improbable the participation of the products of zinc interaction with molecular oxygen in the polymerization initiation.

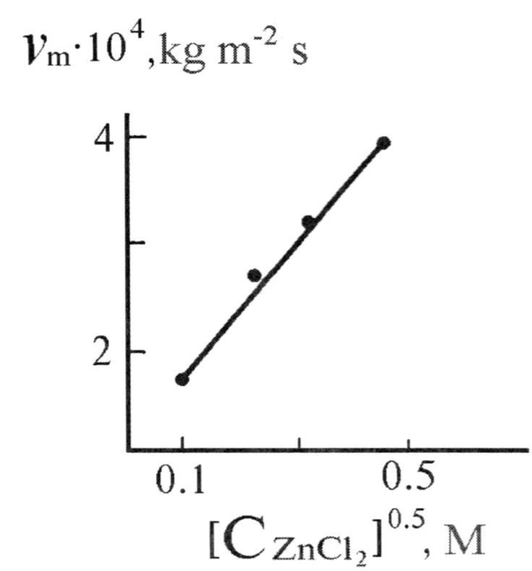

Fugure 18. Dependence of polymer deposition rate on the cathode on zinc chloride concentration. Solution composition (M): acrylamide – 7, formaldehyde – 7. Current density 10 A/m^2.

The second variant of the initiating particle formation also can not be considered absolutely adequate, since our studies demonstrated that zinc chloride substitution by zinc sulfate, albeit somewhat decreased the rate of polymer formation on the cathode, but still maintained it on a level significantly higher than in the absence of zinc salts (Table 4).

Here the rates difference can be explained by the depolarizing effect of the chlorine ion on the reduction of the positively charged complex zinc-acrylamide ion due to manifestation of the ψ'-effect [130]. The most convincing proofs in the favor of initiating radicals $[ZnOHAAH]^{\bullet}$ produced from the hydrolyzed form of the zinc-acrylamide complex $[ZnOHAA]^{+}$ are presented in [65]. However, in this case, at increasing pH the concentration of the hydrolyzed complex and, therefore, the EIP rate must increase monotonously, which is not corroborated by experimental data (Figure 19, curves 1, 2).

If one assumes anyway that the initiating particles have the form $[ZnOHAAH]^{\bullet}$, then the fact of a complicated dependence of the EIP rate on pH can be explained as follows. The pH growth from 1 to 7 increases as the concentration of $[ZnOHAA]^{+}$ as the share of the electrochemically inactive hydrolyzed form of formaldehyde $CH_2(OH)_2$ [126].

Table 4. Salt effect on polymer film formation rate. E = - 1.16 V. pH = 5

Composition, M	$V_m \cdot 10^4$, kg/m^2·s
7 AA + 7 F	0.238
7 AA + 7 F + 0.05 M $ZnCl_2$	4.220
7 AA + 7 F + 0.05 M $ZnSO_4$	0.562
7 AA + 7 F + 0.05 M LiCl	0.062

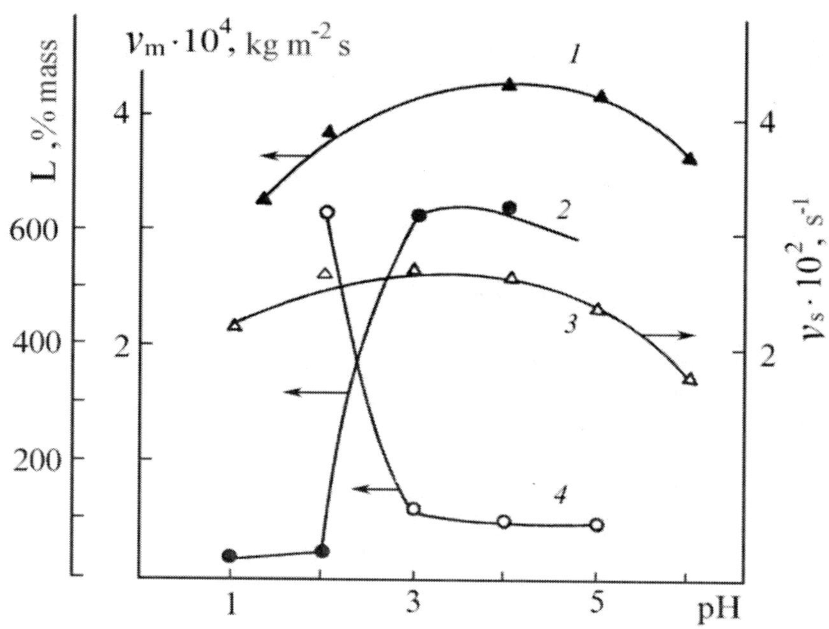

Figure 19. pH effect on polymer deposition rate on the cathode (1, 2), rate of electrode covering by the film (3) and swelling ability (4). Composition (M): acrylamide – 7, formaldehyde – 7, $ZnCl_2$ – 0.05 (1, 3) and 0.4 (2, 4). E = -1.16 V (1, 3); i=10 A/m^2 (2, 4).

So the acidity decrease, on the one hand, accelerates the initiation process, but, on the other hand, suppresses the formaldehyde participation in the methylolation and cross-linking processes. As a result of such an ambiguous pH effect, the polymer film formation rate is a complex combination of factors and, in particular, does not virtually change at pH 3-5.

7.2. Effect of Electrode Potential, Zinc Chloride Concentration, and pH on Film Formation Kinetics

The dependence of the polymer deposition rate on the cathode on the polarization *potential* in the presence of zinc chloride has a complex character (Figure 17). At $E \leq -1.05$ V, when the Zn initiating role is not yet manifested, the rate is small, while at $E \geq -1.1$ V, when initiation is determined by active particles formed at discharge of the zinc-monomer complex, the film formation rate grows dramatically (Figure 5.2, curve 1). Simultaneously, the film becomes solid and strongly adhered to the weakly swelling substrate. Metal zinc is formed under the film in amounts growing in the course of the potential increase (Figure 17, curve 2). At further potential increase ($E > -1.2$ V) and $C_{ZnCl_2} > 0.3$ M, the initiation rate continues to grow, but, due to excessive concentration of initiating particles in the near-electrode area, there increases the share of low-molecular water-soluble fractions, whose removal from the film results in the reduction of the rate of the polymer deposition on the cathode (Figure 5.2, curve 1). Simultaneously, there occurs alkalization of the layer between the film and electrode, which facilitates the dissolution of zinc formed on the cathode (Figure 17, curve 2).

In summary, , providing a high film formation rate at simultaneous preservation of its quality and minimal solution contamination by water-soluble fractions requires application of optimal *concentrations* of $ZnCl_2$ (0.01–0.3) M and *potentials* (-1.1) – (-1.2) V.

A substantial effect on the film quality and formation rate is provided by the solution *pH.* The maximal cross-linking (maximal swelling ability) and film growth rate are provided at pH 3-5 (Figure 19) under as galvanostatic as potentiostatic electrolysis conditions. At pH<2, the film formation rate significantly decreases due to the increase of the share of current spent on the hydrogen emission reaction, near-electrode layer mixing and polymerization product removal into bulk solution.

As follows from Table 5, the film formation rate is related to the cathode material nature and increases in the row Ti<Al<Cu<Fe. Such a dependence results from the fact that metals with an oxide film (Ti, Al) are characterized by an increased overpotential of zinc discharge, which is responsible for the decrease of the initiation rate provided by the zinc-acrylamide complex reduction.

Table 5. Effect of cathode material on polymer film growth rate. Solution composition (M): acrylamide – 7, formaldehyde – 7, $ZnCl_2$ – 0.4; pH 3. i=10 A/m^2

Cathode	Ti	Al	Cu	Steel
$V_m \cdot 10^4$, $kg/m^2 \cdot s$	0.18	0.58	2.96	3.13

7.3. Kinetics of Formation of Poly-AA-F Films

The *dynamics of film formation on the cathode* in the composition AA-F-$ZnCl_2$ under galvano- and potentiostatic electrolysis conditions is shown in Figures 20 and 21. The coating formation starts in both cases almost immediately, after switching on current and goes on initially with a constant rate, which then reduces due to metal coating by the polymer and hindering the transport of discharged ions to the electrode surface.

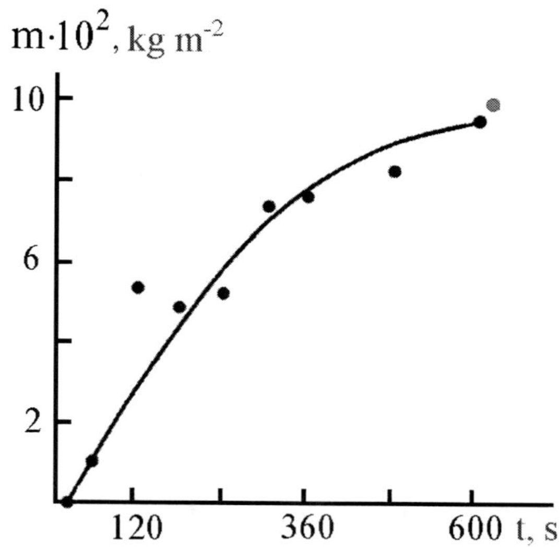

Figure 20. Changing of the polymer deposit mass at galvanostatic electrolysis. i=10 A/m^2. Solution composition (M): acrylamide – 7, formaldehyde – 7, $ZnCl_2$ – 0.01. pH = 5.

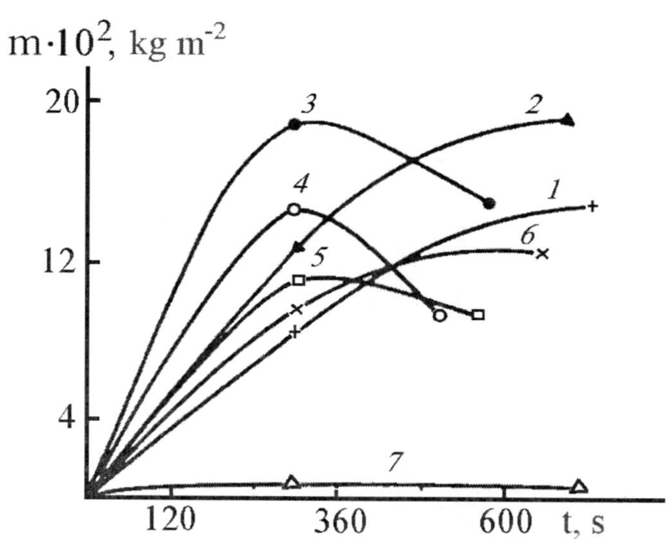

Figure 21. Changing of the film mass at potentiostatic electrolysis. Cathode potential (V): -1.14 (1), -1.16 (2, 7), -1.2 (3), -1.4 (4), -1.5 (5), -1.6 (6). Solution composition (M): acrylamide – 7, formaldehyde – 7, $ZnCl_2$ – 0.05 (1 – 6) and 0.00 (7). pH = 5.

Under the galvanostatic electrolysis conditions, the gaseous hydrogen emission starts after the complete metal surface covering by polymer, which worsens the coating quality. Under the potentiostatic electrolysis conditions, the gas emission and the defects it induces are virtually absent, since in the course of the electrode covering by the polymer the resistance increases, while the working current decreases. The films obtaining under constant potential has one more advantage: after the complete cathode covering by the polymer there occurs the removal of low-molecular fractions from the film and its electroosmotic drying, which results in the reduction of the coating mass (Figure 21., descent on the curves) and the formation of dense insoluble deposits. That is why the potentiostatic electrolysis conditions are optimal for EIP.

Since within the first few minutes of electrolysis ($\tau \leq 5$ мин) the mass of the polymer deposit on the cathode (m) increases linearly over time (Figure 21), the polymer cathode formation rate can be estimated as $dm/d\tau$, limiting the electrolysis period by five minutes.

Dependence of the coating formation rate on the acrylamide/formaldehyde ratio. The electrosynthesized poly-AA-F film is a copolymer product. The polymer IR-spectrum contains a series of bands at 1035, 1330 and 1400 cm^{-1} corresponding to the OH–group vibrations in primary alcohols (Figure 22). The latter confirms the formation of the methylolamide groups –NH–CH_2OH in the polymer in accordance with the reaction:

$$V_m = V_H + k i_0^{1/2} c_{AA}^{a} c_{ZnCl_2}^{1/2} \exp(\alpha n F \eta / RT)$$

Besides, the spectra contains the band 1130 cm^{-1}, which can be assigned to the vibration of alkyl ethers – CH_2–O–CH_2– [131]. The presence of this band indicates the cross-linking of the polyacrylamide film as a result of methylamide groups condensation.

The degree of methylolation and cross-linking estimated on the increase of the swelling ability and solubility in water depends on the electrolysis conditions and composition components and correlates with the change of the intensity of the band 1130 cm^{-1} in IR-spectra.

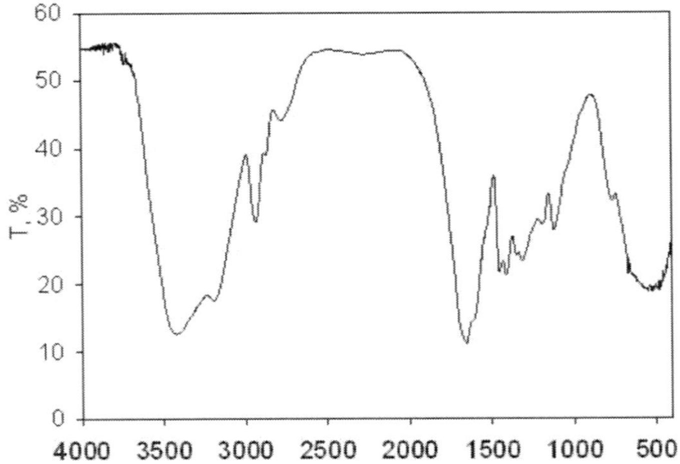

Figure 22. IR-spectrum of the copolymer of acrylamide and formaldehyde.

The most important factor, which determines the film quality, its formation rate and thickness, is the formaldehyde/acrylamide molar concentration ratio. At the molar ratio F:AA<1, the rate of deposit formation on the cathode is virtually the same as in the absence of formaldehyde (Figure 23, linear part of curve 1). Therefore, at small formaldehyde concentrations, its participation in the film formation and cross-linking processes is insignificant. The latter conclusion is confirmed by the polymers IR-spectra. Cross-linking starts to take place only at the molar ratio of the components under study equal or more than 1, which results in an abrupt acceleration of the electrode covering by the polymer (Figure 23, ascent on curve 1). Under the above conditions, the film growth rate is proportional to the acrylamide concentration in solution (Figure 23, curve 2).

A general increase of the acrylamide and formaldehyde concentrations in the solution at the ratio AA:F = 1:1 results in an increase of the rate of the coating formation (Figure 24), homogeneity of the film distribution over the electrode and its average thickness. Only at the acrylamide and formaldehyde concentrations above 7 M, the above regularity gets broken due to the attainment of the acrylamide solubility limit and the start of its crystallization on the cathode. Thus, the highest rate of the formation of a film with a highly cross-linked structure is attained in the solutions containing the formaldehyde amount equivalent to that of acrylamide or exceeding it (at the acrylamide concentration 5–7 M).

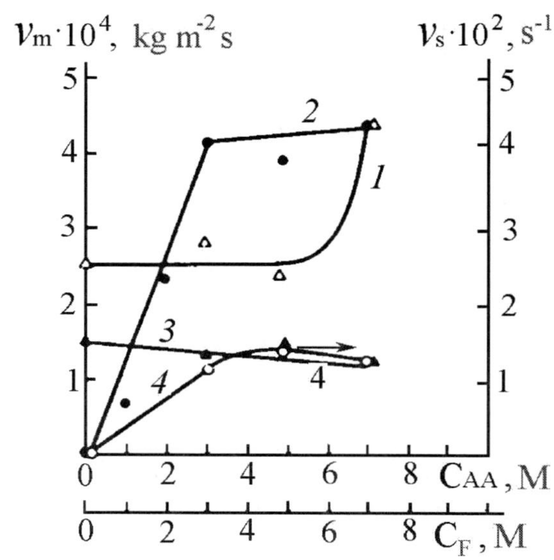

Figure 23. Effect of formaldehyde (1, 3) and acrylamide (2, 4) concentration on the polymer deposition rate on the cathode (1, 2) and the rate of the electrode surface covering by the deposit (3, 4). Composition (M): acrylamide – 7 (1, 3), formaldehyde – 7 (2, 4), zinc chloride – 0.05 (1–4). E = -1.16 V. pH = 5.

Coating formation rate. At the excess of F and $C_{AA} \leq 3$ M, the rate V_m of the polymer mass increment on the cathode increases proportionally to the acrylamide concentration, while at $C_{AA} > 3$ M, it is virtually independent of C_{AA} (Figure 23, curve 2). The process kinetics at the discharge potentials of the zinc-acrylamide complex, with taking into account initiation in accordance with the mechanism (5.1) and (5.2), can be expressed by the equation:

$$V_m = V_H + k i_0^{1/2} C_{AA}^{a} C_{ZnCl_2}^{1/2} \exp(\alpha n F \eta / RT) \quad (11),$$

where V_H – the polymerization rate at the limit discharge current of the hydroxonium ions; a = 1 at ($C_{AA} \leq 3$ M) or 0 (at $C_{AA} > 3$ M); C_{AA} и C_{ZnCl_2} – the concentrations of acrylamide and zinc chloride, respectively; i_0 и η – the exchange current and overpotential of the discharge reaction of zinc-acrylamide complexes.

The film formation is characterized not only by the rate of the polymer mass growth, but also by the rate of the electrode surface coating by the polymer V_s. Both of these rates are related to each other as:

$$V_m = dm / d\tau = d\, S\, h_\tau \rho / d\tau = V_s\, h_\tau \rho \quad (12),$$

where τ – the electrolysis period; *S, h* and ρ – the film surface area, thickness and specific weight, respectively.

Juxtaposition of films V_m, V_s, the swelling ability and electroconductivity provides additional information on the deposit formation regularities. The estimation of V_s and electroconductivity was performed on the curves of current drop over time under potentiostatic electrolysis conditions (Figure 25). In the course of the electrode isolation (due to covering) by the polymer, its free surface and, therefore, the current value, decrease. At small electrolysis periods, when the fraction of the isolated surface is small, the latter enables one to find the values V_s in accordance with the equation:

$$V_s = d\, S / d\, \tau = (i_н - i_\tau) / (i_н - i_к)\, \tau \quad (13),$$

where $i_н$, i_τ, $i_к$ – the initial current density (CD), CD in time and residual CD after complete electrode covering by the polymer, which depends on the film electroconductivity, respectively. That is why the coating electrical conductivity was characterized on $i_к$ at E = const.

Under the potentiostatic electrolysis conditions, whether the current decreases monotonously (Figure 25, curves 1, 2) or the general decrease is accompanied by a temporary growth (curves 3–5). The maximum formation is concerned with the alkalization of the near-electrode layer due to water discharge, temporarily increasing the reaction current:

$$H_2O + \bar{e} \rightarrow \tfrac{1}{2} H_2 + OH^- \quad (14)$$

and decreasing the zinc discharge overpotential. In acidic solutions at pH<3, the necessary alkalization degree is not attained and the curves do not have maximums.

The rates V_m and V_s change in a symbate way with the change of the potential, monomer concentration and pH (Figures 17, 19, 23). As seen from Equation (12), this is possible whether in the case of the unchangeable thickness of the formed film or at its insignificant transformation as in the process of the surface covering by the polymer at the electrolysis conditions modification. On the basis of the above, the coating formation can be represented as proceeding in two stages: first, there occurs the initiation of the polymerization sites of specific thickness; second, the sites spread two-dimensionally over the metal surface until merging into solid film. Here, the variation of the solution composition and electrolysis conditions have little effect on the coating thickness.

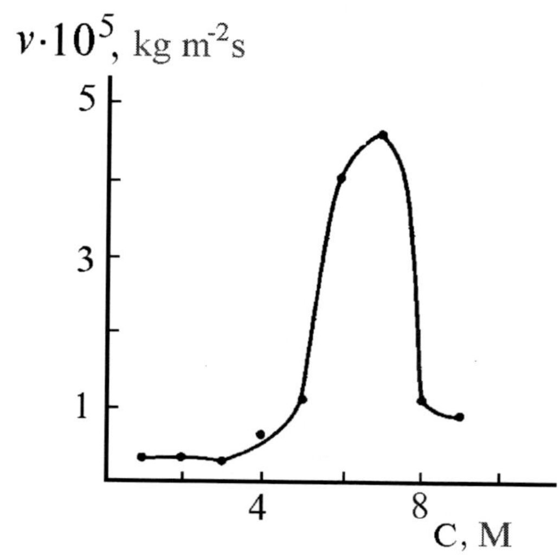

Figure 24. Dependence of the polymer film deposition rate on the cathode on the acrylamide and formaldehyde concentrations: [AA]:[F] = 1:1, i=30 A/m^2.

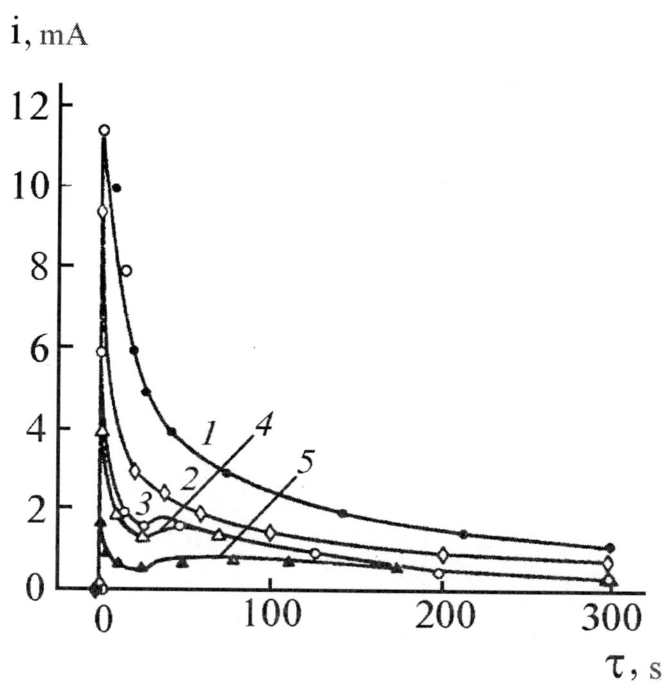

Figure 25. Changing of the current at potentiostatic electrolysis. E = -1.16 V. Solution composition (M): acrylamide – 7, formaldehyde – 7, $ZnCl_2$ – 0.05. pH: 1(1), 3 (2), 4 (3), 5 (4), 6 (5).

8. Formation of Polymer Films from Aqueous Solution of N-Methylolacrylamide [132]

The polymer similar in composition to poly-AA-F is formed in the process of EIP of N-methylol acrylamide (N-MAA), which is also a water-soluble monomer stable in weakly acidic solutions [125]. The monomer is capable to reduce at E = -2.1 V (double bonds reduction). The wave height is proportional to the monomer concentration (Figure 26). The N-MAA reduction process is irreversible, which is corroborated by the absence of a peak on the anode part of the polarogram (*di/d*E) – E.

The polarographic data indicates the possibility of direct EIP of N-MAA, but only at high cathode potentials. However, using metals with hydrogen overpotential below then for mercury results in a turbulent emission of gaseous hydrogen in the range of such high potentials, which prevents the polymer deposition on the electrode.

Though, it was established that the electrolysis of N-methylolacrylamide in an aqueous solution without additives was accompanied by the emerging of the polymer film on the steel cathode as early as at E ≥ -0.5 V. The latter indicates the possibility of indirect initiation of N-methylolacrylamide polymerization by an atomic hydrogen formed at the potentials (-0.5) V – (-1.0) V (Figure 27). However, the rate of the polymer layer formation on the electrode is very low, just like the coating quality, thickness and adhesion. In view of the above, we investigated the possibilities of electropolymerization of N-methylolacrylamide in the presence of initiators: persulfate-ions and $ZnCl_2$. The addition of initiators increases the EIP rate 30-50-fold. However, the use of persulfates results in the complete conversion of the monomer within 24 hours after the current shutoff. At the same time, the EIP process in the presence of $ZnCl_2$ is limited by the cathode surface. Poly-N-MAA films are formed in the presence of $ZnCl_2$ under the electrolysis galvanostatic and potentiostatic conditions. However, the deposits thicknesses and rates of growth on the metal surface get substantially intensified in the potential range of the zinc cathode reduction (Figure 28, curve 1; Figure 29, curve 3).

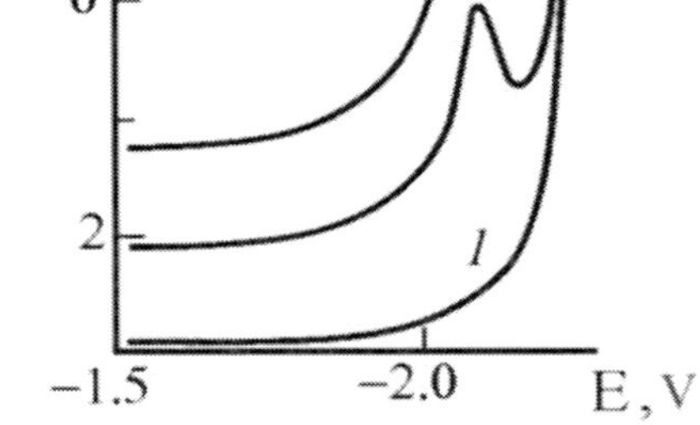

Figure 26. Differential oscillographic polarograms of 0.1 M $LiClO_4$ (1) and N-MAA (2, 3). Concentration of N-MAA (M): $2{\cdot}10^{-4}$ (2) and $5{\cdot}10^{-4}$ (3).

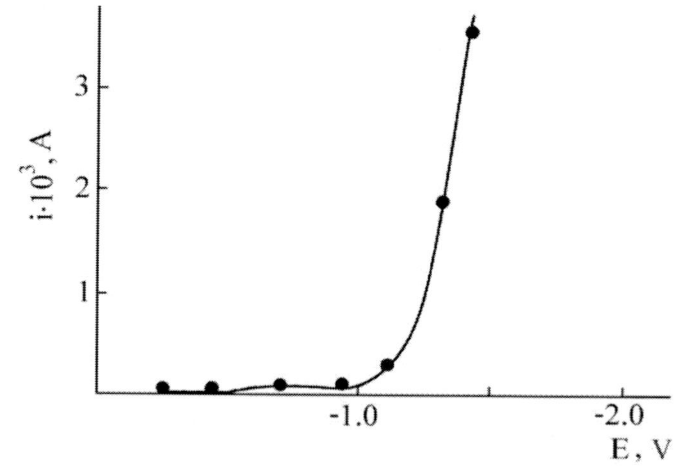

Figure 27. Polarization curve of steel cathode in aqueous solution of 3 M N-MAA.

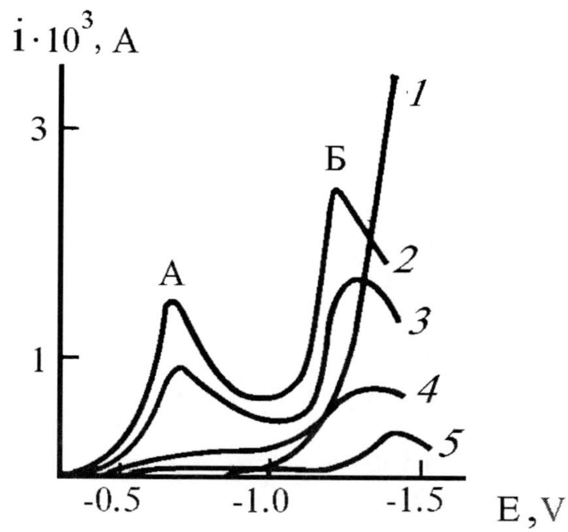

Figure 28. Polarization curves of the steel cathode in the aqueous solution of zinc chloride, 0.25 M (1) and the same solution with addition of N-MAA(M) : 1 (2), 2 (3), 4 (4), 6 (5).

The cathode polarization curves (Figure 28) contain two waves, whose descents are concerned with the formation of the polymer deposit covering the cathode. The first wave is formed independently of the presence of $ZnCl_2$ in the solution and reflects the reaction of hydrogen ions reduction (Figure 28, A), since its height increases with the solution acidification. The potential range of the second wave (Figure 28, B) corresponds to zinc reduction to the metal sub-layer under the polymer film.

The rate of the polymer growth on the electrode at potentials (-0.5 ÷ -1.1) V is equal to $(4–5)\times10^{-4}$ kg/m^2·s. (Figure 29, curve 3). In this case, a thin coating, weakly adhered to the substrate, is formed. In the zinc reduction potential range (E > -1.1 V), the polymerization rate increases 20-50-fold (Figure 29, curve 3), so that the polymer layer formation completes within 2-5 minutes.

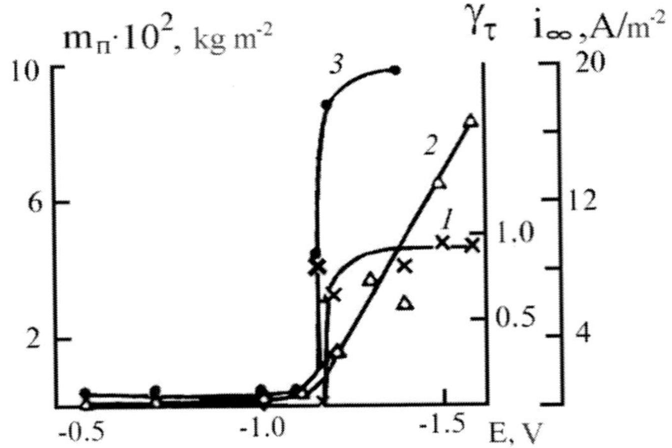

Figure 29. Effect of cathode potential on the soluble fraction content in the film (1), residual current value (2) and polymer coating mass (3). Concentration (M): N-MAA – 3, $ZnCl_2$ – 0.25. pH = 5.

The formed polymer films are distinguished by solidity, heterogeneity, good adhesion to metal and low swelling ability (12–19 weight %). Such a dependence on the polymer deposits quality and the rate of their formation on the cathode polarization value indicate to the existence of two mechanisms of EIP of N-MAA. At E = (-0.5 ÷ -1.1) V, the role of the initiator belongs to the active particles formed at the hydrogen ions reduction, just like in the monomer solution without initiators [54]: N-MAA + H^+ + e^- → N-MAAH$^•$. On the other hand, at E > -1.1 V, the initiation is produced by the active particles formed at the reduction of the zinc-monomer complex.

The complex existence is confirmed by the shift of the zinc semi-wave potential to negative values and decrease of the wave height at the increase of $C_{N\text{-}MAA}$ and constant $ZnCl_2$ concentration (Figure 28, curves 2-5).

At some fixed potential, somewhat lower than the zinc semi-wave potential, the zinc reduction starts only after the attainment of some specific concentration of its salt, which is corroborated by an abrupt increase of the initial current i_{in} (Figure 30, curve 1). Simultaneously, with the start of the zinc active reduction, the polymer film formation rates increases (Figure 30, curve 2).

Since the polymerization rate V_p is determined by the concentrations of the monomer [M] and initiating particles [R$^•$] [54], whose amount depends on the initiation current I, then

$$V_p \sim k\,(I)^n\,[M] \qquad (15),$$

where n = 0.5 or 1.0 for radical and ionic polymerization, respectively.

As follows from Figure 28, the total working current i_w is spent on the reduction of hydrogen and zinc. If i_w is higher than the limiting density of the hydrogen ions discharge current ($i_L^{H^+}$), the latter current can be considered for the composition N-MAA – $ZnCl_2$ as a side one, since the EIP rate for N-methylolacrylamide, determined by the hydrogen reduction, is negligible. Then the effective initiation current (I') is equal to:

$$I' = i_w - i_L^{H^+}$$

I' can be estimated quantitatively, if one knows the specified value i_w and determines $i_L^{H^+}$ from the polarization curves. Then

$$V_p = k\,(I')^n\,[M] \qquad (16).$$

Since $V_p \sim dm/d\tau$, where m – the polymer film mass; τ – the electrolysis time, at which the degree of the electrode surface covering by the polymer film is less or equal to 1, then

$$m \sim k\,\tau\,(I')^n\,[M] \qquad (17).$$

Under the galvanostatic conditions, EIP of N-methylolacrylamide in the presence of $ZnCl_2$, when $i_L^{H^+} < i_w \leq i_L^{Zn^{2+}}$ ($i_L^{Zn^{2+}}$ – the limiting density of the Zn^{2+} discharge current), the rate of the initiating particles feeding into the solution can be considered constant (here. the zinc current yield is accepted as equal to 100%). Under such conditions, the N-MAA polymerization rate is proportional to the monomer concentration (Figure 31), which confirms the validity of Equation (16). The polymer deposit mass depends linearly on $(I')^{0.5}$ (Figure 32), which indicates to the radical nature of the active particles formed at the zinc-monomer complex reduction.

The latter conclusion is confirmed by the absence of post-polymerization and bulk polymerization in the potential range from -0.5 V to -3.0 V and high polymerization rate in aqueous media. The IR-spectra of poly-N-MAA contain the absorption bands in the range 1040 cm^{-1}, thus confirming the polymer cross-linking on –C–O–C– groups.

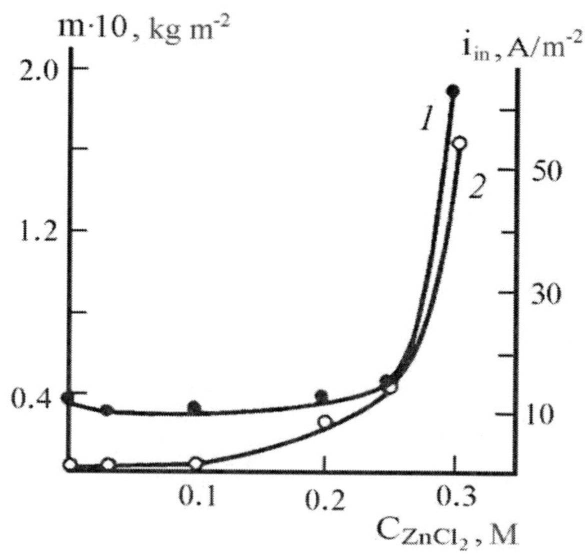

Figure 30. Effect of zinc chloride concentration on the initial current density (1) and polymer coating mass (2). 3 M N-MAA; E = -1.16 V.

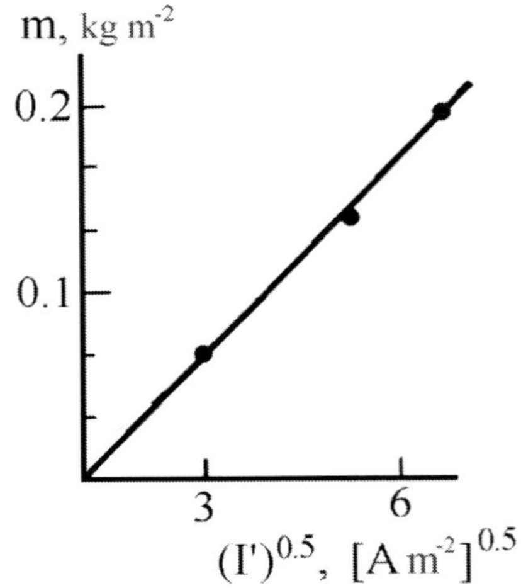

Figure 31. Dependence of polymer coating formation rate on the N-MAA concentration. Zinc chloride concentration 0.2 M. Current density 50 A/m^2.

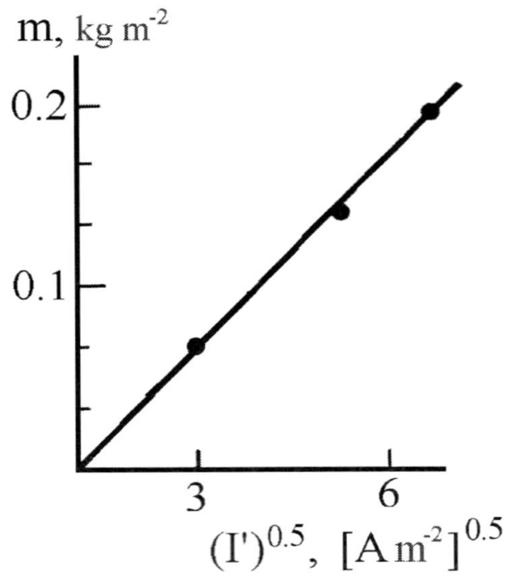

Figure 32. Dependence of the polymer coating mass on the initiation current value. Concentration (M): N-MAA – 3, zinc chloride – 0.25. pH = 5. Electrolysis time 60 s.

9. Principles of the Polymer Layer Formation in the Process of EIP [98]

The film formation process is characterized by the film growth rate V_m, time τ_p, required for stopping the polymerization, and the maximal polymer mass m_p. The higher the film electrolyte-permeability, the higher m_p corresponds to the moment of the complete electrode covering and, as a result, polymerization stopping. Thus, m_p unambiguously characterizes the film's looseness. Besides, since the polymer density (ρ) depends insignificantly on the polymerization conditions, the value m_p characterizes the limiting maximal coating thickness h_p, as $h_p = m_p/\rho$.

The rate of film growth on the electrode V_m at $\tau \rightarrow 0$ can be estimated on the slope angle of the curves $m - \tau$ or the change of the current density in the electrolysis process. In the latter case, the rate of film growth on the metal V_m can be calculated from the equation:

$$V_m = k_1 \frac{i_{in} - i_\tau}{(i_{in} - i_\infty)\tau} = k_1 \; V_m \tag{18}$$

where k_1 – a constant, i_{in} – the initial electrolysis current, i_τ – the current by the time τ, i_∞ – the current at $\tau \rightarrow \infty$.

The residual current i_∞ characterizes the permeability f of the solid polymer film and, therefore, its electroconductivity χ_p:

$$\chi_p = k_2 \cdot f = k_3 \cdot \frac{i_\infty}{E} \quad \chi_p = k_3 \cdot F \tag{19}$$

where χ_p – the solution electroconductivity, E – the voltage drop on the film, k_2 and k_3 – constants.

The current density, which defines the concentration of radicals and, consequently, the rate of EIP and the breakdown in the chain, dramatically alters in the course of electrolysis (Figure 33). At the same time, at a higher current density, the concentration of radicals is greater; the number of growing chains is larger; and their breakdown is more intensive, which leads to synthesis of low-molecular-weight (LMW) products. Indeed, the formation of a polymer coating at the surface of metal during an EIP process occurs in conditions of continuous formation of polymer fractions with different molecular weights and cross-linkage.

Some of these fractions may be soluble and can be washed away from the growing polymer film even while it is still growing. The amount of the soluble product γ_τ in a coating is maximum at the initial period of electrolysis (Figure 34).

The coatings deposited from solutions of different pH but the same EIP conditions differ in their characteristics. As the solution pH is decreased, the overall weight of the film m_p (Figure 35, curve 1) and the film's extreme thickness and electroconduction (i_{res}) (Figure 35 , curve 2) rise, although the share of the soluble fraction is in completely formed films drops (Figure 35, curve 3; Figure 34). This is evidence that properties of films are defined not so much by the magnitude of γ_τ as by the rate of removal of LMW products out of the coating. At larger removal rates, the number of voids and micropores emerging in the film that are not filled by the polymer are likely to be larger and, consequently, the films' looseness and permeability by an external electrolyte increases, as does the longevity of the process of the cathode coverage by a continuous film.

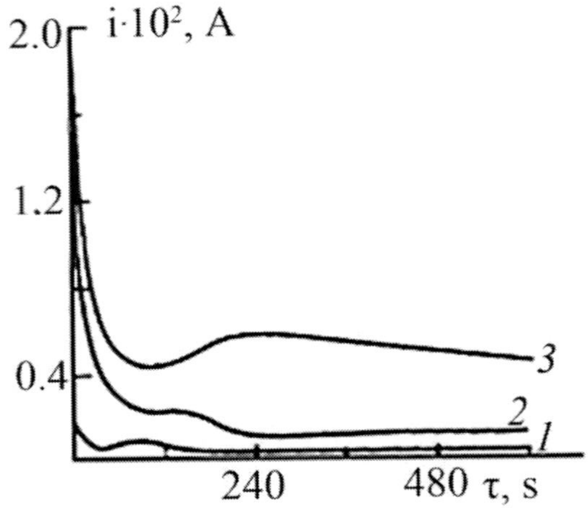

Figure 33. Changing of the current at potentiostatic electrolysis. E = -1.16 V. pH: 5 (1), 2 (2), 1 (3). Cathode area 1.3 cm^2. Solution composition (M): N-MAA 3, $ZnCl_2$ – 0.25.

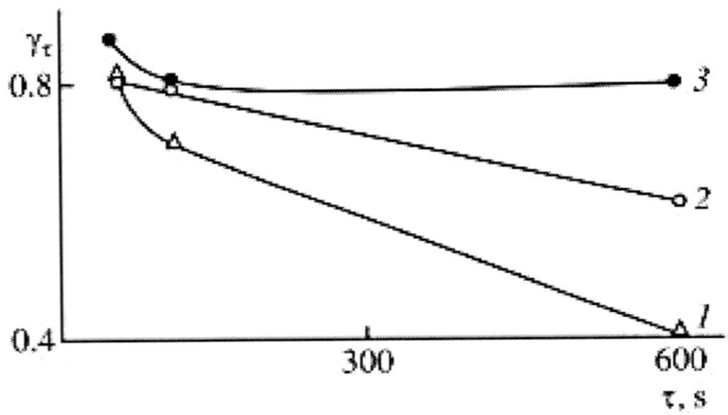

Figure 34. Time dependences of the soluble-fraction content in a polymer film during electrolysis of pH. (1) 1, (2) 2, and (3) 5; E = –1.16 , N-MAA.

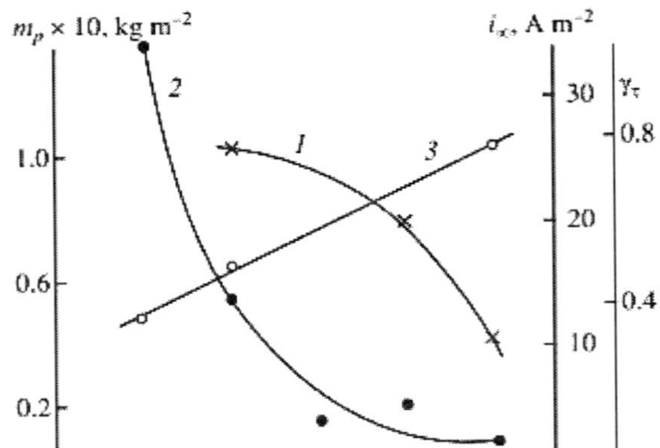

Figure 35. Effect of pH of electrolyte II on (1) weight of a polymer film, (2) residual current in it, and (3) share of soluble fraction in it; E = –1.16 V, deposition time 600 s. N-MAA.

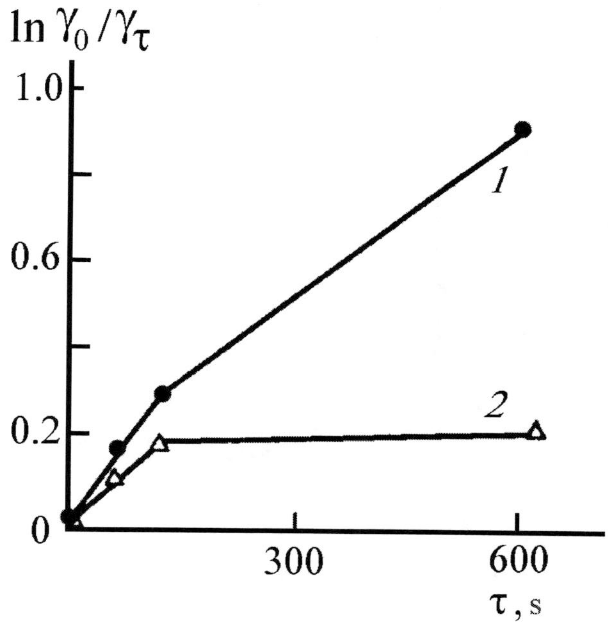

Figure 36. Dependence of the soluble-fraction share in a growing poly- N-MAA film from electrolysis time.

The direct proportional dependence between ln γ_0/γ_τ and τ (γ_0 – the fraction of γ at minimal electrolysis time τ) in the initial period of the film growth (Figure 36) indicates the fact that the concentration of the water-soluble fraction during the coating formation changes according to the equation:

$$\gamma_\tau = \gamma_0 \cdot e^{-k\tau} \quad (20)$$

and, therefore, its rate of dissolution from the film is equal to:

$$V_d = k_{pH} \cdot \gamma_\tau \quad (21)$$

where k_{pH} – a constant increasing with the pH increase.

Effects of N-MAA and zinc chloride concentrations on the rate of film growth on the cathode (Figure 37, curve 1; Figure 38, curve 1), initial current density (Figure 37, curve 2), soluble fraction content (Figure 37, curve 3; Figure 38, curve 2), polymer film mass (Figure 37, curve 4; Figure 38, curve 3) and residual current value (Figure 37, curve 5; Figure 38, curve 4) are submitted in Figures 37 and 38.

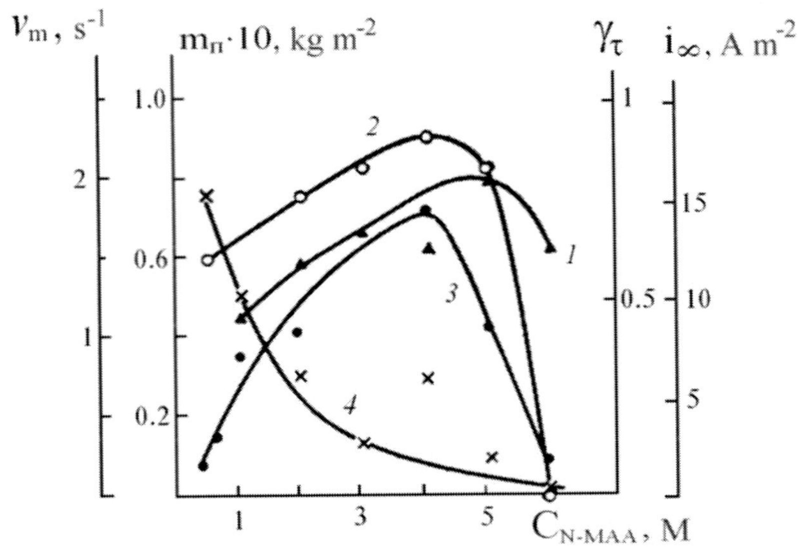

Figure 37. Effect of zinc chloride concentration on the rate of film growth on the cathode (1), initial current density (2), soluble fraction content (3), polymer film mass (4) and residual current value (5). Concentration of N-MAA 3 M. pH = 5. E = -1.16 V.

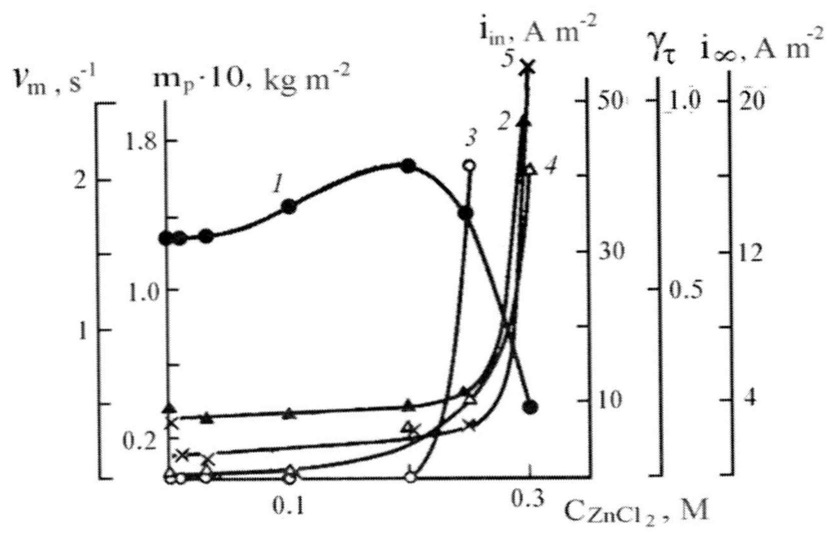

Fugure 38. Effect of N-MAA concentration on the rate of film growth on the cathode (1), initial current density (2), soluble fraction content (3), polymer film mass (4), and residual current value (5). Concentration 0.25 M. pH = 5. E = -1.16 V.

In summary , the formation of the film structure can be represented, in the first approximation, as follows. Initially, a loose network is formed from the copolymer low-molecular fractions emerging as a result of the action of the high rate of discharge of the complex generating the particles – polymerization initiators. Furthermore, this network is filled by denser, high-molecular fractions formed on the slow electrolysis stage. Such filling starts from the electrode/film interface and moves inside the solution. That is why the coating layer directed to the solution usually has a looser structure than the near-electrode layer. In the course of the potential change to more negative values than -1.16 V, the time of the dense film formation, registered on the moment of attainment of stationary values of the electrode capacity and resistance change rate, increases. Therefore, the loose network thickness increases as well.

10. Current-Voltage Characteristics of Electrode Modified by the Polymer Film [43, 133, 134]

The electrochemical behavior of the platinum electrode modified by the film was investigated in solutions containing potassium ferro- and ferricyanides (0.05 M) or tetramethylthionine chloride (TCh) (10^{-3} M) as depolarizers. The selection of test substances was caused by the following factors: (1) they are capable to undergo reversible oxidation-reduction transformations in the electrolysis process; (2) the depolarizers' redox-potentials lie in the range of electrochemical stability of the polymer under study (Figure 39, curve 1, Figure 39, curves 2-9); (3) the signs of the charge of the selected diffusing particles are opposite: $[Fe(CN)6]^{3-/4-}$ and $[C_{16}H_{18}N_3S]^+$; (4) the particles moving under the electrical field effect have different radii r: r ($[Fe(CN)6]^{3-/4}$) ~ (0.36-0.455) nm; r (TCh) ~ (0.75-0.79) nm.

The cyclic current-voltage graphs of the depolarizers under study on pure platinum and the electrode modified by the polymer film, obtained at different potential sweeping rates, have characteristics for this type of substances anode and cathode maximums at 0.340 V and 0.230 V for $[Fe(CN)_6]^{3-/4-}$ (Figure 39, curve 2, 3) and -0.220 V and -0.280 V for TCh (Figure 39, curve 4, 5). The presence of film on the electrode does not virtually change the values of these potentials.

The impedance spectroscopy demonstrated that the electrode modification by the poly–AA–F film affected insignificantly the change of the solution ohmic resistance, which, under identical conditions on Pt and Pt/film electrodes, is equal to 9.9 Ohn cm^2 and 9.0 Ohm cm^2 for $[Fe(CN)_6]^{3-/4-}$ (Figure 40) and 15.3 Ohm cm^2 and 15.0 Ohm cm^2 – for TCh (Figure 41), respectively. Such an effect must be caused by the polymer's high swelling ability. Nevertheless, the diffusion mobility of TCh, $K_3[Fe(CN)_6]$ and $K_4[Fe(CN)_6]$ in the polymer decreases substantially.

The latter is confirmed by the decrease of the density of the peak i_p on current-voltage curves, measured on polymer-modified electrodes with the above mentioned polarizers, as compared to pure platinum (Figure 39). On the other hand, at the potential sweeping rates $V \leq 10^{-2}$ V/s, the values i_p in all the systems under study are almost directly proportional to $V^{1/2}$ (Figure 42). The latter enabled us to consider the processes on pure and modified electrodes quasi-reversible and calculate the diffusion coefficients of depolarizers in solution (in the case of Pt electrode) and in polymer film (in the case of Pt modification) on the Randles-Shevchik equation [135]:

$$i_p = 2.69 \cdot 10^5 \cdot n^{3/2} \cdot A \cdot D^{1/2} \cdot V^{1/2} \cdot C^0,$$

where C^0 – the depolarizer concentration in solution (film).

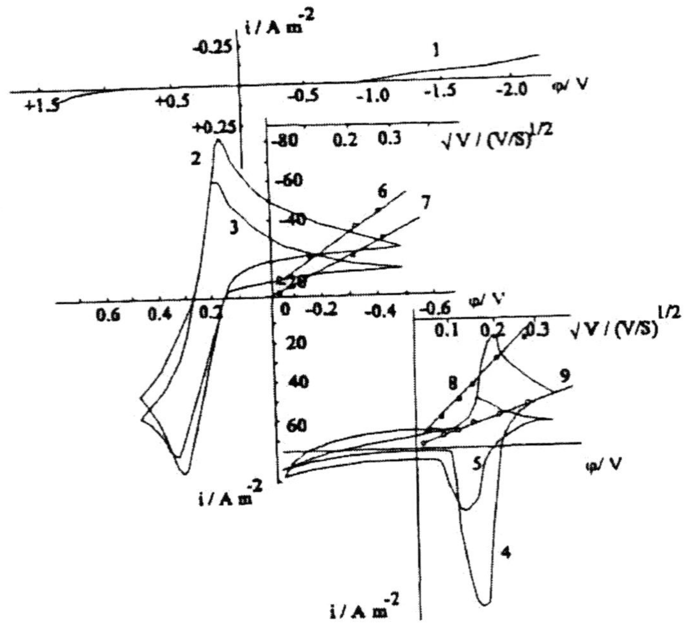

Figure 39. Voltampograms and dependences of current on $V^{1/2}$ on Pt (2, 4, 6, 8) and Pt/poly–AA–F film (1, 3, 5, 7, 9) in dimethylformamide (1); 0.05 M $[Fe(CN)_6]^{3-/4-}$ / 0.5 M Na_2SO_4 (2, 3, 6, 7); 0.001 M tetramethylthionine chloride / 0.5 M Na_2SO_4 (4, 5, 8, 9). The velocity of the polarizing voltage change ($V\ s^{-1}$): 0.001 (1); 0.08 (2-5).

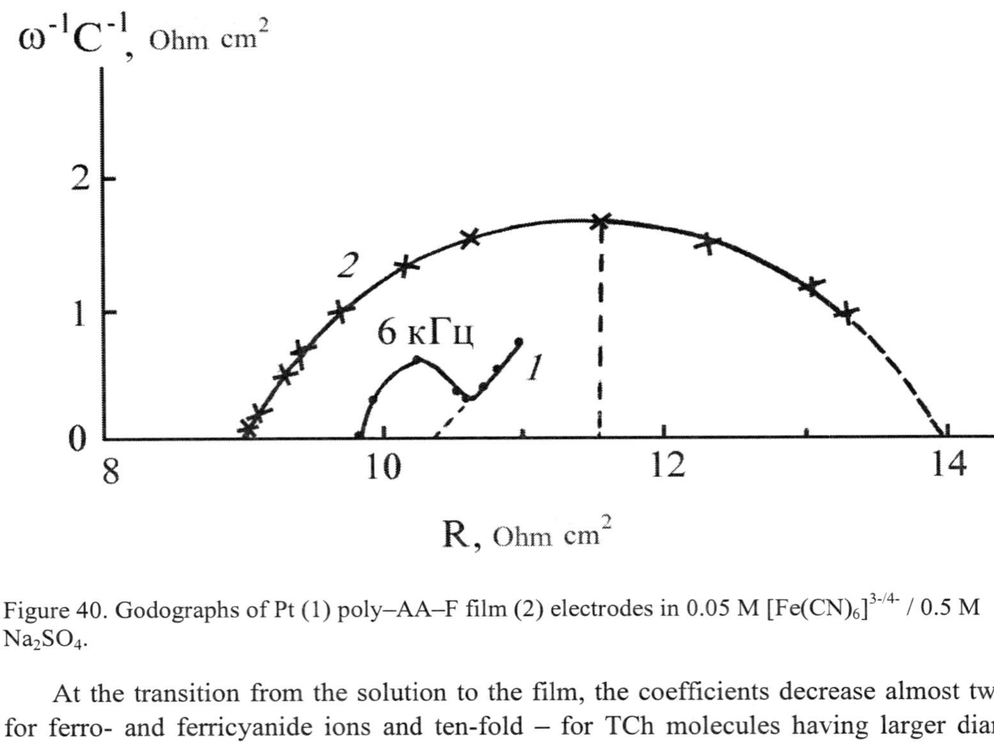

Figure 40. Godographs of Pt (1) poly–AA–F film (2) electrodes in 0.05 M $[Fe(CN)_6]^{3-/4-}$ / 0.5 M Na_2SO_4.

At the transition from the solution to the film, the coefficients decrease almost two-fold for ferro- and ferricyanide ions and ten-fold – for TCh molecules having larger diameters (Table 6). Therefore, the depolarizers' concentration and mobility in the polymer film can be substantially different from those in the solution. That is why the modified electrode must be characterized by proceeding those electrode transformations (among parallel possible ones), which are concerned with oxidation (reduction) of only ions and molecules with good permeability through the film. In other words, such an electrode would manifest selective properties.

Table 6. Depolarizers diffusion coefficients in solution (D) and poly–AA–F film (D_{film}). Temperature 18°C

Depolarizer	$D \times 10^5$, cm^2/c	$D_{film} \times 10^5$, cm^2/c	D / D_{film}
0.05 M $K_4[Fe(CN)_6]$	0.85	0.44	1.93
0.05 M $K_3[Fe(CN)_6]$	0.94	0.53	1.77
1.07×10^{-3} M TX	6.90	0.68	10.20

For TCh, the film hindering effect is not limited only by complications of the molecule transfer from the solution to the reaction area. On the modified polymer electrode, one is also capable to observe the limitations imposed on the depolarizer reduction products removal from the metal/film interface, which are registered neither in the sodium sulfate solution on the film-coated electrode (Figure 43, curves 1, 2) nor in the TCh solution on pure platinum (Figure 43, curve 3).

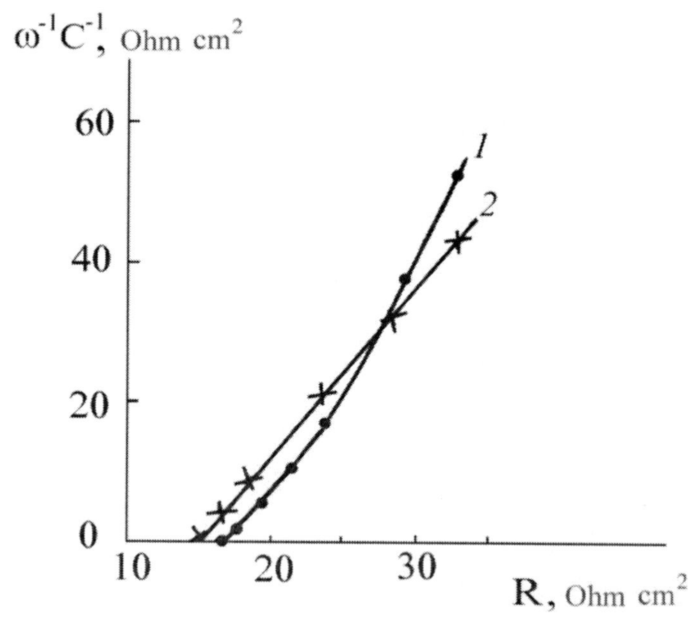

Figure 41. Godographs of Pt (1) poly–AA–F film (2) electrodes in 1.07×10^{-3} M TCh / 0.5 M Na_2SO_4.

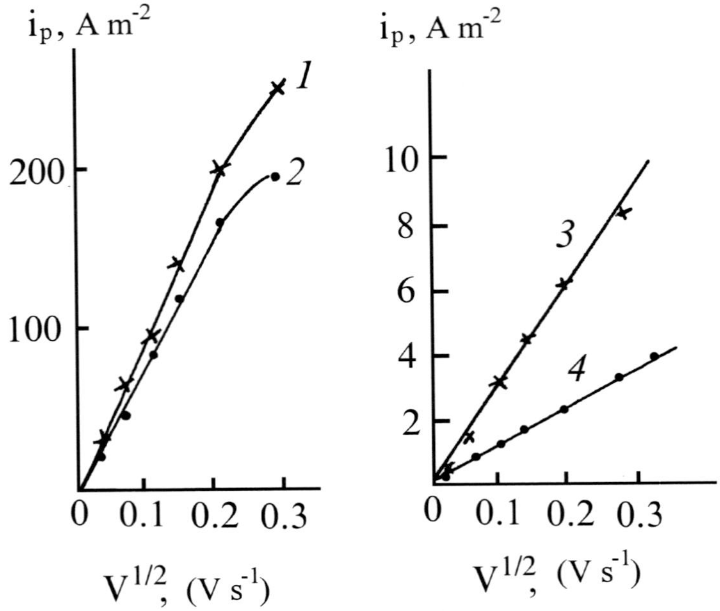

Figure 42. Dependence of the peak current value on $V^{1/2}$ on voltamperograms of Pt (1, 3) and Pt / poly–AA–F film (2, 4) electrodes in solutions: 0.05 M $[Fe(CN)_6]^{3-/4-}$ / 0.5 M Na_2SO_4 (1, 2) and 1.07×10^{-3} M TCh / 0.5 M Na_2SO_4 (3, 4). Potential sweeping rate 0.08 V/s.

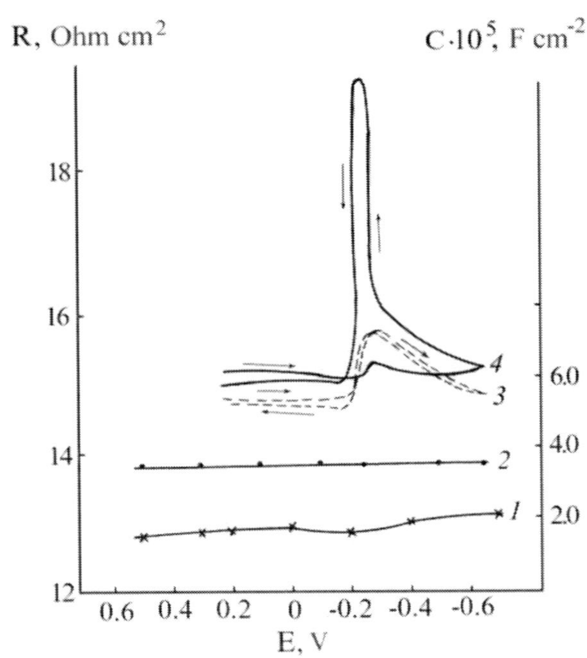

Figure 43. Dependence of capacity (1) and resistance (2-4) of Pt (3) and Pt/poly-AA-F film (1, 2, 4) electrodes on electrolysis potential. Composition: 0.5 M Na2SO4 (1, 2) and 1.07×10-3 M TCh / 0.5 M Na2SO4 (3, 4). Potential sweeping rate 0.01 V/s. 8000 Hz.

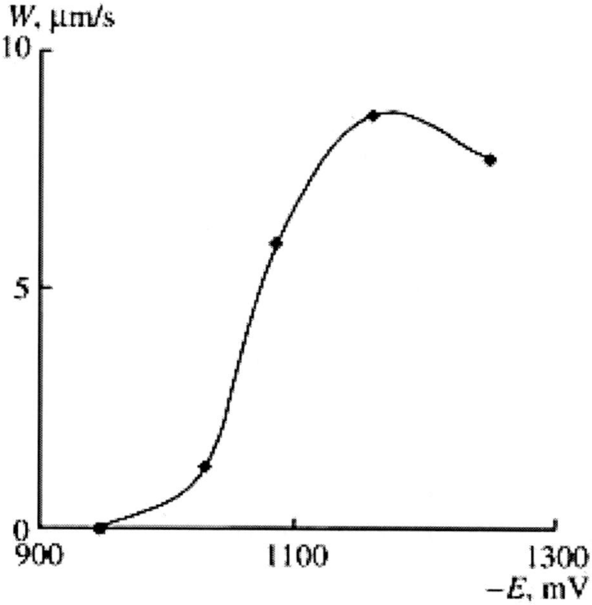

Figure 44. Dependence of the film growth rate *W* on the cathode potential. Electropolymerization time is 20 s, temperature 26°C.

The anomalous behavior of TCh is most clearly seen on different characters of the curves R – E (Figure 43, curves 3, 4) on pure and modified electrodes in the TCh solution. On pure platinum, the direct and reverse curve trends are identical (Figure 43, curve 3). The observed on curve 3 small maximum in the range of TCh redox-potentials is concerned with the formation of low-conducting TCh molecules in the process of their reduction and adsorption on the electrode. However, similar direct and reverse curve trends at the potential cycling indicates the free removal of TCh electroreduciton products from the metal surface.

On the other hand, during the reverse potential sweep on the modified electrode, one can observe an abrupt resistance increase in the range of the reduction of tetramethylthionine chloride (-0.280 V) (Figure 43, curve 4). Such a behavior could be the result of a substantial sub-film near-electrode zone accumulation of the reduced form of TCh molecules, whose removal from the surface is limited by the film, as well as of their high adsorbability on the electrode surface.

11. In Situ Study of Electropolymerization Kinetics of Acrylamide, Formaldehyde, and N,N'-Methylenebisacrylamide [136]

11.1. Using High-Resolution Videomicroscopy for Research of Polymerizations Kinetics

Electropolymerization as a route to polymer materials and coatings on their basis is now the subject of wide discussions in the literature. However, published data on growth kinetics, layer formation mechanism, and the supramolecular structure and physicochemical characteristics of layers is mostly data obtained by traditional methods based on mass and thickness measurements of the polymer deposit on the cathode surface and various physicochemical investigations of the polymer carried out after the experiment, i.e., ex situ. The prolonged time of investigation is one complication of the traditional approach; another difficulty is the deformation and even decomposition of samples, which are inevitable, for example, when films are being separated from the support or when polished sections (cuts) are being prepared for microscopic studies. Furthermore, some subtle features of real-time electropolymerization escape the attention of the researcher. An *in situ* study will evidently be more informative and correct than *ex situ* investigation. The above-mentioned difficulties prompted to seek new methods for investigating electropolymerization and its products not suffering from the above-mentioned disadvantages or having, at least, minimum drawbacks.

One solution to the problem is using the electrochemical quartz crystal microbalance (EQCM) piezoelectric procedure [137, 138]. Another solution for in situ investigations of electropolymerization is using optical microscopy. Visual observations with a microscope can give ample information for rationalizing fundamental processes that form polymer layers during electropolymerization [139].

To study growth dynamics of the polymer deposit, we have developed an experimental apparatus. The apparatus consists of a microcell, optical microscope with differential interference contrast (Axioplan 2 imaging, Carl Zeiss, Germany) and a digital videocamera (NV–GS 250, Panasonic, Japan) [136]. The kinetics (dynamics) of electropolymerization was investigated by the following procedure. An appropriate voltage was supplied to the microcell using a potentiostat to initiate and maintain growth of a polymer film at a given cathode potential. The moment of voltage supply was acoustically fixed on a videorecord. Film growth was visualized with a videocamera. Video information was analyzed and the reaction rate determined on a display by fixing the period of time within which the growth front of the polymer layer has shifted to a certain distance.

The induction period of electropolymerization was determined on a setup with an identical design except that the microscope had a phase contrast feature. The microscope was connected to a light detector (but not to a videocamera) based on an FD-288 V photodiode included in the classical scheme of a current-to-voltage converter. The signal of the light detector was recorded with a four-channel storage oscillograph (ASK 3107). Apart from recording the light detector signal, the oscillograph simultaneously recorded the potential on the reference electrode and cathode–anode voltage on the microcell.

Conventional light microscopy has insufficient potential for investigating electrically synthesized transparent colorless polymer films based on acrylamide. In the visible range, their refraction factor is almost the same as that of the environment (in our case, aqueous solutions of monomers; water). As a result, the polymer layer is almost invisible under a light microscope. To alleviate this difficulty, we used high-resolution videomicroscopy combined with an optical contrasting mode, which makes a light microscope a powerful tool for investigating films with contrast as low as in our case.

An original design of the electrochemical cell underlies the new approach to in situ visualization of the electrochemical formation of a polymer layer on the electrode surface. The cell was placed in a lateral plane in the field of the microscope. The growing polymer layer (the product of electropolymerization) may be visualized, and changes in film thickness may be recorded for different reaction times and conditions. The results of this study afford kinetic data about electropolymerization if taken into account that the amount of the synthesized substance is proportional to the thickness of the electrodeposited layer.

The suggested cell design makes it possible to observe all changes on the electrode surface and in the polymer matrix in a "section," avoiding labor-consuming polishing, cutting, and shearing procedures.

This ensures visualization and real-time evaluation of changes in the physical state of the polymer layer, affording additional information about the mechanism of the processes.

The in situ study of the dependence of the film growth rate on the cathode potential within a strictly fixed period of time of electropolymerization showed that film formation is characterized by complex kinetics (Figure 44).

The shape of the curve presented in Figure 44 agrees well with our previous ex situ data on the kinetics of electropolymerization in an acrylamide–formaldehyde monomer composition in the absence of N,N'-MBAA, which is an efficient cross-linking agent [95].

In Figure 44, it can be seen that in the range of potentials -(950–1030) mV, the film growth rate is low. In the range –(1030–1160) mV, the rate increases drastically and then decreases in the range -(1160–1250) mV. Similar conclusions are obtained when interpreting data on the dynamics of the layer thickness at different potentials of electropolymerization (Figure 45).

The start of active growth of the polymer film (cathode potential is ≤-1085 mV) is accompanied by zinc deposition on the cathode surface (Figure 46). At a lower cathode potential (−1030 mV), the zinc sublayer is not visualized. Consequently, one can assume that the discharge of the [$ZnCl_2$/acrylamide] complex occurs at potentials from the range –(1030–1085) mV and is accompanied by zinc deposition on the cathode and generation (into the monomer solution) of active species that initiate polymerization and polymer deposition on the zinc surface.

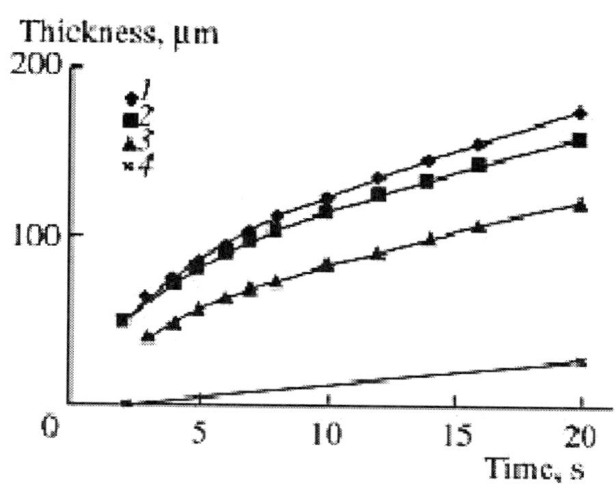

Figure 45. Dynamics of polymer layer growth for different electropolymerization potentials, mV: –1160 (1), –1250 (2), −1085 (3), –1030 (4).

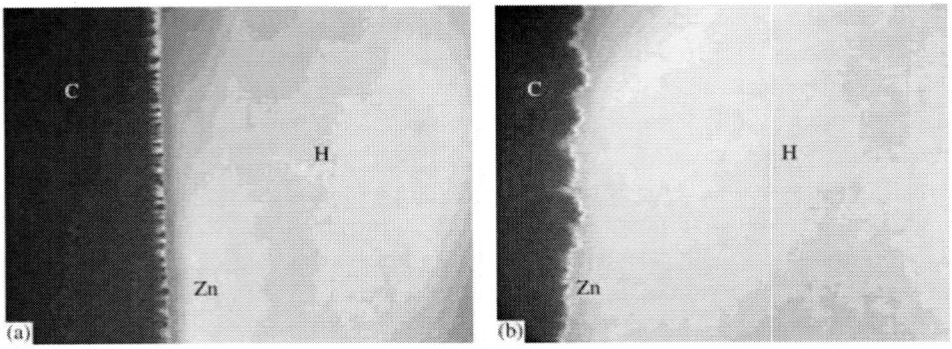

Figure 46. Electropolymerization of the monomer mixture AA–F–N,N'-MBAA–$ZnCl_2$: initial stage of zinc deposition (a), growing zinc dendrites (b), cathode (C), polymer layer (H).

In situ studies using optical microscopy allowed us to visualize for the first time, the growth of both zinc and polymer layers and to refine certain features of this process. As can be seen from Figure 46, at the initial stage, the zinc deposit is a very thin layer, uniformly distributed over the cathode surface. The thickness and roughness of the zinc sublayer increase with time. Quick growth of the polymer layer is accompanied by the slow growth of zinc dendrites. It can be seen that the polymer layer grows among zinc dendrites, fully repeating the microrelief of the zinc deposit. This leads to high adhesion between the polymer coating and the metal cathode surface. The mechanism of polymer layer formation revealed in this study also accounts for the fact that the near-cathode surface of the polymer membrane is rougher than the surface that contacts the solution [140].

Raising the temperature under identical conditions (components of the composition, electrode potential, cathode material) in the range of potentials –(1085–1160) mV leads to increased growth rate of the polymer layer, indicating that the temperature coefficient of the electropolymerization reaction is positive (Figure 47).

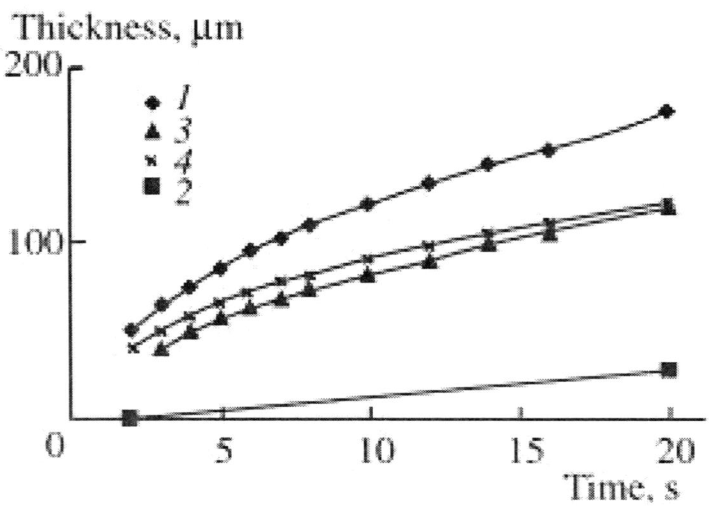

Figure 47. Dependence of layer growth dynamics on the electrode potential and temperature: –1160 mV (1), 26°C; –1085 mV, 13°C (2), –1085 mV, 26 °C; –1160 mV, 13°C (4).

11.2. Postpolymerization Effect

It was established that the polymer layer continues to grow on the cathode even after the cell is disconnected (Figure 48). The dynamics of layer thickness, reflected by the curve, is qualitatively the same as the dynamics reflected in Figure 45. As can be seen in Figure 48, *postpolymerization* damping is faster at lower temperatures. However, comparative analysis of data presented in Figures 45 and 48 indicate that under identical conditions (20 s, –1085 mV, 26°C) the growth rate of the polymer layer is 3.5 times higher during electrolysis than after disconnection (6 and 1.75 μm/s, respectively).

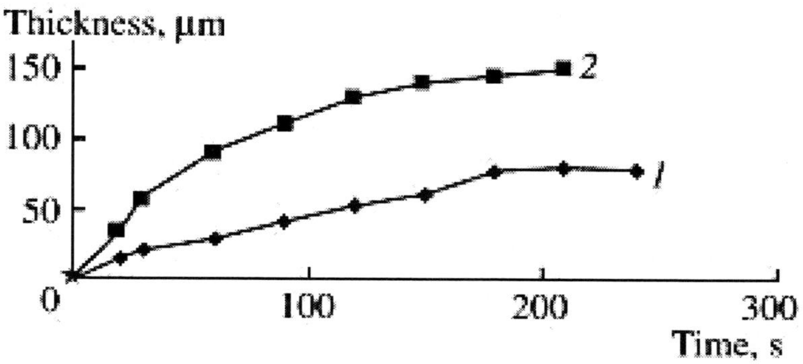

Figure 48. Growth of the polymer layer after electrolysis has stopped. Electropolymerization potential is –1085 mV; temperature, °C: 13 (1), 26 (2). {Leave the bottom figure}

The observed postpolymerization effect may be explained from different standpoints. It may be due to the high concentration and relatively high lifetime of polymerization-initiating radicals, and to the slow stage of polymerization chain break. Another factor may be direct contact between the polymer deposit and the aqueous solution of the monomer composition during electropolymerization and after it. Therefore, growth of the polymer deposit after cell disconnection may be attributed to gradual swelling of the polymer in the aqueous solution and as a consequence, to increased thickness of the polymer film. The third reason for postpolymerization growth of the polymer layer may be uncontrolled polymerization involving the zinc deposit and the monomer composition. We have detected the formation of a transparent elastic film that swells in water and is deposited on the surface of a zinc plate submerged into the monomer composition under study. The three factors may operate all together.

11.3. Contact Polymerization of Acrylamide on the Zinc Surface

A peculiarity of the electropolymerization of acrylic monomers with the addition of chloride of zinc is that the electrolysis process produces simultaneously a polymer film layer and a metal zinc sublayer on a cathode. Comparative studies of the kinetics of electropolymerization and zinc-initiated polymerization by the gravimetric method showed that zinc initiation is more efficient (Figure 49). It is evident that electropolymerization results in the formation of sufficiently fine-porous dense film, so that subsequent polymer deposit mass growth is limited by the porous blockage in the same manner as at switched-on current.

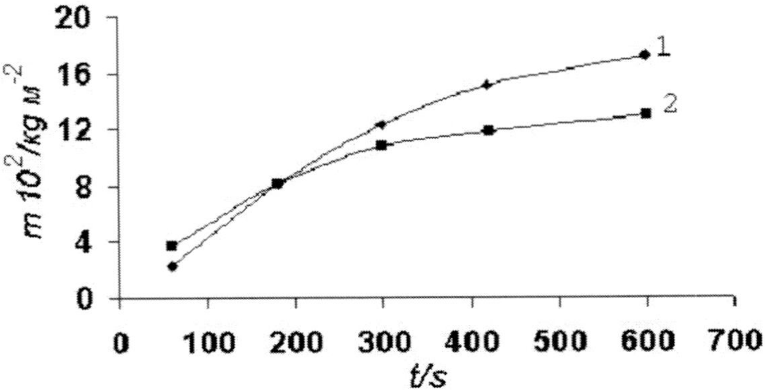

Figure 49. Kinetics of electroinitiated (2) and zinc-initiated (1) polymerization. E = -1.16 V; 25°C.

It is probable that at zinc-initiation, the formation of active particles inducing polymerization is also the result of the zinc-acrylamide complex electroreduction. The point is, after putting the zinc rod into the polymerization composition under study it gains the potential of about -(0.95-1.00) V. This value is sufficient to induce efficient electropolymerization. The most efficient cathode EIP potential is equal to -1.16 V (the highest polymerization rate, the most optimal film quality) [140]. The used overstated, as compared to the zinc rod potential, E = -1.16 V value is explained by the contribution of the solution resistance and the polymer porous film itself into the potential value. During initiation by zinc, the current does not go from cathode to anode and, therefore, a stray potential drop does not occur in the interelectrode space. The highest potential was attained using the monomer solution without $ZnCl_2$. Furthermore, the potential gradually decreases in the course of $C_{Zn^{2+}}$ increase. As a result, the polymerization rate and polymer product yield are decreasing (Figure 50).

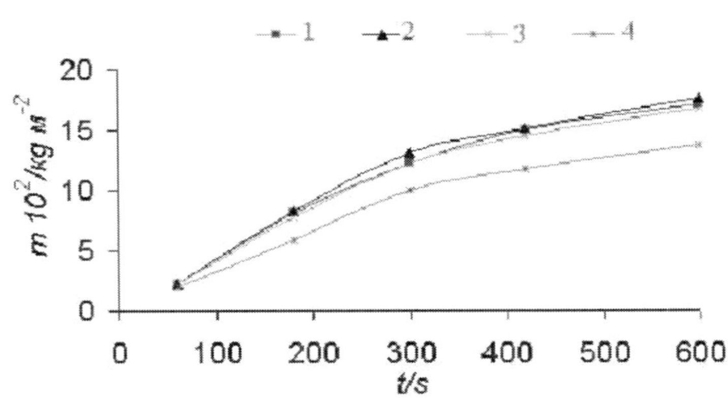

Figure 50. Formation of the polymer deposit on zinc electrodes in dependence on polymerization time and $ZnCl_2$ (M) content in polymer composition: 0.2 (1), 0.1 (2), 0.5 (3), 0 (4). T = 25°C.

11.4. Condensation of Polymer Film during EIP

In optical microscopy in situ study of electropolymerization, we recorded a new effect of changed optical density of the growing polymer layer, which is presumably due to the physical condensation of the polymer. As shown by experiments, the condensation front wave moves simultaneously toward the cathode and the solution. In the latter case, the front motion is not limited, but in the former case, it is limited by the metal surface. One can assume that this effect forms a very dense barrier layer on the near-cathode surface of the polymer film whose properties are described in [140].

The data of Figure 51 is proof of polymer matrix condensation in the near-cathode region. A gas bubble lying on the surface of the electrode serves as a reference point. From Figure 51, one can see that the polymer envelopes the bubble and grows further on it. During further electropolymerization, the bubble gradually increases in volume in the polymer (Figure 52). As the bubble is flattened, one can conclude that the polymer material has mechanical anisotropy since expansion predominantly occurs across (but not along) the polymer layer.

In the course of postpolymerization, one can observe an increase of volume and change of shape of the gas cavity. The bubble is being flattened and becomes close to ellipsoid in its shape (compressed spheroid). Moreover, the L_1 axis of the bubble (perpendicular to the electrode surface) does not practically change (Figure 53, curve 3) while the length of the L_2 axis (parallel to the electrode surface) (Figure 53, curve 2) first increases in proportion to the postpolymerization time, than this process is retarded, and at last, is practically ceased. Dynamics of change of the bubble big and small axes length ratio in the course of postpolymerization is shown in Figure 53, curve 1. It follows from this figure that the size of the bubble and, therefore, its volume must rise at first and then be stabilized in a certain time after switching out of current.

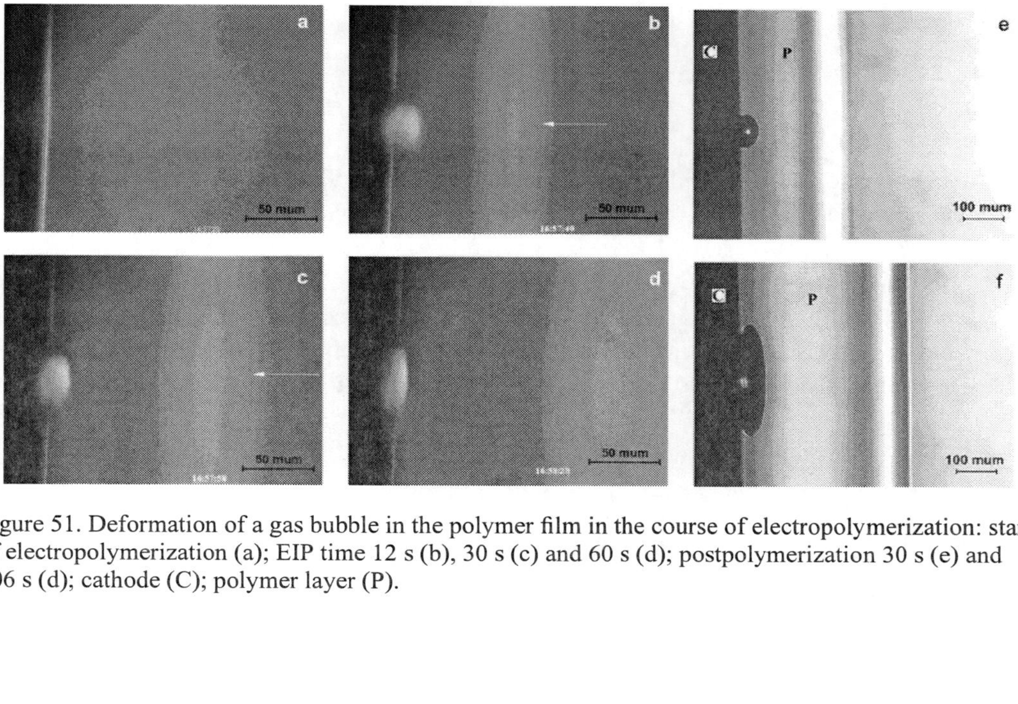

Figure 51. Deformation of a gas bubble in the polymer film in the course of electropolymerization: start of electropolymerization (a); EIP time 12 s (b), 30 s (c) and 60 s (d); postpolymerization 30 s (e) and 406 s (d); cathode (C); polymer layer (P).

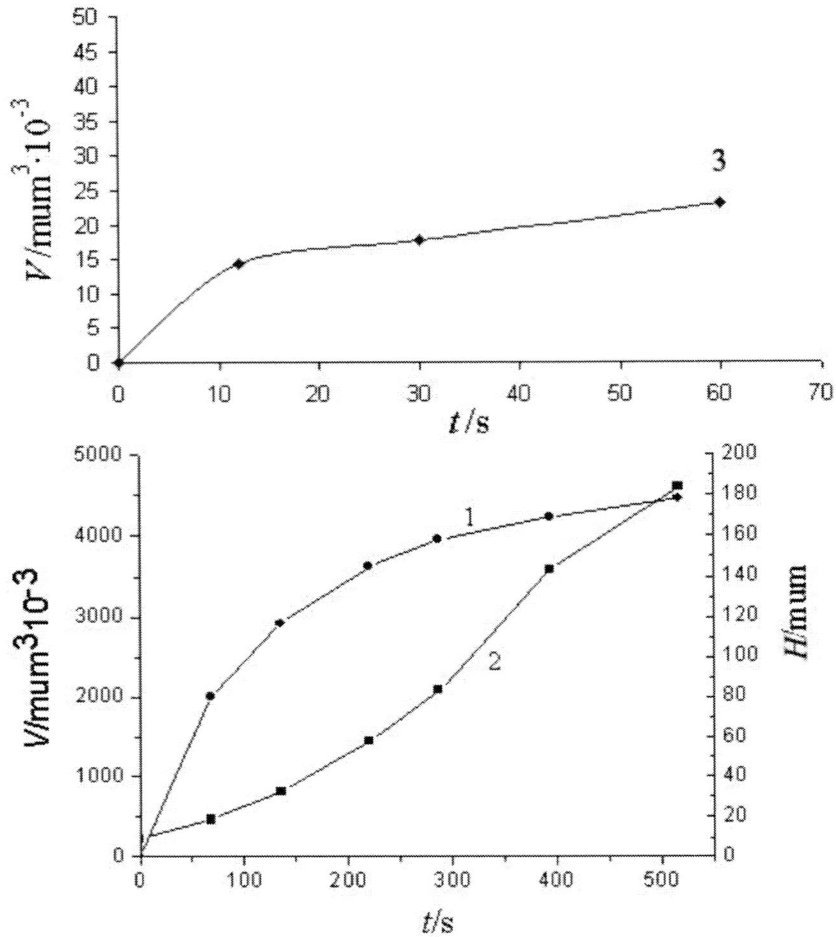

Figure 52. Change of the polymeric layer thickness (1) and the volume of the gas cavity (2) in the process of post-polymerization (2) and polymerization (3). Composition is standard. E = -1085 mV. 26 °C.

This conclusion well correlates with the data of Figure 9. Moreover, an increase of the gas cavity volume (Figure 52, curve 2) proceeds together with that of the polymeric layer thickness (Figure 52, curve 1).

Horizontal flattering of the gas bubble along the electrode surface, i.e. perpendicular to direction of the polymeric film growth, means that the process of post-polymerization is accompanied by a rise of limiting forces in the polymer volume which are orthogonal to the electrode surface. One can try to estimate their value. It takes to assume the following:

- The bubble is initially formed in liquid.
- The bubble is flattened due to pressure of the flat polymeric layer.

The rounding radius of the bubble free surface, adjacent to liquid, r = h/2, where h – the bubble height (Figure 54). The surface tension value γ may be determined from the condition of equality of pressure which the polymeric layer renders on the gas bubble (P_p) and

Laplacian pressure (P_L) which is conditioned by the curvature of its free surface. $P_p = F/\pi R^2$, provided the bubble is subjected to the action of the force F and the area of its contact with the flattened bubble $S = \pi R^2$. The value of Laplacian pressure $P_L = \gamma/r = 2\gamma/h$. $P_p = P_L$ in the steady-state regime. Then, the force acting on the bubble equals $F = 2\pi R^2\gamma/h$. Respectively, $\gamma = Fh/2\pi R^2$. R and h values may be determined experimentally with high accuracy.

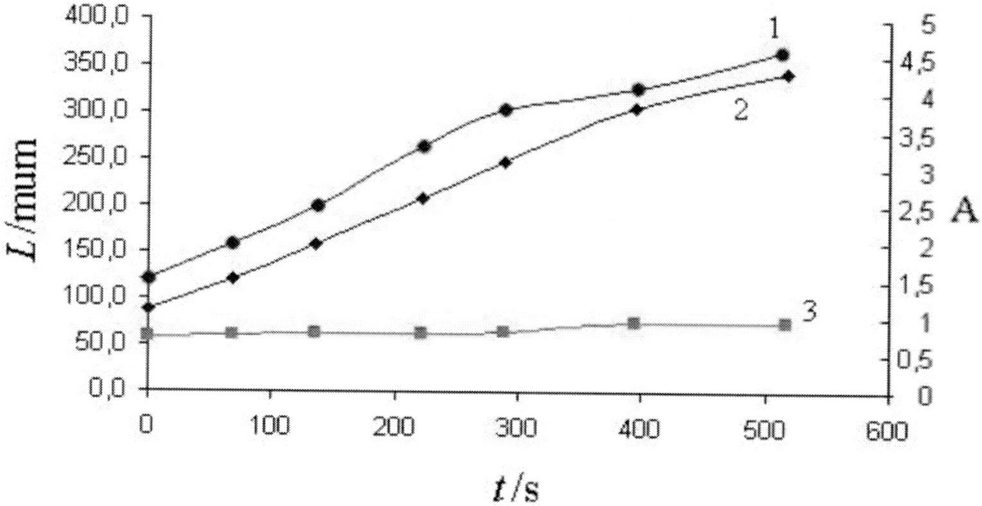

Figure 53. Change of the bubble's big (2) and small (3) axes lengths of the gas cavity in the process of postpolymerization and their ratio. Conditions of synthesis are analogous to those in Figure 52.

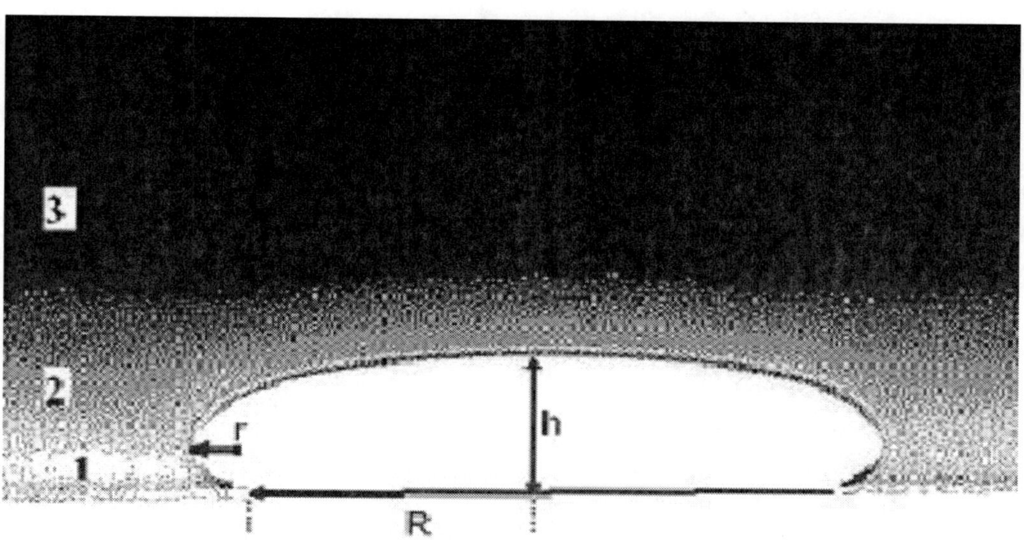

Figure 54. Schematic image of gas bubble on the cathode. In a growing polymer, layers of various density from liquid (1) to more dense layers (2, 3) corresponding to formation of polymeric layer may be distinguished.

Although this model is extremely simplified and it does not take into account possible compression of the gas into the bubble, the addition of gas into the bubble is a result of electrolysis of aqueous solution (rise of pressure in the bubble) and possible dependence of the surface pressure value on the polymerization time, nevertheless, it gives a general conception on changes in the composition polymeric layer / gas cavity.

The character of the bubble expansion shows the presence of an anisotropic deformation of the polymeric material which can arise as a result of structural transformations into its volume caused by forces densifying the near-cathode layer of the electrosynthesized polymer. Rapid insulation of the cathode with the layer of the non-conducting polymer results in a uniform coating of the external surface of the gas bubble with the polymer. Dynamics of its growth is roughly restricted, while an increase of volume of the gas bubble observed after switching off the current and change of its shape may be a consequence of the following processes. Due to its high swelling capacity, the polymer is saturated with electrolyte containing HCl (obtained in the process of hydrolysis of zinc chloride). The acid interacts with electrodeposited zinc.

As a result of the chemical reaction, an additional amount of hydrogen is formed that increases the gas pressure in the cavity. In turn, zinc dissolving promotes to the disturbance of adhesion between the polymer layer and cathode. A partial separation of the polymeric layer from the cathode on the border of the earlier formed bubble takes place. Then the gas pressure moves apart the polymer edges, loosing their cohesion, near the spherical cavity to opposite sides along the electrode. As a result, the gas cavity turns into spheroid. It follows from the experimental data that the gas pressure into the bubble is not sufficient to lift the polymer layer upwards. But it is sufficient to press the walls of the swelled film channels which are placed lateral to the electrode surface [45]. Compression of the channels results in the restriction of the electrolyte access to the electrode surface as well as retardation and, as a consequence, the following cessation of reaction. Practical invariability of L_1 and output of L_2 and volume of the gas cavity to stationary values are the confirmation of such a type of the process.

11.5. Induction Period of Electropolymerization

Analyzing the videorecord of the dynamics of electropolymerization, we have found that in all cases, the polymer layer did not start to form immediately at the moment of voltage supply to the microcell; rather, it started to form within a period of ~1 or 2 s after the potentiostat had been turned on. This period of time defines the *induction period* of polymerization. We have developed the device and procedure for exact determination of this time period. A microscope is used for projecting a microcell section within a circle 40 μm in diameter to a photodetector (Figure 55a). The light flux from the illuminator goes through the stated region and the microscope to the photodetector. In the absence of electrolysis, when the photometrically scanned section of the microcell is filled with electrolyte alone, the response of the photodetector is conventionally taken for the reference (zero baseline). Turning on the potentiostat initiates electrolysis and further polymerization. Once a solid polymer has formed in a section S previously filled with electrolyte, light intensity changes relative to its initial value. Polymer growth leads to monotonously changing light intensity.

Figure 55b shows a typical plot obtained using the suggested procedure for determining the induction period.

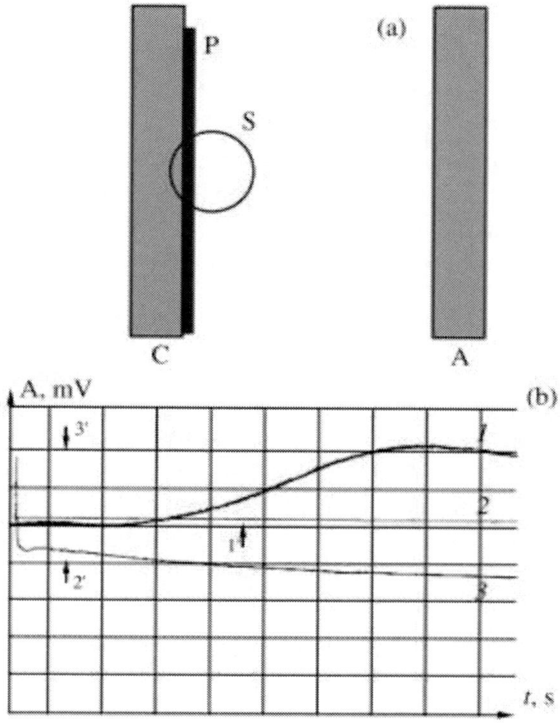

Figure 55. Microcell for investigating the induction period (a): cathode (C), anode (A), electrodeposited polymer layer (P), microcell section projected on the photodetector (S); determination of the induction period (b): signal from the photodetector (1), cathode potential (2), cathode–anode voltage (3). Cathode potential is –1160 mV, temperature 22°C. One square on the abscissa corresponds to 1 s for all curves. **A** is signal amplitude. One square corresponds to 1000 mV for curves 2 and 3 and to 50 mV for curve 1. The origin (horizontal lines) does not coincide in the curve amplitude for curves 1–3; for clarity, it is indicated by the corresponding primed figures.

It can be seen that the cathode potential (curve 2) reaches the given value (–1160 mV) within a very short period of time, practically simultaneously with the cathode–anode voltage supply to the microcell (curve 3). Once the cathode potential has established at a given value, light intensity slightly fluctuates, which is detected by the photodetector. After 2.3 s (this period of time can accurately be determined from the plot), the signal from the photodetector tends to change monotonously (curve 1). Thus the polymer layer starts to grow in 2.3 s after the cathode potential has reached the specified value at which polymerization can be effectively initiated.

Thus, in situ investigation of electropolymerization permitted to look into the process and obtain new information.

1. It allows real-time investigation of electropolymerization both on the electrode surface and in the electrolytic cell based on the information simultaneously visualized and videorecorded.

2. The method is almost contact-free if the interaction of light from the microscope illuminator with the substance of the monomer mixture and the polymer product is neglected. This approach rules out sample deformation in the course of the study.
3. The procedure allows the investigation of low-contrast, colorless, optically transparent polymer samples whose refraction index differs insignificantly from that of the solution.
4. The method permits direct real-time in situ investigation of the kinetics of electropolymerization from changes in growth dynamics of the polymer layer on the cathode in the lateral plane.
5. Our procedure permitted us to detect an induction period of electropolymerization and to evaluate it.
6. It was found that the optical and mechanical density of the polymer film changes during electropolymerization. It is concluded that the polymer material has mechanical anisotropy and mechanical stresses whose front is directed along the normal to the cathode surface and toward the polymer/solution interface.
7. Formation of a Zn/polymer double layer has been visualized, and reasons for high adhesion of the polymer layer to the metal surface have been revealed.
8. Spontaneous growth of the polymer film after electrolysis has been detected. The kinetics of postpolymerization has been determined.

The approach to an in situ study of the mechanism and kinetics of electropolymerization is universal and may be used for investigating different electrochemical processes.

12. Synthesis of Ultrafiltration Membranes by Electrochemically Initiated Polymerization of Acrylic Monomers [22, 23, 141]

Membrane methods for the separation and concentration of organic and mineral substances are promising for the refinement of chemical substances and recovery of valuable components from production wastes. Diversity of application fields requires the development of membranes with a wide range of characteristics and simple preparation procedures. However, presently, the main material for fabricating ultrafiltration (UF) membranes is cellulose acetate, which readily degrades even in weakly aggressive media. Therefore, active searches for alternative materials is in progress. At the same time, methods for production of all presently known UF and reverse-osmosis membranes are very labor-consuming, difficult, and multistage and require expensive and toxic organic solvents [142, 143]. Therefore, the development of new, simpler technologies and new membrane materials for hyperfiltration is one of the most promising and important ways to ensure progress of physicochemical and biological engineering.

Our studies allowed to propose an original and principally new method for preparing UF membranes by electrochemically initiated polymerization (EIP) of monomers [44]. We showed [95, 98, 118, 132] that EIP of vinyl monomers in aqueous solutions yields loose or dense cathodic films, uniform in thickness. By the looseness of the film, we mean that the structural matrix of the polymer precipitate contains micropores and various voids, which result from the leaching of soluble, low-molecular-weight fractions and sorbed electrolyte from the polymer. The fraction of the soluble fraction in an electrochemically prepared polymer and the rate of its leaching are determined by the potential and time of electrolysis, pH and initiating components These films swell in water, but do not dissolve because of cross-linking of the polymer. Water absorption of these films suggests the presence of pores and channels and allows these films to be considered as a filtering material.

A number of undisputable advantages of the EIP method make it very promising for the production of UF membranes:

1) the preparation is simple, since synthesis of the polymer and formation of the polymer film are combined in a single fast stage, which takes no more than 2-30 min;
2) uniform films differing in composition, structure, thickness, and porosity can be prepared by varying the composition of electrolytic baths and electrolysis conditions;
3) the process conditions can be improved, and the environmental impact can be reduced, since no organic solvents are used, and no complicated procedures are required that are usually used when forming, drying, transporting, and storing membrane materials;
4) wide prospects are opened for development of an efficient method for fabricating membrane units, since polymer films can be formed on conducting supports of any configuration.

The compositions of electrolytic baths are listed in Table 7.

Table 7. Conditions of preparation of polymer films

Membrane no.	Composition of solution for EIP (M)	Cathode	Anode	Potential, V	Current, A m^{-2}	Time, min
1	3 AA, 7 F, 0.05 $ZnCl_2$	Ni	Pt	-1.40		10
2	3 AA, 3 F, 0.45 AAC, 0.3 $ZnCl_2$	Ni	Al	-1.20		30
3	3 AA, 3 F, 0.1 *N,N'*-MBAA 0.1 $ZnCl_2$	Ni	Al	-1.16		10
4	3 AA, 3 F, 0.45 AAC, 0.55 VA, 0.4 $ZnCl_2$	Ni	Al	-1.20		20
5	7 AA, 5.4 F, 0.1 $ZnCl_2$	Ni	Al	-1.70		5
6	3 AA, 3 F, 0.5 $ZnCl_2$	Ni	Al	-1.16		2
7	3 AA, 3.7 F, 0.4 $ZnCl_2$	Ni	Pt		61.7 400.0 10.0	2 0.5 15
8	3 AA, 7 F, 0.05 $ZnCl_2$	PTEM	Pt	-1.20		3
9			Pt	-1.16		3
10			Pt	-1.1-		5
11		PTES	Pt	-1.20		3
12			Pt	-1.16		3
13			Pt	-1.10		3
14			Pt	-1.10		2
15	3 AA, 7 F, 0.25 $ZnCl_2$, sucrose, saturated solution	Ni	Al		38.7	1
16	3 AA, 7 F, 0.25 $ZnCl_2$, UV irradiation, 10 min	Ni	Al	-1.16		3

The electrolysis was performed under potentiostatic (E = -1.1...-1.7 V, vs. SCE) and galvanostatic (i = 10-400 A m^{-2}) conditions. The combined mode involved brief pulses of high current densities, preceded and followed with longer intervals with lower current densities (Table 7). The electrolysis time (unless otherwise stated) was determined by the condition of preparing films with similar thickness.

The specific membrane performance G was determined volumetrically with respect to distilled water and calculated as follows [82, 142, 143]:

$$G = V / S\,t,$$

where V is the filtrate volume (l), S is the membrane surface area (m^2), and t is the time (days).

Studies showed that polymer films that had been prepared as described in [95, 118] and removed from the metal, do not filter water under atmospheric pressure, or their permeability is very low, so that they cannot be considered as dialysis membranes. However, upon applying pressure, the films become permeable. This suggests the possibility of using these films as UF and reverse-osmosis membranes. However, requirements imposed upon polymer coatings and semipermeable membranes substantially differ in many respects. The property that is advantageous for films (e.g., good adhesion to the metal, absence of pores, or low swellability) can be disadvantageous for membranes. Therefore, conditions for electrochemical preparation of high quality polymer coatings and permeable membranes can differ.

In accordance with the foregoing and with general principles of preparation of hyperfiltration membranes [142, 143], we have formulated necessary conditions that should ensure the formation of hydrophilic permeable polymeric films using the EIP method:

1) use of water-soluble monomers that form cross-linked, insoluble, and water-swelling polymers upon EIP;
2) high polymerization rate, leading to enrichment of the polymer in low-molecular-weight soluble fractions, which are washed out from the polymer and provide a loose structure of the remaining film;
3) the presence of blowing agents;
4) appropriate EIP conditions that would allow separation of the film from the metal without altering mechanical and physicochemical properties of the polymer.

The monomer compositions that meet these requirements include as major components acrylamide, formaldehyde, and zinc chloride; the appropriate electrolysis conditions are listed in Table 7. As noted [95, 118], with such composition polymer films from methylol derivatives of acrylamide and formaldehyde, uniform in thickness, can be obtained on the cathode under both galvanostatic and potentiostatic conditions. Their swellability and looseness are determined by the nature of copolymerizing monomers, the degree of cross-linking of the polymer matrix, and the ratio of the initial substances; these properties are optimal when the concentrations (C) of acrylamide and formaldehyde are 3-7 M, with C_{AA}/C_F=1. With lower concentrations of acrylamide and formaldehyde, the polymerization rate decreases, and higher-molecular weight products giving dense and impermeable films are formed. Higher concentrations are inaccessible because of limited solubility of acrylamide.

Also, the maximal rate of the synthesis of polymer coatings is attained at potentials E = -1.1…-1.3 V is solutions containing the above-given amounts of acrylamide and formaldehyde and 0.01-0.3 M $ZnCl_2$ and having pH 3-5 [95]. However, the effect of these parameters on ultrafiltration properties of films was not studied; the optimal characteristics of ultrafilters can be attained under different EIP conditions.

Our studies showed that the main components of the formulation have several functions. For example, formaldehyde, on the one hand, initiates EIP via intermediate radicals $CH_2OH^{\bullet}$ formed in protonic medium upon electrochemical reduction [95, 126]. On the other hand, formaldehyde acts as a hydroxymethylating and cross-linking agent [64, 95]. Thus, formaldehyde favors the formation of products with a network structure. In addition, being an antiseptic, formaldehyde improves the membrane resistance against biological degradation with microorganisms. This is extremely important for the functioning and storage of filtering materials and to a large extent determines the economical characteristics of the filtration process and even the feasibility of its commercial implementation.

Zinc chloride is also a multifunctional component. Firstly, it indirectly initiates electropolymerization of acrylamide via complexation in aqueous solution [65, 95]. Since the reduction of the zinc acrylamide complex occurs easier than the reduction of free acrylamide (-1.1 and -1.97 V vs. normal hydrogen electrode, respectively), the potential of EIP of acrylamide in aqueous solutions can be substantially decreased. This not only decreases the energy consumption, but also improves the film quality, since at lower potentials no gas evolution due to electrochemical decomposition of water will occur at the cathode, and hence, perforation of the growing film, which can deteriorate the filtering properties of the film, will be eliminated. Second, zinc chloride is a blowing agent [143]. Third, it allows cathodic deposition of a two-layer zinc/polymer coating [118], which is extremely important for membrane synthesis, since dissolution of the zinc sublayer between the metallic support and the polymer film allows stripping of the film from the support with the aim of testing the film as a membrane.

Since stripping of the film is a very demanding stage in electrochemical synthesis of UF membranes and must not alter the properties of the film, we have developed special procedures for the removal of films from metallic supports.

1) Chemical dissolution of zinc is performed in 0.1 N HCl or 1 N Na_2SO_4 during 1-2 min. Because of low concentrations and slow diffusion of the electrolyte through the film, chemical dissolution of zinc occurs slowly and causes uniform and smooth exfoliation of the polymer from the support without mechanical damage and chemical degradation [95].
2) Electrochemical dissolution of zinc is performed in the initial monomer formulation, in 0.1-1 N Na_2SO_4, or in KCl at potentials from 0.5 to -0.7 V. In this range, the polymer film does not undergo electrochemical degradation (Figure 39, curve 1). The electrolysis time at E = 0.1 V did not exceed 1-2 min.

The main service characteristic of hyperfiltration membranes is their permeability, which was quantitatively estimated by the specific performance G. Since G depends not only on the pressure differential across the membrane (P), but also on the composition of the solution being filtered and differs by 1-2 orders of magnitude in various [143], we compared the performances of membranes prepared by EIP with respect to distilled water at P = 0.15 MPa.

Tables 7 and 8 show that performance with respect to water of all polymer films obtained under various EIP conditions is comparable to that of a number of known membranes.

Apparently, the structure of polymer films and hence, their permeability can be varied taking into account the mechanism and kinetics of electrosynthesis of polymer films and coatings from acrylamide, formaldehyde, and methylolacrylamide [95, 98, 118, 132]. Indeed, the membrane performance can be controlled by various methods: by varying concentrations of the main components of the electrolytic bath or adding copolymerizing agents, by using various electrolysis modes, or by modifying films through chemical or physical action on the polymer (Tables 7 and 8). For example, low-performance membranes were prepared with the galvanostatic combined mode of electrolysis. Since under conditions of galvanostatic EIP, the rate of generation of initiating species is constant and their concentration is determined by the current density (i) and increases with growing i [54], increase of the current results in the accumulation of low-molecular-weight soluble fractions in the copolymer due to the acceleration of chain termination. These fractions, as shown in [98], are easily washed off and loosen the structure of the film. In contrast, low current densities afford a dense layer. By alternating low and high values of i during electrolysis, we obtained polymer films with a complex structure, i.e., with loose middle and dense boundary layers. As seen from Table 8 (membrane no. 7), specifically dense layers, owing to their poor permeability, limit the rate of filtration through membranes prepared from such films and decrease G by a factor of 2-8 as compared to other membranes.

Table 8. Performances of membranes

Membrane no.,	Membrane performance (l m^{-2} day^{-1})	
	0.15 MPa	0.20 MPa
1	160	
2	214	
3	591	
4	255	
5	217	
6	257	
7	-	76
8	14	
10	12	
11	33	
12	23	
13	35	
PTEM	2771710	
PTES	6860160	
UEM-200	72-140	
UAM-100	140-390	
UAM-150	400-790	
15	-	557
16	-	835

Under conditions of potentiostatic electrolysis, at potentials of reduction of the zinc-acrylamide complex (E ≥ -1.1 V, vs. SCE) the EIP rate substantially increases with increasing C_{ZnCl_2} [95, 98, 118, 132]. In this case, acceleration of initiation and an increase in the total concentration of active species also accelerate chain termination and increase the content of the soluble fraction in the polymer and hence, the film's looseness. As a result, the permeability of the films must increase. As seen from Table 7 and 8, the most dense, weakly swellable (90%), and weakly permeable films are formed at C_{ZnCl_2} = 0.05 M; with increasing the zinc chloride concentration to 0.5 M, the specific performance of the membrane increases by a factor of 1.6 (membranes nos. 5 and 6). A similar effect is observed with increasing the electrolysis potential and acrylamide concentration: the membrane performance increases (membrane nos. 1 and 5) because of looser structure of the film.

Moderately polar functional groups in the membrane, such as –COOH, NH_2, -OH, and –CHO, because of competition, decrease the tendency of water molecules to form clusters and thus, favor water permeation through the membrane [144]. As seen from Table 7, the monomer formulations studied allow the introduction of the above functional groups into the polymer matrix of electrochemically prepared membranes. This can be performed both by adding other monomers into the acrylamide-formaldehyde composition (membranes nos. 2-4) and by taking acrylamide in an excess relative to formaldehyde, which ensures incomplete hydroxymethylation (membrane no. 5). The presence of corresponding groups in the polymer is confirmed by the IR data. Comparison of the specific performances of membranes nos. 1, 2, and 5 demonstrates a correlation between the amount of functional groups incorporated into the membrane and its performance *G*. For example, the performance of membrane no. 2 (-OH, -COOH) or 5 (-OH, $-NH_2$) is higher than the performance of membrane no. 1 (-OH).

It was found that filtration characteristics of membranes also largely depend on the properties of copolymerizing monomers. Comparison of the specific performances of membranes nos. 2 and 4 shows that the addition of vinyl acetate VA (whose polymer exhibits low swellability, 12.7% [145]) onto the electrolytic bath not only changes the qualitative composition of the membrane, but also decreases its swellability (70-80 wt %) and increases the performance by 20%. However, the greatest increase in the specific performance (by a factor of 2.3-8 relative to other membranes) is attained when the bifunctional cross-linking agent *N,N'*-MBAA is used as a comonomer (membrane no. 3). Additional rigid cross-linking of the polymer matrix decreases the film's swellability to 35 wt %, makes it less loose, and, which is most important, decreases the pore deformation during service. All of these factors have a positive effect on the filtration rate.

Further studies showed that pore size and hence, the membrane permeability can be increased by using the blowing agent sucrose, whose molecules are larger than those of zinc chloride (membrane no. 15), or by combining EIP with UV irradiation of the cell (membrane no. 16) [146].

We found that polymer films can be formed on both all-metal and metal-ceramic cathodes (porous nickel, titanium, etc.). In this case, the membrane performance is determined not only by EIP conditions, but also by the properties of the reinforcing material. Filling of coarse (5-50 μm) pores of the ceramics with a polymer and deposition of a film with required thickness on the cathode surface allow single-stage fabrication of composite membrane units which are ready to use immediately after removal from the electrolysis bath (Table 7 and 8, membranes nos. 8-14).

As seen from Table 8, metal ceramics themselves do not limit the filtration, since their performance is 5 orders of magnitude higher than of the film. However, the specific performances of composite membranes formed on PTES and PTEM under similar EIP conditions differ by a factor of 2.3-2/5, i.e., the ratio of membrane performances is the same as the ratio of performances of corresponding uncoated porous materials (Table 8). With increasing thickness of the electrodeposited polymer (e/g/, due to increasing the EIP potential), the performance of composite membranes decreases (membrane no. 12, 90 μm; membrane no. 13, 40 μ). At the same time, the performance of the non-reinforced film is by an order of magnitude higher than that of the composite film/PTEM membrane and by a factor of 5 higher than that of the film/PTES membrane (membranes nos. 1, 8, and 11, respectively). This data shows that the permeability of composite membranes can be controlled by using metal ceramics with various pore sizes and by varying the degree of pore filling with the polymer. This allows preparation of mechanically-strong membranes for both ultrafiltration and reverse osmosis. For example, the performance of membrane no. 14 under conditions of reverse osmosis at a pressure differential of 4 MPa was 4000-5000 $l\ m^{-2}day^{-1}$ with respect to 0.5% NaCl.

13. Properties of Electrosynthesized Ultrafiltration Membranes and Fields of Their Practical Application [42, 44]

The main operation parameters of ultrafiltration membranes determining the expediency of their practical application comprise their *productivity* and *selectivity*. The permeability and specific productivity *G* of electrosynthesized (both film and composite) poly-acrylamide-formaldehyde membranes increase with the pressure growth, which is caused by the pore deformation expansion (Figure 56). The observed dependence is of a linear character in the pressure range 0.05-0.6 MPa. At the same time, the membranes of this type are characterized by a slight hysteresis even after attaining high pressure of 0.8-1.0 MPa (Figure 57). Such a trend of the curve is evidence of the polymers good elastic properties. In addition, the membranes elastoplastic properties are confirmed by the results of their days-long tests under periodic pressure drop conditions. After the pressure withdrawal, the membrane returns to the state close to the initial one, while the character of productivity change, stationary value of *G* and the time of achieving the plateau value do not virtually change (Figure 58). This high stability of operation parameters gives the electrosynthesized membranes extra benefits in comparison with cellulose acetate membranes.

Under constant pressure conditions, the membrane productivity decreases and achieves plateau values, which is concerned with the polymer matrix compacting (Figure 59, curves 1 and 2). The permeability decrease is accompanied by the membrane selectivity growth (Figure 59, curve 3 and Figure 60).

The electrosynthesized membranes are stable in organic solvents, solutions of mineral acids and alkali in the pH range 1-14. Such membrane properties enabled one to use them in purification of wastewaters from the lubricating-cooling liquid “Aquol-10”, in which cellulose acetate membrane decompose (Table 9). Here, the selectivity could attain 90%, depending on the EIP conditions and membrane modification.

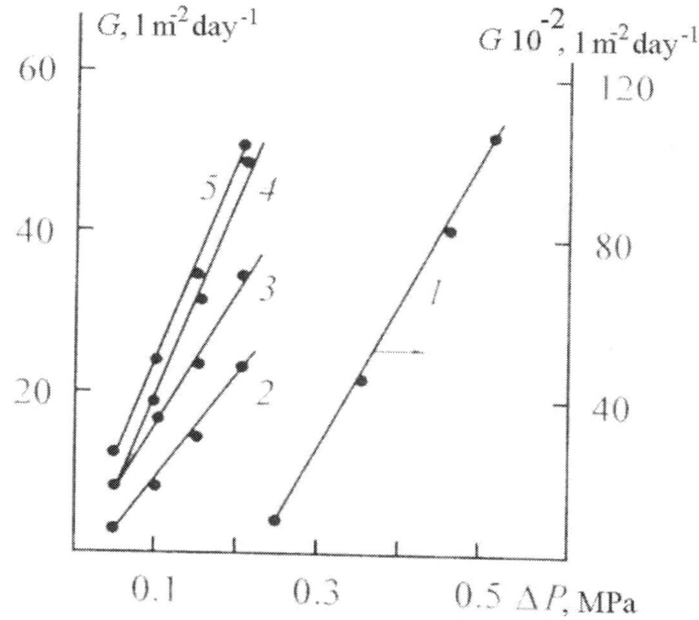

Figure 56. Dependence of productivity of film (1) and composite (2-5) membranes on pressure. (1) – membrane UV-treatment. Filtered solution: lubri-coolant (1), water (2-5). Electrolyte composition: 3 M AA, 7 M F, 0.25 M $ZnCl_2$ (1); 3 M AA, 7 M F, 0.05 M $ZnCl_2$ (2-5). E (V): -1.16 (1 3), -1.2 (2, 4), -1.1 (5).

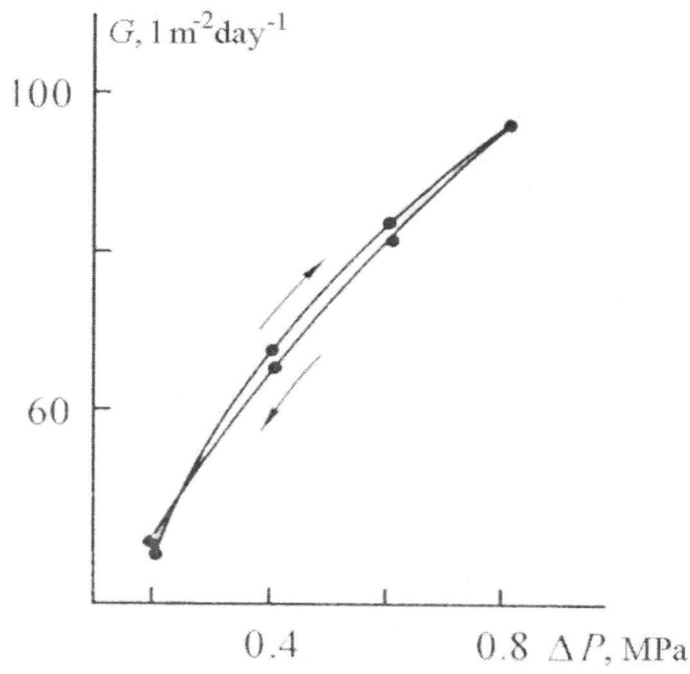

Figure 57. Hysteresis of poly-AA-F membrane. Composition (M): 3 AA, 7 F, 0.4 $ZnCl_2$. $i = 61.7$ A m^{-2}, 1 min. Membrane modification by the solution (triethanolamine + $FeCl_3$).

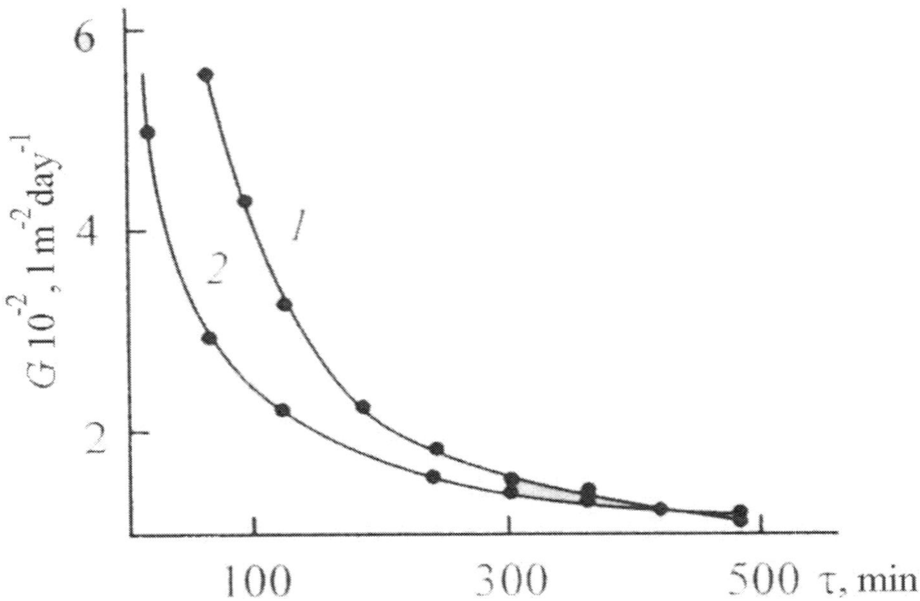

Figure 58. Dynamics of membrane productivity under periodic pressure drop conditions. ΔP 0.2 MPa. Membrane formation conditions – as in Figure 57.

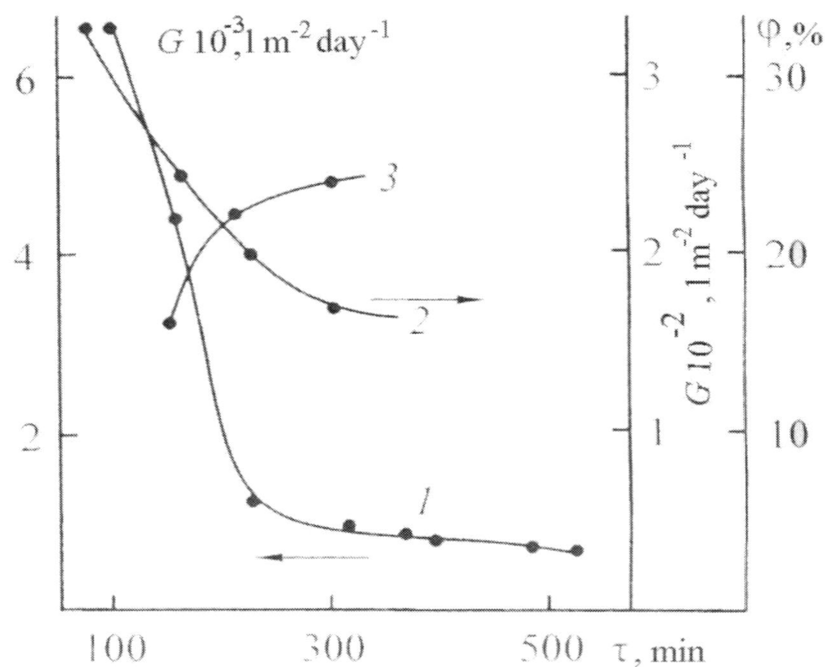

Figure 59. Change of productivity (1, 2) and selectivity (3) of membranes over time. Membrane formation conditions – as in Figure 56, curve 1 (2, 3), the same for the membrane UV-modification. (1).

Table 9. Ultrafiltration of lubri-coolant "Aquol 10". Electro;lyte composition (M): 3 AA, 7 F, 0.25 $ZnCl_2$

№	EIP regime	Modification of the film	Pressure, MPa	Penetration, l/m^2 day	Selectivity, %
1	-1.16 V; 3 min	UV, 10 min	0.184	835	27
2	-1.12 V; 2 min	UV, 60 min and 2 M $Na_2S_2O_3$, 2 min	0.184	634	37
3	38.7 A/m^2; 1 min	2 M $Na_2S_2O_3$; 30 min	0.24	557	90

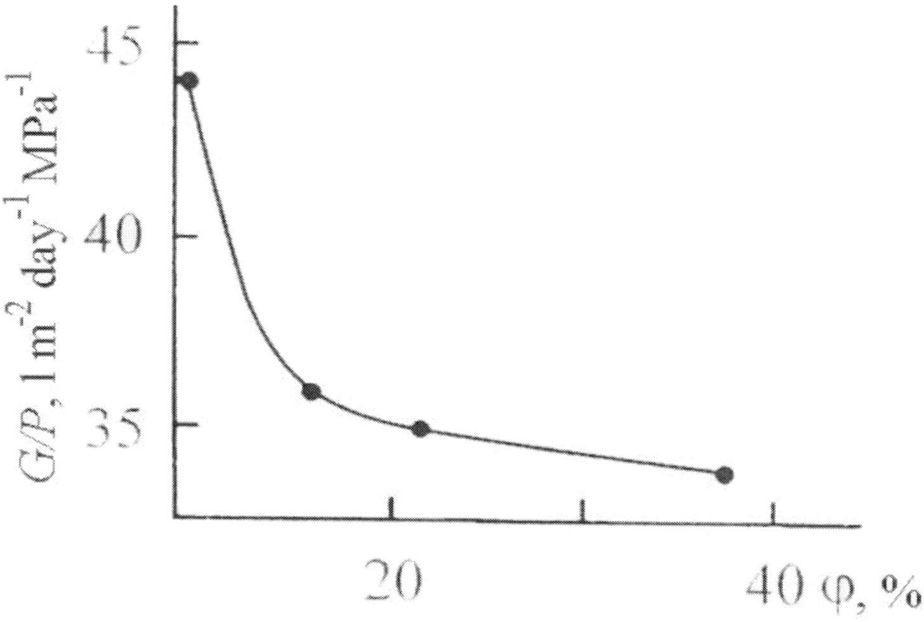

Figure 60. Dependence of membrane selectivity on specific productivity. Membrane formation conditions – as in Figure 59, curve 1.

The membranes of this type demonstrate high selectivity in regard to colorants. As was shown on the example of Congo red, the membrane selectivity depends on the EIP conditions (Table 10). The lowest selectivity was found for the membrane synthesized at high potential E=-1.7 V, when the membrane pore size must be maximal [95].

On the other hand, the use of the optimal EIP rate (E=-1.16 V), the addition of N,N'-MBAA, and the increase of the formaldehyde content in the monomer composition and galvanostatic electrolysis improve the membrane selectivity. However, the highest degree of removal of Congo red molecules can be attained by the introduction of negatively charged carboxyl groups into the polymer, for example, by adding acrylic acid or vinylacetate to the electrolysis bath. Mutual repulsion between the membrane and similarly charged Congo red particles results in hindering the dye transport through the ultrafilter, thus increasing the membrane selectivity up to 99.9%. The pressure increase from 0.15 up to 0.4 MPa results in an almost 60-fold ultrafilter productivity increase with almost no selectivity deterioration. The pore estimation on Congo red, whose molecule radius and occupation site area are known, demonstrated that the pore diameter of electrosynthesized membranes was in the range from 7 up to 50 Å depending on the electropolymerization conditions.

Table 10. Electrosynthesized membranes selectivity on Congo red

№	Electrosynthesis conditions	Selectivity, %		
		0.15 MPa	0.2 MPa	0.4 MPa
1	(3 AA, 7 F, 0.05 $ZnCl_2$) M; -1.4 V; 10 min	-	-	74.0
2	(3 AA, 3 F, 0.45 AAc, 0.3 $ZnCl_2$) M; -1.2 V; 30 min	99.9	-	99.1
3	(3 AA, 3 F, 0.1 N,N′-МБАА, 0.1 $ZnCl_2$) M; -1.16 V; 10 min	95.8	-	94.6
4	(3 AA, 3 F, 0.45 AAc, 0.55 VA, 0.4 $ZnCl_2$) M; -1.2 V; 20 min	99.9	-	99.4
5	(7 AA, 5.4 F, 0.1 $ZnCl_2$) M; -1.7 V; 5 min	-	-	58.0
6	(3 AA, 3 F, 0.5 $ZnCl_2$) M; -1.16 V; 2 min	-	-	76.0
7	(3 AA, 3.7 F, 0.4 $ZnCl_2$) M; 61.7 A/m^2 (2 min), 400 A/m^2 (0.5 min), 10 A/m^2 (15 min)	-	91.0	-
8	UAM-150	99.4	-	-

The principal possibility was demonstrated for the application of electrosynthesized ultrafiltration membranes in the technologies of concentration and purification of *arabinogalactan*. The technology enables one to maximally remove water and dihydroquercetin admixtures with minimal power consumption without using extra reagents. Such membranes are comparable, in terms of separation factor and concentrating degree, with the commercially available membranes UAM-150P and UAM-50PT. The electrosynthesized membranes' advantage in the case of the mentioned process consists in a simplified technology of their production.

The possibility of the application of the modified films on the basis of cross-linked polyacrylamide as pervaporation membranes to separate the ethanol-water mixtures was established. The ways of improving membranes pervaporation properties was suggested.

The possibility of using ultrafiltration in the technology of extraction of phytic acid and its salts from rice production waste was also demonstrated. Ultrafiltration enables one to decrease 1.2-fold total reagents consumption and 60-fold – the time of phytic acid derivatives extraction, perform deep purification of solutions from proteins and other organic and inorganic components, thus improving 15-fold the salts whitening and obtaining high-purity products. The advantages of electrosynthesized membranes, as compared to commercially available filters, consist in their sterilizing effect, increase of the phytic acid hydrolyzate purification degree, 10-100-fold increase of the filtration rate without deteriorating main positive characteristics of the final product [42]. The technological setup for extraction of phytic acid using ultrafiltration has been developed.

14. The Overall Porosity and Pore Sizes of Electrosynthesized Ultrafiltration Membranes [141]

The above assets of electrosynthesized membranes point to the potentiality of applying both the electrochemical method and the produced ultrafilters. However, to solve some theoretical and practical problems related to the use of membrane materials, it is necessary to fully know the structure and parameters of capillary systems. Hence, there is no doubt whatsoever that the above characteristics of membranes produced by the non-traditional electrochemical method must be studied in detail.

Altering the electrolysis modes and the monomer composition allows one to control the volume structure of formed polymer films and vary their friability and, consequently, the overall number and size of pores [98]. However, these quantities were not determined quantitatively, nor was their dependence of the synthesis conditions.

The most important characteristics of ultrafiltration membranes, which define their productivity and selectivity, are the overall porosity and pore radii. These characteristics may be determined by different techniques, each of which has its own advantages and drawbacks. Therefore, to most correctly determine these characteristics, it is of importance to correctly select the membrane test method.

The overall porosity W is the volume of pores in a unit volume of a membrane. Depending on the nature of the membrane material and technique of its formation, the overall porosity may vary quite widely from 5 to 80% [142, 143, 149].

The volume structure of ultrafiltration membranes formed by an electropolymerization method is highly ordered [45]. The pores in the ultrafiltration membranes are positioned normally to the system, and the number of blind and closed pores are at a minimum. Therefore, we expect that the overall porosity of the ultrafiltration membranes will have large values.

To determine the overall porosity W of electrosynthesized ultrafiltration membranes, we employed the Manegold method [147, 148]. The volume of a water-saturated membrane was calculated with the equation

$$V_n = \frac{p_{n_{H_2O}} - p_{n_{CCl_4}}}{\gamma_{CCl_4} - \gamma_{H_2O}}$$

and the overall porosity,

$$W = \frac{V_n - V}{V_n} = \frac{V_n - \dfrac{p - p_{n_{H_2O}}}{\gamma_{H_2O}}}{V_n} \qquad (22)$$

Here, V is the volume of the solid substance or the membrane core, V_n and p_n are the volume and weight of water-saturated membrane in air, $p_{n_{H_2O}}$ and $p_{n_{CCl_4}}$ are the membrane weights in water and carbon tetrachloride, while γ_{H_2O} and γ_{CCl_4} are specific weights of water and carbon tetrachloride.

It was preliminarily established, with the aid of thermogravimetry, that the polymer undergoes no decomposition in the temperature region 100–105 °C (Figure 61), which allowed using this temperature region for the determination of the weight of dry samples.

The results of the measurement of the overall porosity, obtained using equation (22), appear in Table 11.

Table 11. Overall porosity of electrochemically synthesized membranes

Membrane no.	Synthesis conditions		Overall porosity, %
	Electropolymerization solution composition, mol l^{-1}	Electropolymerization mode	
1	3 AA, 3 F, 0.1 $ZnCl_2$	–1.16 V; 10 min	99.3
2	3 AA, 3 F, 0.1 $ZnCl_2$	–1.3 V; 5 min	98.4
3	3 AA, 7 F, 0.1 $ZnCl_2$	–1.2 V; 5 min	95.6
4	3 AA, 3 F, 0.1 (N,N'-MBAA), 0.1 $ZnCl_2$	–1.16 V; 3 min	92.8
5	3 AA, 3 F, 0.1 (N,N'-MBAA), 0.1 $ZnCl_2$	–1.16 V; 5 min	92.3
6	3 AA, 3 F, 0.1 (N,N'-MBAA), 0.1 $ZnCl_2$	–1.16 V; 10 min	91.25
7	3 AA, 3 F, 0.1 (N,N'-MBAA), 0.1 $ZnCl_2$	–1.2 V; 3 min	90.7
8	3 AA, 3 F, 0.1 (N,N'-MBAA), 0.1 $ZnCl_2$	–1.2 V; 5 min	90.6
9	3 AA, 3 F, 0.1 (N,N'-MBAA), 0.1 $ZnCl_2$	–1.2 V; 10 min	88.65
10	3 AA, 3 F, 0.1 (N,N'-MBAA), 0.1 $ZnCl_2$	–1.3 V; 3 min	90.76
11	3 AA, 3 F, 0.1 (N,N'-MBAA), 0.1 $ZnCl_2$	–1.3 V; 5 min	90.3
12	3 AA, 3 F, 0.1 (N,N'-MBAA), 0.1 $ZnCl_2$	–1.3 V; 10 min	89.0

As follows from the data presented in Table 11, the maximum overall porosity of 98.4–99.3% is exhibited by weakly cross-linked membranes 1 and 2. Increasing the degree of the polymer cross-linkage by raising the formaldehyde concentration in the source monomer or by adding into solution N,N'-methylene-bis-acrylamide (N,N'-MBAA) reduces the overall porosity to 90–95%, as the data for membranes 3 and 8 indicate. A similar dependence is observed following an increase in the electrolysis potential (membranes 4, 7, 10) and the membrane synthesis time (membranes 4–12, W = 89–93%). Such a mutual interrelation between the overall porosity and the polymer cross-linkage nicely conforms to the mechanism of formation of a porous structure in electrosynthesized films proposed earlier. According to this mechanism, the volume of voids in a polymer matrix is intimately dependent on the share of soluble low-molecular weight fractions that form in the course of synthesis and the rate of their removal from the film [98]. Thus, the studies performed showed the overall porosity of electrochemically synthesized ltrafiltration membranes to exceed that of the known membranes. The large values of the overall porosity are still one more trait of ultrafiltration membranes, which improves their characteristics.

Another important parameter of membrane filters is the pore size. To determine the average radius of pores in ultrafiltration membranes, one can resort to a variety of methods. A direct technique for the determination of the average pore radius is electron microscopy. However, we deem it necessary to underline that, firstly, the size of pore radii may be measured sufficiently accurately with the aid of electron microscopy only in the trek membranes and, secondly, to process electron micrographs is quite a daunting task. No direct measurement of the average pore radius is possible in traditional membranes due to their involved structure. Hence, the necessity to evoke indirect methods, which include computation with the aid of mathematical models. There are several methods of this type : (1) *Method of immobilization of particles;* (2) *The bubble point method*; (3) *Method of diffusion;* (4) *The water extrusion method;* (5) *The mercury extrusion method.* The basic disadvantage of the above methods is that all of them are usually based on a model of cylindrical pores with circular or elliptical cross-sections. However, a model of the type is acceptable only for isotropic membranes. In the case of anisotropic membranes, it leads to considerable errors.

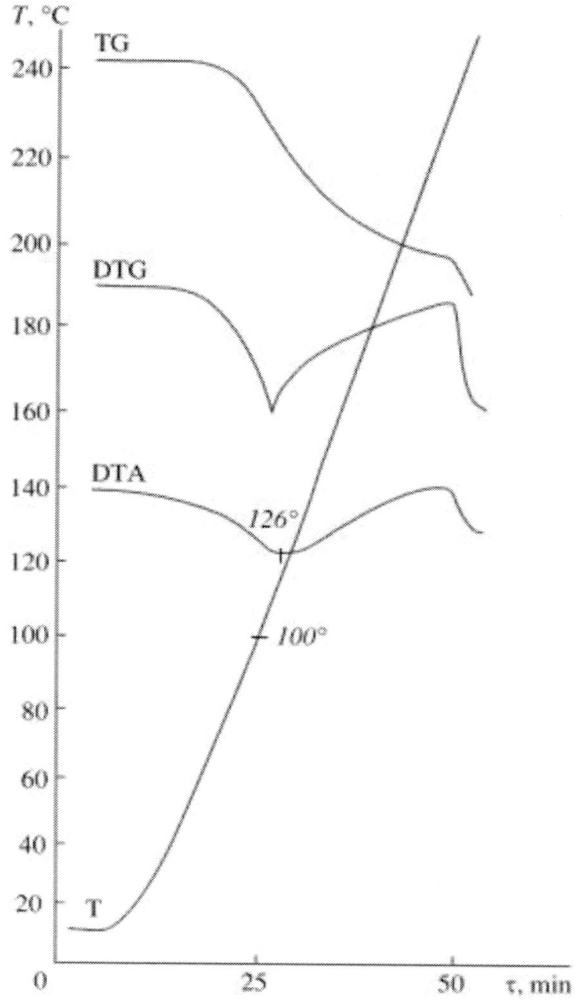

Figure 61. Termogravigram for a copolymer of acrylamide, formaldehyde, and N,N'-methylene-bis-acrylamide; heating rate 5°C per minute.

Besides, when determining the magnitude of pores, one has to bear in mind that the real size of pores in a working ultrafilter differs from that determined by the above methods. One of the two reasons for this fact is that the membrane undergoes deformation under the pressure. The other is that a layer of bonded solvent, which exists on the surface and inside the pores of a lyophilic membrane in solution, makes the cross-section of the pores narrower; hence, the filtrate moves over capillaries of smaller sizes [150]. Nevertheless, the estimates of pore size obtained by different methods coincide well enough. The electron microscopy data happens to be somewhat overestimated as compared with the results of hydrodynamic methods.

The average pore size may be calculated on the basis of properties of the fluid flow in membranes. To this end, one can make use of the Poiseuille equation, according to which the water flux through a membrane is

$$J_1 = N \pi r^4 P / 8 \eta l = W r^2 P / 8 \eta l \quad (23)$$

Here, J_1 is a flux per unit surface area of the membrane; N the number of pores per unit surface area; r a pore radius; P the external pressure gradient; l the membrane thickness; η the water viscosity; and $W = N \pi r^2$ is the porosity, which is frequently equaled to the water content in the membrane. As follows from (2),

$$r_1 = \sqrt{\frac{8\eta l G}{W}} \quad (24)$$

where $G = V/S\,\tau$ is a leakage coefficient.

This is the form of a formula for the average pore radii under the assumption that all the pores are cylindrical and perpendicular to the membrane surface and have the same length equal to the membrane thickness. If the pores are distributed uniformly in three mutually perpendicular directions, the fluid will pass only through the pores that cross either side of the membrane, i.e. through one third of the overall number of pores. In this case,

$$r_2 = \sqrt{\frac{24\eta l G}{W}} \quad (25)$$

It is absolutely clear that equations (23)–(25) are too simple by half to adequately describe real membranes, as they use too many assumptions and ignore the pore tortuosity, blind pores, and the pore scatter by radii. Besides, it will be remembered that the results are additionally distorted by the procedure used for the determination of the overall porosity in the absence of a pressure head, i.e. in conditions different from real ultrafiltration processes.

Pore radii r_n of EIP membranes calculated with equation (25) allowing for the overall porosity of membranes (Table 12), may be compared with some other results. For example, with those established with the aid of the ultrafiltration of test particles of the known size (congo red, methylene blue; $r_n \sim$ 5–26 Å [22, 23, 44]) or scanning tunneling microscopy ($r_n \sim$ 10–25 Å) and electron microscopy ($r_n \sim$ 50–100 nm) [45]. The comparison shows that, basically, average pore radii of membranes under study determined by different techniques have comparable values. However, the Poiseuille method [151], as opposed to the tunneling and electron microscopy methods, yields averaged results and completely fails to estimate the fraction of large pores and their contribution to the filtration process.

The results presented in Tables 11 and 12 allowed us to follow the general regularity of variation in the pore radii as a function of the electrolysis conditions. In the weakly cross-linked membranes 1 and 2, increasing the electrolysis potential and, consequently, the EIP rate, loosens the polymer structure and increases the pore radii. A similar action is exerted by an increase in the degree of cross-linkage of the membrane matrix at the expense of an excess concentration of F (membrane 3) and due to adding N,N'-MBAA, which is a bifunctional cross-linking agent, into the monomer (membranes 4–12).

In the latter case, the effect manifests itself most vividly, because cross-link-age of macromolecules makes the gaps between them rigid, thus reducing the pore scatter by size. When using the cross-linkage N,N'-methylene-bis-acrylamide molecules, whose linear size exceeds that of formaldehyde molecules, the gap between the chains being cross-linked and the pore radii of the membrane being formed will be greater than in the case of the cross-linkage done by formaldehyde molecules only. According to experiments, when the polymer is rigidly cross-linked by N,N'-methylene-bis-acrylamide molecules, the film structure directly depends on the synthesis time. With the EIP time equal to or under three minutes, r_n is independent of the electrolysis potential (membranes 4, 7, 10). With the EIP time increased from 3 to 10 min, the pore radii increase by 3–12 times (membranes 4–6, 7–9, 10–12).

According to [98], five to ten minutes into the electrolysis at potentials of discharge of a zinc–monomer complex, the content of the soluble fraction in a polymer reaches 0.8–0.85. With increasing the EIP duration, conditions for the dissolution of the soluble fraction inside the electrolyzer bath improve, thus raising the fraction of larger pores. The maximum effect occurs upon increasing the electrolysis potential from –1.16 to –1.2 V and the electrolysis duration to ten minutes (Tables 1, 2; membranes 6, 9). Experiments also show that the overall porosity of large-pore rigidly cross-linked membranes 6, 9, and 12 is lower than that of small-pore membranes 1 and 2 (Tables 11, 12).

For all the membranes studied, the narrowing of pores (r_τ) was observed in the course of ultrafiltration, which correlated with the decrease in the productivity of filters and testified to the film deformation under the pressure (Table 12). However, studies showed that the thickness of membranes remained practically unchanged after the ultrafiltration process, which confirms elastic-elastic properties of polymer established in [23].

Table 12. Average radius of pores in membranes, determined from the overall porosity

Membrane no.	G_n x 10^6, m^3 m^{-2} s^{-1}	G_τ x 10^6, m^3 m^{-2} s^{-1}	l_n x 10^3, m	l_τ x 10^3, m	r_n ,Å	r_τ, Å
1	1.388	0.115	0.112	0.108	8.6	2.4
2	7.870	0.925	0.069	0.069	16.1	5.6
3	6.805	0.578	0.083	0.074	16.8	4.5
4	2.120	0.115	0.760	0.740	17.3	6.57
5	9.050	0.925	1.000	0.980	51.5	17.3
6	106.810	27.20	0.835	0.820	207.6	103.8
7	2.920	0.115	0.457	0.420	17.3	5.0
8	5.350	0.173	0.790	0.760	34.6	8.3
9	28.300	2.314	0.900	0.900	363.0	17.3
10	3.472	0.347	0.500	0.490	17.3	9.5
11	4.630	0.231	0.560	0.520	34.6	8.0
12	28.94	0.520	0.610	0.500	86.6	11.8

Note: Membrane numbers and synthesis conditions same as in Table 1. The subscript n denotes the beginning of the ultrafiltration; t, after ultrafiltration.

Determining the average pore radius r from the consumption of the flow of the fluid through a membrane and from its *electroresistance* is much simpler than other techniques, as it allows us to compute r without preliminarily determining some quantities that are difficult to measure experimentally [144, 152, 153]. The method is based on the assumption that the electroconductivity of a membrane made of an insulator is defined by the electroconductivity of electrolyte in its pores. Assuming that the conditions that determine the passing of an electric current are identical to the conditions of the fluid flow through the membrane, we can find the average pore radius from measurements of the coefficient of leakage of the membrane G (see equation (25)) [142, 143, 149] and its electric resistance R_e [144, 152, 153]:

$$R_e = \rho_e\, l \,/\, \pi r^2 n S_0,$$

where ρ_e is the electroresistivity. Then, the equation for the calculation of radii of pores will assume the following form:

$$r = \sqrt{\frac{8\eta R_e \chi S_2 V}{\Delta P S_1 \tau}} \qquad (26)$$

here, η is the fluid viscosity, expressed in units of Pa s; R_e the membrane electroresistance, expressed in ohm; χ the solution electroconductivity, in S m^{-1}; S the membrane area when determining the electroconductivity, in m^2; V the fluid volume passing through the membrane within time period τ, in m^3; P the pressure head, in Pa; τ the ultrafiltration time of volume V of the fluid, in s; and S_1 the membrane area during the ultrafiltration, expressed in m^2. The membrane electroresistance was measured on a thermostated cell of a conical type, as measurements in cells of a cylindrical type in dilute solutions of electrolytes give a considerable error due to a large resistance of a cell filled with electrolyte. The cell was filled with a KCl solution, which continuously circulated through either chamber of the cell.

In order to determine the resistance of membranes, we used the alternating current method [151, 154]. The measurements were carried out on plane membranes preliminarily saturated with 0.1 M KCl. The electroresistance of membranes in solutions of electrolytes, R_e, was computed as the difference between the electroresistance of the cell filled with an electrolyte, R_c, and the cell with an electrolyte and a membrane, $R_{c\text{-}m}$, i.e. $R_e = R_{c\text{-}m} - R_c$.

Equation (26) is valid in the case of equality of thicknesses of membranes used in the ultrafiltration cell, l_1, and in the cell intended for the determination of the electroconductivity of the polymer film, l_2. Therefore, we used samples cut out of the same membrane and having the same thickness both in the cells intended for ultrafiltration and in the cells intended for the determination of the electroconductivity of the polymer film, i.e. $l_1 = l_2$. In addition, to exclude the deformation action of pressure on the membrane structure in the course of a filtration process and to approximate the conditions of hydrodynamic tests to the determination of the electroresistance of polymer films as much as possible, the ultrafiltration process was performed at a minimum pressure head of 0.05 MPa. As follows from the data presented in Table 12, once these conditions are met, the thickness of the membranes after the ultrafiltration process, l_τ, decreases but slightly as compared with the initial value, l_n. All of these measures allowed us to use the simplified equation

$$r = \sqrt{\frac{8\eta R_e \chi S_2 G}{\Delta P}} \quad (26)$$

for calculating the average pore radius. Results of experimental estimation of the electroresistance of membranes under study, R_e, and values of the pore radii, calculated with equation (26), appear in Table 13.

Table 13. Dependence of pore radii in membranes on electropolymerization conditions (determination time G_τ is equal to 7.92 10^4 s)

Membrane no.	G_τ 10^6, m^3 m^{-2} s^{-1}	Re, ohm	Pore radius r, Å	Electropolymerization mode	
				E, V	τ, min
3 M (AA + F), 0.1 M $ZnCl_2$					
1	0.405	1.6	-	-1.16	5
2	0.347	3.7	31.8	-1.20	5
3	0.925	5.1	73.9	-1.30	5
4	0.115	5.5	32.7	-1.16	10
3 M AA, 7 M F, 0.1 M $ZnCl_2$					
5	0.041	5.0	73.8	-1.16	5
6	0.578	4.3	39.6	-1.20	5
7	1.273	3.6	66.6	-1.30	5
8	0.115	8.2	39.3	-1.16	10
3 M (AA + F), 0.1 M N,N'-MBAA, 0.1 M $ZnCl_2$					
9	0.115	16.8	48.5	-1.16	3
10	0.925	15.7	123.0	-1.16	5
11	27.200	22.1	788.5	-1.16	10
12	0.115	17.2	40.0	-1.20	3
13	0.173	21.6	50.0	-1.20	5
14	2.314	21.6	241.9	-1.20	10
15	0.347	12.5	71.5	-1.30	3
16	23.100	18.8	60.0	-1.30	5
17	52.000	19.4	94.7	-1.30	10

As follows from the tabulated data, the electroresistances and pore radii of membranes synthesized from monomer solutions of different compositions at different intensities and durations of electrolyses differ. The pore dimensions calculated from the electroresistance with allowances made for hydrodynamic parameters are on average 1.5–3 times (deviation coefficient *k*) those determined from the overall porosity (Tables 12, 13). The maximum deviations (k ~ 4) occur in small-pore membranes (Table 12, membranes 1 and 2; Table 13, membranes 3 and 4). The reason for these deviations is probably that such membranes (where the degree of cross-linkage of the polymer matrix by short formaldehyde molecules is low) consist practically of linear molecules, which are elastic and readily undergo transformation upon a change in the operation conditions, which leads to the opening of pores and the increase in their diameter. For more rigidly cross-linked polymers, the deviation coefficient is smaller, specifically, *k* ~1.1–2.3 (Table 12, membranes 3-12; Table 13, membranes 6 and 9-

17). Structures of this type, which are cross-linked by N,N'-methylene-bis-acrylamide molecules, are less prone to deformation when swollen and when a flow of fluid is passing.

The character of dependence of pore radii on the electropolymerization time remains the same. Specifically, r increase with the time of electrosynthesis of a polymer film (Table 13, membranes 9-11). A comparison of results, obtained after estimating the pore dimensions by different methods, shows that the combined method, which combines electrochemical and hydrodynamic measurements in conditions as identical as possible, yields data that is closest to the results of scanning electron microscopy and tunneling microscopy.

15. Structural and Morphological Investigation of Electrochemically Synthesized Acrylic Polymeric Films [45, 140]

15.1. Structure of the Polymer Film

Methods that yield maximum information on the characteristics of polymer films and membranes are the microscopic methods, which permit the resolution of structural and morphological parameters of the top and bottom surfaces and cross-sections of membranes. However, it is necessary to exercise much care when dealing with such methods and even more so when trying to interpret the results. In the first place, not all microscopic methods are applicable for investigations of membrane materials in actual conditions of synthesis and exploitation (when swollen in water). Secondly, it is necessary to take into account the fact that these methods may severely distort the real polymer structure, because of specifics of the sample preparation. For example, scanning electron microscopy does allow one to explore the fine structure of polymer films, but under the action of an electron beam, the sample can undergo damage and even burn up. Another daunting problem in electron microscopy is how to dry up moist samples, for capillary forces also may damage or alter the polymer structure.

Atomic force microscopy also exhibits high resolution. However, as opposed to SEM, it requires neither vacuum drying of samples nor sputtering of an synthesis corresponding layer. Thus, one can avoid structure distortions connected with severe conditions of preparation of a sample for investigations. The mechanical damage inflicted on the polymer by the rigid needle in the case of a semicontact AFM method is minimum. And yet, the AFM method allowed us to obtain satisfactory micrographs only when examining films dried in air. As the swollen samples had a non-uniform relief over the profile height, the needle damaged them and produced no clear-cut picture of the image of the surface.

It follows from the above that parameters, which reflect the structure and properties of a polymer film and a membrane prepared from it and which are obtained in the SEM and AFM conditions, may differ from those that characterize the polymer and the membrane in the working (swollen in water) state. Hence, it is absolutely obvious that it makes much sense to apply such a contactless and hardly damaging method as transmission microscopy, using which one can work with both dry and swollen-in-water films. To be frank, however, only a combination of various microscopic methods makes it possible to determine structural and morphological peculiarities of polymer membranes that would approach the real parameters to a maximum degree.

The electron microscopy images of a cross-section of a polymer film presented in Figures 62a and 62b show that the layer, which forms in the course of electropolymerization, has a uniform thickness but the film structure is asymmetric [45, 133]. In the cross-section, we distinguish three layers united into one (Figure 62a). The film consists of a very thin layer ~0.1 μm thick (the surface that faces the electrode during the synthesis), which is much denser than the substructure. This is precisely the barrier layer that defines the permeability and selectivity of films when used as ultrafiltration membranes. The barrier layer is combined with a roughly porous layer ~40-50 μm in depth consisting of the same material. A gradual increase in the pore size ensures a high filtration capacity and high barrier capability of such a combined bilayer. The third layer (the surface that faces the solution during the synthesis) is ~1 μm thick and consists of an ordered phase whose profile is shaped as a cord. Such a structure, which is not unfamiliar in the physical chemistry of polymers, is typical for the aggregation of a supramolecular structure of an amorphous copolymer comprising acrylic acid and sodium ethylenesulfonate [133]. This part of the polymer film kind of "tightens" the surface, imparting mechanical strength to it.

The difference between the density of the structure of layers may be explained by specific features of electropolymerization. The initial current pulse creates conditions that favor the formation of high-molecular-weight polymer products. These conditions include (1) the formation of radicals directly on the electrode and (2) a high coverage of its surface by the monomer due to specific adsorption of acrylamide and its derivatives, which hinders recombination of active centers. Both factors lead to the formation of a thin dense surface layer. Later on, the growth of the film thickness becomes affected by diffusion limitations that are caused by the shift of reagents to the electrode surface through pores or voids formed by the fluctuations of macromolecules. As a result, once the monolayer film forms, the reaction may pass from the kinetic region into the diffusion one. The processes of synthesis and washing-away of soluble fractions out of the film being formed, which occur in parallel, cause the loosening of the membrane structure and the formation of larger pores and channels. As the monomer diffusion to the electrode and the migration of LMW products into the bulk solution are possible only on the film side facing the solution, the diameter of channels on this surface is much larger than on the side facing the electrode (Figures 62c, 62d).

A micrograph of a cross-section of a film clearly shows that the pores in a membrane are normal to the surface in accordance with the orientation of monomer molecules in an electric field (Figure 62a). Such an ordered structure of a polymer film ensures a maximum porosity and a minimum number of cul-de-sac pores [149], which is an advantage of electrochemically synthesized ultrafiltration membranes as compared with those produced by traditional methods. In the latter case, the membranes are characterized by a chaotic distribution of pores, a multitude of their profile shapes, and a large number of cul-de-sac pores, owing to which the substance being filtered travels over a tortuous path, which considerably reduces the filtration rate. The electropolymerization is devoid of most of these drawbacks.

An ordered orientation of pores in a film is confirmed also by the anisotropy of the resistance of a polymer swollen in an Na_2SO_4 solution.

The resistance along the layer thickness, i.e. along the pores, is not great (10–20 kohm for a film ~1 mm thick) and is defined by the resistance of the electrolyte that fills the pores. The resistance of the film along its surface, i.e. normally to the position of pores, is much higher (150 ohm on a portion 10 mm long). This data suggests that the channels and the pores in a film are isolated from one another and the high surface resistance of a film is connected to the resistance of the electron-non-conducting polymer matrix.

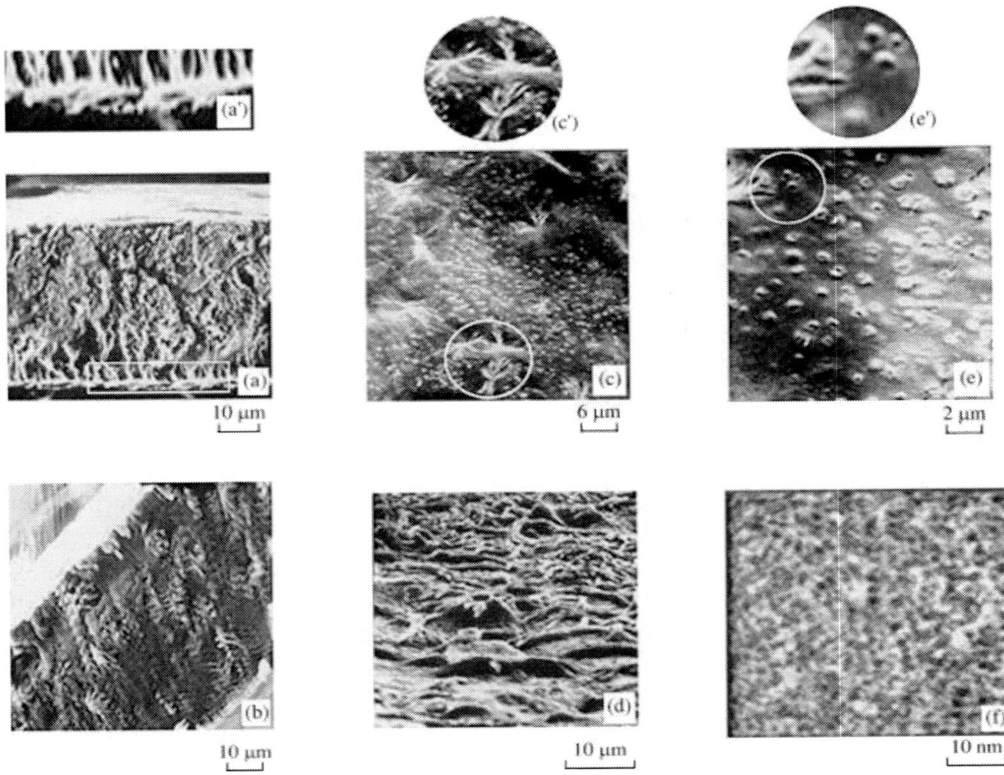

Figure 62. SEM images of (a, b) cross-section of a polymer film, (c) surface facing electrode during synthesis, (d) surface facing solution, and (e) amorphous portion and (f) STM image of film surface; [acrylamide] : [formaldehyde] = 3 : 7, copper cathode, E = -1.0 V, deposition time 5 min.

Thus, the above data suggests that the method of electropolymerization of monomers in an aqueous environment holds some promise as to the creation of highly selective asymmetric (anisotropic) membranes for ultrafiltration and reverse osmosis.

It is a known fact that polymers with a more ordered structure are more stable when employed. Under their "ordering" one should understand not only an ordered deployment of various pores and channels in a polymer matrix but also the presence of crystalline and amorphous phases in it.

The ordering of the supramolecular structure of polymers depends first and foremost on the orientation of macromolecules in the bulk phase. The ordering increases during crystallization when there emerges long-range order, i.e. when mutual arrangement of macromolecules is not random even if the distance from one another is arbitrarily long. In the case of EIP, an electric field will presumably exert an orienting effect on the position of monomer molecules and growing polymer chains along the force lines. The resultant polymer will have a strictly ordered structure containing a crystalline phase.

As seen in Figures 62c and 62d, the membrane surface is nonuniform and microrough. The film comprises amorphous and crystalline parts, i.e. it is a semicrystalline polymer containing crystalline regions incorporated into the amorphous matrix. The crystalline parts protrude above the amorphous ones; hence, the amorphous portions lie in the pits between the crystallites. It follows from Figures 62b, 62c, and 62d that the crystallization process makes a stop at the stage of the formation of fibrils and spherulites. The length of fibrils and the diameter of fibrillar spherulites equal ~4.0 to 4.5 mm. The spherulites are piercing the entire body of the membrane matrix (Figure 62b). As expected, they are indeed normal to the membrane surface, which means that they are oriented along the force lines of the electric field.

As a rule, the crystalline phase strengthens the polymer and raises its ultimate tensile strength, for the crystallites behave as effective cross-links. At the same time, the strongly cross-linked crystalline polymers are brittle, with the brittleness being proportional to the density of cross-links. The amorphous regions, on the other hand, are characterized by elasticity. However, these regions are most vulnerable, are easily deformed, and are the first to undergo destruction during hydrolysis or a microbial decomposition of the polymer. That is why a polymer material with optimum characteristics must have a certain ratio between crystalline and amorphous phases and, consequently, strictly fixed electrolysis modes, which define such an ordering.

Amorphous parts consist of densely packed globules pierced by pores (Figure 62e). The diameter of such parts is 12–18 mm relative to the centers of spherulites. The radius of globules is ~100 to 250 nm. The average separation between the centers of large globules amounts to 0.3–1.2 mm. As seen in the micrograph, the cross-section of the pores is rounded. The radius of pores of the film on the side facing the electrode during the synthesis is equal to ~50–100 nm (Figure 62c), as determined from the micrographs. The density of such pores is ~8 ´ 107 pores per centimeter squared. According to calculations, they occupy an insignificant part (0.6–2.5%) of the overall membrane area.

The membrane side that faces the solution during electropolymerization (Figure 62d) exhibits the same complex amorphous–crystalline structure as that is adjacent the metal surface. However, the radius of pores on this side is much larger and may reach~700 nm. Such a difference between the pore size on the opposite sides of a membrane characterizes its high anisotropy. Determined from micrographs, the anisotropy (the ratio between the pore sizes) is equal to ~10–14.

The electron microscopy data nicely conforms to the EIP results of studying the formation kinetics for a polymer layer on a metal. These studies showed that the molecular weight of a polymer and, as a result, the polymer structure are substantially dependent on the rate of discharge of the zinc–monomer complex. As this process is a non-steady-state, the formed film exhibits a non-uniform structure.

According to the data of *scanning tunneling microscopy*, the morphology of a membrane surface is much more complicated than what follows from the electron microscopy data. The amorphous portions contain a large number of tiny pores ~1.0 nm in diameter (Figure 62f) whose overall density reaches ~54 10^{12} pores per cm^2 and which occupy ~35% of the membrane surface. As follows from the baromembrane measurements, it is precisely these pores that provide for the major characteristics of permeability and selectivity of electrosynthesized ultrafiltration membranes. As seen in the figure, in some places the pores overlap, two to three of them merge, and then their diameter correspondingly increases to ~2.5 nm. That the pores overlap or merge follows also from the fact that the pore circumference becomes other than circular. However, one must bear in mind the complexity of analyzing micrographs, which has been noted by many researchers. The fact that clear images are difficult to obtain is due to the film roughness. As a result, several small pores situated close to one another may be interpreted as one large pore. As the image of the lower-situated layers of the film is not quite clear, the size of tiny pores sometimes cannot be estimated correctly, because their profiles are not quite clear.

The first task was to find out the reason for the formation of a dense near-cathode layer, i.e. whether it forms directly in the course of electropolymerization or results from the deformation processes inherent in the sample preparation for the SEM investigations. *Optical microscopy* allowed us to answer this question. As the film under study is colorless and transparent in water, in the case of optical microscopy, we used a contrasting regime (differential interference contrast). Our reasoning was that sometimes, one fails to see an object not because it is small but because its refraction coefficient barely differs from that of the environment (in our case, water) [155, 156]. To expand the possibilities of the optical contrasting method, we combined it with some elements of video magnification, for example, at the expense of a camcorder.

Figure 63 presents micrographs of cross-sections of polymer films synthesized at different electrolysis durations and swollen in water. The micrographs were obtained with the aid of optical microscopy. The cross-sections of all membranes that were synthesized at the electropolymerization time equal to 3–10 min exhibit at least two layers, which have boundaries between them and are contrasted optically.

Investigations showed that a thin, morphologically isolated and optically dense layer always forms on the film side adjacent to the cathode (near-cathode layer). It precedes a thicker layer, which faces the solution during electrolysis (matrix layer). The thickness of the near-cathode layer in a swollen film depends on the electrolysis duration and varies in the interval 2 to 10 μm.

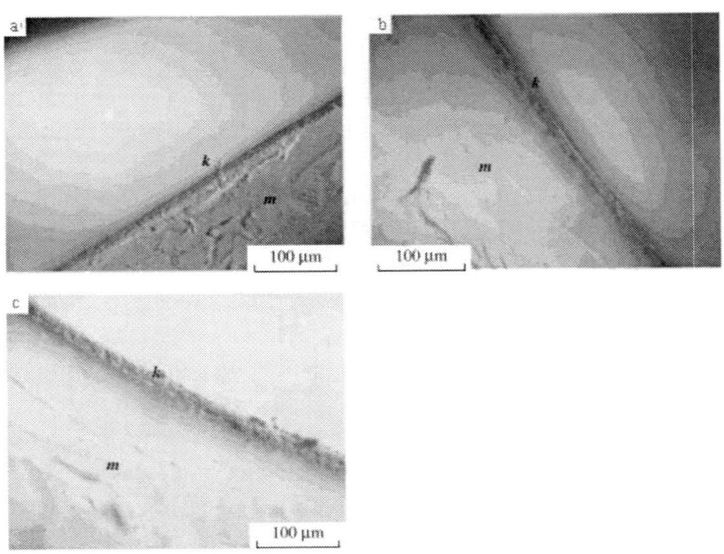

Figure 63. The optical micrographs (differential interference contrast) of cross-sections of UF membranes, with *k* referring to the near-cathode layer and m, to the matrix layer; standard electrolyzing-bath composition, E = –1.16 V, synthesis duration (a) 3, (b) 5, and (c) 10 min.

The near-cathode layer in the membranes synthesized for the duration of 0.5–1 min of electrolysis is barely pronounced and has small dimensions of thickness. This is due to the fact that the layer did not yet form completely. As follows from Figure 64 (curve 3), the near-cathode layer grows largely within the first 1–5 min, whereas in the course of a more prolonged electrolysis, its thickness varies but little. Comparing these results with the data that reflects the way the overall thickness of a polymer film depends on the synthesis duration (Figure 64, curve 1) allows us to unambiguously conclude that the larger increment of the film thickness is realized at the expense of the matrix layer. The growth of both the near-cathode layer and the entire film dramatically slows down in the interval of electropolymerization durations 5 to 10 min and practically terminates. Thus, the optimum duration of the film formation amounts to 3–5 min.

The multilayer arrangement of the membrane and the ratio between layer thicknesses we established are in good agreement with the model of electrochemical formation of polymer films presented in [133]. According to that model, the difference in the density of the structure of layers is defined by specific features of electropolymerization. An initial current pulse creates conditions that favor synthesis of polymer products with a high molecular weight. The conditions we mean include in the first place, the formation of radicals directly on the electrode and the high electrode coverage by the monomer due to specific adsorption (i.e. a maximum monomer concentration on the metal). These factors favor the recombination of active centers and provide for an analysis of high growth rate and minimum breakdown of chains. The resultant surface layer on the metal is thin and dense, it insulates the working electrode, thus reducing the current density and, correspondingly, the electropolymerization rate. Diffusion limitations, created by the shift of reacting substances to the electrode surface, start telling on the growth of the film thickness.

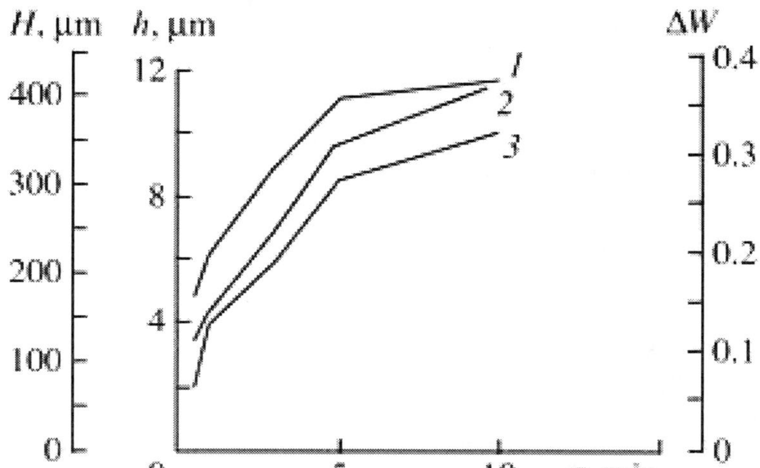

Figure 64. Dependences of (1) film thickness H, (2) coefficient of linear thermal expansion ΔW, and (3) thickness of near-cathode layer h on the electrolysis duration; standard electrolyzing-bath composition, E = –1.16 V.

As a result, the subsequent process of the film formation turns into a non-steady-state and proceeds in conditions of simultaneous formation of insoluble high-molecular-weight fractions and soluble low-molecular weight fractions. The washing of the latter out of the coating leads to the appearance of voids and cavities in the film and to the formation of a porous membrane matrix. This process creates conditions for capillary penetration of electrolyte into the film. As a result, the electropolymerization process continues and the matrix layer grows to the required size. By varying the polymerization rate, one can produce dense insulating polymer layers, which may be used as special coatings. Alternatively, one can synthesize polymer structures with properties of ultrafiltration membranes. When busy with the investigation, we made the assumption that the difference in the brightness of the two layers we fixed was due to the difference in their densities, i.e. the darker layer is simultaneously the denser layer. The data we obtained with the aid of optical photometry can serve as a roundabout confirmation of this assumption. We established that the near-cathode layer attenuates the light stronger than the matrix layer does. One of the possible explanations for this phenomenon (provided the chemical composition of the polymer is strictly homogeneous) is that the higher density of the near-cathode layer stems from the presence of pores of a far smaller size in it, whereas the pores in the matrix layer are conversely larger. Consequently, the same volume contains different amounts of substance contained in it. Yet another confirmation of our assumption as to the different density of layers is the fact that the membranes we electrosynthesized have a tendency to form into coils with the near-cathode side inside. Such a behavior of the membrane may be explained from the position of a denser and more rigid, finely porous and less swollen near-cathode layer.

Thus, the data we presented demonstrates that the polymer films we electrosynthesized are asymmetric from the viewpoint of a structural–morphological classification [82]. It is obvious that it is precisely the near-cathode layer, which is denser, that will be determining selective properties of a formed UF membrane. The near-cathode layer is sometimes called a barrier layer [149]. The matrix layer, on the other hand, is presumed to play the role of a substrate of a kind, on which the selective layer is affixed. These results are in good agreement with the filtration characteristics established in [23] and the capillary and structural characteristics established in [45] of the membranes under investigation.

15.2. The Crystalline Structure of Acrylamide

One of the peculiarities of the polymer films (membranes) formation by EIP of acrylic monomers consists in a multilayer structure of electrosynthesized films [41, 157]. Here, the densest layer is formed directly on the metal substrate. One of the reasons of the asymmetric structure could be concerned with high specific acrylamide adsorption on the electrode. It is generally known that in the aqueous medium, acrylamide can exist as in the monomer as in the associated state, in particular, in the forms of cyclic and acyclic dimers, while at $C_{AA} >$ 0.75 M, the dimeric form is predominant [158]. However, to determine the thickness of the dense near-cathode layer, one should know accurate parameters of the crystalline structure of acrylamide and the specific behavior of its molecules, in particular, the ability for dimerization. Since, in the literature data, the accuracy of the acrylamide crystalline structure determination does not exceed ~19% [159] and the complete data on bond lengths between adjacent atoms is absent, it appears impossible to estimate the adsorption layer thickness.

Table 14. Crystallographic data, experimental parameters and structure corrections

Gross formula	C_3H_5NO
Molecular mass	71.08
Temperature, K	133(2)
Wavelength, Å	MoK_α (0.71073)
Crystal system	Monoclinic lattice, $P2_1/c$
a, b, c, Å	8.228(1), 5.759(1), 9.760(1)
β, degree	120.04(1)
V, Å^3 ; Z	400.3(1), 4
ρ_{calc}, g/cm^3	1.179
μ, mm^{-1}	0.090
F(000)	152
Crystal size, mm	0.08×0.16×0.32
Data acquisition, θ(°)	4.20-29.01
Index interval	$-9 \leq h \leq 11, -7 \leq k \leq 7, -13 \leq l \leq 13$
Number of measured reflection	2691
Independent reflection	1057 (R_{int} = 0.029)
Reflection with I > 2σ(I)	867
Improvement method	Full matrix LS method
Improvement of variable	66
GOOF	1.091
R-factors for $F^2 > 2\sigma(F^2)$	R1 = 0.0543, wR2 =0.1263
R-factors for all reflections	R1 = 0.0664, wR2 = 0.1319
Δρ(min) and Δρ(max), e/Ă^3	– 0.199 и 0.343

To clarify the AA crystalline structure, the studies using the X-ray structural analysis at a temperature of 133(1) K on the diffractometer BRUKER SMART 1000 CCD (MoK_α-irradiation, graphite monochromator; accuracy 5.43%) were performed. In the experiments we used the acrylamide monocrystal (Reanal) in the form of a plate of size 0.8 x 0.16 x 0.32 mm. The data collection was conducted by groups of 906 frames at the angle values φ= 0, 90 and 180° and the crystal-detector distance 45 мм, ω-scanning with an increment 0.2° and exposition 20 s per frame. The crystal structure of acrylamide (I) was determined by the direct method and corrected by the least square method at the anisotropic approximation for non-hydrogen atoms. The hydrogen atoms were localized from the differential synthesis and corrected in the isotropic variant. The collection and editing and correction of the elementary unit parameters were performed in accordance with the programs SMART and SAINT *Plus* [160]. All of the calculations on the structure determination and correction were performed in accordance with the programs SHELXTL/PC [161]. The main crystallographic data is presented in Table 14, the atom coordinates and their equivalent isotropic thermal parameters – in Table 15, the interatom distances and valent angles – in Table 16.

In the acrylamide crystal structure, the molecules C_3H_5NO are flat (Figure 65) (maximal atoms shift from the plane is equal to 0.06 Å), while the carbon and nitrogen atoms coordination has a shape of the triangle with valent angles close to 120° (Table 3), which confirms the suggested localization of hydrogen atoms. In the acrylamide molecule, the valent bonds C(1)-C(2) and C(3)-O correspond to multiple bonds (Table 15).

Table 15. Atom coordinates and their isotropic thermal parameters (Å^2) in the structure I

Atom	*x*	*y*	*z*	*U*(eq)
C(1)	0.3486(3)	-0.1350(3)	0.1250(2)	0.0379(4)
C(2)	0.2826(2)	0.0219(3)	0.0128(2)	0.0273(4)
C(3)	0.1751(2)	0.2263(3)	0.0172(2)	0.0215(3)
N	0.1076(2)	0.3668(2)	-0.1072(2)	0.0271(3)
O	0.1509(2)	0.2632(2)	0.1313(1)	0.0297(3)
H(1)	0.331(3)	-0.125(4)	0.216(3)	0.041(5)
H(2)	0.418(3)	-0.265(4)	0.121(3)	0.048(6)
H(3)	0.301(3)	0.012(4)	-0.077(3)	0.047(6)
H(4)	0.037(3)	0.482(4)	-0.112(2)	0.030(5)
H(5)	0.127(3)	0.338(3)	-0.187(2)	0.029(5)

In the acrylamide structure, the molecules C_3H_5NO are linked by the hydrogen bonds N-H(4)...O into centrosymmetric dimers (Figure 66), which, in their turn, form bimolecular layers parallel to the plane (100) through the hydrogen bonds N-H(5)...O (Table 17). The distance between layers is equal to 1.5Å, while the layer thickness is about 6 Å. Binding between the layers of the structure is realized only by weak molecular interactions of the type C...C (C1-C1 3.790 and C2-C2 3.717 Å), since the molecule heads are directed outside the layers.

Table 16. Bond lengths *d*, Å and valent angles ω, grad in the molecule C_3H_5NO

Bond	*d*	Bond	*d*
C(1)-C(2)	1.309(2)	C(3)-O	1.244(2)
C(1)-H(1)	0.97(2)	C(3)-N	1.328(2)
C(1)-H(2)	0.95(2)	N-H(4)	0.87(2)
C(2)-C(3)	1.486(2)	N-H(5)	0.89(2)
C(2)-H(3)	0.96(2)		
Angle	ω	Angle	ω
C(2)-C(1)-H(1)	122(1)	O-C(3)-N	122.3(1)
C(2)-C(1)-H(2)	121(1)	O-C(3)-C(2)	121.7(1)
H(1)-C(1)-H(2)	117(2)	N-C(3)-C(2)	116.0(1)
C(1)-C(2)-C(3)	121.8(2)	C(3)-N-H(4)	118(1)
C(1)-C(2)-H(3)	122(1)	C(3)-N-H(5)	121(1)
C(3)-C(2)-H(3)	116(1)	H(4)-N-H(5)	121(2)

Table 17. Geometric parameters of hydrogen bonds in the structure of I (*Operation codes of equivalent atoms symemtry: #1-*x*,-*y*+1,-*z*; #2 x,-*y*+1/2, *z*-1/2)

N-H…O*	*d*(N-H), Å	*d* (H…O), Å	*d* (N…O), Å	∠ NHO, angle
N-H(4) …O#1	0.87(2)	2.07(2)	2.935(2)	175(1)
N-H(5) …O#2	0.89(2)	1.97(2)	2.848(2)	172(1)

The latter implies that the occurring solid-phase reaction of acrylamide polymerization must proceed with the breakage of double bonds C1=C2 in monomers and the formation of bonds between adjacent carbon atoms C1-C1 and C2-C2 from different layers (the polymerization route is shown in Figure 66 by the solid line).

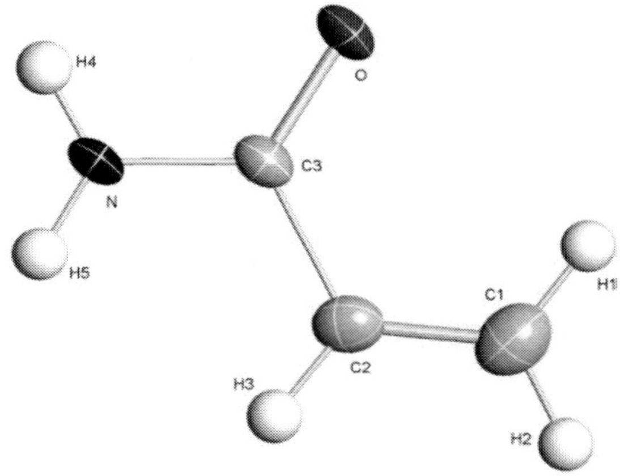

Figure 65. Structure of acrylamide I.

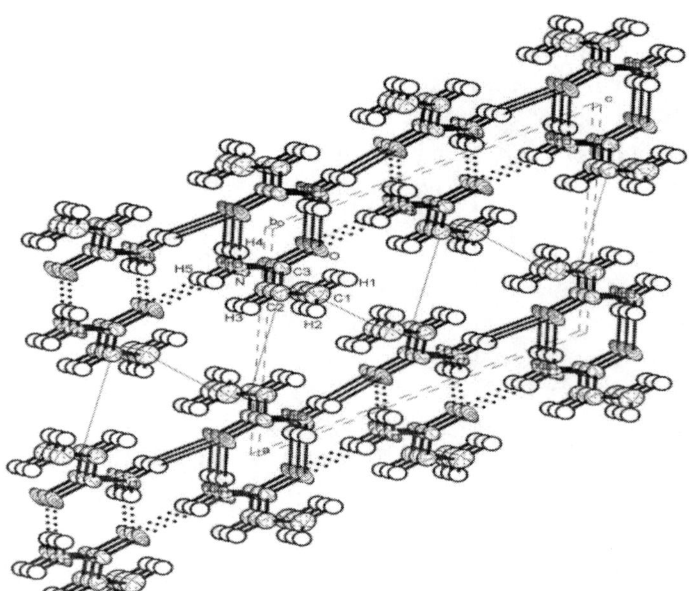

Figure 66. Projection of structure of acrylamide I onto the plane (010).

As follows from the description of the acrylamide crystal structure, in the course of acrylamide dissolution in the aqueous medium, its crystal structure must dissociate, due to the breakage of intermolecular and hydrogen bonds into molecular centrosymmetric dimers with two hydrogen bonds. Then the electrochemical formation of a thin polyacrylamide polymer film from the aqueous solution on a metal substrate must proceed in a following way. Molecular dimers, in the process of deposition on the substrate, form, under the effect of directed hydrogen bonds N-H5...O, bimolecular layers of a thickness around 6 Å, parallel to the substrate plane. The polymer film formation takes place at the participation of two bimolecular layers necessary for the polymerization process. The thickness of such a molecular film must be a multiple of the bimolecular layer. Taking into account the gap between layers, this parameter must have the value of at least ~13 Å. This is much less than the thickness of the barrier layer (2-20 micrometers) we found by microscopic methods. Therefore, a dense selective layer has a more complex, polylayer structure. Here the barrier layer thickness demonstrates a good correlation with the thickness of the double electrical layer, which, in the presence of a specific adsorption, can attain the value of 6-16 micrometers [162].

15.3. The Morphology of the Membrane Surface

To examine the morphology of the near-cathode layer and the matrix layer of the membrane, we used the SEM method (Figure 67). The inference from the figure is that these two layers have markedly different morphology (Figures 67a, 67b). The near-cathode layer exhibits clearly pronounced roughness. At the same time, when strongly magnified, both the near-cathode layer (Figure 67c) and the matrix layer (Figure 67d) exhibit globular structure. On the whole, the globules are distributed uniformly over the entire surface of either layer. The density of large globules (the number of globules per unit area) is relatively low. The explanation for this is that, as was established in [22], copolymers of AA with F and N,N'-MBAA are cross-linked. Because of this, such a material contains both rigidly-linked and relatively free polymer chains. In this case, the probability of self-organizing in the presence of other favorable conditions, into such a supramolecular structure as a globule (and a globule of a size sufficient to be fixed by the SEM method) is high only for the chains that are more mobile.

Figures 68a and 68b present micrographs of surfaces of the near-cathode and the matrix layers of swollen ultrafiltration membranes. The micrographs were obtained by a method of optical microscopy. One can discern that the observed pattern is, on the whole, similar to the SEM images. Specifically, the near-cathode layer exhibits clearly pronounced roughness, while the matrix layer is practically even.

It is of interest to mention that we observed large polymer globules only at the surface of the matrix layer. However, their density was rather low, as compared with the density of dried polymer samples. Using an AFM method, we obtained an ultimate confirmation of the globular structure of the polymer (Figure 68c). As seen in this figure, the membrane's polymer matrix has globular composition, at least in the near-surface layer. To a certain degree of reliability, we deem it possible to assume that the size of the observed globules is equal to 50–300 nm.

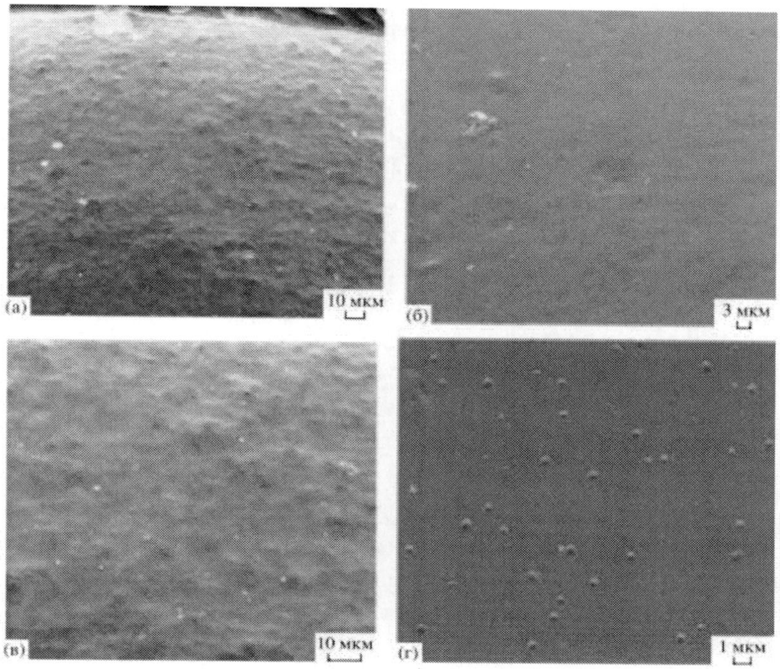

Figure 67. The SEM micrographs of the surface of UF membranes: (a, c) near-cathode layer and (b, d) matrix layer; standard electrolyzing-bath composition, E = –1.16 V, synthesis duration 5 min.

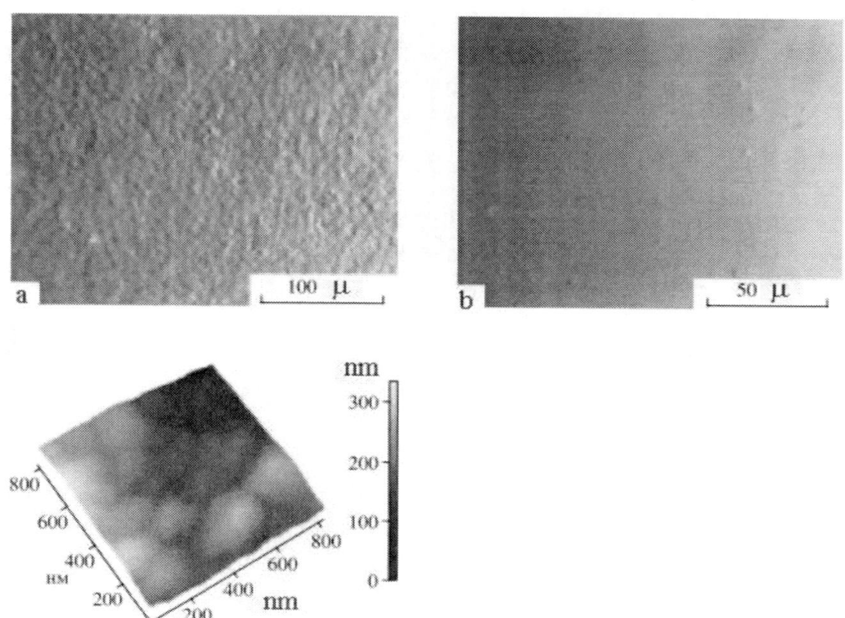

Figure 68. Micrographs of the surface of polyacrylamide–formaldehyde films: (a, c) near-cathode layer and (b) matrix layer; composition of electrolyzing bath is standard; E = –1.16 V; panels *a* and b: optical microscopy, differential interference contrast; panel *c*: AFM.

At the sites where globules get in touch with one another, we observe pores with a broad mouth and narrowing channel, which is precisely the place where particles divide into large and small. The mouth size depends on the globule diameter and is defined by the interglobular distance inside the pack; it amounts to approximately 1 to 50 nm. The factor that is limiting the transport rate is the diameter of the narrowest parts of channels.

The results we obtained allow us to draw the conclusion that the structure of pores in the electrosynthesized membranes represents a system of densely packed closed spheres. Such a structure is similar to the organization of organic and inorganic membranes that are manufactured either by sintering or by a method of inversion of phases with a nodular (granular) composition of an external layer [82].

We established that the surface topography (roughness) of the near-cathode polymer layer directly depends on the concentration of zinc chloride in the monomer composition and the electrolysis potential. The roughness increases with the increasing content of zinc chloride in the electrolyte. Raising the cathode potential from –1.09 to –1.2 V (the region of discharge of the zinc–acrylamide complex, which yields metallic zinc) exerts a similar influence. On the other hand, this effect is practically not visualized at a potential of less than or equal to –1.0 V, at which no zinc intermediate deposited on the electrode. In other words, it is absolutely obvious that the observed roughness of the near-cathode layer of the polymer intimately depends on the microgeometry of crystals of the sublayer of metallic zinc deposited on the electrode. The formed polymer deposit covers the protrudings and pits of the zinc layer, faithfully copying its profile. Once peeled off the substrate, the microgeometry of the polymer surface partially levels off at the expense of the swelling of the film and the increased linear dimensions of the film (stretching). The fact that the matrix layer has a more even relief is due to the layer's formation at the polymer/solution interface. This implies that the relief is defined solely by the morphology of the polymer layer. The data is of utter importance for the determination of optimum conditions of electrochemical synthesis of the UF membranes. The point here is that, in the case of membrane materials, it is necessary to monitor the roughness of the near-cathode layer, in order to ensure that the surface is as smooth as possible: otherwise the membranes would be prone to a faster pore-clogging. Nevertheless, one must not reduce the zinc chloride concentration or the electrolysis potential too much, for this would lead to a decrease in the electropolymerization rate and give rise to difficulties associated with the peeling of the membrane off the electrode.

Tthe morphology of the membranes we electrosynthesized is dependent on the degree of cross-linkage of the polymer. The results we presented in the foregoing were obtained when studying polyacrylamide membranes. In that case, we used N,N'-MBAA as an additional cross-linking agent. In the absence of MBAA, when the cross-linking agent is less efficient formaldehyde [44], we observed more drastic differences between dry samples and those swollen in water. In the latter case, the morphology of the near-cathode layer of the poly-AA–F and poly-AA–F–N,N'-MBAA membranes was similar. However, after drying the poly-AA–F samples, the surface morphology and the phase composition of the polymer alter, because, as we discern in Figure 62c, on the whole, areas emerge in an amorphous polymer matrix, which are represented by spherulites, whose size is on the order of a few micrometers. Spherulites are crystalline formations. All this suggests that the conditions that are created in the drying process favor self-organization of polymerchains into a corresponding supramolecular structure. Our explanation for this phenomenon is as follows. The conditions of synthesis of polymers under consideration entail a different degree of their cross-linkage.

Consequently, the polymer chains that make up the polymers have different degrees of cross-linkage and mobility (we assume that the pliability of monomer units in the chains of these two polymer types differs not too much and, under these conditions, the process of one aggregation or another is limited by the degree of mobility of the polymer chain). As a result, the polymer chains in a less cross-linked polymer are more mobile and prone to self-organization and ordering. However, when swollen in water, both polymer types exhibit a poor tendency to self-organization with the formation of crystalline structures. As a result, their supramolecular structure is represented only by rare globules (low organization degree).

The forces that act between polymer chains and lead to their physicochemical interaction (for example, to mutual attraction, aggregation, ordering, and so on) have an electrostatic origin [150, 151]. Consequently, the capability of such polymer chains to undergo aggregation is infinitesimally small, even if the polymer chains in question have sufficient pliability and are not rigidly fixed anywhere but intermolecular forces are practically absent. We attribute the weakening of the intermolecular interaction in a polymer swollen in water precisely to the presence of water, for water is a substance with a large value of permittivity ε'. Consequently, it weakens the electrostatic forces that are acting between polymer chains by ε' times. Upon drying, water evaporates little by little and the magnitude of the force increases. As a result, the larger mobility of polymer chains (in the case of a weakly cross-linked polymer) in combination with large intermolecular forces results in the formation of such a highly-organized supramolecular structure as the spherulites.

The data concerning the effect that the constitution of a monomer composition and electrolysis regimes exerts on the thicknesses of a polymer film and the near-cathode layer appear in the table. It follows from the data that raising the zinc chloride concentration from 0.05 to 0.2 M leads to a perceivable increase in the overall thickness of the film. To our minds, this increase is connected with the acceleration of electropolymerization.

Reducing the content of bipolar cross-linking agent N,N'-MBAA in the electrolyte from 0.1 to 0.01 M leads to a similar result. The reason for this effect is that, at a low N,N'-MBAA concentration, the share of poorly cross-linked fractions in the polymer is larger than that in the case of electropolymerization performed in solutions containing larger quantities of the cross-linking agent. The resultant effect is two-pronged. In the first place, the synthesized polymer swells in water to a larger degree. Secondly, the formed layer is more porous and insulates the working surface of the cathode in the course of electrolysis to a smaller degree. Hence, both the supply of reactants to the electrode surface and the formation of active species become easier, resulting in the creation of conditions favoring the continuation of the growth of the polymer layer on the metal.

The elevating electrolysis potential in the interval –1.09 to –1.2 V yields thinner films. In all probability, increasing the cathode polarization boosts electropolymerization and raises the degree of cross-linkage of the polymer, as a result of which the formed films are thinner and weakly swell in water [152, 153]. However, the result which has been obtained at $E \leq -1.0$ V, when one could have expected producing a film of a maximum thickness, was diametrically opposite. It is possible to assume that this phenomenon is due to a change in the initiation mechanism, for the discharge potential of the zinc complex is not reached [144], resulting in a lower electropolymerization rate and smaller layer thickness.

We discovered no other regularities that would govern variations in the near-cathode layer thickness as a function of the electrolysis conditions. Following some variations in the composition of the compound and regime of electrolysis or others, the thickness varies from

5.5 to 9 μm. And it is only in the case of a reduced N,N'-MBAA content that the thickness of the near-cathode layer reaches 16 μm, but then the film in this case is the thickest (Table 18).

Table 18. Effect of electrolysis conditions on polymer film thickness H and near-cathode layer thickness h; synthesis duration 5 min

Electrolysis solution, M	$-E$, V	H, μm	h, μm
0.2 $ZnCl_2$ + 3 AA + 3 F + 0.05 N,N'-MBAA	1.16	400	9.0
0.1 $ZnCl_2$ + 3 AA + 3 F + 0.05 N,N'-MBAA	1.16	315	7.3
0.05 $ZnCl_2$ + 3 AA + 3 F + 0.05 N,N'-MBAA	1.16	218	6.4
0.2 $ZnCl_2$ + 3 AA + 3 F + 0.1 N,N'-MBAA	1.16	300	7.0
0.2 $ZnCl_2$ + 3 AA + 3 F + 0.01 N,N'-MBAA	1.16	500	16.0
0.2 $ZnCl_2$ + 3 AA + 3 F + 0.05 N,N'-MBAA	1.20	380	5.5
0.2 $ZnCl_2$ + 3 AA + 3 F + 0.05 N,N'-MBAA	1.09	458	7.0
0.2 $ZnCl_2$ + 3 AA + 3 F + 0.05 N,N'-MBAA	1.00	342	6.4

The linear dimensions of polymer films substantially increased upon their peeling off of a metallic cathode. Such a behavior is connected with the ability of the synthesized polymer to swell in water. Due to a strong adhesion, the dimensions of a film rigidly fixed on a metal are restricted by the dimensions of the substrate. Once the film is peeled off, the restricting factors vanish and the film spreads proportionally to the swelling capability of the polymer. The increase in the linear dimensions of a film is dependent on the electrolysis conditions. Figure 64 (curve 2) shows a typical example of the way the coefficient of linear thermal expansion W alters with the electropolymerization duration. The maximum rise of the coefficient occurs in the interval 0.5–5 min, whereas the dynamics of this process slows down in the duration interval 5 to 10 min. The progress of curve 2 correlates well with the curve that describes variations in the film thickness (Figure 64, curve 1).

This fact proves yet once again that the polymer layer largely grows precisely within the first 0.5–5 min of electrolysis. As the competing processes in the course of the formation of a polymer film are the supply of an electroactive substance to the cathode surface and the insulation of the cathode surface, we deem it possible to assume that the process predominant in the interval 0.5–5 min is the supply of the substance to the cathode, whereas the cathode insulation tells more strongly only after 5–10 min of electrolysis.

Thus, a complex application of the methods of scanning electron microscopy, atomic force microscopy, and optical microscopy, has allowed us to explore the structure and morphology of polyacrylamide–formaldehyde films and ultrafiltration membranes formed by a method of electropolymerization. It is shown that the optical methods of contactless non-destructive investigation of polymer membranes hold much promise for studying their structural characteristics in conditions that approach to a maximum degree the conditions of synthesis and exploitation of ultrafilters.

16. Biodegradation of the Polymer Film

At the present time, when polymers have an overwhelming presence in our everyday life, their waste management problem is also of a great importance and urgency. That is why, in the course of the development of novel technologies for polymer utilization, one should pay special attention to the waste processing as well.

The studies focused on revealing the possibilities to process electrosynthesized polyacrylamide-formaldehyde material using microorganisms allowed discovering the earlier unknown characteristic feature of the obtained material – the ability to biodegrade. It was found that such a polymer, after washing out of the initial solution containing toxic formaldehyde, induced a specific bacteria behavior consisting of a complex of reactions directed to search, capture and consume nutrients [163]. The elements of the bacteria motion activity were registered on an oscillogram (Figure 69).

It was established that the moving bacterial forms changed the motion character near the polymer film, approached the 'attack' object (film), attached to the surface using special organelles and started slow rotation, thus making a cone with a vertex in the attachment point (A). The figure shows the sinusoid character of the moving organelles throbbing. To get unattached from the substrate, the bacteria start to rotate faster, and the flagellar apparatus works with higher frequency (C). Furthermore, they stop moving , but the flagellar vibrations remain, which is evidence of the start of the nutrition process (D).

All of the above bacterial reactions can be recorded for qualitative and quantitative estimations of the applicability of various substrates, which is important in the optimization of a purposeful synthesis of biodegrading polymers. The characteristics of the microorganisms' behavior are crucial for establishing their species type in order to produce microorganisms of specific features in view of biodegradation.

The main characteristic feature of biodegradation of the polymer under study consists of the fact that bacteria attack the polymer film side less intensively (Figure 70), which is directed to the electrode in the process of synthesis (selective layer, Figure 70a). Mostly, the attack proceeds on the opposite side – the matrix layer (Figure 70b). The latter must be attributed to the difference in the film density and the topography of the membrane selective and matrix layers. The membrane selective layer appears to be, due to higher density, more 'rigid' for the bacteria feeding system: this fact serves as an additional proof of an asymmetric structure of electrosynthesized membranes. On the other hand, a loose and smooth matrix layer is more accessible for microorganisms.

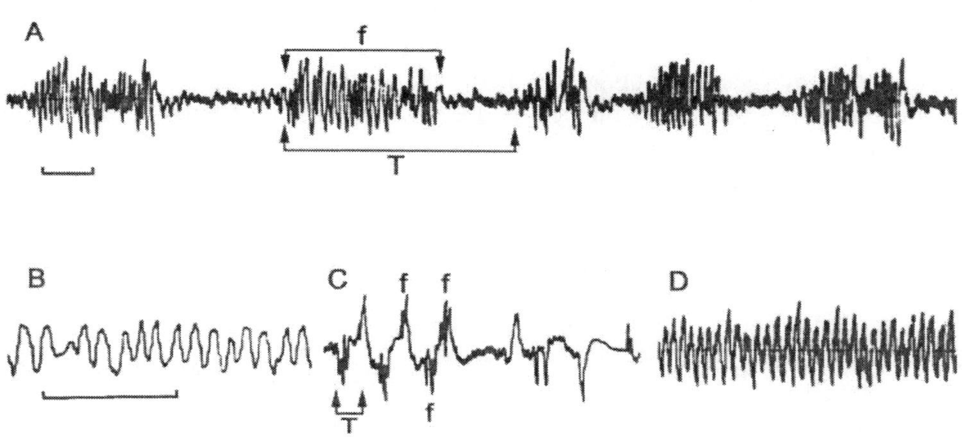

Figure 69. Oscillogram of the registration of bacteria feeding behavior during the polymer 'attack'. T – rotation period; f, V – registering of bacteria flagella vibrations at different sweeping rates.

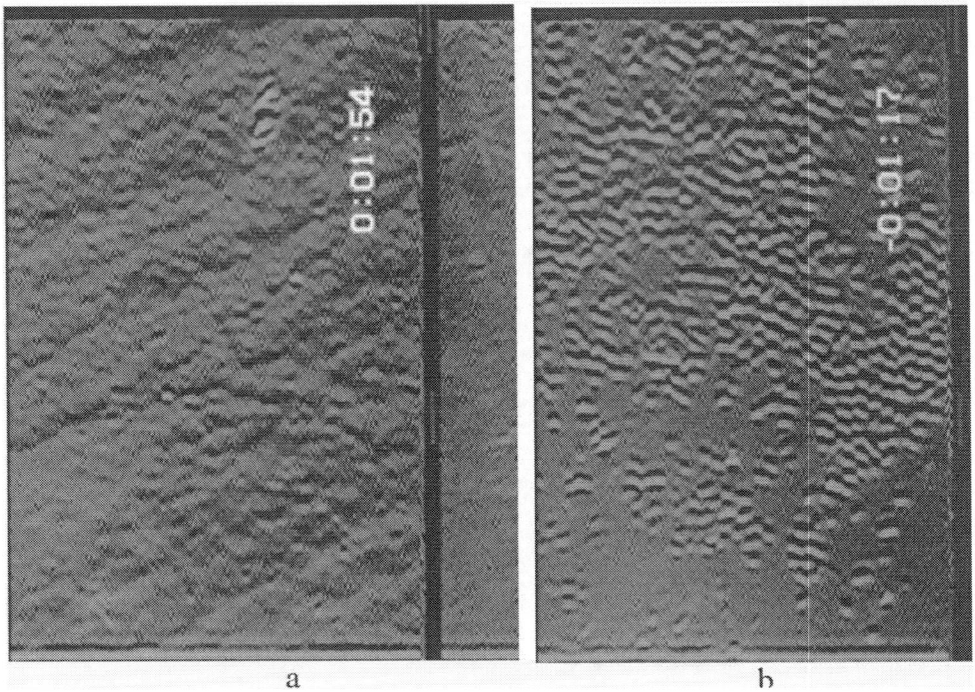

Figure 70. Microscopic microphotographs of the polymer film populated by microorganisms colonies: a) selective layer surface; b) substrate surface (multiple bacterial colonies are seen).

The established fact of the feeding-attractive effect of the electrosynthesized acrylamide-formaldehyde polymer was corroborated by studies in [164], in which it was demonstrated by classic seeding methods that the (AA-F-MBAA) produced using electropolymerization could serve as a nutrition substrate for a number of bacterial species inducing biocorrosion. One should emphasize that one of the main advantages of the applied methods of studying the membrane biodegradation consists of the fact that all of the results were obtained by direct observation under the real-time conditions. Other method advantages include simple visualization, video-documenting of each process stage, and qualitative and quantitative process registration. One more crucial positive feature of the method consists of the fact that it allows obtaining fast sought-for results , unlike the gravimetric method, in which accumulation of quantitative data requires days-long or even months-long experiments.

In summary , the suggested approach not only extends the number of methods of studying electrosynthesized films, but also makes its own substantial contribution into the development of chemical processes based on microorganisms' application.

Conclusion

The peculiarities of EIP, which have been revealed up to present, enable one to develop and regulate the conditions of the electrochemical synthesis of polymer films with pre-determined characteristics. Modification of the electrolysis conditions and variation of monomer compositions allow the producing of the protective, electro-insulating, decorative, and antiseptic coatings deposited on metals [101] or forming porous polymer films of the membrane type [23, 42, 165]. The electrosynthesized polymer layers can be, simultaneously with EIP or on subsequent stages, electrochemically or chemically modified by the formation of surface complexes or immobilization of organic substances or metal powders into a polymer matrix [32]. Such an approach enables one to create novel composite materials with special properties, in particular, the artificial implants which have become an inseparable part of modern medicine science and practice. The estimated present demand in novel materials for tissue engineering comprises of dozens of tons. The materials of this type must be porous and swelling to provide the tissue growth and dosed output of implanted medicinal preparations. The primary focus of attention in these materials' development is concerned with their biological and physical properties which depend to a great extent on the method of production. From the latter point, the EIP has real prospects of practical application.

Another prospective filed of the electropolymerization application constitutes the development of different types of selective electrodes and sensors, in particular, biosensors as well as composite materials for enzymatic catalysis. A large number of the last decade's publications are concerned with this very field [23, 165-204].

The enzymatic methods of catalysis, which are now under intensive development as interdisciplinary analytical chemistry/biochemistry projects, are distinguished by the combination of high-sensitivity characterizing voltamperometry and expanded selectivity and stability of enzymes [182]. The latter methods are characterized by the express features, simplicity of the used equipment and experimental techniques, and relative economic efficiency. However, a wider spreading of these methods is limited by the enzymes low stability and impossibility of multiple uses. The above disadvantages can be overcome by the enzyme immobilization into a polymer matrix. Immobilization increases the enzyme stability at prolonged storage, stability to external impacts and allows providing the biocatalysts multiple uses. In this case, the enzyme catalytic activity in EIP-coatings is significantly higher than in films produced by other methods, for example, by the photochemical initiation [167-170, 173, 175, 176, 178, 180, 181, 183, 185-193, 196-199, 202].

There also exists the demand in biosensors for the analysis of the activity of biological and chemical substances. The largest number of studies are devoted to the development of glucose sensors [170, 173, 175, 188, 191]. Also, a great attention was paid to producing sensors for the determination of trace amounts of metals and organic compounds [201, 203].

Electropolymerization is successfully applied in development of novel materials for chemical batteries, for example, separators and catalysts for oxygen reduction on the cathode and methanol oxidation on the anode [205, 206]. It is of special interest to apply EIP to produce cathodes and anodes of biofuel cells [167]. In [177], the possibility of the application of electropolymerization for creating acoustic sensors was demonstrated.

One more prospective field of EIP application is the electrode modification by polymer fullerene films obtained with the help of EIP C_{60} fullerene [207]. Such films can be modified by the introduction of different redox-substances as mediator systems.

A rather interesting EIP method of producing composite films based on implanting film-forming particles into the polymer molecules matrix has been patented recently [208]. The polymer molecules adsorb on the surface of the above particles by forming an ionic bond. The EIP conditions ensure polymerization near the electrode with capturing these particles, and, as a result, a matrix composite film material is formed on the electrode. Introduction of dyes into the particles enables one to impart films with the required optical or fluorescent properties.

In summary, , the recent studies have demonstrated high scientific and practical value of the electropolymerization method. It is highly probable that not all of the method advantages have been revealed, so that the development of works in this field is of substantial urgency. From this point of view, the electrochemical method of producing novel polymer materials has real prospects of finding the place among the leading ones in this field of science and technology.

References

[1] Tang, Y.; Pan, K.; Wang, X.; Liu, Ch.; Luo, Sh. *Microchim Acta*. 2010, *vol* 168, 231.

[2] Byers, J.C.; DiCarmine, P.M.; Mahmoud, M.; Abd Rabo, Moustafa; Wang, X.; Pagenkopf, B.L.; Semenikhin, O.A. *J Phys Chem*. 2009, *vol* 113 (48), 15715.

[3] Raja, V.S.; Venigopal, A.; Saji, V.S.; Sreekumar, K.; Suseelan, Nair R.; Mittal, M.C. *Progress in Organic Coatings*. 2010, *vol* 67, 12.

[4] Abedin, S.Z.El.; Borissenko, N.; Endres, F. *Electrochem Communication*. 2004, *vol* 6, 422.

[5] Blau, A.; Weinl, C.; Mack, J.; et al. *J Neuroscience Methods*. 2001, *vol* 112, 65.

[6] Bieniasz, L.K.; Bureau, Ch. *J Electroanal Chem*. 2000, *vol* 481, 152.

[7] Cram, S.L.; Spinks, G.M.; Wallace, G.G.; et al. *J Applied Polymer Science*. 2003, *vol* 87, 765.

[8] Deniau, G.; Charlier, J.; Alvado, B.; et al. *J Electroanal Chem*. 2006, *vol* 586, 62.

[9] WO 02098926. 2002.

[10] Russell, D.; Meyer, R.; Jubran, N.; et al. *J Electroanal Chem*. 2004, *vol* 567, 19.

[11] Bin, Yu.; Soo, B. Kh. *Electrochimica Acta*. 2005, *vol* 50, 1917.

[12] Bergamini, J.F.; Jouini, M.; Aeiyach, S.; et al. *J Electroanal Chem*. 2005, *vol* 579, 125.

[13] Vagin, M.Yu.; Trashin, S.A.; Karyakin, A.A. *Electrochemistry Communications*. 2006, *vol* 8, 60.

[14] Hu, J.-M.; Liu, L.; Zhang, J.-Q; et al. In *Progress in Organic Coatings*. 2007; Vol. 58, 265.

[15] Tanguy, J. *J Electroanal Chem*. 2000, *vol* 487, 120.

[16] Deniau, G.; Viel, P.; Bureau, Ch.; et al. *J Electroanal Chem*. 2001, *vol* 505, 33.

[17] Decker, C.; Vataj, R.; Louati, A. In *Progress in Organic Coatings*. 2004; Vol. 50, 263.

[18] Liu, Guo-kin L.; Ren, Bin; Wu, De-yin; et al. *J Electroanal Chem*. 2006, *vol* 594, 73.

[19] Okano, M.; Takeda, K.-I.; Toriumi, T.; et al. *Electrochim Acta*. 1998, *vol* 44, 659.

[20] Okano, M.; Toriumi, T.; Hamano, H. *Electreochim Acta*. 1999, *vol* 44, 3475.

[21] Guenbour, A.; Kacemi, A.; Benbachir, A. In *Progress Organic Coatings*. 2000; Vol 39, 151.

[22] Kolzunova, L.G.; Kalugina, I.Yu.; Kovarskii, N. Ya. *Rus J Appl Chem*. 1996, *vol* 69, 135.

[23] Kolzunova, L.G.; Suponina, A.P. *Russ J Appl Chem*. 2000, *vol* 73, 1542.

[24] Kolzunova, L.G. *Bull FEB RAS*. 2002, No. 4, 49.

[25] WO 2007007343. 2007.

[26] Mertens, M.; Calberg, C.; Baute, N.; et al. *J Electroanal Chem*. 1998, *vol* 441, 237.

[27] Yilmaz, F.; Cianga, L.; Guner, Y.; et al. *Polymer J*. 2004, *vol* 45, 5765.

[28] Bureau, Ch.; Chong, D.P. *Chemical Physics Letters*. 1997, *vol* 264, 186.
[29] Bureau, Ch.; Soudan, J.-M.; Lecayon, G. *Electrochim Acta*. 1999, *vol* 44, 3303.
[30] Bureau, Ch. J. *Electroanal Chem*. 1999, *vol* 479, 43.
[31] Jonnard, P.; Vergand, F.; Bonnelle, C. *Int. J. Adhesion and Adhesives Result List*. 1997, *vol* 17, 263.
[32] *Electrochemistry of Organic Compound to XXI century*; Eds. Gultiay, V.P.; Krivenko, A.G.; Tomilov, A.P.; Sputnik$^+$: Moscow, 2008; 578 p. [In Russian]
[33] Deniau, G.; Thomé, T.; Gaudin, D.; et al. *J Electroanal Chem*. 1998, *vol* 451, 145.
[34] Charlier, J.; Bureau, Ch.; Lecayon, G. *J. Electroanal Chem*. 1999, *vol* 465, 200.
[35] Viel, P.; Bureau, Ch.; Deniau, G.; et al. *J Electroanal Chem*. 1999, *vol* 470, 14.
[36] Guenbour, A.; Kacemi, A.; Benbachir, A.; et al. In *Progress Organic Coatings*. 2000; Vol. 38, 121.
[37] Kuramitz, H.; Nakata, Y.; Kawasaki, M.; et al. *Chemosphere*. 2001, *vol* 45, 37.
[38] Kuramitz, H.; Saitoh, J.; Hattori, T.; et al. *Water Research*. 2002, *vol*. 36, 3323.
[39] Wu, K.; Hu, Sh. *Carbon*. 2004, *vol* 42, 3237.
[40] Zaid, B.; Aeiyach, S.; Lacaze, P.C.; et al. *Electrochim Acta*. 1998, *vol* 43, 2331.
[41] Kolzunova, L.G.; Karpenko, A.A.; Karpenko, M.A.; Udovenko, A.A. *Russ Chem J*. 2005, *vol* 49(5), 137. [In Russian].
[42] Kolzunova, L.G.; Zemnukhova, L.A.; Fedorishcheva, G.A.; Kurilenko, L.N.; Sergienko, V.I. *Russ J Appl Chem*. 2000, *vol* 73(10), 1726.
[43] Kolzunova, L.G.; Suponina, A.P. *Russ J Electrochem*. 2000, *vol* 36(4), 392.
[44] RF Patent 1560280
[45] Kolzunova, L.G.; Barinov, N.N. *Anal Bioanal Chem*. 2002, *vol* 374, 746.
[46] RF Patent 2171808
[47] *Polymer Electrochemistry* Eds. Tarasevich, M.P.; Orlov, S.B.; Shkolnikov, E.I. et al. Nauka: Moscow, 1990; 238 p. [In Russian].
[48] Vorotyntsev, M.A.; Levi, M.D. In *Advances in Science in Technology. Electrochemistry;* VINITI: Moscow, 1991; Vol. 34, p. 154. [In Russian].
[49] Kolzunova, L.G.; Kovarsky, N.Ya. *Polymer Coatings and Electrophysical Deposition Methods;* Nauka: Moscow, 1976; 86 p. [In Russian]
[50] Bezuglyi, V.D.; Alekseeva, T.A. In *Progress in Electrochemistry of Organic Compounds;* Nauka: Moscow, 1975; p. 156. [In Russian]
[51] Bezuglyi, V.D.; Alekseeva, T.A. *Electrochemistry of Polymers*; Osnova: Kharkov, 1990; 184 p. [In Russian]
[52] Bezuglyi; V.D.; Alekseeva; T.A.; Shapovalov; V.A. In *Progress in Electrochemistry of Organic Compounds*. Nauka: Moscow, 1980; p. 161. [In Russian]
[53] Alekseeva, T.A.; Shapovalov, V.A., Bezuglyi, V.D. In *Progress in Electrochemistry of Organic Compounds*. Nauka: Moscow, 1983; p. 265. [In Russian]
[54] Shapoval, G.S.; Lipatova, T.E.. *Elektrokhimicheskoe initsiirovanie polimerizatsii* (Electrochemical Initiation of Polymerization). Naukova Dumka: Kiev, 1977; 235 p. [In Russian]
[55] Shapoval, G.S.. In: *Electrode Reactions with Organic Compounds*. Nauka: Moscow, 1990; p. 151. [In Russian]
[56] Shapoval, G.S.; Gorodysky, A.V. *J Adv Chem*. 1973, No 5, p. 854. [In Russian]
[57] *Electrochemistry of Organic Compounds;* Eds. Tomilov, A.P.; Feoktistov, L.G. Mir: Moscow, 1976; 732 p. [In Russian]

[58] Makarov, K.A.; Zytner, Ya.D.; Myshlennikova, V.A. *Electrochemical Polymer Coatings;* Khimia: Leningrad, 1982; 128 p. [In Russian]
[59] Kovalchuk, E.P.; Aksimentyeva, E.I.; Tomilov, A.P. *Electrosynthesis of Polymers on Metal Surface*. Khimia: Moscow, 1991; 224 p. [In Russian].
[60] Bezuglyi, V.D.; Karpinetz, A.P. In *Advances in Science in Technology. Electrochemistry;* VINITI: Moscow, 1993; Vol. 39. p. 65. [In Russian]
[61] Zytner, Ya.D.; Tikhonova, L.S.; Makarov, K.A. In: *Advances in Science in Technology Electrochemistry;* VINITI: Moscow, 1993; Vol. 39. p. 94. [In Russian]
[62] Bruno, F.; Dubois, J.E. *Electrochim Acta.* 1972, *vol* 17, 1161.
[63] Turin, Yu.M.; Smirnova, L.M.; Naumov, V.I. *Electrochemistry*. 1979, *vol* 15, 445.
[64] Kolzunova, L.G.; Kovarsky, N.Ya.; Inberg, A.S. *Russ Chem Bull.* 1994, *vol* 43, 227.
[65] Collins, G.L.; Thomas, N.W. *J Polymer Sci. Polymer Chem.* 1977, *vol* 15, 1819.
[66] Lopyrev, V.A.; Kashik,T.N. *Vysokomol Soed.* 1984, *vol* (B) 26, 594. [In Russian]
[67] US Patent 3434946
[68] Kolzunova, L.G.; Kovarsky, N.Ya. *Plasticheskie massy [Plastics*]. 1974, No. 5, 28. [In Russian]
[69] Molodov, A.I.; Losev, V.V. In *Advances in Science in Technology. Electrochemistry*; VINITI: Moscow, 1971; Vol. 7. p. 65. [In Russian]
[70] RF Patent 713896
[71] Chang, W.-P.; Whang, Wh.-Tz.; Lin, P.-W. *Polymer*. 1996, *vol* 37, 1513.
[72] Mengoli, G.; Daolio, S; Giulio, U. et al. *J Appl Electrochem*. 1979, *vol*. 23, 2117.
[73] Mengoli, G.; Musiani, M.M. *Makromol Chem*. 1980, *vol* 181, 1909.
[74] Mengoli, G.; Daolio, S.; Musiani, M.M. *J Appl Electrochem*. 1980, *vol* 10, 459.
[75] Bagotskaya, I.A.; Boitsov, V.G.; Kazarinov, V.E. *Russ J Electrochem.* 1989, *vol* 25, 111.
[76] Borkowska, *Z. J Electroanal Chem*. 1988, *vol* 244, 1.
[77] Brodsky, A.M.; Urbah, M.I. *Electrodynamics of interface metal/electrolyte*; Nauka: Moscow; 1989; 296 p.
[78] Pillai, K.C.; Waghorne, W.E.; Wilson, O.J. *J Electroanal Chem*. 1988, *vol* 252, 151.
[79] Margaryan, K.S.; Sargisyan, S.A.; Vasil'eva,U.B.; et al. *Russ J Electrochem.* 1989, *vol* 25, 748.
[80] Kargin, U.M.; Vorintsova, L.M.; Manapova, L.Z.; et al. *Russ J Electrochem.* 1987, *vol* 23, 415.
[81] *Su, W.; Iroh, J.O. Electrochim Acta*. 1999, *vol*. 44, 3321.
[82] Mulder M. *Basic Principles of membrane technology*. Kluwer: Dodrecht, 1995; 513 p.
[83] Karpinets A.P., Lizenko N.V. *Features of association of organic radicals with cation background into non-protonic medium*. GPNTB: Kiev, Ukraine, 1995; No. 184-Uk95. 17 p.
[84] Karpinets, A.P.; Bezuglyi, V.D. *Vysokomol Soed.* 1991, *vol* (B) 33, 123.
[85] US Patent 3417003.
[86] US Patent 3554882.
[87] Garg, B.K.; Raff, R.A.; Subramanian, R.V. *J Appl Polymer Sci.* 1978, *vol* 22, 65.
[88] Kronin, V.V.; Maximov, Yu.M.; Podlovchenko, B.I. *Adsorption of pyridine and aniline on platinum electrodes;* VINITI: Moscow: 1988; No. 7987-V88, 13 p.
[89] Bagrii, V.A.; Shapoval, G.S. *Ukr Chem J* 1987, *vol* 53, 1293.
[90] Bagrii, V.A.; Shapoval, G.S. *Ukr Chem J* 1986, *vol* 53, P. 1065.

[91] Tidswell, B.M.; Mortimer, D.A. *Eur Polym J.* 1981, *vol* 17, 735.
[92] Gierst, L.; Franck, C.; Quarin, G. et al. *J Electroanal Chem.* 1981, *vol* 129, 353.
[93] Mengoli, G.; Munari, M.T.; Bianco, P. et al. *J Appl Polym Sci.* 1981, *vol* 26,. 4247.
[94] Tanguy, J.; Deniau, G.; Zalczer, G et al. *J Electroanal Chem.* 1996, *vol* 417, 175.
[95] Kolzunova, L.G.; Kovarsky, N.Ya. *Russ J Electrochem.* 1984, *vol* 20, 144.
[96] Hillman, A.R.; Mallen, E.F. *J Electroanal Chem.* 1987, *vol* 220, 351.
[97] Liu, Ch.; Chen, W-J; Martin, Ch. R. *J Membrane Sci.* 1992, *vol* 65, 113.
[98] Kolzunova, L.G.; Kovarsky, N.Ya.; Novichkova, L.M. *Vysokomol Soedin. Ser A.* 1985, *vol* 27, 2370.
[99] Pomogailo, A.D.; Savost'anov, V.S. *Uspekhi khimii (Progress Chem)* 1991, *vol* 60, 1513. [In Russian]
[100] Lipatova, T.E.; Matyushova, V.G. *Carbon.* 1985, *vol* 23, 59.
[101] Kolzunova, L.G. In *Electrochemistry of Organic Compound to XXI century;* Eds. Gultiay, V.P.; .Krivenko, A.G.; Tomilov, A.P.; Sputnik+: Moscow, 2008; p. 429. [In Russian]
[102] Houzé, F.; Boyer, L.; Noël, S.; et al. *Synthetic Metals.* 1994, *vol* 62, 207.
[103] Rault-Berthelot, J.; Granger, M.M. *J Electroanal Chem.* 1995, *vol* 382, 169.
[104] Tanguy, J.; Deniau, G.; Zalczer, G.; et al. *J Electroanal Chem.* 1996, *vol* 417, 175.
[105] Baute ,N.; Martinot, L.; Jerôme, R. *J Electroanal Chem.* 1999, *vol* 472, 83.
[106] Tanguy, J.; Viel, P.; Lécayon, G. *Electrochimica Acta.* 1993, *vol* 38. 1501.
[107] Bureau, Ch.; Defranceschi, M.; Delhalle J.; et al. *Surface Science.* 1994, *vol* 311, 349.
[108] Azzem ,M.A.; Yousef, U.S.; Pierre, G. *Europ. Polym J.* 1996, *vol* 32,111.
[109] EP 0 665 275. 1995.
[110] US Patent 5232574. 1993.
[111] US Patent 3567612. 1971.
[112] Mironov, V.S.; Pleskachevsky, Yu.M.; Klimovich, A.F. *Russ J Appl Chem.* 1981, *vol* 54, 87.
[113] El-Hamouly, S.H.; Azze J n, M.A; Yousef, U.S. *Europ Polym J.* 1993, *vol* 29, 1271.
[114] Otero, T.F.; Cantero, I. *Electroanal Chem.* 1995, *vol* 395, 75.
[115] US Patent 3464960.
[116] Makarov, K.A., Zytner, Ya.D., Omel'chenko, I.Yu.; et al. *Russ J Appl Chem.* 1980, *vol* 53, 698.
[117] Japan Patent 48-36402.
[118] Kolzunova, L.G.; Kovarskii, N.Ya. *Vysokomol Soedin. Ser A.* 1983, *vol* 25, 1702 [*Polymer Sci USSR.* 1983, 25A Engl. Transl.]
[119] Gorbunova, K. M. and Polykarov, Yu. M. In *Itogi Nauki. Elektrokhimiya [Advances in Electrochemistry]*; VINITI: Moscow, 1966 [in Russian].
[120] Kolzunova, L.G.; Kovarsky, N.Ya. *Russ J Electrochem* 1978, *vol* 14, 861.
[121] Vetter, K.J. *Electrochemische kinetic*; Springer-Verlag: Berlin-Göttingen-Heidelberg, 1961; 856 p.
[122] Kolzunova, L.G.; Kovarsky, N.Ya. *Vysokomol Soedin. Ser. B.* 1983, *vol* 25, 468.
[123] US Patent 876198.
[124] *Encyclopedia of inorganic materials*; Higher School: Kiev, 1977; Vol. 1. p. 582. [in Russian].
[125] Rostovsky, E.N.; Novichkova, L.M. *Russ J Appl Chem.* 1968, *vol* 41, 346.

[126] Mairanovskii, S.G.; Stradyn', Ya.P.; Bezuglyi, V.D. *Polyarografya v organicheskoi khimii* (Polarography in Organic Chemistry); Khimiya: Leningrad, 1981; 325 p. [in Russian].

[127] Mengoli, G. *Adv Polymer Sci Berlin.* 1979, *vol* 33, 1.

[128] US Patent 3464960.

[129] Tikhonova, L.S.; Makarov, K. A.; Zytner, Ya. D.; et al. *Russ J Electrochemistry.* 1981, *vol* 17, 880.

[130] *Double layer and electrode kinetics* Ed. Kazarinov V.E.; Nauka: Moscow; 1981, p. 376.

[131] *Establishment of structure of organic connections by physical and chemical methods;* Khimia: Moscow, 1967, *vol.* 2, 190 p.

[132] Kolzunova, L.G.; Kovarsky, N.Ya.; Novichkova, L.M. *Vysokomol Soedin. Ser. A.* 1986, *vol.* 28, 227.

[133] Kolzunova, L.G. *Russ J Electrochrem.* 2004, *vol.* 40, 337.

[134] Kolzunova, L.G.. Kovarskii, N. Ya. In *Electrically-Based Microstructural Characterization;* Eds. Gerhardt, R.A.; Taylor, S.R.; Garboczi, E.J.; MRS: Pittsburgh, 1996; *vol.* 411, 425.

[135] Galus, Z. *Teoretyczne Podstawy Electroanalizy Chemicznej.* Panstwowe Wydawnictwo Naukowe: Warszawa, 1971; 131 p.

[136] Karpenko, M.A.; Karpenko, A.A.; Kolzunova, L.G. *Russ J Electochem.* 2007, *vol.* 43(10), 1144.

[137] Charlier, J.; Bureau, C.; Lecayon, G. *J Electroanal Chem.* 1999, *vol.* 465, 200.

[138] Trabanelli, G.; Zucchi, F.; Fonsati, M. *Electrochim Acta* 1998, *vol.* 44, 311.

[139] Sukhotin, A.M.; Rotinyan, M.A. *Tekhnika eksperimental'nykh rabot po electrokhimii, korrozii i poverkhnostnoi obrabotke metallov;* (Experimental Techniques in Works on Electrochemistry, Corrosion and Surface Treatment of Metals); Khimiya: St. Petersburg, 1994, p. 166.

[140] Karpenko, M.A.; Kolzunova, L.G.; Karpenko, A.A. *Russ J Electrochem.* 2006, *vol.* 42(1), 89.

[141] Kolzunova, L.G.; Greben', V.P.; Suponina, A.P. *Russ J Electrochem.* 2003, *vol.* 39(1), 1452.

[142] Brock, T.D. *Membrane Filtration: a User's Guide and Reference Manual*, Madison, WI: Science Tech., 1983.

[143] Dytnerskii, Yu. I., *Baromembrannye protsessy;* (The Baromembrane Processes); Khimiya: Moscow, 1986.

[144] Dytnerskii, Yu. I., *Obratnyi osmos i ul'trafil'tratsiya;* (Reverse Osmosis and Ultrafiltration); Khimiya: Moscow, 1978, p. 18.

[145] Kolzunova, L. G.; Podorvanova, N.F.; Kovarskii, N. Ya., Russ J *Electrochem.* 1982, *vol.* 18(3), 425.

[146] Nikolaev, A.F. *Sinteticheskie polimery I plasticheskie massy na ikh osnove;* (Synthetic Polymers and Plastics Based on Them); Khimiya: Moscow, 1966, 603 p. [In Russian]

[147] Grigorov, O.N.; Karpov, I.F.; Koz'mina, Z.P.; Tikhomolova, K.P.; Fridrikhsberg, D.A.; Chernoberezhskii, Yu.M. *Rukovodstvo k prakticheskim rabotam po kolloidnoi khimii;* (A Laboratory Course on Colloidal Chemistry); Khimiya: Moscow, 1964. [In Russian]

[148] Dubyaga, V.P.; Perepechkin, L.P.; Katalevskii, E.E. *Polimernye membrany*; (Polymer Membranes); Khimiya: Moscow, 1981. [In Russian]

[149] Kesting, R.E. *Synthetic Polymeric Membranes*, Wiley: New York, 1985.

[150] Mank, V.V.; Lebovka, N.I. *Spektroskopiya yadernogo magnitnogo resonansa vody v geterogennykh sistemakh;* (Nuclear Magnetic Resonance Spectroscopy of Water in Heterogeneous Systems); Nauka Dumka: Kiev, 1988; 152 p.
[151] *Processing with Membranes;* Eds. Lacey, R.; Loeb, S.; Wiley: New York, US, 1972.
[152] Selin, M.M.; Tambovskii, S.V. *Khim-farm Zh.* 1987, *vol* 21, 1256 p. [In Russian]
[153] Selin, M.M.; Tambovskii, S.V. *Khim-farm Zh.* 1989, *vol* 23, 1384 p. [In Russian]
[154] Karpenko, L.V.; Demina, O.A.; Dvorkina, G.A.; Parshikov, S.B.; Larshe, K.; Okler, B.; Berezina, N.P. *Russ J Electrochem.* 2001, *vol.* 37, 328.
[155] Salmon, E.D.; Shaw, S.L.; Waters, J.; Waterman-Storer, C.M.; Maddox, P.S.; Yeh, E.; Bloom, K. *Methods Mo. Cell Biol.* 1998, *vol.* 56, 185.
[156] *Svetovaya mikroskopiya v biologii: Metody* (Light Microscopy in Biology: Methods); Ed. Leisi, A. Mir: Moscow, SU, 1992.
[157] Udovenko, A.A.; Kolzunova, L.G. *Russ J Structural Chem.* 2008, *vol.* 49(5), 993.
[158] Grigoryan, M.V.; Grigoryan, J.D.; Chshmarityan et. al. *Russ J Phys Chem.* 2004, *vol.* 78(4), 651.
[159] Isakov, I.V. *Russ J Struct Chem.* 1966, *vol.* 7(6), 898.
[160] Bruker (1998). SMART and SAINT-*Plus. Versions 5.0. Data Collection and Processing Software for the SMART Sistem.* Bruker AXS Inc., Madison, Wisconsin, USA.
[161] Sheldrick, G.M. (1998), SHELXTL/PC. *Versions 5.10. An Integrated System for Solving,* Refining and Displaying Crystal Structures From Diffraction Data. Bruker AXS Inc., Madison, Wisconsin, USA.
[162] *Electrochemistry of Organic Compounds*; Eds. Tomilov, A.P.; Feoktistov, L.G.; Mir: Moscow, SU, 1976, p. 53.
[163] Karpenko, M.A.; Kolzunova, L.G. In *Microbial indication and remediation.* 2004, FENU: Vladivostok, p.177.
[164] Suponina, A.P.; Kolzunova, L.G.; Karpenko, M.A. *Russ J Appl Chem.* 2006, *vol* 79, 934.
[165] Zang, W.; Plieth, W.; Kossmehl, G. *Electrochim Acta* 1997, *vol* 42, 1653.
[166] Bureau, Ch., Deniau, G., Valin, F.; et al. *Surface Science.* 1996, *vol.* 355, 177.
[167] WO 2005093888, 2005.
[168] US Patent 5286364, 2006
[169] US Patent 5540828, 2006
[170] US Patent 6413396, 2006
[171] McMahon, C.P.; Killoran, S.J.; Kirwan, S.M.; et al. *Chem. Commun.* 2004, 2128.
[172] Noroozifar,M.; Khorasani-Motlagh, M.; Taheri, A. *Talanta.* 2010, *vol* 15, 1657.
[173] US Patent 7041209, 2006.
[174] Long, D.D.; Marx, K.A.; Zhou, T. *J Electroanal Chem.* 2001, *vol.* 501, 107.
[175] Chung, T.D.; Jeong, R-A.; Kang, S.K.; et al. *Biosensors and Bioelectronics.* 2001, *vol.* 16, 1079.
[176] Compagnone, D.; Isoldi, G.; Moscone, D.; et al. *Analytical Letters.* 2001, *vol.* 34, 841.
[177] Peng, H.; Yin, F.; Zhou, A.; et al. *Analytical Letters.* 2002, *vol.* 35, 435.
[178] Chen, X.; Hu, Y.; Wilson, G.S. *Biosensors and Bioelectronics.* 2002, *vol.* 17, 1005.
[179] Kim, J.-H.; Kim, B.-G.; Yoon, J.-B.; et al. *Sensors and Actuators A: Physical.* 2002, *vol.* 95, 108.

[180] Yang, H.; Chung, T.D.; Kim, Y.T.; et al. *Biosensors and Bioelectronics*. 2002, *vol.* 17, p. 251.
[181] Gajovic-Eichelmann, N.; Ehrentreich-Förster, E.; Bier, F.F. *Biosensors and Bioelectronics*. 2003, *vol.* 19, 417.
[182] Ciszewski, A.; Milczarek, G. *Talanta*. 2003, *vol.* 61, 11.
[183] Cirpan, A.; Alkan, S.; Toppare, L.; et al. *Bioelectrochemistry*. 2003, *vol.* 59, 29.
[184] Yin, F. *Talanta*. 2004, *vol.* 63, 641.
[185] Ojani, R.; Raoof, J.-B.; Salmany-Afagh, P. *J Electroanal Chem*. 2004, *vol.* 571, 1.
[186] Yang, M.; Yang, Yun.; Yang, Yu; et al. *Analytica Chimica Acta* 2005, *vol.* 530, 205.
[187] Tao, W.; Liu, Y.L.; Nie, L.; et al. *Bioelectrochemistry*. 2004, *vol.* 65, 51.
[188] Lakard, B.; Herlem, G.; Labachelerie, M.; et al. *Biosensors and Bioelectronics*. 2004, *vol.* 19, 595.
[189] Weetall, H.H.; Ragers, K.R. *Talanta*. 1998, *vol.* 62, 329.
[190] Broncová, G.; Shishkanova, T.V.; Matějka, P.; et al. *Analytica Chimica Acta*. 2004, *vol.* 511, 197.
[191] Curulli, A.; Valentini, F.; Orlanducci, S.; et al. *Sensors and Actuators B: Chemical*. 2004, *vol.* 100, 65.
[192] Defever, T.; Deniau, G.; Palacin, S.; et al. *J Electroanal Chem*. 2006, *vol.* 589, 46.
[193] Obirai, J.; Nyokong, T. *Electrochimica Acta*. 2004, *vol.* 49, 1417.
[194] Gong , J.-L.; Gong, F.-Ch.; Zeng, G.-M.; et al. *Talanta*. 2003, *vol.* 61, 447.
[195] Suprun, E.V.; Budnikov, H.C.; Evtugyn, G.A.; et al. *Bioelectrochemistry*. 2004, *vol.* 63, 281.
[196] Loh, K.P.; Zhao Sh.,L.; Zhang, W.D. *Diamond and Related Materials*. 2004, *vol.* 13, 1075.
[197] Zhou, H.; Chen, H.; Luo Sh.; et al. *Sensors and Actuators B: Chemical*. 2004, *vol.* 100, 224.
[198] Jeong, H.; Kim, H.; Jeon, S. *Microchemical J*. 2004, *vol.* 78, 181.
[199] Reyes De Corcuera, J.I.; Cavalieri, R.P.; Powers, J.R. *J Electroanal Chem*. 2005, *vol.* 575, 229.
[200] Ebarvia, B.S.; Sevilla, III. F. *Sensors and Actuators B: Chemical*. 2005, *vol.* 107, 782.
[201] Jin, G.; Zhang, Y.; Cheng, W. *Sensors and Actuators B: Chemical*. 2005, *vol.* 107, 528.
[202] Lomillo, M.A.A.; Ruiz, J.G.; Pascual, F.J.M. *Analytica Chimica Acta*. 2005, *vol.* 547, 209.
[203] Zhang, L.; Li, W.; Shi, M.; et al. *Talanta*. 2006, *vol.* 70, 432.
[204] Han, J.H.; Taylor, J.D.; Kim, D.S.; et al. *Sensors and Actuators B: Chemical*. 2007, *vol.* 123, 384.
[205] Bettelheim, A.; Soifer, L.; Korin, E. *J Powwer Sources*. 2004, *vol.* 130, 158.
[206] WO 0124291, 2001.
[207] Bettencoutr-Dias, A.; Winkler, K.; Fawcett, W.R.; et al. *J Electroanal Chem* 2003, *vol.* 549, 109.
[208] US Pat 6803097, 2006.

In: Polymer Films: Properties, Performance and Applications ISBN 978-1-61209-201-0
Editors: Stefan A. Romano and George P. Somners

Chapter 2

POLYMER SURFACE MODIFICATION BY ELECTROMAGNETIC RADIATION PROCESSING

Daniel Eduardo Weibel
Departamento de Físico-Química, IQ, UFRGS,
Porto Alegre, RS, Brazil

ABSTRACT

Polymers have been applied successfully in many different fields such as membrane and thin-film technologies, biomaterials and biocompatibility, sensors, adhesion, protective coatings, microelectronic devices, composites, etc. Success or failure in a particular application depends many times on special surface properties of the material with regard to chemical composition, hydrophilicity, roughness, crystallinity, conductivity, lubricity, and cross-linking. Polymers very often do not possess the surface properties needed for these applications. However, surface modification techniques, which can transform these inexpensive materials into highly valuable finished products, have become an important part of the plastics and many other industries. In recent years, many advances have been made in developing surface treatments to alter the chemical and physical properties of polymer surfaces without affecting their excellent bulk properties. In particular, the past two decades have seen the rapid development of new strategies for the design of polymer functional surfaces. As one of the main techniques developed to achieve efficient surface modification of polymeric materials, electromagnetic radiation-induced surface modification has been widely applied as simple, useful and versatile approach to improve the surface properties of polymers. This chapter surveys the recent advances in electromagnetic radiation–induced surface modification of polymers, predominantly focusing on the research work carried out in the last decade and covering the modification and functionalization of polymer surfaces using: (1) Ultraviolet (UV); (2) Vacuum-UV (VUV); (3) Laser and (4) Synchrotron radiation sources. Each radiation source combined with traditional or new experimental approaches has the ability to tailor specific applications.

The academic interest in this area is driven for two reasons: the first one is the advance in the understanding and manipulation of nano and microstructures designs at surfaces together with the chemistry involved at the molecular level. The second is associated with the unique potential in use light to transform and pattern surfaces. The extensive base of knowledge of photochemistry accumulated in the last half century, in particular the advances in surface photochemistry, offers exciting potential for modification of surfaces with control of both the density and nature of surface chemical groups. Electromagnetic radiation permits a high selective control of the changes produced at the surfaces due to quantum allow transition rules. Simultaneously the use of ultrafast laser sources, where ablation takes place because of non-resonant process by multi-phonon absorption, can be used even in materials normally transparent to the laser wavelength. Additionally, electromagnetic radiation treatments offer solutions in terms of industrial applications, which need for materials to compete in the increasing low-cost consumer market. The materials for these applications should be cheap, easy to make, and, last but not least, be environment-friendly, combustible, or better yet biodegradable or photodegradable. These characteristics, which are not usually found in traditional wet and dry surface modification techniques of polymer surfaces, offer great potential for expansion of modification methods based in electromagnetic radiation.

1. Introduction

The field of polymer photochemistry is recognized as an active and very important area in applied photochemistry, with rapidly extending applications in industrial development. Light source-driven technologies such as UV-hardering, photolitography, photo-polymerization, photohardering, photochemical grafting, etc. became very common as surface modifications techniques. In particular, photopolymerization and photocuring science and technologies are being developed toward designing novel and specific initiators and materials for special applications. Photocrosslinking of polymers is also very attractive in terms of enhancing the physical and mechanical properties of conducting materials.

Optical properties of polymers remain an active area of strong development, with a continued growth in photochromic and liquid crystalline materials. New polymer-based composite materials containing nanoparticles is a rapid growing field [1], particularly in relation to fast optical switches, magneto-optics data storage, interferometers, nonlinear optical properties, etc. Photoactive polymer films by immobilization of photoactive complexes in polymer matrices enable the study of the electro- and photochemistry of species in non-diffusive conditions. Research in polymer based solid state electrochemoluminescent [2] and dye based solar cell systems [3] have shown the widespread applications of those approaches in recent years. Those and other examples [4] demonstrate that processes, such us, electrochemoluminescence seen in solution can be achieved also in immobilized systems.

Surfaces that can respond to a specific external stimulus in a specific manner, the so called smart surfaces [5], are of increasing interest in areas as diverse as organic electronics, cell culture, microfluids and coatings. The combination of those smart surfaces with the flexibility provided through molecular synthesis, supramolecular chemistry, and surface science, open a series of exciting opportunities for the development of new materials and devices. Photoswitchable molecules immobilized in polymers are a recent attempt for the generation of photoswitchable polymers. For example, a photo/electrochromic dithienylcyclopentene can be immobilized on a surface as a thin film (<10 nm) through the electropolymerization of a methoxystyryl unit attached to the photoactive unit [6]. The limited conductivity of the polymer backbone limits the oxidative electropolymerization of the methoxystyryl unit. The polymer film is formed consistently as a thin film and can be switched both optically and electrochemically between the open and closed state of the photochromic dithienylethene moiety, although problems with oxidation and extended UV irradiation leads to rapid destruction of the photochromic functional unit.

Photodegradable polymer is receiving increasing attention in the last decade, due there are compelling economic and social reasons for using degradable plastics in certain applications and, therefore, considerable research is now devoted to devising new polymers with improved performance. The ideal photodegredable polymer would be such material that should end up as litter rather quickly after light activation. Basically, there are two methods to obtain polymers photochemically degradable [7, 8]. One method is to introduce chemically a chromophore into the polymer chains. Indeed, the most commercially successful chromophore is the carbonyl group [7, 8]. Ultraviolet radiation absorption leads to typical photoreactions of the carbonyl chromophore, *i.e.*, the well known Norrish type I and II processes or by an atom abstraction process [9], all of which produce the degradation of the polymer material. The second general method for making polymer materials photochemically degradable is to mix a radical initiator into the polymer [10, 11]. Although direct irradiation of polymers might be the simplest method, in some cases, the use of photosensitive structures has some advantages. The photosensitization method should be preferred when there is a need of controlling the photodegradation, as well as when the irradiation is performed at wavelengths that are not absorbed by the polymer. The excitation of the sensitizers may result in the production of free radicals that can initiate polymer degradation processes. In general, these radicals abstract hydrogen atoms from the macromolecule forming polymeric alkyl radicals that react with oxygen, initiating an oxidative chain reaction. Both the carbonyl-containing and the radical initiator require ultraviolet light for degradation. Recently, some research groups tried to synthesize visible light photodegradable polymers based in a new class of polymers that contain metal–metal bonded organometallic dimmers interspersed along the polymer backbone [12]. These polymers are photodegradable because the metal–metal bonds can be cleaved with visible light and the resulting metal radicals captured with oxygen or other traps.

In the polymer photochemical processes described above, the surface of the solid material is a very important property in successful applications and among all parts of a material, the outmost surface of the solid will probably interact physical and chemically with its environment in the largest extend. The microstructure and chemical functional groups present at the surfaces are usually different from those in the bulk. To achieve materials with desired surface properties for specific applications, the surface of solids are modified by a wide range of surface treatment techniques. Surface modification has many advantages, such as, no need to redesign the bulk material to achieve a target surface performance, improvement of the surface properties without affecting its bulk properties and in many cases reduction in cost, since all chemistry is confined to the surface. In recent years, many advances have been made in developing surface modification techniques for polymers to tailor specific applications which change their original chemical and physical surface properties and keep their bulk properties intact. The more common surface modification techniques now a day available are treatments by plasma, corona, photons, electron beams, ion beams, X-rays, γ-rays, chemical grafting and combinations of different methods, including modern hybrid techniques. By using those treatment technologies different polymer surfaces can be modified to achieve the following objectives:

1. Increase hydrophilicity.
2. Increase hydrophobicity.
3. Introduce special functional groups at the surface for a specific application.
4. Modify surface free energy.
5. Improve chemical inertness.
6. Introduce surface cross-linking.
7. Modify surface morphology increasing or decreasing roughness.
8. Remove weak boundary layers or contaminants.
9. Increase separation efficiency in membranes.
10. Increase surface electrical conductivity.
11. Avoid or decrease corrosion rate.
12. Decrease friction.

Surface functionalization of polymeric materials by using polychromatic or monochromatic electromagnetic radiation is a relatively new research field, in particular when monochromatic radiation is used as a tool to introduce new surface functionalities. UV-light treatment of polymer surfaces provides similar effects with other techniques, such as plasma treatment modifying their hydrophilicity, chemical properties on surfaces, and so on. The use of UV-light for surface treatment is a classical application of this methodology for surface modification. However, there are significant differences, advantages and disadvantages of UV-treatments compared with plasma modification techniques. Plasma treatments are generally limited to surfaces, but photochemical reactions can be surface-limited or can take place deep inside the bulk depending on the UV absorption coefficient at the specific UV-wavelength. Plasma can treat wide surface areas, while UV-treatment can process wide areas as well as very small spots. Another difference is the power-intensity control. Controlling plasma intensity is generally limited although with some methods like a magnetic field confinement it can be enhanced locally. In UV-photo treatment, one can use continuous wave (CW) UV-lamps with a moderate light intensity to very high power output of a pulsed laser. In general, UV processes do not require costly equipment and offer the advantage of patterned surface modification by use of lithographic techniques. In the last decade, new applications of UV surface functionalization have been published combining reactive gas or vapours atmospheres with simultaneous UV irradiation of the polymer [13-18].

The developments of excimer lasers and vacuum-ultraviolet (VUV) sources based on dielectric barrier discharges, DBDs, is inciting increasing interest as means to oxidize polymer surfaces. Although low-pressure radiofrequency (r.f.) plasmas were used to study VUV since the early 1970s [19], the field of materials processing with VUV radiation has only recently received much attention [20-27]. Organic materials with multiple bonds or with atoms which have non-bonding electron pairs such as oxygen or nitrogen absorb light with a wavelength shorter than 200 nm very strongly [28]. This property limits the penetration depth of the photons and 90% of the radiation is absorbed within a surface layer between 50 and 100 nm. The penetration depth contrast with UV light, where, depending on the intensity of the source, the wavelength or the polymer structure, the radiation is absorbed non-uniformly within microns (1-6 μm). Principal advantages of VUV excimer radiation sources are high monochromatic radiant exitances, simplicity of construction, large emission area with high energy photons, low cost, and availability of different wavelengths. Excimer VUV lamps are very promising for surface modification for large area industrial applications. Excimer radiation sources may be preferred to laser and plasma techniques for the treatment of large surface areas as well as for the simplicity of the irradiation equipment used, taking into account that plasma systems require most often vacuum lines and controlled gas mixtures.

The understanding of the interaction of a laser radiation with a material has contributed to the increase application of lasers to surface modification, machining, laser ablation deposition, and chemical analysis. The availability of eximer laser sources in the VUV spectral region has increased the number of publications in polymer surface modification using lasers as excitation sources. In particular, laser technology has found interesting applications in the fabrication of polymer based sensor arrays by applying conventional patterning steps on appropriate photosensitive materials [29]. Further polymer engineering should maximize the number of areas with different sensing properties on the same substrate producing a novel class of polymer sensor arrays with enhanced detection efficiency of liquid/gas analytes [30]. The photo induced changes in the VUV in one hand modulate the physical/chemical functionality of the polymeric surfaces such as hydrophilicity, wetability, or biocompatibility and on the other hand, lead to surface modification with nanoresolution. The high value of the absorption coefficient and the high molecular photo-dissociation probability allow atomic resolution depth control. In addition, modification of chemical composition and roughness of the irradiated part of the organic films is achieved, enhancing thus the surface-probe binding strength and the detection sensitivity in a predetermined way.

Within the wide range of photochemical sources used to modify polymer surface properties, synchrotron radiation (SR) are the less explored way of treatment. A synchrotron yields a continuous spectrum of radiation extending from the far IR to the hard X-ray regime. Third and projected fourth SR sources have much more intense photon flux than can be obtained by most conventional sources. UV and X-ray monochromators provide an adjustable spectral resolution for all photoemission experiments. Further advantages of SR are its 100% polarization in the plane of the ring, its high degree of collimation, its high stability and well-defined time structure for time-resolved measurements. In spite, the unique properties of this source a few studies have been carried out to modify the surface properties of polymeric materials [31-36]. Surface modification by SR exposure in the soft X-ray region under gas atmosphere is effective in controlling surface properties; however, it has not been extensively reported because of its experimental difficulties originating from the lack of a suitable window material in this energy region and the use of ultra high vacuum conditions. Probably, new kind of studies will start to appear in the future using SR radiation to modify selectively polymer surfaces due to the increasing importance of this radiation to characterize, modify and synthesized new materials.

This book chapter is intended to review the status of worldwide research in surface modification of polymeric materials using mainly electromagnetic radiation as excitation source and covering from the UV up to soft X-ray region. Second section reviews various techniques to perform surface modifications of polymeric materials and typical experimental techniques used to characterize the changes produced at the treated surfaces are presented in Section 3. Thereafter the electromagnetic radiation treatment of polymer surfaces are discussed depending on the source used, UV, VUV, laser and SR, in sections 4, 5, 6 and 7 respectively. Finally, a few conclusions and outlooks in surface modification by electromagnetic radiation are summarized in section 8.

2. Techniques in Surface Modification

2.1. Wet Chemical

In wet chemical surface modification, a material is treated with liquid reagents to generate reactive functional groups on the surface. This classical approach to surface modification does not require specialized equipment and thus can be conducted in most laboratories. It is also more capable of penetrating porous three-dimensional substrates than plasma and other energy source surface modification techniques [37], and allows for in situ surface functionalization of microfluidic devices. For example, cyanuric chloride was used as a coupling molecule between hydroxyl groups on the surface of an epoxy resin and polyamines [38]. In this way two polyamine, diethylenetriamine and branched polyethylenimine, were successfully coupled to the surface via chemical modification. Recently, an interesting application of TiO_2 self cleaning coatings on polycarbonate (PC) substrates was reported [39]. TiO_2 coatings on plastics may find widespread application in auto and construction industries if the coating produces a polymeric material with new differentiated photocatalytic and mechanical properties. A chemical surface treatment method was used to create hydrophilic groups on the PC surface. Then TiO_2 was deposited based on wet coating using an anatase sol of TiO_2 nanoparticles of 30 nm size. PC with a self cleaning TiO_2 layer exhibited better hardness and scratch resistance, as well as good photocatalytic and mechanical properties.

Chromic acid and potassium permanganate in sulfuric acid have been used to introduce reactive oxygen-containing moieties to poly(ethylene) (PE) and poly(propylene) (PP) [40, 41]. PP, like PE and other polyolefins, has become an increasingly important material. The ability to tailor the bulk properties of polyolefins through catalyst design makes these already important polymers derived from relatively inexpensive feedstocks even more important. However, these polyolefins are generally hydrophobic materials and in many practical instances, improved adhesion, wetability, printability, or biocompatibility is desired. As a result, there is continuing and widespread interest in new chemistry capable of modifying these polymer surfaces. For example, PP was initially oxidized by etching to produce a modified surface [41]. Then, a series of repetitive grafting experiments using a diamine derivative of poly(tert-butyl acrylate) were used to produce surfaces containing significant amounts of poly(acrylic acid) obtaining water contact angles of approximately 20-30°. On the other hand, treatment of the modified PP surfaces by chemical etching with alkali produced a more hydrophilic carboxylate surface. Treatment of these surfaces first with ethyl chloroformate followed by pentadecylfluorooctylamine produced a hydrophobic fluorinated surface. Modifying a pristine polyolefin surface allows for further processing of the polymer increasing the application range and impact of the material in the market.

Fluorinated polymers, like poly(tretrafluoroethylene) (PTFE) has widely used in several fields of industry because of its excellent properties, such us thermal and chemical stability, low surface energy and high resistivity. PTFE membranes are frequently used for filtration, as gas permeable membranes in gas electrodes and diffusion cells and can be used in the construction of biosensors based on an oxygen sensor or an ammonia gas electrode. Membranes and powders prepared from PTFE were investigated for their potential use as multifunctional supports for proteins, enzymes and sugars [42]. PTFE surfaces have been modified by refluxing for 8 h with elementary sodium in toluene to generate double bonds. In a next step, samples were oxidized at 120 °C in a 1:1 mixture of trifluoroacetic acid and hydrogen peroxide (38%) which improved membrane wetting and allowed immobilization of the enzyme alliinase.

Unfortunately, these wet chemical methods are non-specific, introducing a range of oxygen-containing functional groups. Chain surface orientation, as in PMMA ester modification, for example, adds additional complexity in the treatment processes and the degree of surface functionalization may therefore not be repeatable between polymers of different molecular weight, crystallinity, or tacticity. These wet chemical methods also generate hazardous chemical waste and can lead to irregular surface etching [43]. Many of these techniques also require extended treatment in concentrated corrosive solutions. For these reasons, while useful in the laboratory environment, these wet chemical processes may not be suitable for larger scale, industrial applications.

2.2. Plasma

Broadly, a plasma can be defined as a gas containing neutral and charged species, including electrons, positive and negative ions, radicals, atoms and molecules in ground and excited estates and electromagnetic radiation (mainly UV and VUV). In a plasma, the average electron temperature ranges between 1 and 10 eV, the electron density varies from 10^9 to 10^{12} cm^{-3}, and the degree of ionization can be between 10^{-6} and 0.3[44]. There are a wide range of scientific published reviews [44-47] and books (see the review of Denes, et al. [45] and references therein) concerning plasma treatment of polymers and only a brief account will be given in this chapter section.

Plasma treatments in its different forms have moved from the academy to become important industrial processes for modifying the surface properties of polymers. The main advantages of plasma modification techniques can be summarized in the following:

1. Bulk properties of the polymer remain intact after treatment because the surface modification can be confined to the surface layer, typically less than ~ 100 nm.
2. Choosing the gas in the plasma, it is possible to graft particular chemical functionalities at the polymer surface.
3. Common problems encountered in wet chemical techniques (section 2.1) such as residual solvent on the surface and swelling of the substrate can be avoid with the use of a gas plasma.
4. Excited species in gas plasma can modify the surfaces of all polymers, regardless of their structures and chemical reactivity, and this modification is fairly uniform over the treated surface.

The disadvantages of the plasma processes are as follows:

1. A vacuum system is required in traditional plasma treatments increasing the operation cost.
2. Optimal parameters developed for one system usually cannot be adopted for another system. Because the plasma process is extremely complex; it is necessary a very

good control of the plasma parameters such as Radio Frequency (r.f.), power level, gas flow rate, gas composition, gas pressure, sample temperature, and reactor geometry. In consequence, it is difficult to achieve a good understanding of the mechanisms involved between the plasma species and the surface.

3. It is difficult to control the concentration of a particular functionality formed on a polymer surface.
4. Usually, polymers substrates modified by plasma have to be further processed after treatment, between hours and up to a few days, because polymers recover pristine properties, commonly in a short period of time.

Plasma occurs naturally on the sun and other stars [45] and it is also artificially produced in fluorescent lights, plasma displays and flat panel TV. The generation and sustenance of a plasma requires an energy source to produce ionization of the gas species. The use of direct current (DC) in a plasma requires that electricity has to be conducted through the electrodes, where in one of them, the sample is set for treatment. Frequently, polymer films are insulators which it would charge up the electrode and terminate the discharge. To solve the problem an AC power source is used because positive charges accumulated during one half-cycle can be neutralized by electron bombardment during the next cycle. To provide a continuous discharge a frequency of 50-100 kHz is sufficient to obtain many of the qualitative features observed in a DC glow discharge. A r.f. plasma can be initiated and sustained by external electrodes outside the reactor vessel or the coil wound around the reactor vessel.

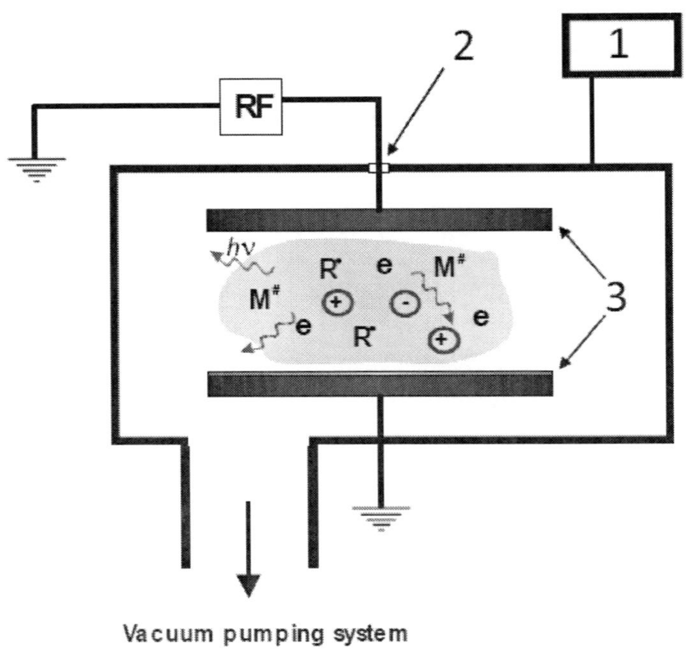

Figure 1. Schematic diagram of a parallel plate cold-plasma reactor representing the plasma-generated species. 1: gas supply and flow control; 2: electric insulator; 3: electrodes; RF: r. f. generator; R^*: radical species; $M^\#$: metastabel excited species; $h\nu$: UV and VUV emitted radiation.

A typical plasma system is constituted by a vacuum pumping system, a reactor vessel, a matching network, a power source and a gas handling or flow control system. A schematic diagram of a typical r.f.-plasma reactor including the generated species in a plasma is shown in Figure 1.

In general, the interaction of the reactive species generated in a gas plasma can react with polymer surface according the following processes [44]:

1. Surface reactions: Reactive plasma gas-phase species can react directly with the polymer surface. Alternatively, surface species can produce new functional groups after exposition to oxygen or other gases. In addition, those reactive species can give place to crosslinking reactions at the surface. Examples of these reactions include plasma treatment by argon, ammonia, carbon monoxide, carbon dioxide, fluorine, hydrogen, nitrogen, nitrogen dioxide, oxygen, and water.
2. Plasma polymerization: A thin film on the surface of a polymer is formed via polymerization of an organic monomer such as CH_4, C_2H_6, C_2F_4, C_3F_6, acrylic acid, etc. in a plasma. These complex processes involve reactions between species in the gas phase, reactions between surface species and reactions between gas-phase species and surface species.
3. Etching and surface cleaning: Undesirable materials at the polymer surface or atmospheric contaminants are removed from a surface by chemical reactions and physical etching at the surface to form volatile products. Organic contaminants presented at the surface are usually removed by oxygen-containing plasmas Etching differs from cleaning only in the amounts of materials that are removed from the surface.

Depending on the plasma conditions, they are generally divided in two main categories: near-equilibrium or non-equilibrium plasmas that are defined as hot plasmas or cold plasmas respectively. Very high temperatures of electrons and heavy particles, charged and neutral, are typically found in hot plasmas which also present maximal degrees of ionization (close to 100%). On the contrary, cold plasmas are composed of low temperature particles (charged and neutral molecular and atomic species) and relatively high temperature electrons. They are characterized with very low or much lower degrees of ionization than hot plasmas. Electrical arcs, plasma jets of rocket engines, thermonuclear reaction generated plasmas, etc. are examples of hot plasmas. Cold plasmas are formed in low-pressure DC and RF discharges, discharges from fluorescent illuminating tubes and in corona discharges. A detailed description of the different plasma types can be found in a recent review by Denes, *et al.* [45].

2.3. Pulsed Plasmas

Modulated discharges, *i.e.*, pulsed plasma can be created by using an r.f. carrier, switched on and off at a low frequency. One common type of modulation is the so called 'chirp' modulation where during a given pulse interval a carrier is swept over a wideband. Pulse is a variation of voltage or current where specific electronic circuits are required for the generation and control of precisely timed waveforms. A rise and a decay time of infinitesimal duration characterize that variation and can be easily modified during a particular experiment.

The main parameters that characterize modulated discharges are: frequency, pulse period (the time it takes to complete a pulse cycle) and the duty cycle (a percentage of ON period with respect to a total period ON plus OFF). They can be defined by:

$$\text{Period} = t_{ON} + t_{OFF} \tag{2.1}$$

$$\text{Duty cicle} = \frac{t_{ON}}{t_{ON} + t_{OFF}} \text{x}\, 100 \tag{2.2}$$

By selecting a particular duty cycle in pulsed plasma processes it is possible to allow recombination mechanisms that occur in the absence of the plasma state, in time or in space. These reactions are characterized by minimization of the typical dissociation plasma processes and as a result molecular precursors can be physically or covalently incorporated through charged or free radicals active species, into the surface. The partial or total absence of plasma state allows the development of recombination processes which are not present during the plasma limiting significantly the intensities of plasma-induced fragmentation mechanisms.

In pulsed plasma experiments, negative ions are not trapped in the glow by the plasma sheaths to form particulates and 'dust' particle formation can be diminished altering the discharge chemistry. This allows to maintain the etch/deposition rate despite lower power. For example, in Si etching with pulse durations less than 10 ms by using SF_6, the etch rate is increased to the level of continuous discharge when the duty cycle is kept constant at 20% [48].

Another interesting characteristic of pulsed plasmas is that gas phase chemistry can be altered during the discharge by modification of plasma parameters. For example, the CF/CF_2 ratio and the radical densities of CF_3 radicals can be controlled through variation of the duty cycle for 100 ms modulation period [49]. Modulating an Electron Cyclotron Resonance (ECR) plasma enables the control of the CF_2 radicals and the F atom density ratio in CHF_3 [50].

An important pulse plasma treatment process for surface modification is known under a general name of plasma immersion technique which include plasma source ion implantation (PSII), plasma immersion ion implantation (PIII or PI3), plasma ion implantation (PII or PI2), plasma ion plating (PIP), plasma immersion ion implantation and deposition (PIIID), metal plasma immersion ion implantation and deposition (MePIIID), plasma doping (PLAD), plasma ion immersion processing (PIIP), plasma-based ion implantation (PBII) [51, 52]. The family of techniques covers a wide range of related techniques ranging from pure plasma ion implantation, to implantation–deposition hybrid modes, to modes that represent plasma film deposition with substrate bias. A historical review with many examples and future trends of PII can be found in reference [51]. In the most general sense, all plasma immersion techniques have in common that the surface of a substrate (target) is exposed to, or immersed in, a plasma and that relatively high substrate bias, usually pulsed, is applied. The pulse applied to the substrate has different objectives:

1. to avoid arcing on the substrate
2. to allow the near-substrate region to be 'refilled' with ions during the bias-off time

3. to limit the sheath size when operating with very high voltages (>10 kV) and low pressure plasmas;
4. to have additional degrees of freedom such as pulse length and duty cycle during a particular treatment process.

Briefly, when a negative high voltage is applied to a conducting substrate immersed in a plasma, plasma electrons are repelled from the near-substrate region in a time scale of the inverse plasma electron frequency [51]. Ions have larger inertia than electrons and on this time scale; they are still at their original location, forming the ion matrix sheath. On the longer time scale of the inverse ion plasma frequency, ions are accelerated to the negatively biased substrate and impact the surface with an energy that depends on their initial position within the sheath. PII is not an ideally mono-energetic process. Usually, the voltage pulse applied to the substrate, relatively to the ion travel time, has slow rise and fall time, thus ions have less energy at the beginning and end of each pulse producing a sheath expansions due to the ions motions. Details for various situations such as other-than-plane geometry, collision sheath, supersonic plasmas, can be found in the specific literature [51, 53, 54].

2.4. UV/O_3

Surface treatments by gas phase ozonation under irradiation of UV light, as well as plasma and chemical vapors deposition, have been extensively used for improving the wetability and adhesion properties of polymer surfaces and removing organic contaminants on inorganic substrates [55-65]. Dry surface treatment methods, such us oxygen plasma, corona discharge or UV radiation in combination with ozone, can modify the physical and chemical properties of substrate surfaces facilitating the interactions and contacts between two dissimilar phases, without affecting the properties of the bulk material.

In contrast to plasma techniques, UV/O_3 treatment is milder than the oxygen plasma due to the absence of high kinetic energy particles. Compared to other dry surface treatment techniques, UV/O_3 methodology is important in that the process can be conducted under atmospheric pressure, and hence the equipment and running costs are relatively low. These advantages are used in semiconductors and electronics industries that are one of the major areas where the UV/O_3 treatment has been finding ever-increasing applications in recent years. In addition, the use of dry medium eliminates the need for immersing the whole device/component in a solution for treatment. A review showing characterization parameters and range of applications of the UV/O_3 treatment was recently published [60].

The photochemical and kinetic reactions of gas phase ozone are well known and were first proposed by Chapman, who in 1930 suggested a static photochemical steady sate model, as follow [28]:

$$O_3 + h\upsilon \longrightarrow O + O_2 \qquad 2.3$$

$$O + O_3 \longrightarrow 2\, O_2 \qquad 2.4$$

$$O + O + M \longrightarrow O_2 + M \qquad 2.5$$

Oxygen plasma and UV-Ozone (UV-O_3) treatments are often used industrially for cleaning (long exposure times) or introducing oxygen species to organic surfaces (short exposure times). Short-time oxidation treatments can be extremely effective to improve the adhesion between polymer surfaces and other surfaces. UV-O_3 is especially important for improving adhesion between multilayer films for packaging and modifying the surface of biologically compatible polymers [64, 65]. The ozonation treatment with UV light can easily be carried out with different media, such us gas, solvent and solution without the need for a vacuum system and complex experimental equipment. UV-O_3 treatment is also suitable for organic substrates which usually are unstable materials when the temperature of the treatment increases. Many different polymers have been studied over the years by UV-O_3 treatment, but one that was investigated in extend is polystyrene (PS). A simple and widely used polymer such as PS was probably chosen because new functional groups containing oxygen, for example, would be easy detected by surface sensitive techniques. By this way the surface modification processes would be better tracked.

X-ray photoelectron spectroscopy (XPS) and atomic force microscopy (AFM) techniques have been used to investigate the surface and near-surface properties of PS substrates which have been subjected to controlled surface modification processes performed in situ [66]. Modification of the PS surfaces were carried out using UV at 184.9 and 253.7 nm, in oxygen (UV-ozone), in nitrogen (UV-only) and by an exposure to an atmosphere of ozone in the absence of UV (ozone-only). Both UV-ozone and ozone-only modifications incorporate oxygen directly, but the surface irradiated with UV-only in an oxygen-free environment takes up oxygen after exposure to air. XPS analyses revealed that the oxygen content of the modified surfaces increased in the order of ozone-only < UV < UV-ozone for a 60 s exposure (0.05, 0.7 and 5 at.%, respectively). Friction force imaging of patterned surfaces revealed chemical heterogeneity. Statistical analysis of the friction force data shows that the resolution attainable with UV irradiation in nitrogen was higher than surfaces patterned by UV-ozone treatment.

A combination of ozone aeration and UV irradiation was used to study the introduction of amine and amide groups on PS surface in aqueous ammonia solution [67]. Treatment in aqueous ammonia solution introduced amide or amino-groups onto the PS surface, together with hydroxyl and carbonic acid groups. Treatment in pure water introduced only hydroxyl and carbonyl groups onto the surface. Ozone can be dissolved in water by aeration of the gas and if it is excited by UV light ($\lambda < 254$ nm), the excited ozone decomposed in water and produces very reactive oxygen species, such as hydroxyl ($OH^{\bullet}$), peroxide ($OOH^{\bullet}$), and superoxide radicals ($O_2^{\bullet -}$) radicals, according to the following scheme [67]:

$$O_3 + h\upsilon \longrightarrow O + O_2^{*} \qquad 2.6$$

$$O + H_2O \longrightarrow 2\ OH^{\bullet} \qquad 2.7$$

$$O_3 + OH^{\bullet} \longrightarrow HO_2^{\bullet} + O_2 \qquad 2.8$$

$$HO_2^{\bullet} \rightleftharpoons O_2^{\bullet -} \qquad 2.9$$

Aromatic compounds are especially reactive when they are in contact with those active oxygen species and can generally decompose to produce water-soluble compounds involving hydroxyl groups and carbonic acid. In water, the active oxygen species produced by UV-ozone reacted with the aromatic ring of surface molecules on the PS film adding directly hydroxyl groups to the aromatic rings, some of which were opened and oxidized to carboxylic acid. The reactive surface then can react with ammonia to form NH groups and amide on treatment in aqueous ammonia solution. The treatment was suitable for modifying PS films to introduce at the surface NH, NH_2, and amide functional groups [67].

Modification of poly(ethylene terephthalate) (PET) surfaces were recently reported using a simple method by brief exposure of PET surfaces to UV/ozone treatment [59]. The methodology used generated high surface concentration of hydrophilic moieties that facilitated subsequent chemisorption of organosilane precursors. The physico-chemical properties of the organosilane layers attached to PET were studied with several surface sensitive techniques, including near-edge X-ray absorption fine structure (NEXAFS) spectroscopy, XPS, contact angle (CA), AFM, and ellipsometry. Atomic oxygen concentration and surface energy of the treated PET samples increased with increasing UV-O_3 treatment time up to 4–5 min, as evidenced from CA, XPS and NEXAFS measurements. When the treatment time increased the surface roughness measured by AFM also increased. The oxidized PET surfaces were then used as a substrate for depositing an hydrophobic fluorinated organosilane self-assembled monolayers (SAM). The quality of the SAMs generated for different UV-O_3 treatment times were monitored with CA, XPS and NEXAFS. Surface analyses confirmed that 2 min of UV-O_3 treatment time represents an optimum condition for creating stable PET-UVO/fluorinated silane samples. In addition, NEXAFS measurements showed that the fluorinated chains were oriented nearly perpendicularly to the substrates.

The effects of UV/ozone treatment of a polymethyl methacrylate (PMMA) derivative polymers dielectric surface on the properties of pentacene thin films for organic transistors were recently investigated [58]. Organic thin film transistors (OTFTs) are drawing considerable interest for the low-cost and large-area flexible electronic device applications and many studies were devoted to the investigation of the interfacial properties between pentacene (Pn) and inorganic gate insulators. The UV-O_3 treatments of the films were performed using a very simple set up using a low pressure mercury vapors grid lamp under ambient condition. The treated surface exhibited higher offset values for the highest occupied molecular orbital levels between Pn and the polymer measured by Ultraviolet Photoelectron Spectroscopy (UPS). This result would indicate that a higher threshold voltage need to be applied for the device and the dipole field would be reduced in the Pn film on the treated surface. The hydrophilic nature of the treated surface, observed by water CA measurement, allowed for a larger grain size of the Pn film, as confirmed by the AFM measurements, which will also favorably contribute to device mobility.

3. Surface Characterization Techniques

Improvement in surface treatment techniques cannot be made without an understanding of the chemical and physical properties of polymer surfaces after each step in the modification process. Different surface treatment techniques will introduce diverse functional groups and concentrations that have to be measured in detail. This understanding and specially the control of the chemical-physics changes produced at the polymer surface will allow to study the mechanisms involve in the surface modification and simultaneously to improve the treatment process for a particular application. For example, in a printing process of a hydrophobic polymer film, the polymer surface has to be modified to hydrophilic allowing the ink to wet the surface and in consequence transferring it to the film. At the same time, the wetting of the treated surface does not have to induce spreading of the ink in printing images. Another example that has seen rapid growth in the past decade in such industries as biomedical, textiles, microelectronics, bio-processing, and food packaging is the covalent immobilization of bioactive compounds onto functionalized polymer surfaces. In the field of hydrolytically degradable polymers, *i.e.* polymers that have hydrolytically labile chemical bonds in their back bone, the presence of functional groups susceptible to hydrolysis, such us esters, orthoesters, anhydrides, carbonates, amides, urethanes, ureas, etc. is the fundamental importance. Therefore, it is necessary to obtain an in-depth knowledge about concentration, morphology and chemical functionalities at the surface of treated polymers for successful applications. This chapter section focuses only on selected surface sensitive analytical techniques used in polymers analyses. The techniques covered here is obviously not complete and this section cannot cover all existing surface analysis techniques. For the interested reader much more details about surface analytical techniques and applications can be found in specialized literature [68-72].

Within the range of surface analytical techniques actually available, the concept of "surface" can be misleading depending on the technique used. For example, a surface defined by Fourier transform infrared spectroscopy-Attenuated Total Reflection (FTIR-ATR), may correspond to bulk information for Time-of-flight secondary ion mass spectrometry (ToF-SIMS). The most appropriate term to be used in surface analysis is information depth or sampling depth that can vary even for a same technique that is operating in a particular mode, such us XPS and angle resolved XPS (ARXPS). A typical sampling depth for XPS is between 10 and 50-100 Angstroms depending of the kinetic energy of the secondary electrons and on the operation mode (ARXPS been much more sensitive to surface species). On the contrary, other spectroscopy techniques, such as energy or wavelength dispersive X-Ray spectroscopy (EDS or WDS respectively) have sampling depth in the micrometer range.

When surface sensitive techniques has to be involve in the study of a particular sample, it is necessary to consider at first what kind of surface information is required. Each technique will supply different and usually complementary information on the surface and in general, the more exact surface information will came from a combination of several surface sensitive techniques. The choice of a suitable technique(s) is affected by diverse factors, and usually when the sample surface condition is not well defined, the right procedure should be start with simple techniques; even they may have deeper sample information, and move to more sensitive surface techniques. When it is necessary to carry out chemical analyses, ToF-SIMS technique can be appropriates and even a simple technique, such us water contact angle (WCA) can be used to study hydrophilicity or hydrophobicity of a treated surface. If quantitative and chemical-state information are important, XPS will be a good choice. When high-resolution images (topology) of the surfaces are needed, AFM, Scanning electron microscopy (SEM) or Scanning tunnel microscopy (STM) can be invoke. If very low concentrations of species, impurities, contaminants, etc. have to be determined at the surface, static SIMS (ToF-SIMS) can be selected. Also ToF-SIMS should be the technique to choose if a chemical image of the surface or a chemical depth information image is necessary, especially when the primary ion is a cluster (3D *cluster*-SIMS). The importance of interfaces and specially the interface solid-gas or solid-liquid in all kind of applications is reflected in a myriad of surface science techniques available (see Table 1). Definitely, more critical than choosing a particular surface science technique to analyze a sample, is selecting the right investigator or technical stuff. In spite, they may have excellent skills in a particular technique; they have to be able to identify the information that is relevant to the study and suggest alternatives techniques to improve the analysis of the samples.

Table 1. Summary of main surface analyses techniques used for chemical and morphological analyses

Technique	Main information obtained
Work function	Surface spatial charge
UPS: Ultraviolet Photoelectron Spectroscopy EELS: Electron Energy Loss Spectroscopy	Energy level and density of states
IRRAS: Infrared Reflection Absorption Spectroscopy HAS: Helium Atom Scattering. (HRHAS: High Resolution HAS) FTIR-ATR: Fourier transform infrared spectroscopy-Attenuated Total Reflection	Vibrational energy levels
TPD: Temperature-Programmed Desorption TDS: Thermal Desorption Spectroscopy ESD: Electron Stimulated Desorption PSD: Photon Stimulated Desorption	Adsorption, desorption and chemical kinetic studies
LEED: Low Energy Electron Difraction STM: Scanning Tunnel Microscopy STS: Scanning Tunnel Spectroscopy ATM: Atomic Force Microscopy ICISS: Impact Collision Ion Scattering Spectroscopy ESDIAD: Electron Stimulated Desorption Ion Angle-Resolved (angle-resolved ESD)	Estructural and topographic. Image capacity.
AES: Auger Electron Spectroscopy XPS: X-Ray Photoelectron Spectroscopy (also known as ESCA: Electron Spectroscopy for Chemical Analysis) NEXAFS: Near edge X-ray Absorption Fine Structure SIMS: Secondary Ion Mass Spectroscopy ToF-SIMS: Time-of-flight-SIMS	Chemical composition analizes. Imagen capacity

3.1. Water Contact Angle

The spreading of a liquid on a solid surface and its wetability is an old subject that has been addressed in the early nineteenth century by very famous physicists as Gibbs (1875-1878), Laplace (1806, 1807), Plateau (1869) or Young (1805) who set the theoretical basis of the thermodynamic and mechanic of a pendant water drop, capillarity, shape of a meniscus, etc. They were successful in explain a large number of experimental situations at least as long as only macroscopic phenomena were involved.

Wetability is directly related to surface energy and it is one of the primary surface properties of solid materials. It is mainly governed by chemical composition, microstructure and recently, with the help of powerful electron microscopes, the importance of the nanostructures superimposed on micrometer-scale in nature surfaces was realized [73]. Several excellent examples are well known and studied, such as lotus leaves [74], cicada orni's wings [74, 75], water strider's legs [76], etc. Those natural produced surfaces are water repellent in that the water droplet roll off the surface and simultaneously remove contaminants from the surface (self-cleaning). New theoretical models were introduced in the last decades trying to explain those new observed phenomena [77, 78]. The surface solid wetability is a very important property of a material and a research topic of fundamental interest with a variety of technological implications. The control of the surface wetability is crucial in many practical applications, such as, thin film technology, lubrication, antifouling paints, windows self-cleaning, microfluids, textiles, anti-snow-sticking surfaces, etc.

The behavior of a liquid droplet on a solid surface depends on two main solid properties: the surface energy that is determined by the chemical nature of the top most molecular layers of the solid, and the surface roughness. Materials with low surface energy such as carbon fluorine compounds have a maximum contact angle measured of 120° [79]. A simple experimental method to measure the wetability of a surface is the water contact angle (WCA) technique.

WCA measures surface hydrophilicity or hydrophobicity by measuring how much a droplet of water spreads on a surface. As shown in Figure 2, the lower the contact angle, the more hydrophilic the surface is. As a surface becomes more oxidized, or has more polar groups introduced to it, hydrogen bonding with the water becomes more facile and the droplet spreads along the hydrophilic surface, resulting in a lower contact angle. On the contrary, if the interaction of the water droplet with the surface is low, the WCA is hydrophobic or superhydrophobic.

If a small liquid drop is deposited on a solid, three interfaces come into play, and three interfacial tensions are involved: γ_{SL}, γ_{SV} and γ_{LV}, respectively the solid-liquid, solid-gas and liquid-gas interfacial tensions [69]. The γ_{ij} can also be considered as forces per unit length of the triple line which forms where the three phases meet (see Figure 3).

The interfacial tension γ_{ij} represents the free energy necessary to increase by one unit the area of contact between two different phases *i* and *j*. The physical origin of this free energy is easy to understand: inside a dense phase, *i*, the molecules attract each other, and they can better interact with neighbors of the same species *i* when located in the bulk material than when located close to the interface.

Figure 2. Schematic of contact angles representing different wetting conditions. θ: < 90°: water wet the surface; 150° > θ ≥ 90°: poor wetting conditions; θ ≥ 150°: superhydrophobic surface.

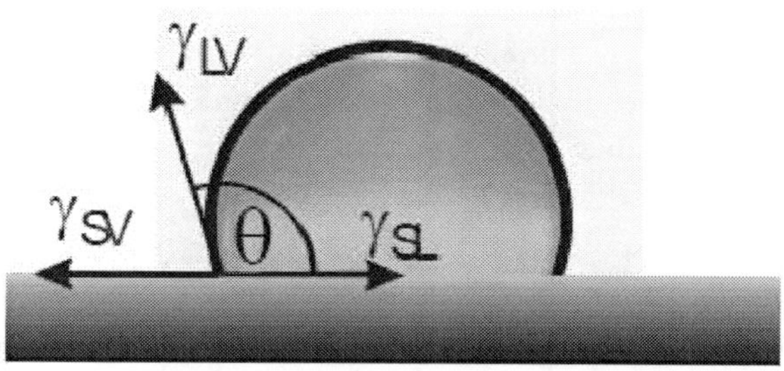

Figure 3. Equilibrium contact angle representation.

The contact angle, as shown in Figure 3, is governed by the force balance at the three-phase boundary and is defined by the Young's equation [80]:

$$\gamma_{LV} \cos\theta = \gamma_{SV} - \gamma_{SL} \quad 3.1$$

For a small liquid drop on the solid surface, the important quantity is the spreading parameter (S), *i.e.* the free energy difference between a bare solid, directly in contact with the vapours and a solid covered by a flat, thick liquid layer:

$$S = \gamma_{SV} - (\gamma_{SL} + \gamma_{LV}) \quad 3.2$$

Clearly, if S > 0 the drop spontaneously spreads and tends to cover the solid surface. Such a situation is called total wetting. The case where S < 0, is referred to as partial wetting: the liquid remains as a drop on the solid and reaches an equilibrium shape. It is possible to combine equation 3.1 with 3.2 to obtain:

$$S + \gamma_{LV} = \gamma_{LV}\cos\theta \quad 3.3$$

Figure 4. Schematic representation of hysteresis. Left: hysteresis is measured tilding the sample. Right: hysteresis can also be measured by the sessile method. θ_A: advancing CA. θ_R: receding CA.

A real solid surface is extremely hard to clean and is never ideal; it always has a small roughness or small spatial inhomogeneities. Any wetting experiment is extremely sensitive to heterogeneities or chemical contamination and one of the simple results of the inhomogeneity is contact angle hysteresis. On a real world solid surface one measures a static contact angle that depends on the history of the sample. If the liquid-vapors interface has been obtained by advancing the liquid, (after spreading of a drop for example) the contact angle has a value θ_A larger than the equilibrium value; if on the contrary the liquid-vapors interface has been obtained by receding the liquid (by retraction or aspiration of a drop) the measured contact angle θ_R is smaller than the equilibrium contact angle. Even when the solid surface is only slightly heterogeneous, the difference θ_A - θ_R can be as large as a few degrees. Many phenomena observed in everyday life can be explained by contact angle hysteresis, such us a raindrop attached to a vertical real window that flows down under the action of its weight. In the upper parts of the drop the liquid has a tendency to recede and the contact angle is the receding contact angle, in the lower parts of the drop, the liquid has a tendency to advance and the contact angle is the advancing contact angle; the difference in contact angles creates a capillary force directed upwards that can balance the weight. A schematic representation of the hysteresis is shown in Figure 4.

The important of hysteresis has recently been realized. Surface solid wetability is a very important property of a material and a research topic of fundamental interest with a variety of technological implications. The control of surface wetability is crucial in many practical applications, such as thin film technology, lubrication, antifouling paints, windows self-cleaning, microfluids, textiles and anti-snow-sticking surfaces. In recent years, a great deal of experimental and theoretical research has been devoted to the so-called superhydrophobic self cleaning surfaces, those that exhibit WCA higher than 150° and low-hysteresis (drops roll off easily even at small inclinations) [74, 81-84]. Wetability is mainly governed by chemical composition, microstructure and more recently with the help of powerful electron microscopes, the importance of the nanostructure superimposed on micrometer-scale topology in nature surfaces was understood [73]. Nature evolution over million years has produced unique products that challenge human mind in every knowledge field. Several excellent examples are well known and studied, such as lotus leaves [74], cicada orni's wings [74, 75], and water strider's legs [76]. Those natural produced surfaces are water repellent in that the water droplet roll off the surface and simultaneously remove contaminants from the surface (self-cleaning). The increase in the contact angle was explained by the use of surfaces of proper roughness [81, 85, 86] and based mainly in two theoretical models (Wenzel [87] and Cassie [88]) many methods have been used in recent years to prepare controlled rough surfaces. Figure 5 represent the conditions corresponding to Wenzel and Cassie-Baxter states. Theoretical models were also introduced in the last decades trying to explain those new observed phenomena [77, 78].

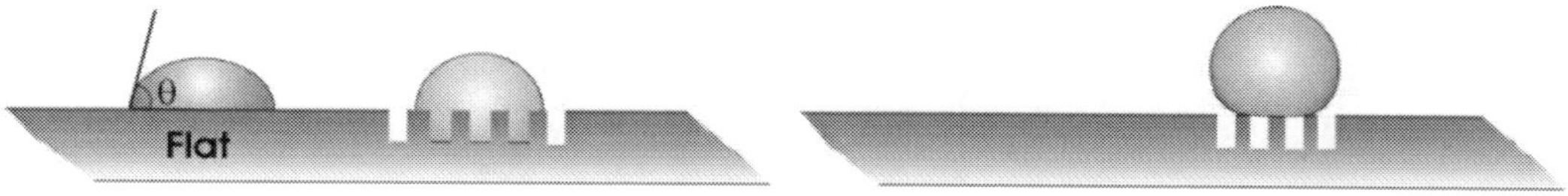

Figure 5. Schematic representation of superhydrophobic surface in Wenzel (left) and Cassie-Baxter states (right).

While contact angle is a simple and rapid measure of the change of a surface's hydrophilicity, it is limited by its inability to distinguish between different hydrophilic functional groups and it is affected by a number of factors, including the followings: surface preparation, contamination, inconsistent water pH and hardness, changes in environmental temperature and humidity and drop size. A detailed description about contact angle measurements and important factors to be considered in the analysis can be found in the book of C.M Chan [69].

3.2. Spectral Analysis Methods

Surface analysis involves probing a sample with several chemical or physical probes and detecting the various characteristic emitted species to yield information about a surface. Typical probes include photon, electron or ion beams. When a particular probe is used a spectrum, i.e. an array of entities, such as light waves or particles, ordered in accordance with the magnitudes of a common physical property, as wavelength, electrons or mass is produced. Depending on the secondary entities produced by the probe, the technique is usually called as spectroscopy (spectrum intensity as a function of photon energy, electron kinetic energies, photon wavelengths, etc.), or spectrometry (spectrum intensity as a function of particles masses, commonly mass/charge ratio). For the more discerning reader, excellent books are available which describe in detail each surface analytical technique [68-72, 89].

3.2.1. X-Ray Photoelectron Spectroscopy (XPS)

XPS, also commonly known as electron spectroscopy for chemical analysis (ESCA), is a very versatile surface analytical technique with wide-ranging applications. XPS determines the atomic composition of a solid's top several nanometers. Upon exposure to X-ray photons, a surface emits photoelectrons whose binding energies can be compared to known values to identify the element and its oxidation state [68]. The resulting spectrum is a plot of intensity (arbitrary units or counts per second) versus binding energy (eV). The intensity of the ejected photoelectrons relates directly to the material surface atomic distribution and can therefore be used to quantify percent atomic composition and stoichiometric ratios. In addition to quantifying change in surface atomic composition, XPS can be used to estimate extents of reaction by dividing measured atomic concentrations by theoretical values calculated by assuming complete conversion. Figure 6 shows schematically the basis processes in XPS spectroscopy.

The kinetic energy E_K of the photoelectrons is determined by the difference between the X-ray photon energy, $h\nu$ (ν is the X-ray frequency), and the binding energy E_b of that inner-shell electron. Since the photon energy is known and the kinetic energy of the photoelectrons is measured in the experiment, the binding energy can be easily calculated by the following equation:

$$E_b = h\nu - E_k \qquad 3.4$$

Including the work function (ϕ) of the spectrometer the above equation has to be corrected changing E_k by the term $E_k + \phi$.

The binding energy of a particular shell of an atom is unique for each element; therefore, knowing the binding energy allows identification of the elements. All elements with binding energy lower than the excitation energy of the X-ray source can be excited. Figures 7 shows a typical survey spectrum obtained from a clean film of polyurethane (PU) while Figure 8 presents the corresponding molecular structure of PU. Signal belonging to O 1s (532 eV), N 1s (400 eV), C 1s (285 eV), Si 2s (154 eV) and Si 2p (103 eV) can be seen clearly in the XPS survey spectrum. When the surface of a PU film is modified by a particular technique, changes in a typical XPS pristine spectrum can be easily monitored following modifications in peak intensity, as well as the appearance of new signals. These modifications can be measured qualitatively and quantitatively.

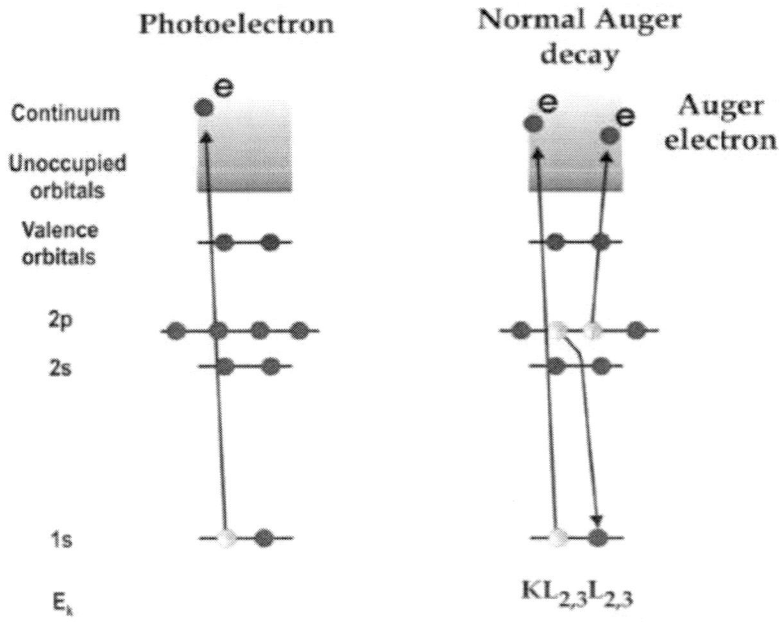

Figure 6. Schematic representation of XPS and normal Auger processes.

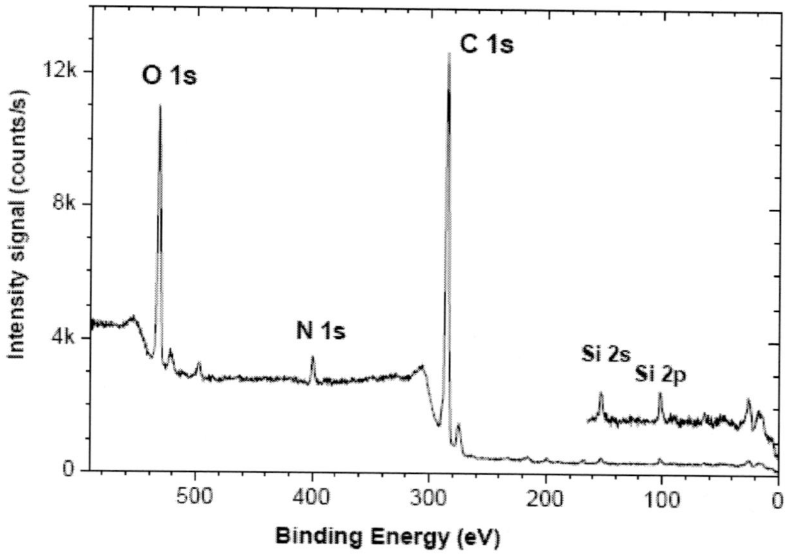

Figure 7. Survey spectrum for a ~ 5 μm clean film of polyurethane showing different core levels excitations that belong to the polymer atom composition.

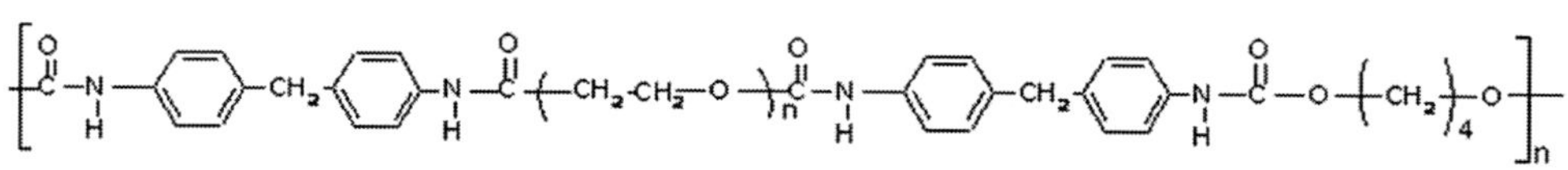

Figure 8. Molecular structure of the polyurethane used to obtain the XPS survey spectrum of Figure 7.

In polymer surface modification, it is of interest to identify the presence of specific functional groups. Curve synthesis can be used on high resolution scans to better understand the nature of a bond, but curve fitting models must be chosen carefully, functionalities are typically present in low concentration, and fitted curves overlap, making quantification complex. A method that is very useful for polymeric systems is the use of an internal standard, which depends on the invariance of a chosen chemical functional group in different molecules. The knowledge of the molecular structure of the polymer is required for the application of this method. For example, the C 1s spectrum of pristine polyurethane is shown in Figure 9.

The C 1s region yields six mayor components, indicating that there are six different types of carbon atoms in the structure (see Figure 8) corresponding to the aromatic C=C (284 eV), aliphatic C-C/C-H (285 eV), C-N (285.8 eV), C-O (286.5 eV), C=O (287.4 eV) and COO (289.2 eV) bonds in the polymer chain [17, 90]. The lowest binding energy components, speciphically, the position of the C-C/C-H are usually specified as an internal standard for the correction of shifts caused by charging. Other peaks of different carbon environments are fixed relative to this peak, set at 285.0 eV. This method is particularly useful for polymers because carbon atoms, which can be used as internal standards, are readily available.

A different approach to identifying the presence of specific functional groups is through the use of functional group labeling experiments (derivatization) using chemical derivitizing agents[70]. In some cases there is a lack of specificity in core level binding energies and many functional groups give rise to similar chemical shifts. This problem is even more pronounced when several such groups are present in the surface, since individual peaks overlap and the spectrum show broadening due to substituent secondary effects.

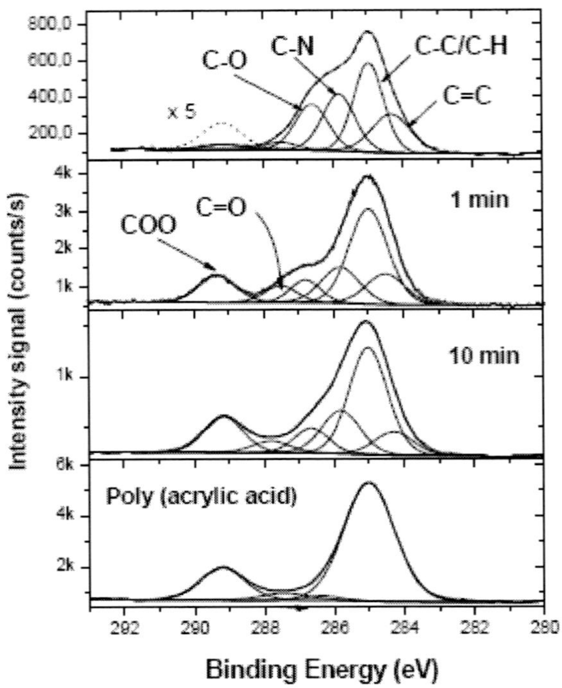

Figure 9. XPS spectra of the C 1s peak of untreated polyurethane (top), Acrylic Acid plasma treated polyurethanes at 1 and 10 minutes of plasma treatment times and synthesised poly(acrylic acid)[90]. The C=O and COO groups are showed in different spectra for better presentation.

Many researchers have attempted to overcome these problems by using specific derivatizing agents which react with one functional group and label it with distinctive element. This procedure may have the additional advantage of increasing detection sensitivity if the new element has higher cross-section than carbon, oxygen or nitrogen. A complete list of reagents for derivatization reactions can be found in the book of Briggs, et al. [70].

Much higher surface sensitivity can be obtained by Angle-resolved XPS measurement s (ARXPS). ARXPS depth-profiling is a non-destructive technique that provides information about the chemical modifications over a 5–10 nm range of depth. ARXPS analyses provide the ‘‘apparent’’ atomic composition of the surface according to the different photoemission angles used (defined as the angle between the axes of the entrance slit of the photoelectron detector and the surface normal). The atomic composition can be measured over a certain depth by integration of experimental data. The total photoelectron flux *I*, arising from a given chemical element from the sample, is given by the following expression [91]:

$$I_\theta = k\int_0^{+\infty} c_z \exp\left(\frac{-z}{\lambda\cos\theta}\right)dz \qquad 3.5$$

where k depends on the photoemission cross-section as well as instrumental and geometrical factors, c(z) is the concentration of the given element as a function of depth z, λ is the photoelectron inelastic mean free path (IMFP), and θ is the photoemission angle.

Ideally, XPS analysis should consist of the determination of the complete concentration depth profile within the analyzed volume. In many samples, such as metal, oxides and semiconductors this can be accomplished by means of ion etching. In the case of polymers ion etching is inappropriate, for example when the sample would undergo chemical degradation. In such cases ARXPS has to be used without the help of ion etching, in which the sample is tilted so as to vary the angle between the axis of the photoelectron analyzer and the normal to the sample surface (θ). XPS data is usually collected at a series of angles θ from 0° to 75° photoemission angle. Five or six angles are usually sufficient. Cumpson [92] has recommended concentrating the data acquisition at the higher photoemission angles, where the analysis is most sensitive to the sample surface, such us at 0°, 40°, 55°, 63° and 70°.

XPS is a very powerful technique for surface analysis of polymers but also it includes an inability to detect lighter atoms, principally hydrogen and Helium. For their detection different techniques have to be used [72, 93] (see also section 3.2.2 in this chapter). Because XPS analysis is limited to the top few nanometers of the surface, samples must be handled carefully as even minor surface contamination is pronounced in the resulting spectrum. In addition, polymers are fairly stable under X-ray irradiation, and damage caused by X-rays can be minimal during a typical XPS experiment. In spite of that, the operator has to be caution when peak intensity or peak-shape changes occurs during analysis or discolored regions on a specimen appear following analysis [94, 95]. These effects are among the most readily observed signs of a probably presence of damage. There are at least three possible causes for the degradation of a polymer under prolongated exposure to X-ray: secondary electrons generated at the X-ray source which reach with the sample, interaction of X-ray photons with the polymer and secondary electrons generated within the polymer. As new generations of instruments often include more finely focused beams and increased incident fluxes, analysts need to be increasingly aware of the potential for beam damage during analysis of samples using XPS. Special care has to be taken when synchrotron radiation facilities are used, due to the high photon flux of new generation synchrotron sources.

3.2.2. Near Edge X-Ray Absorption Fine Structure (NEXAFS)

Core-level excitation spectroscopy in its various forms is useful as a probe of the short-range structure and bonding in molecules and solids. In Near Edge X-Ray Absorption Fine Structure, NEXAFS, also called X-Ray Absorption Near Edge Structure (XANES), a core-level electron is excited to empty states near or above the ionization limit usually by X-ray absorption or the impact of high-energy electrons. Today, the term NEXAFS is typically used for soft x-ray absorption spectra and XANES for hard x-ray spectra. In a beginning, the most widely used core level excitation spectroscopy was EXAFS (Extended X-ray Absorption Fine Structure) which includes the spectral region extending from approximately 30 eV to several hundred eV above a core level excitation edge. In EXAFS the region within the first 30 eV above the excitation edge is excluded, but from a practical point of view always there has been considerable interest in understanding this excluded spectral region close to the edge for several reasons. First, the peaks in the near-edge region are usually an order of magnitude more intense than those in the EXAFS region which makes possible the analysis of smaller concentrations of specific atoms in bulk samples and on surfaces whose EXAFS signal is too weak. Second, experimental studies of the near-edge region have shown in the last 20 years that the technique is especially sensitive to the bonding in molecular complexes with an unique property that can be used as "fingerprints" of similar polymers made with different formulations. Third, new generation of synchrotron sources with much higher intensity and the possibility to use ultra fast pulses for time resolved studies, have given an even larger impulse to NEXAFS spectroscopy. Scanning transmission X-ray microscopy [89, 96-98] and more recently by the first demonstration on femtosecond NEXAFS [99], it was realized the increasing importance of NEXAFS spectroscopy within the field dominated by traditional spectroscopy techniques.

In NEXAFS, the X-ray energy is scanned and the absorbed X-ray intensity is measured; contrary to the related XPS or ESCA technique, where the photon energy is fixed and the electron intensity is measured as a function of electron kinetic energy. As X-ray photon energy is scanned, a synchrotron facility is the ideal place to use this technique. The most common methods to record NEXAFS spectra are transmission and electron yield measurements [89]. The transmission technique requires thin foils while the electron yield technique, often called total electron yield (TEY) detection, can be used for conventional samples. The photo absorption spectrum in TEY mode usually is obtained by measuring the total electron yield, (determining the sample drain current under X-ray illumination), simultaneously with a photon flux monitor (Au grid is generally used). The final data are normalized by this flux spectrum to correct for fluctuations in beam intensity and monochromator transmission. For energy scale calibration of a NEXAFS spectrum, CO_2 gas can be added and the transmission spectrum of the mixture of the polymer and CO_2 gas can be recorded [100]. By this way, the energies of the CO_2 Rydberg transitions from the high-resolution NEXAFS spectra are used to calibrate the spectra. Grafite is also used as useful calibration method for thin polymer films [101].

NEXAFS is a unique technique for examining the surface of organic films, providing elemental, bonding, and structural information [89]. Concerning polymers having phenyl groups, NEXAFS spectroscopy has shown its benefits with respect to XPS to extract phenyl ring sub-peak information [102-104]. Because in NEXAFS spectroscopy the C 1s $\rightarrow \pi^*$ phenyl ring resonance is one of the most prominent spectral features, its changes can be easily monitored during a surface modification treatment. In the 250-750 eV range, the *K*-edges of carbon, nitrogen, oxygen, and fluorine, are commonly studied. At higher energies, higher-order transitions of heavy elements can be investigated. Within each *K*-edge, peaks appear on the basis of the electronic energetic of specific chemical bonds. NEXAFS is element specific because the X-ray absorption edges of different elements have different energies. This is illustrated in Figure 10 for a polyurethane thin film (compare with the information provided by XPS in Figure 7 and 8). For illustration proposes, NEXAFS spectra of Figure 10 shows two important transition of polyurethanes: an intense C 1s (C-H) $\rightarrow \pi^*_{C=C}$ transition at 285.2 eV [100] and a broader signal at 293.5 eV that can be assigned to a C 1s $\rightarrow \sigma^*_{C\text{-}C}$ transition [98, 100]. The core excitation is indicated between parenthesis and the nature of the upper level of a given transition is indicated as the final subscript after the arrow. NEXAFS is very sensitive to the bonding environment of the atom. This is already visible in Figure 10. The NEXAFS spectrum exhibits considerable fine structure above each elemental absorption edge. This fine structure arises from excitations into unoccupied molecular orbitals.

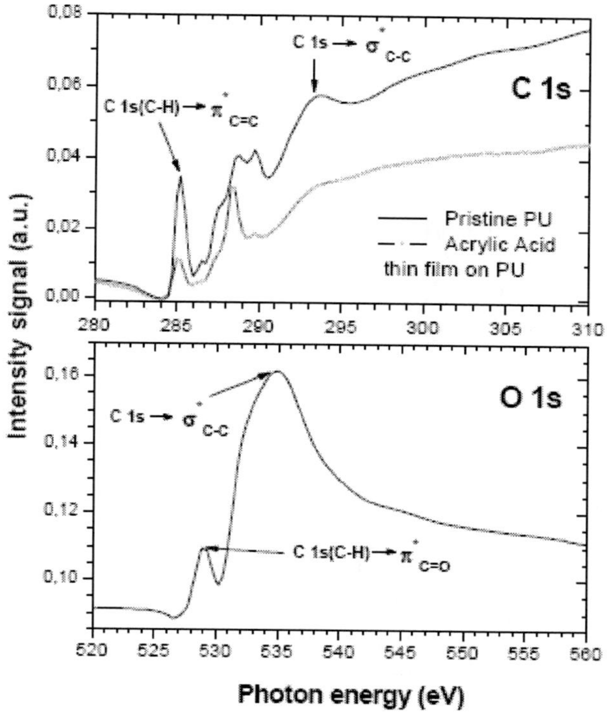

Figure 10. Carbon (top) and oxygen (bottom) *K*-edge NEXAFS spectra of a polyurethane film showing a few main core shell transitions. Carbon *K*-edge NEXAFS spectrum of thin film of poly(acrylic) acid polymerized on the polyurethane film is also shown on top (see text for details).

NEXAFS spectrum is dominated by multiple scattering of a low-energy photoelectron in the valence potential set up by the nearest neighbors. Often one can use a spectral "fingerprint" technique to identify the local bonding environment. The spectra exhibit chemical shifts within each group similar to XPS spectra but more importantly considerably different fine structure, such us for carbon in different molecular groups (see Figure 10, top). Due to a combination of larger chemical shifts in NEXAFS and the high energy resolution of synchrotron sources (a few hundred meV or less than 100 meV in third generation sources), the overlapping, for example, of C-C and C=C signals separated in ~1 eV in XPS (see Figure 9) is not a problem in the NEXAFS spectrum. Figure 10-top shows that the separation between C=C and C-C functional groups is more than 8 eV. In addition, when a thin film of poly(acrylic acid) (PAA) is polymerized on top of the polyurethane film, XPS spectra show a change in shape with the increasing treatment time (see Figure 9), resembling a PAA like film at longer plasma treatment times. When a similar films are analyzed by NEXAFS, even for less than 10 nm PAA thickness, the NEXAFS spectrum changes completely (see Figure 10-top). The new shape of the spectrum is similar to that obtained by Swaraj *et al.* in plasma polymerized experiments using Acrylic Acid at high duty cycle or in continuous modes [105]. Figure 10-bottom shows O 1s NEXAFS spectrum of a PU film. When a particular polymer does not have a specific functionality, such us polystyrene, NEXAFS spectroscopy can give detailed information about new functionalities grafted on the polymer surface after the treatments. C 1s and O 1s edges can be recorded, for example, comparing the spectra before and after the treatments to know the surface changes produced [35, 104].

NEXAFS can also provide information about molecular orientation and ordering of surface species. NEXAFS relies on a polarized beam of high-intensity X-rays from a synchrotron source, which excites a core electron to create a core hole. The area of each peak is proportional to the concentration of bonds and the edge jump for any particular elemental *K*-edge is proportional to the total concentration of that element. Both of these quantities, the near edge and the edge jump are dependent on the incidence and the emission angles, allowing for the determination of surface bond orientation. Since the incident X-ray beam is strongly polarized and absorption by specific bonds is polarization dependent, the net orientation of molecules can obtained. Thus, the variation in the intensity of a particular peak as a function of the electric field vector orientation of the photon beam is a sensitive measure of the molecular orientation on the surface. From simple molecules, such as CO, N_2, and NO to polymers, NEXAFS spectroscopy has been used to determine the orientation of surface-bond molecules [106] or three-dimensional structures [107-109]. Additionally, utilizing partial electron yield variations as a function of detector entrance grid voltage bias allows for rough depth profiling, Auger electrons escaping from deeper within the film experience multiple inelastic scattering events which decrease their energy and depth information can be obtained. Depth profiling was demonstrated applying a series of retarding voltage biases on polystyrene films treated by oxygen plasma [104]. By changing the retarding voltage it was possible to infer the effect of treatment as a function of polymer depth. NEXAFS, even at relatively high retarding bias fields, still probed much deeper into the surface than water surface tension.

NEXAFS spectra, as in XPS, can be fitted to extract more information about the different transitions involved. In the case of NEXAFS, the spectra present much lower overlapping in signals than XPS [89, 103] and, therefore, there is far less peak de-convolution in NEXAFS spectra than when using XPS, thereby distinguishing chemical changes more easily. By comparison with reference polymers it is possible to assign the shift and growth of peaks to specific chemical environments. All the NEXAFS properties discussed in this chapter section clearly illustrate the power of NEXAFS to distinguish chemical bonds and local bonding. In many ways it is superior to XPS, which does not provide local structural information.

3.2.3. Time-of-Flight Secondary Ion Mass Spectrometry (TOF-SIMS)

Most materials used in technological applications are organic in origin with different degrees of complexity and in the bioscience field the analysis and interpretation of the results given by a particular technique increases even more. Powerful techniques are available for high-sensitivity surface analysis (some of them already introduced in the present chapter), such as XPS, NEXAFS, surface mass spectrometry (MS), and the topological techniques of scanning tunneling microscopy (STM) and its related counterpart, atomic force microscopy (AFM). These methods must be evaluated with respect to the general analytical question, "What are the nature, concentration, and location of all atomic and molecular species present in the surface region?" To answer this question, the analyst must consider the unambiguous identification of atoms as well as molecules or molecular clusters (unknown surface species); information on the relative or absolute coverage by these species (quantification of these surface species); and information on their lateral and depth distributions (2D and 3D location of these surface species in the sample). It should be possible to identify all types of elements and molecular species and to distinguish between isotopes. Quantification and locating species should be also possible, even for very low surface concentrations and with high lateral and depth resolution. In addition, all types of materials and all sample shapes should be accessible, including insulators.

In XPS, information on the elemental composition of the uppermost atomic layers is obtained with sensitivities down to 0.1% of a monolayer together with important information on the chemical environment of the identified elements (chemical shift). However, identification of unknown complex molecules by this chemical shift is not possible, and the achievable lateral resolution is limited to a few micrometers. NEXAFS suffers with similar problems when molecular structural information and chemical image is needed. STM and AFM allow lateral resolution in the sub-nanometer range so that single atoms can be probed. However, these techniques cannot be used to identify unknown surface species.

The most important feature of surface MS, in addition to high sensitivity, is its ability to provide detailed molecular information and information at shallow depths, allowing the identification and quantification of all elements, isotopes, and molecular species. Therefore, surface MS should be an excellent means for surface analysis. One efficient way to carried out surface MS is the use of ions as primary particles that hit the sample under analysis. The resulting technique receives the name of SIMS (Secondary Ion Mass Spectrometry) and can be defined as the mass spectrometry of ionized particles which are emitted when a surface is bombarded by energetic primary particles, usually ions, such us Ar^+, Xe^+, Cs^+, Ga^+, clusters, etc. After the primary impact, different sputtered secondary particles are emitted: electrons, neutrals species atoms or molecules (in ground or excited states), atomic and cluster ions. A schematic diagram of the SIMS process can be seen in Figure 11.

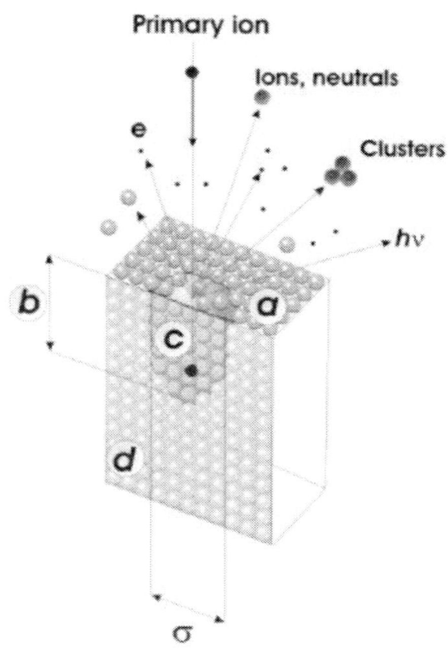

Figure 11. A schematic representation of the SIMS process. a: surface layer; b: penetration depth; c: bulk damage volume; σ: damage surface area described by the damage cross section; d: bulk.

The vast majorities of those species are neutral, but are the secondary ions formed during or after the primary impact that are extracted and analyzed by a mass spectrometer. It is all that process which provides a mass spectrum of a surface and enables a detailed chemical analysis of the surface material. The interested reader can find more information about the SIMS technique in specialized books [71, 72, 110] or reviews [111-115].

SIMS technique has evolved during the years from the initial "dynamic" mode (several hundred monolayer sputtered per second); the "static" mode where the Time-of-flight SIMS is the dominant variant of SIMS which maintains the integrity of the surface layer within the time-scale of the analytical experiment; up to cluster-SIMS [116] that in the last decade has transformed molecular analysis by SIMS. Cluster-SIMS has allowed minimizing the bombardment induced damage, reducing dramatically the static limit condition, such that for many materials the static limit requirement can be abandoned. A completely new analytical regime has opened up so that molecular depth profiling and 3D chemical imaging became possible [114-119].

The TOF-SIMS technique is based on the principle that heavier secondary ions take longer to travel as compared to smaller, lighter ions. Mass determination is achieved by measuring the flight time of the desorbed secondary ions in a drift path of known length after accelerating them in an extraction field to a common energy E of some keV, which is illustrated by the equation:

$$E_k = \frac{1}{2} m v^2 \qquad 3.6$$

where E_k is the kinetic energy, m is the mass, and v is the velocity. Equation 3.6 shows that the flight time is proportional to the square root of the mass of a secondary ion. For an accurate determination of flight times, the start and arrival times of the secondary ion must be well defined. To meet these requirements, either the primary ion beam or the extraction field is pulsed and the arrival time of the secondary ion is determined by a fast detection system. The sensitivity of the instrument may further be enhanced using a pulsed sub-nanosecond ion beam [116, 120], using ion reflectors and triple sector analyzers. In the case of linear TOF and reflectors the time-of-flight equation are more complicated than equation 3.6. For example, the flight time of an ion leaving the surface can be calculated following the innovations of Wiley-McLarem [121]. The detector arrival time of an ion generated at a time t_0 is:

$$t = t_s + t_D + t_0 \qquad 3.7$$

where: t_s is the acceleration time, t_D is the drift time region and the ion flight time is calculated by:

$$t = \frac{\sqrt{(2m)}}{eE}\left(\sqrt{(U_0 + eEs_0)} + \sqrt{U_0}\right) + \frac{\sqrt{2m}D}{2\sqrt{(U_0 + eEs_0)}} + t_0 \qquad 3.8$$

where $(U_0 + eEs_0)$ is the initial and final kinetic energy of an ion having an initial kinetic energy of U_0 and accelerating from the initial source position, s_0.

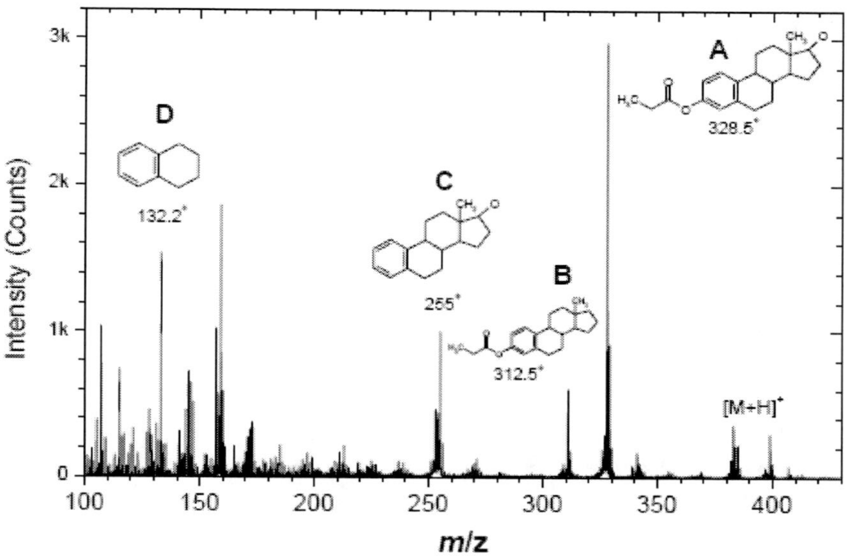

Figure 12. Positive TOF-SIMS spectrum of *β*-Estradiol dipropionate on Si (100) using 25-keV impact Ga^+.

The output from a SIMS machine may also be interpreted as an image, which coined the name of chemical microscope [122] (microprobe imaging) by applying a narrow primary ion beam across the analyzed surface, thus obtaining a comprehensive mass spectrum of the surface. The detection limit of TOF-SIMS is more sensitive than XPS, on the order of parts per billion; however ionization of surface molecules is dependent on sample matrix, and is therefore not quantitative. TOF-SIMS can be used to complement XPS results by offering identification of chemical species and also be used to differentiate samples that have similar XPS spectra.

Figure 12 shows typical features of a TOF-SIMS spectrum of a thin layer of *β*-estradiol dipropionate on Si (100) setting a 5 μL drop of a 2 x 10^{-4} M solution by spin-coating. The TOF-SIMS spectrum of Figure 12 was acquired using 25 keV Ga^+ primary ions and 10^6 pulses with a total acquisition time of 100 seconds. Typical fragmentation pattern of the molecular ion appears in the spectrum together with a low intensity of the protonated molecular ion. The high degree of fragmentation observed in Figure 12 is characteristic of excitation with monoatomic primary ions.

The capacity of TOF-SIMS applied to polymer surfaces has been demonstrated extensively in the last two decades [69, 71, 72, 112, 114, 123-125]. TOF-SIMS has been shown to differentiate analogous polymers structures, such as polyethylene, polypropylene and polystyrene which have very similar XPS spectra. A wide range of examples of TOF-SIMS applications to polymer characterization can be found in the literature and will not be described here. Instead, brief remarks of the use of ToF-SIMS in polymer surface modification and the benefit of cluster-SIMS in polymer characterization will be given.

TOF-SIMS has been employed in many studies on surface polymer modifications using different techniques, such us plasma, chemical grafting, radiation treatment, UV-treatment, etc. and in general TOF-SIMS is used in combination with other surface sensitive technique to obtain detailed information about the modifications after the treatments. Briggs and coworkers have been shown the benefit to combine two powerful techniques, such as TOF-SIMS and XPS to study the surface modification changes produced by different treatments [125]. For example, plasma-deposited polymeric films (PDF) prepared from carbonyl precursors were carefully studied by a combination of FTIR, derivatization XPS and TOF-SIMS [126]. It was demonstrated that oxygen-rich PDF that were stable both in air and in water could be reproducibly created from mixtures of acetone and oxygen. The surface enrichment of oxygen in these films was related to the addition of oxygen gas to the acetone feed. TOF-SIMS was mainly used to compare the surface chemistry of carbon-oxygen PDF created from acetone-O_2 mixtures with a commercial tissue culture polystyrene (TCPS) surface. Both techniques, XPS and static SIMS, suggest that the surface chemistry of the TCPS surface differs significantly from both the pure acetone PDF and the acetone-O_2 PDF.

A multi-technique approach was used by the group of Bradley [127-129] in UK in the fabrication of Acrylic Acid plasma polymers by pulsed plasma in the millisecond pulse regime. A comparison between continuous plasma and pulsed plasma was carefully investigated, including for the first time, a mass spectrometry study to monitor the neutral and charged species in pulsed plasma (gas-phase) as a function of plasma time off period. The results showed that solid-phase plasma polymers obtained from nominally the same power plasmas of acrylic acid (pulsed or CW) were found to be remarkably similar. Based in previous studies the authors concluded that this not surprising result may highlight the limitations of the spectroscopic tools employed (XPS and SIMS) for the characterization of the films.

In another interesting study, a novel plasma functionalization process based on the pulsed plasma polymerization of allyl glycidyl ether was reported for the generation of stable and highly reactive epoxy-functionalized surfaces with well-defined chemical properties [130]. The effect of the plasma parameters on the final epoxy-functionalized surface was characterized systematically using a combination of several surface science techniques: TOF-SIMS, XPS, FTIR-ATR, AFM and ellipsometry. It was found that the composition of the allyl glycidyl ether plasma polymers (AGEpp) films varies with the plasma duty cycle (plasma on/plasma off) in pulsed plasma polymerization and, to a lesser extent, with the applied plasma power level, as shown by XPS and TOF-SIMS data. Further surface analysis showed that on storage in air, the AGEpp films were sufficiently stable toward ambient post plasma oxidation and retained the capability for immobilizing molecules via nucleophilic addition to epoxy groups. Interestingly, after blocking of epoxy surface groups with ethanolamine, no nonspecific adsorption of lysozyme took place, indicating that allyl glycidyl ether plasma polymers could be used for biodiagnostic and biotechnology applications where both covalent bioconjugation capability and facile passivation of unused surface groups, by ethanolamine blocking instead of albumin blocking, are required, such as for protein microarrays.

TOF-SIMS was the dominant mode of the technique within the field of SIMS until beginning of this decade. Although the static mode showed impressive results in the bio and polymer science, those studies also illustrated some of the challenges that faced the application of static SIMS to the study of complex systems using a monoatomic liquid metal ion source: low or very low secondary ion yield, large damage cross sections and the static limit (Primary ion dose < 10^{13} ions/cm^{-2}). The static limit restrings the minimum area available for sample analysis. For example, considering an area of 1 μm x 1 μm (10^{-8} cm^2) and a molecular area of 5 Å x 5 Å (N_2) means that there are 4 x 10^6 molecules/pixel available to analyze. Taking into account the static limit (less than 1% of the surface has to be analyzed) and a typical low secondary ion yield of 10^{-3}, there are less than about 40 molecular ions available to detect. If the analyst is interested in detecting heavier masses, the fragmentation process decreases even more that number. Figure 12 illustrates clearly this effect for a simple molecule. The yield of molecules above about 300 amu using atomic ion bombardment is very small and many significant bio-molecules are very much larger. These issues clearly limited the potential application of SIMS to studies of bio-surfaces.

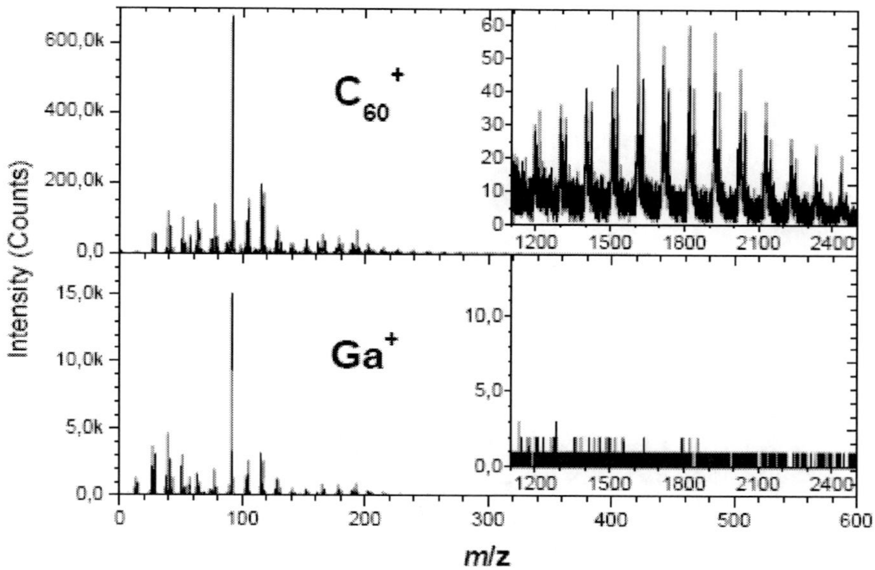

Figure 13. Positive TOF-SIMS spectrum of PS-2000 using 10-keV impact C_{60}^+ and Ga^+. Dose (ions): $C_{60}^+/Ga^+ = 0.94$.

The situation in favor of a more powerful technique to analyze bio-surfaces (2D and 3D) with detection of heavier masses, even intact molecular ions, started to change in the last decade with the move from monoatomic to polyatomic primary ions. The development of stable cluster ion beams allows cluster-SIMS to approach the same or greater degree of sophistication now associated with dynamic SIMS. Many research groups have shown that the static limit can be ignored using cluster bombardment with simultaneous molecular depth profiling [116, 120, 131-133]. C_{60}^+ was especially successful as primary source in increasing yields from organic materials [116, 120] and open the way for new analytical approaches who promise to enable the technique to deliver significant new data to surface studies of complex materials and particularly to research into biological systems. Fundamental measurements suggest that the sputtering yield of organic materials can be very high—several hundred to several thousand molecules per projectile, and that the molecular ion yield can also be enhanced by a built-in chemical ionization mechanism. For example, Figure 13 shows a comparison of secondary ion yield and fragmentation data under 10.0 keV Ga^+ and 10.0 keV C_{60}^+ ion bombardment for a traditional polymer, a very low molecular weight polystyrene, PS-2000. A large increase in secondary ion yield is apparent using the C_{60}^+ primary ions, and this is especially evident in the higher mass region. More than 10^3 times increase was measured for the cluster source compared to the monoatomic Ga^+. However, the relative yields increase proportionately more in the high-mass region [116, 120, 134]. Disappearance (damage) cross sections under C_{60} bombardment in thick films of molecular solids showed extremely low levels of damage as if the molecules are being peeled off the surface, leaving very little subsurface damage. For such materials, the static limit could be abandoned.

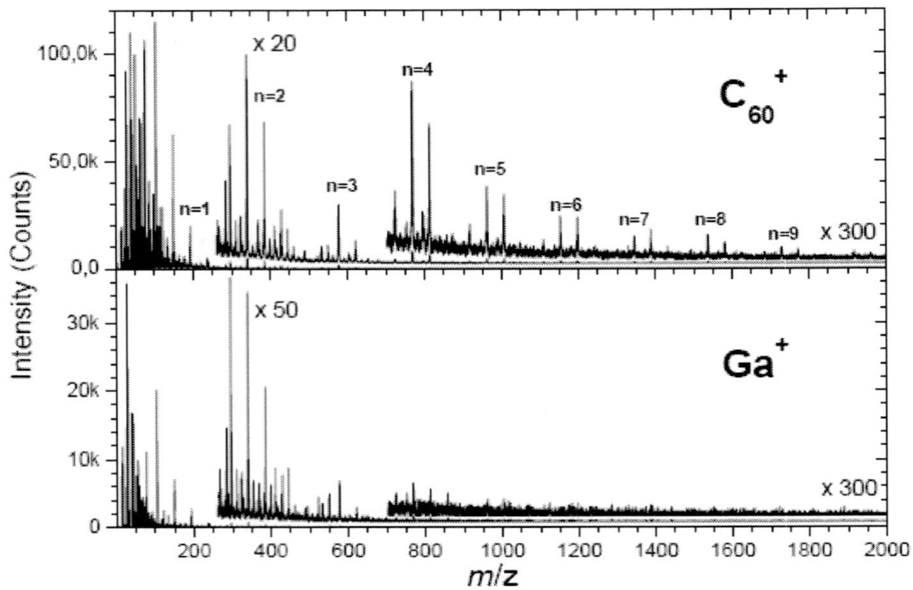

Figure 14. Positive TOF-SIMS spectrum of PET thick bulk film using 10-keV impact C_{60}^+ and Ga^+. Dose (ions): $C_{60}^+/Ga^+ = 0.18$.

Figure 14 shows another example of the potential of cluster-SIMS to characterize another widely used polymeric material, poly(ethyleneterephthalate), PET, compared with a monoatomic Ga^+ source. Again the overall yields efficiencies are 10^3-10^4 higher using $C60^+$, as compared to atomic primary ions [116]. Desorption os clusters characteristic of the PET monomer structure are clearly seen when a cluster source was used as primary particle. Definitely, cluster bombardment delivers no more molecular fragmentation than any other primary projectile used in SIMS. Indeed, in most cases studied, there is less fragmentation and a very dramatic increase in “molecular” ion yield.

The new surface analytical capacity of cluster SIMS presents unique opportunities for high sensitivity and high spatial resolution experiments that combined can allow to performed three-dimensional chemical analysis of molecular films. A number of recent publications have detailed analyses of a range of test systems alluding to the possibility o 3D molecular analysis of biocellular and polymer systems. A first account of 3D molecular imaging of a biological cell, using ToF-SIMS was recently published [135]. By using C_{60}^+ ions a Xenopus laevis oocyte cell was sputtered and images were constructed characterizing the chemical changes in successive layers of a biological cell with minimal sample preparation and intervention. Recently, new research was published combining measurements of AFM and Cluster-SIMS to perform quantitative depth profiles [115, 136]. The ability to perform three-dimensional molecular imaging experiments results in much longer analysis times. This will require new methods for improving the duty cycle of the instruments and to fully capitalize on the vast amount of data generated in such an experiment, multivariate analysis may play an important role in data analysis and interpretation.

3.2.4. Fourier Transform Infrared Spectroscopy-Attenuated Total Reflection (FTIR-ATR) and Raman Spectroscopy

Infrared and Raman spectroscopy are two very closely related spectroscopic techniques that are highly complementary to each other and which belong to the group of molecular vibrational spectroscopies. Over the last 20 years, a substantial amount of literature concerning the use of Fourier transform infrared (FTIR) attenuated total reflectance (ATR) spectroscopy for the in situ investigation of interfacial phenomena has been published. Usually FTIR-ATR technique is used combined with others surface sensitivity techniques already review in this chapter, such as XPS, NEXAFS, TOF-SIMS, etc. to complement and increase the interpretation of the data. FTIR-ATR has been developed specifically for the analysis of surfaces and uses infrared radiation to determine the chemical functionalities present in a sample. When an infrared beam in a defined wavenumber is incident in a sample, stretching, contracting, and bending of chemical bonds cause the absorption of the IR radiation. In the ATR mode, the sample is brought into close contact with the surface of a prism with a high refractive index [137, 138]. An infrared beam approaching the interface from the optically denser medium at a large enough angle of incidence is totally reflected, but some penetration occurs (in the μm range). The penetration depth depends on several factors, such us wavelength of radiation, refractive indices of sample and prism material, and the angle of incidence. If the sample absorbs infrared radiation a transmission-like infrared spectrum can be observed, resulting in a plot of absorbance (or transmittance) versus wavenumber. An example of an FTIR-ATR spectrum is shown in Figure 15 with each absorption band showing up as peaks on the spectrum. Although FTIR-ATR has a relatively deep sampling depth, it does not require ultra high vacuum conditions, as do XPS, NEXAFS and TOF-SIMS, and an analysis can therefore be conducted much easier and in a very short period of time. Due to the relatively simple set up of FTIR-ATR, this technique can also be used to monitor migration of functional groups to the polymer bulk or near surface region or differentiate several reaction pathways that occur in a surface functionalization of polymers. For example, in the vacuum-ultraviolet (VUV) irradiation of polystyrene films in the presence of oxygen it was found that the VUV treatment produced not only oxidatively functionalized surfaces, but generated also morphological changes [139]. The pathways of oxidative functionalization or fragmentation could be differentiated by FTIR-ATR analysis of irradiated polystyrene surfaces before and after washing with acetonitrile.

Raman spectroscopy is less widely used that its counterpart FTIR mainly to study effects on sample degradation and fluorescence, but recent technological advances has simplified the equipment and reduce considerably the experimental problems. Because Raman spectroscopy is an in situ technique, any kind of samples can be measured by Raman: solids, liquids, vapors, adsorbed molecules on surfaces, surface properties changes, etc. On the other hand is based on the fact that when energy is incident on a surface, most photons are scattered elastically causing the Rayleigh effect which is a common technique used for measuring particle size and size distribution down to less than 1 μm. However, every once in 10^6-10^8 times, these photons are in-elastically scattered with output photon energy less than input energy. The energy changes that are detected in vibrational spectroscopy are those require causing nuclear motion.

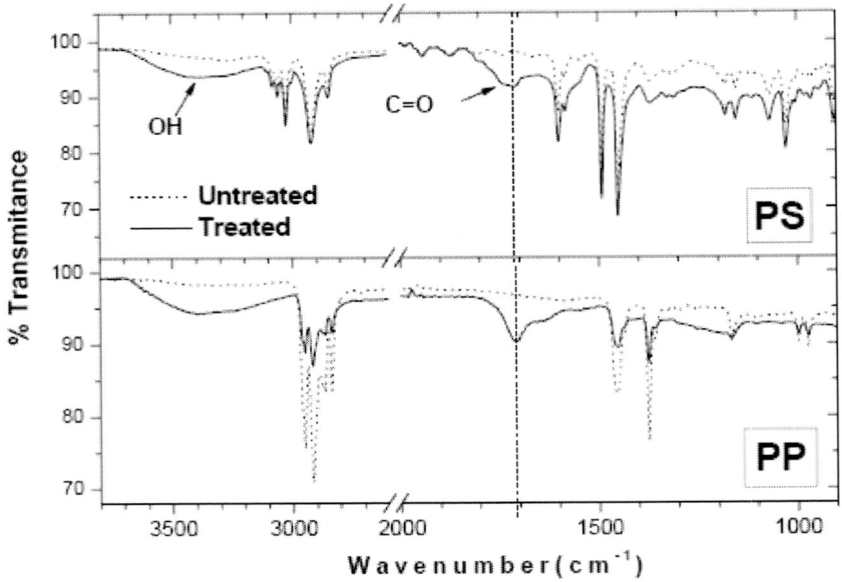

Figure 15. FTIR-ATR spectra of untreated (dotted line) and treated (solid line) polystyrene (PS) and Polypropylene (PP) films with 120 minutes of UV irradiation in presence of Acrylic Acid vapours. The spectra show the characteristic absorption peak of the new functionalities grafted at the polymer surface (OH and C=O bonds) [140].

Because electrons are comparatively light, the interaction of the incident photon with the electron cloud will be very weak and the photon will be emitted with the same energy or very small frequency changes. However, if nuclear motion is induced during the scattering process, energy can be transferred from the incident photon to the molecule or from the molecule to the scattered photon resulting in an inelastic scattered process. This is the Raman scattering, that it is inherently a weak process, but it can be exploited for surface analyze when laser beam are used in small samples areas. Raman spectroscopy highly complements IR spectroscopy as the former is based on polarizability and the latter, on dipole moments. If degradation of the polymer can be avoid during the analysis time, important information can be obtained from treated polymer samples [18, 54, 141]. For example, carbon double bond is a very strong Raman absorber and this signal intensity has been used to follow the concentration changes in several surface modified polymers by plasma [141] or UV treatment [18].

3.4. Microscopic Analyses Methods

3.4.1. Atomic Force Microscopy (AFM)

AFM has evolved into a useful tool for direct measurements of intermolecular forces with atomic-resolution that can be employed in a broad spectrum of applications such as electronics, semi-conductors, materials and manufacturing, polymers, biology and biomaterials [142]. AFM can provide in situ imaging ability to be used for nanoindentation without the necessity of moving the sample, switching tips, relocating the area for scanning, or using an entirely different instrument to image the indentation. In AFM, a tip attached to a flexible cantilever moves across the sample surface to measure surface morphology on an atomic scale. The change in surface height is then measured by the location of the reflected laser beam in the quadrant photodetector, and a surface topographical map is generated from which surface roughness values can be calculated [142]. The AFM consists of three different types, namely contact (includes tapping which is intermittent contact), non-contact, and phase-imaging modes. It involves placing a nano-sized tip (silicon nitride) mounted on a cantilever onto the polymer surface. When the AFM tip is brought into contact with a surface it experiences a very small force (of the order of a few nN) as a result of the interaction with the surface atoms. This leads to the tip being either attracted or repelled as it is scanned over the surface. In contact mode, the tip is in the repulsive region of the inter-molecular force curve and a laser beam is then reflected off the back of the cantilever and onto a segmented photodetector. The deflection of the cantilever is monitored or kept constant using a feedback loop to keep the force at a preset value. In the non-contact AFM mode, the probe hovers 10 nm above the material's surface and by virtue of its attraction to the polymer, recordings are measured. This is particularly effective for hydrophobic polymers. In tapping-mode AFM, the probe is vibrated at its resonance frequency at an amplitude of 20–100 nm whilst tapping the polymer surface. When the phase difference between piezoelectric oscillations is measured, the AFM is in the phase-imaging mode. Because AFM has subnanometer resolution and is capable of conducting measurements in physiological conditions, it has been used to study a range of biological molecules (proteins, DNA, RNA, tissues, etc.).

AFM is a very powerful tool to study modifications on polymer surface by different treatments and the technique has been used as an important complement to chemical techniques [59, 143-145]. In particular, AFM is very useful technique in plasma modifications of material because the etching process due to the plasma treatment can be carefully studied by AFM. Figure 16 shows the effect of Acrylic Acid low pressure plasma treatment on the plasma polymerization process of Poly(Acrylic Acid) (PAA) and etching of the Si substrate. The AFM results illustrate important morphological modifications associated with chemical changes by the plasma power increase. Figure 16 reveals the formation of a thin film of PAA on the Si substrate at a plasma power lower of 20 W (Figure 16-top). When the plasma power was equal or higher than 30 W, sputtering of the Silicon substrates was observed (see Figure 16-bottom).

In addition to evaluating surface topography, AFM can produce two-dimensional chemical affinity maps by modifying the cantilever tip with anything from specific functional groups to biomolecules.

The development of nanotube tips, which can be functionalized with as little as one biomolecule, gives the potential for even greater affinity map resolution. Several new application of AFM has shown the potential in quantification of available active sites in immobilized enzymes, measuring the force of antibody- antigen interactions in immunogenic diagnostic tools, and observing conformational changes of biomolecules under different pH and temperatures [142].

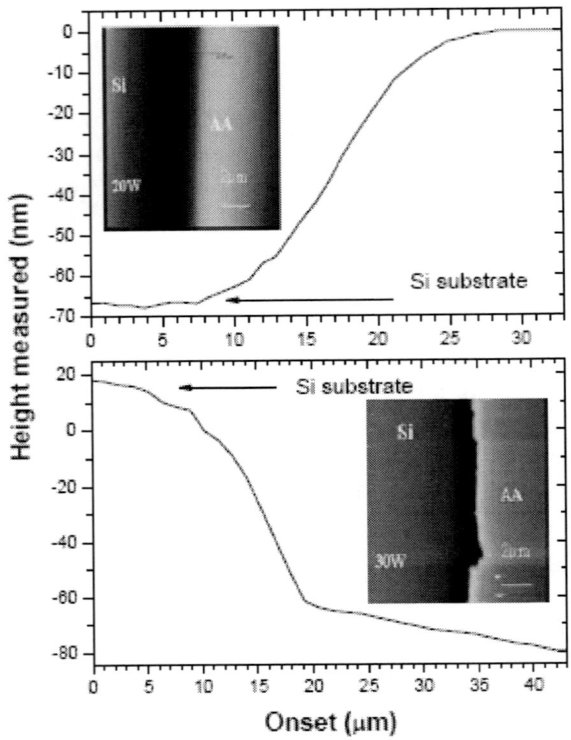

Figure 16. Plot of the film thickness dependence of Acrylic Acid (AA) plasma polymerization treatment using two plasma powers: 20 W (top) and 30 W (bottom). Substrate: Si(100). The reference corresponds to the virgin Si substrate and it is indicated in the figure. Treatment time: 40 min. Image size: 10 x 10 μm. Insets show the actual AFM image.

3.4.2. Scanning Electron Microscopy (SEM)

When a sample is bombarded with electrons, it emits secondary electrons and X-rays. The intensity of the secondary electrons is detected to generate a high resolution three dimensional surface image. X-rays can be detected to conduct elemental analysis. SEM is not as surface sensitive as other techniques, and non-conducting polymers must be sputter-coated prior to analysis. Nevertheless, it is one of the more widely available tools in surface analysis, and it is thus often used to measure surface topography. It has been used to examine cell adhesion on biocompatible and hemocompatible materials and the efficacy of tissue engineering scaffolds. More recently, low-temperature preparation and SEM imaging became available by using cryo-specimen holders at liquid nitrogen temperatures [146]. These new applications of SEM allows the study of biosystems with very high imagen resolution. The types of signals produced by an SEM include secondary electrons, back-scattered electrons (BSE), characteristic X-rays, light (cathodoluminescence), specimen current and transmitted electrons. Secondary electron detectors are the most common detector in SEM. Because of the wide use and the existence of many excellent books in SEM the details of the technique will not be included in this chapter and they can be found in specialized references [147].

4. Photochemical Processing Techniques for Surface Polymers Modification

Given the enormous importance of surfaces, researchers have gone to great lengths to develop a capability for the design and control of surface properties, both on natural and synthetic materials. In this sense, photochemical polymer surface modifications can be seen as light source-driven technology. Only after powerful microwave mercury lamps had become commercially available, UV-hardening of photoresist, curing and painting became practical. From that time, new photochemical sources are now accessible for surface cleaning, grafting and functionalization, such as lasers, dielectric-barrier discharges, excimer lamps, UV LED spots and synchrotron radiations facilities. Surface photochemical modification and selective grafting on polymer surfaces opens up varieties of new products for practical applications. Park, et al. pointed out in a recent review [148] that the ultimate target of surface modification is to control surface properties quantitatively by precise manipulation of surface chemical structure at the molecular level, going beyond to foreseeing that it should be possible to develop surfaces that can be modified using only photons as reagents. In their very interesting review, the authors describe a new paradigm for photochemical surface modification based on the recent knowledge of how to manipulate the surface chemistry at the molecular level coupled with the broad data base of photochemistry. This combination presents great potentials for quantitative surface modification allowing the control of the nature and density of functionalities at the treated surfaces. By this paradigm, ''photochemical surface delivery vehicles'' would involve the grafting of ''smart'' molecules at the surface. Then upon irradiation with proper wavelength and intensity, these complex molecules can be assembled onto the substrate and designed to transform into a desired surface functionality. In the following sections of this chapter new advances and breakthroughs in the last decade on polymer surface photochemistry will be reviewed.

4.1. Ultraviolet–Assisted Polymer Surface Modification

In section 2 were presented a number of effective technologies that have been developed to improve the surface properties of polymers, including physical deposition, wet chemistry or chemical/wet methods. Dry surface treatment techniques, such as corona discharges, plasma, and UV radiation in the presence of ozone or without it, are effective for improving polymer-polymer as well as polymer-ceramic adhesions. All these techniques can modify the physico-chemical properties of surface substrates for intimate interfacial contacts between two dissimilar phases, without affecting the properties of the bulk material. Physical deposition of compounds on the substrates surfaces usually produce bounded coatings in a noncovalent interaction, which are unstable and easy to remove from the substrate. Grafting-to methods have several limitations, such as incomplete surface coverage and difficulty in uniform diffusion of the polymer chains to the substrate by their steric chain repulsion[149]. Compared with other modification methods, surface treatments and graft polymerizations induced by UV irradiation exhibit some advantages, e.g., low cost of processing, simple equipment set up, fast reaction rates, easy industrialization, near surface modification region and no damage of the bulk material.

UV-light irradiation of polymers surfaces leads to a series of reactions, such us photo-crosslinking, photo-oxidation in air, or photochemical reactions in reactive atmosphere. The extent of reactions depends on the reactants, the light intensity, and the photon absorption as a function of photon penetration depth. Light intensity is one of the crucial factors in UV-driven technologies and has to be carefully chosen depending on the application. Low-pressure mercury lamps generally have low intensity, a few mW/cm^2, while medium-pressure and high-pressure mercury lamps have much higher light intensity with their output ranging from deep UV to visible red-light region[9]. Microwave-driven electrode-less mercury lamps have an energy output of as high as 800 mW/cm^2 [150]. A more recent development is the commercialization of constant-wavelength (CW)-excimer laser lamps based on dielectric barrier discharge mechanism [150].

These lamps do not have a high output (about tens of mW/cm^2 for the XeC1 lamp) and they cannot match a pulsed excimer laser source, but it is sufficient for surface modifications. Dielectric-barrier discharges (silent discharges) are used on a large industrial scale and they combine the advantages of non-equilibrium plasma properties with the ease of atmospheric-pressure operation. In comparison with these CW-lamp sources, pulsed UV-laser sources have extremely high energy outputs (~ 10^7 W/cm^2), which is sufficient for laser-induced photoablation opening up a series of important applications. With new powerful light sources becoming available, more new applications will be found for polymer surface modifications in the near future.

Another distinctive characteristic of UV-assisted treatment is the wavelength dependence and penetration depth in polymer materials.

The penetration of the radiation at a particular wavelength depends on absorption coefficients besides mobility of polymer chains, glass transition temperature and diffusion coefficient of gases in polymer films. The Lambert-Beer law shows how UV-light is absorbed as a function of film thickness, or as a function of photon penetration depth, once the absorption coefficient is known. The intensity at a depth L, I, within a film is related to the incident wavelength intensity (I_0), by the equation, $I = I_0 \exp(-\beta L)$ in which β is the absorption coefficient. For example, for polystyrene films irradiated at 254 nm the radiation is strongly absorbed by the films. Polystyrene has a moderate UV-absorption in the longer wavelength region beyond 250 nm due to its aromatic component, e.g., with an absorption coefficient of ~ 4240 cm^{-1} at 254 nm. Figure 17 shows UV-absorption spectra of pristine polystyrene films of different thickness. It can be seen that the rapid attenuation of intensity has the effect of confining the effective reaction zone to the surface layers. The concept of surface modification of those films has a relative meaning. If a polystyrene film of 1 mm thickness is irradiated using 254 nm, more than 60% of the photons are absorbed within 0.15 mm (~ 15% of the volume assuming homogenous irradiation) and the treatment can be said as "surface" modification. On the contrary, if the polystyrene film has 0.15 mm thickness all the polymer will be modified by the 254 nm radiation. In this case the treatment modified the surface and also the bulk of the film. Depending of the polymer used and the functional groups presents in their structures (substituent), the absorption spectrum as function of thickness, such us Figure 17 has to be measured to know the actual depth penetration of the radiation at the selected wavelength.

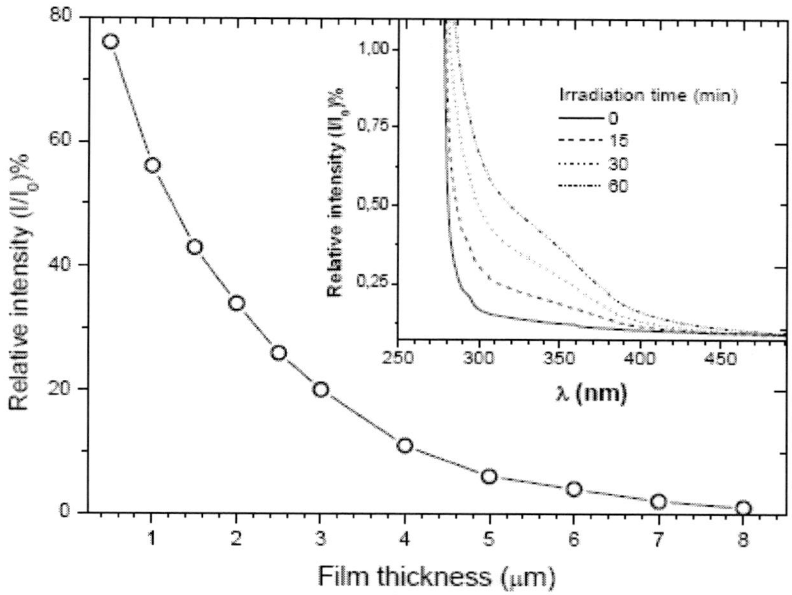

Figure 17. Absorption curve of pristine polystyrene films of different thickness irradiated at 254 nm. Inset: UV absorption spectra of a spin-coating film of polystyrene (~1 mm thickness) irradiated with unfiltered UV light at several photolysis times in ar.

Photo-degradation or photo-crosslinking reactions are typical processes that occur when polymer films with a suitable chromophore are exposed in vacuum or in an inert atmosphere to UV-light of moderated light intensity. In some cases, both reactions take place simultaneously. Even though polystyrene is a crosslinking-type polymer, it degrades under UV-irradiation in air due to photo-oxidation. Inset of Figure 17 shows those processes clearly with the merge of new absorption bands in the 300-400 nm region that were missing in pristine polystyrene. The increase of absorption intensity is proportional to the illumination time. As an illustrative example of the complex photochemical process involved in UV polymer irradiation can be presented here for styrene polymers. More detailed information on UV treatment of diverse polymers can be obtained in specialized books [8, 151, 152] and it is out of the scope of the present chapter. The phenyl groups of styrene polymers have strong absorption of 254 nm ($\pi \rightarrow \pi^*$ transition) which results in the formation of a singlet state (S_1^*) with a low degree of vibrational excitation. Radiation-less transitions (internal conversion, $S_1^* \rightarrow S_0^*$) are very efficient in these polymers [151]. Further deactivation occurs by monomer and excimer fluoresces with decay lifetimes of about 1 and 19 ns respectively [153]. Photochemical processes are associated with fissions (α or β) of bonds adjacent to the phenyl groups [9, 151, 154]. Probable polystirene α fissions are shown in Figure 18. β fissions correspond to C-C and C-H scissions in β positions.

Although polystyrene is a crosslinking-type polymer, it degrades under UV-irradiation in air due to photo-oxidation, as shown in the scheme of Figure 19. The simplify view shown in Figure 19 of PS photodegradation will be exemplify in the following pages with more complex examples and the interested reader can found more experimental details in the references given.

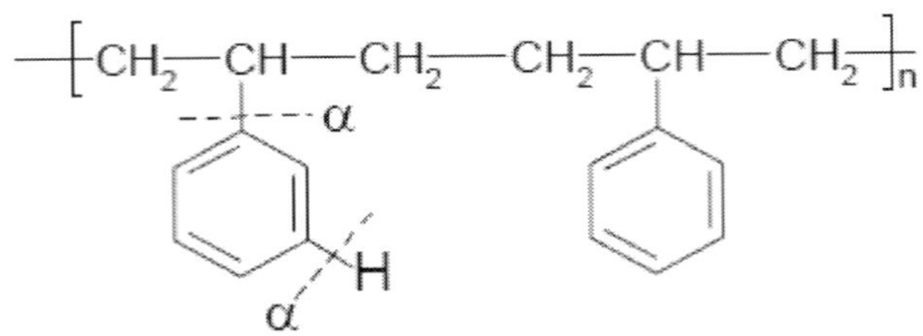

Figure 18. Representations α fissions in polystyrene.

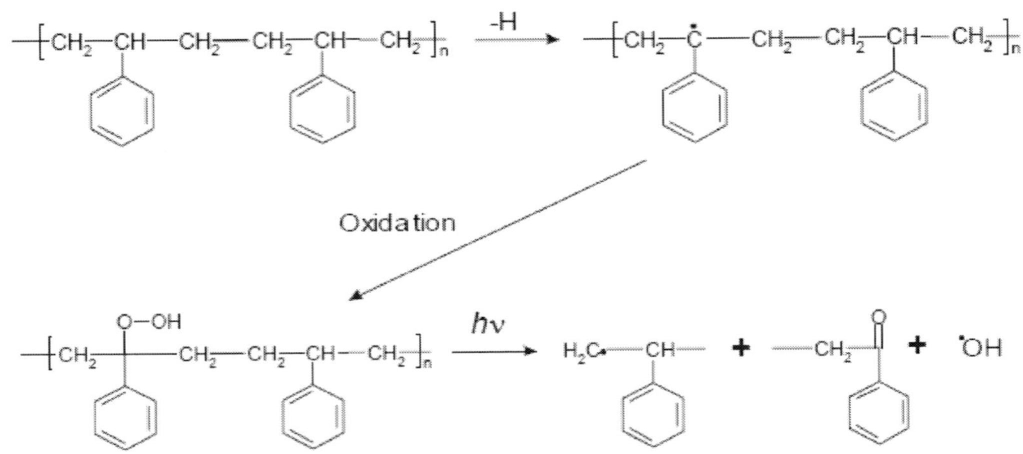

Figure 19. Simplified scheme of UV-degradation of polystyrene in air by photo-oxidation.

4.1.1. UV-Induced Surface Graft Polymerization

UV-grafting photochemistry of polymers has been recently reviewed by Deng *et al.*[16]. Despite the rapid progress in this promising research field, only a few review papers have been published to address the surface photografting polymerizations and related applications [16]. In view of the significance of the technique and the rapidly extending uses, the authors expected to inspire the interests of readers and researchers from the academy and the industry summarizing the latest advances in this research area. The review surveys the advances in UV light induced surface graft polymerizations, mainly focusing in:

i. Photoinitiators, controlling grafting, self-initiated grafting (without photoinitiators), graft polymerizations, liquid, vapor and bulk phase grafting, and the substrates used for grafting;
ii. Characterization of the topography of grafted surface layers.
iii. Application of techniques to produce functionalized polymer surfaces to tailor specific problems.

The review of Deng *et al.* [16] is thus an excellent update of this area and therefore a few illustrative examples will be included in this section.

Surface graft polymerization is extensively used using different methodologies, such us chemical, plasma or UV light irradiation. When UV light is employed, often the grafting is carried out in the presence of a photoinitiator or photosensitizer. Two types of photoinitiators are generally used: Norrish type I or Norrish type II, although the second one photoinitiators are more frequently used, predominantly because their use results in higher grafting efficiency [16]. Among the existing Norrish type II photoinitiators, probably the most widely selected have been benzophenone (BP) [155] and its derivatives [156-158]. These types of photoinitiators have been shown to be very efficient to initiate a number of radical-induced surface photografting polymerizations. A typical BP or BP-based photoinitiator mechanism consists in the absorption of a photon to an excited singlet state, which then by intersystem crossing reach a triplet state. This triplet state undergoes efficiently hydrogen-abstracting reactions from substrates producing surface radicals which initiate the surface graft polymerization reactions.

Asano, M. et al. [159] have prepared proton-conducting membranes for a polymer electrolyte fuel cell (PEFC) by UV-induced photografting of styrene into poly(tetrafluoroethylene) (PTFE) films and subsequent sulfonation. These membranes serve as a separator to prevent mixing of the reactant gases and simultaneously as an electrolyte for transporting protons from the anode to the cathode. The importance in the preparations is that the membrane has to be stable under fuel-cell operation and proton-conductible. An important step in the polymerization process requires both the coating of photosensitizers on the PTFE surface and UV irradiation in a mixture of styrene, acetone, and water. Under these conditions, the excited photosensitizer induces C—F bonds breaking and PTFE radicals can be formed, leading to the graft polymerization of styrene. The effects of various photosensitizers and the variation in the thickness of the PTFE films were investigated. In addition, the UV-photografted proton-conducting membranes were compared with γ-ray-radiation grafted membranes and the commercially Nafion membrane. The procedure carried out is presented in Figure 20.

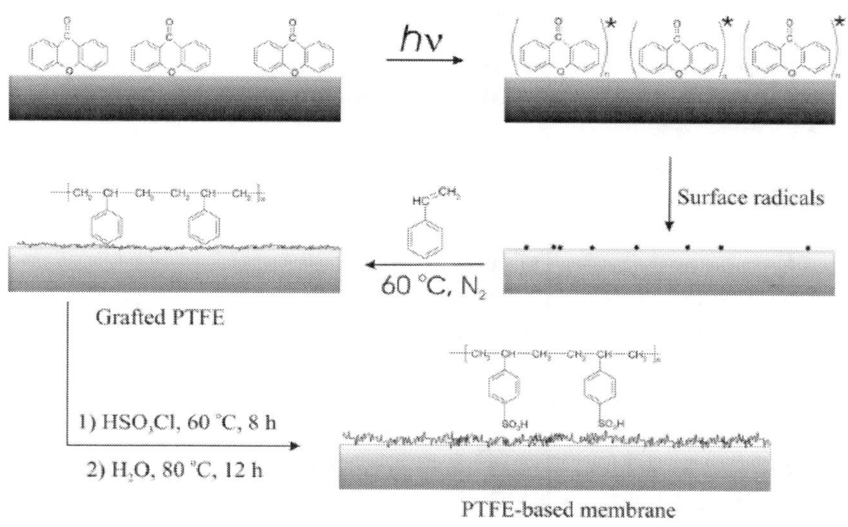

Figure 20. Preparation process of proton-conducting membranes by the UV photografting method. Adapted from Asano, M. et al. [159]. *: Photoexcited sensitizer.

The UV photografting method for the preparation of PTFE-based proton-conducting membranes showed high proton conductivity and better mechanical properties than commercial membranes. UV styrene-photografted membranes showed much lower ion exchange capacity (IEC) than the IEC Nafion membrane. On the other hand, higher proton conductivity and lower water uptake could be obtained. Apparently, UV photografting also proceeded according to the interface grafting mechanism, that is, the graft polymerization occurs on the interface of grafted/un-grafted regions where the styrene penetrates from the grafted region to the interface. This mechanism was also observed in γ-ray radiation grafting, but UV photografting avoids the radiation degradation caused by γ-rays, maintaining the mechanical properties of the resultant proton-conducting membrane. Surface and cross-section morphologies imaged by SEM indicated that completely sulfonated graft chains were effectively constructed to form the network for proton transport. The authors concluded that novel membranes with high proton conductivity with a low degree of grafting could be prepared by UV photografting.

Polypropylene (PP) films were grafted with Poly(acrylic acid) (PAA) chains using UV radiation, benzophenone as photoinitiator and water as solvent [157]. The objective was the study gas transport and permeability coefficient of oxygen, nitrogen, carbon dioxide, carbon monoxide, argon, methane, ethane, ethylene, and propane across the grafted films. Also, the pressure and temperature effects on gas transport were investigated. The photografting process of hydrophylic monomers onto polyolefinic film surfaces is usually conditioned by the solvent. For example, the grafting of a monomer onto PP films using benzophenone as initiator may take place not only on the polymer surface but also in the polymer bulk if the solvent is compatible with PP. The hydrophilic character of the solvent prevents the diffusion of the monomer inside a hydrophobic film. In this work the photoinitiator and the growing chains are soluble in water facilitating radical reaction propagation outside the polymer surface. In addition, the solvent used (water) is inert to the triplet excited state of the photoinitiator benzophenone and in consequence the grafting reaction is not inhibited. Costamagna et al. [157, 160] proposed a general scheme for the grafting mechanism of PAA onto the surface of PP films that is shown in Figure 21. Absorption of a UV photon by the benzophenone generates in a second step a radical which removes a hydrogen atom from a tertiary C-H bond of PP. This free radical mechanism initiates the grafting reaction. Diffusion of the photoinitiator inside the films is inhibited due to the hydrophobic character of PP and the hydrophilic nature of the reaction medium maintains the free radical generation confined mainly to the films surface. In spite the authors did not comment about the monomer Acrylic Acid (AA) photochemistry, important reactions may occur during irradiation due to absorption of UV photons by AA monomers[9].

The results obtained by Costamagna et al. [157] showed that the grafting percentage of PAA seems to be a linear function of the reaction time for grafting times larger than 5 min and that the PA grafting onto the PP surface has induction times larger than 2.5 min. SEM images highlight a grafted PAA layer on the surface of PP films, without observing the presence of PAA inside the PP films. Measurements of the permeability coefficient of several gases, such us oxygen, nitrogen, carbon dioxide, carbon monoxide, argon, methane, ethane, ethylene, and propane across the grafted films showed a sharp decrease compared with a pristine PP membrane. The authors interpreted their permeation results suggesting that radicals created in the tertiary carbons of the grafted chains by effect of UV light or by chain transfer reactions may highly crosslink the PAA grafted layer. When the grafting percentage was higher than 3.5% a rigid layer involving both strong hydrogen bonding and chains crosslinking strongly hinders gas permeation across that layer. The permeability coefficient of gases with 3.5% PAA grafting was found in average nearly 16 times lower than that measured in untreated PP.

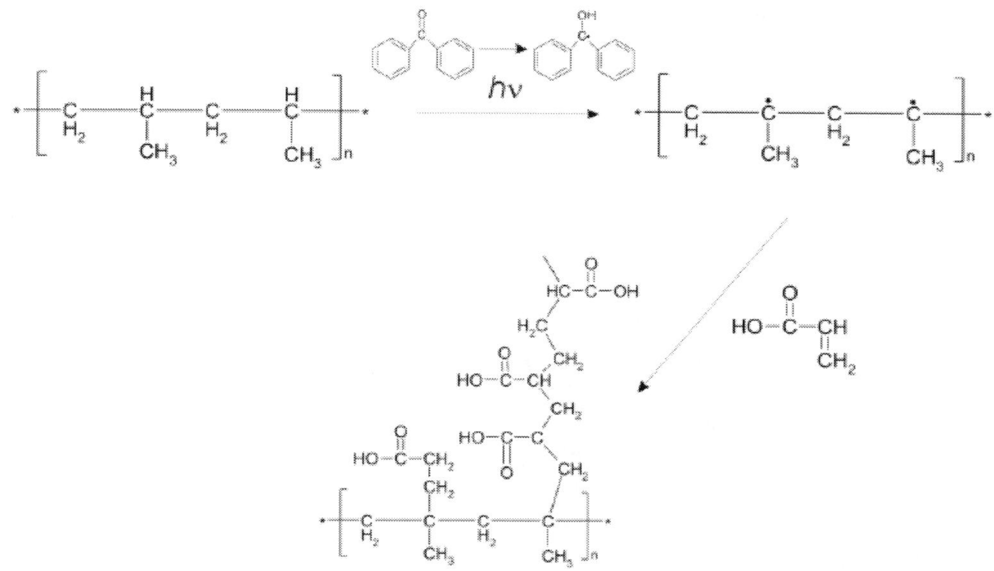

Figure 21. Scheme of the photografting reaction of acrylic acid onto the surface of isotactic polipropylene films. A solution containing the initiator (benzophenone) dissolved in acrylic acid and using distilled water as solvent was irradiated with UV light at room temperature under a nitrogen atmosphere [157].

Depending on the solvent used, surface photografting polymerization conducted in liquid phase depends in many cases on the solvent, which has an important influence on the grafting effects. For example, graft polymerization of (meth)acrylic acid using aliphatic solvents, such as chloroform, cyclohexane was very efficient [161], but on the contrary, when polar solvents such as ethanol were used the graft polymerization reaction was not favorable. Other solvents, such as toluene, rather acts as a screening agent for UV light and thus was not a useful solvent. In addition, different solvents also result in different surface topography of the grafted substrate.

Polypropylene microporous membranes surfaces were modified with 2-hydroxyethyl methacrylate (HEMA) to increase hydrophilicity, biocompatibility, and functionality [158]. Graft polymerization of HEMA onto polypropylene microporous membrane (PPMM) was induced by UV irradiation using ferric chloride ($FeCl_3$) and benzophenone (BP) as co-photoinitiators with different methods. Grafting HEMA onto PPMM surface was almost impossible, with BP as photoinitiator only, no matter which method was adopted and which concentration of BP was used. It is well known that when the reaction mixture containing BP is exposed to UV radiation, photons cause the excitation of the photoinitiator, BP, to a short-lifetime single state from where it relaxes to a triplet state [155]. Then, BP can abstract hydrogen atoms from PP chains by inelastic collision, thereby create free radicals on the solid surface which can serve as active sites for the grafting of different vinyl monomers (see schema of Figure 21). But for the studied monomer, HEMA, it was found that the graft polymerization was hard to progress and the conventional photoinitiator, BP, could not initiate the grafting polymerization of HEMA on the PPMM. By contrary, HEMA was successfully grafted onto PPMM when only $FeCl_3$ was introduced in the reaction system. Incorporation of $FeCl_3$ into the reaction system remarkably enhanced the grafting of HEMA onto PPMM, which could be ascribed to the "synergistic effect" between Fe^{3+} and BP. At low Fe^{3+} concentration it was produced a lot of HEMA homopolymer which was easily removed with acetone and simultaneously the grafting degree was very considerable. But for the cases with high Fe^{3+} concentration, little homopolymer was observed in the reaction system and relative low grafting degree was obtained. This phenomenon was attributed to the presence of metal ions or their complexes. These ions could also act as chain terminators by oxidizing or reducing the chain radicals, depending on the relative oxidation potential of the ion and the polymeric radical.

The work carried out by Costamagna et al. [157] and Hu et al. [158] stress other point in the selection of monomer, wavelength, intensity, solvent and photoinitiator in a particular grafting experiment. Both works use polypropylene polymer as substrate and try to graft different monomers (AA and HEMA respectively) using a common photoinitiator, BP. But whereas Costamagna et al. were successful in graft PAA on PP using BP as photoinitiator, Hu et al. [158] could not initiate the grafting polymerization of HEMA on the PPMM using the same photoinitiator. Only after the addition of small concentrations of $FeCl_3$ into the reaction system the grafting of HEMA onto PPMM was highly enhanced. Complex fast photochemical mechanisms and chemical reactions that proceed at lower rates determine the final grafting results observed.

Recently, Colavita, et al. [162] have investigated the mechanism of covalent grafting of functional organic molecules on amorphous carbon surfaces using the photochemical reaction of terminal alkenes. They have used infrared reflection-absorption spectroscopy (IRRAS) and XPS to compare the reactivity of n-alkenes of similar length with different terminal functional groups at the opposite end. The results were complemented using ultraviolet photoelectron spectroscopy (UPS), photocurrent measurements and computational methods in order to investigate whether electron photoemission is an important process in these reactions. The experimental set up was very simple. H-terminated amorphous carbon (a-C) samples were placed inside a nitrogen-purged reaction chamber sealed with a quartz window to prevent evaporation. A few drops of the neat alkene compound were placed on the substrates and the cell was illuminated with a low-pressure mercury lamp (254 nm). Finally, the samples were then rinsed using chloroform, ethanol and methanol. It was found that the photochemical functionalization of carbon surfaces with terminal alkenes proceeds through a complex mechanism in which, the reaction is initiated by the absorption of a photon and the ejection of a photoelectron into the liquid phase. The obtained results suggested that the reaction involves injection of photoelectrons into the alkene acceptor levels, leading to the formation of radical anions in the liquid phase. As a consequence, the reaction efficiency depends on two factors, the photoemission properties of the a-C surface and the ability of the liquid to accept electrons. In addition, the authors demonstrated that grafting of reactive alkenes can be enhanced by seeding the surface with a small amount of electron accepting groups. These results has provided information into the role of electronic excitations in controlling rates and mechanisms of olefin reactions at surfaces in UV initiated grafting of olefins. In addition, it may shed light over the complex mechanism involved in UV grafting reactions on polymer surfaces.

4.1.2. UV-Assisted Surface Modification Using Reactive Atmospheres

In Section 2, a series of dry surface treatment methods were described, such us plasma or UV radiation in combination with ozone that can modify the physical and chemical properties of substrate surfaces facilitating the interactions and contacts between two dissimilar phases, without affecting the properties of the bulk material. By using those methods, common problems encountered in wet chemical techniques (section 2.1) such as residual solvent on the surface and swelling of the substrate can be avoid with the use of a gas plasma for example. Another similar approach, without the need of any expensive set-up is the modification of the polymer surface with the help of UV light in the presence of a vapour atmosphere. In general, this is low-cost method, easy to apply and in many cases the methodology has comparable results to that obtained by plasma treatments. Additionally, the use of a dry medium eliminate the needs for immersing the whole device/component in a solution for treatment, such as in UV grafting reactions presented in section 4.1.1.

A series of interesting articles on UV-assisted or induced modification of polymer surfaces have been recently published using gaseous organosilanes compounds[13-15, 163]. The authors developed a new process for the photochemical modification of different polymers surfaces employing organosilane compounds. They were interested to see if organosilicon groups can be introduced onto polymer surfaces by photochemical techniques. The hydrosilylation process is well known in organic chemistry and over the past years hydrosilylation of alkenes, carbonyl compounds and alkynes in the presence of free radical initiators (e.g. azo-bis-isobutyronitrile) has been investigated thoroughly. The Si–H bond dissociation energy of trialkylsilanes $R_3Si–H$ and the reactivity of trialkylsilyl radicals $R_3Si^•$ has been subject of numerous studies. Considering free radical chemistry, trialkylsilyl radicals $R_3Si^•$ belong to the most reactive species towards various functional groups. Radicals $R_3Si^•$ add to olefins and also to aromatic rings at high rate constants [15]. As e consequence of the properties of trialkylsilanes a study on surface modification of C=C unsaturated polymers such as polybutadiene was carried out by Spanring, et al. [15]. Common silanes reagents used in hydrosilylation of organic chemicals have low vapour pressure and it is too low for a surface photomodification process based on gaseous reagents. Instead, ethyldimethylsilane (EDMS), a low-boiling liquid, and trimethylsilane (TrMS), available as a liquefied gas, were chosen as reagents [15]. In this work, cis-1,4-polybutadiene (PBU) with 98% cis-content was selected as target polymer. The compounds EDMS and TrMS have high vapor pressures, are easy to handle, non-toxic and almost odorless. Although they are flammable compounds, they do not tend to self-ignition upon contact with moisture or oxygen (in contrast to silane SiH_4). These beneficial properties facilitate handling of the reagents both in laboratory experiments and in technical processes. The irradiation of PBU with a medium pressure Hg lamp in the presence of vapors of EDMS results in an attachment of alkylsilyl groups. Similar results were obtained when monochromatic UV light irradiation by 193 nm ArF* laser was carried out on polybutadiene in the presence of gaseous TrMS. The incorporation of silyl groups was confirmed both by FTIR and XPS. After 10 min of UV irradiation a clear XPS signal corresponding to the Si 2p (102.8 eV) excitation was detected. Changes in the FTIR spectra were also evident after UV irradiation under TrMS or EDMS atmosphere. Signals at 1248 cm^{-1} and 838 cm^{-1} were observed after 60 min of UV lamp irradiation. Contact angle measurements by the sessile drop method were carried out to follow changes in surface tension γ of polybutadiene during UV-assisted surface modification. Pristine cis-1,4-polybutadiene is a hydrophobic polymer showing a water contact angle $\theta = 99.2^o$. As a result of UV irradiation in presence of EDMS, θ increased up to 102.5° within 60 min. When diiodomethane was used to measure changes in contact angle, θ increased from 45.2° (prior to irradiation) to $\theta = 70.8^o$ after 60 min of irradiation. From contact angle data surface tension γ and its dispersive and polar components (γ^d and γ^p) were calculated following the method of Owens and Wendt. Pristine cis-1, 4-polybutadiene has a surface tension of 37.0 mJ cm^{-2}. The surface tension of polybutadiene mainly consists of the dispersive component γ^d, while the polar component γ^p of the surface energy is comparably small. The results showed that UV treatment of polybutadiene with trialkylsilanes mainly affects the dispersive component, γ^d, which decreases significantly; while the polar component, γ^p, remains almost unchanged during this process of surface modification as no polar groups are introduced. The generation of low-energy polymer surfaces by UV treatment in the presence of silanes may be useful to improve their antiadhesive properties. Uptake and permeation of organic solvents and vapours can be reduced by such techniques, although UV surface modification with trialkylsilanes is highly sensitive to traces of oxygen in the process gas.

Spanring, et al. [13] have also carried out the irradiation of polystyrene (PS) in presence of trialkylsilanes (EDMS or TrMS). UV irradiation of PS surfaces with a medium pressure Hg lamp in the presence of vapors of EDMS does not only result in the incorporation of alkylsilyl groups ($R_3Si.$) but also Si–H groups were covalently bond onto the polymer surface. The authors claim that the introduction of the Si–H groups opens new possibilities for further functionalizing polymer surfaces. As in the case of PBU study [15], contact angle measurements were carried out to monitor changes in the surface tension γ of PS during the photo-assisted surface modification. After UV irradiation in presence of EDMS the contact angle θ increased from 91.4° to 99.6° within 60 min. The water contact angle did not increase significantly even after prolonged irradiation. For diiodomethane changes in contact angle data were more pronounced. After 10 min of treatment with EDMS the θ using CH_2I_2 increased from 36.1° to 55.3°, reaching 67.2° after 60 min of irradiation. Similar to PBU treatment, UV treatment of PS in presence of EDMS mainly affects the dispersive component, γ^d, which is reduced to almost 50% of its pristine value after 60 min of UV surface modification. Because no polar groups are introduced during this type of treatment, the polar component, γ^p, remains almost unaffected. In summary, surface modification of PS and PBU treated with EDMS or TrMS results in a low-energy surface. During the UV-assisted treatments the bulk polymer remains unaffected while surface properties can be modified with either hydrophobic or hydrophilic groups bearing 1-alkenes as anchor groups. The results obtained by Spanring, et al. have shown that a common and simple UV methodology can be used universally to treat different polymers.

When a polar or oxidant atmosphere is used in conjunction with UV irradiation hydrophobic polymers can be modified to hydrophilic [17, 164]. For example, Bormashenko et al.[164] have studied the UV irradiation of polysulfone (PSF) and polyethersulfone (PESF) films with a Hg lamp at a wavelength of 254 nm for photolysis times ranging from 10 to 90 min apparently in air atmosphere. The films exposed to these treatments demonstrate an important decrease in their water contact angle. From the initial contact angles of PSF and PESF substrates established as θ_{PSF} in 84° and θ_{PESF} in 79°, respectively; UV irradiation decreases strongly the contact angle to about 20° for both polymers after 90 min of irradiation. The contact angle on the irradiated substrates demonstrates typical aging effect. For example, when a PESF film was irradiated for 90 min and held afterward at ambient conditions for 60 min, the contact angle increased to 35° and returned to their initial values after rinsing the film with bi-distilled water and drying. Based on these results, the authors [164] suggested that the revealed phenomenon may provide a means of controlling the spreading of liquids on PSF and PESF films which can be promising for use in microfluidics applications.

Surface functionalization of polyurethane (PU) films has been obtained by UV-assisted modification in the presence of oxygen or acrylic acid (AA) vapors [17]. The changes produced after the treatments were analyzed by water contact angle measurements, XPS and NEXAFS spectroscopies. When AA vapors were used during irradiation a photo-polymerization process was observed which led to a formation of a thin film of poly (acrylic acid) (PAA) over the PU films. By using a simple set up (see Figure 22), an AA monomer vapour atmosphere was created by bubbling nitrogen through a reservoir containing pure monomer located next to the gas entrance of the reactor. The reservoir was heated at 50 °C to increase its vapour pressure to about 30 mbar. In experiments carried out in the presence of oxygen, its constant flux was introduced during irradiation of the sample [17]. UV irradiation was carried out with unfiltered light of a commercial medium pressure mercury lamp.

Pristine PU is a slightly hydrophobic polymer that shows a water contact angle of about 85°. After UV irradiation in the presence of oxygen, the contact angle decreased linearly down to about 28° within 90 min of photolysis. As can be seen in Figure 23, tuning of photolysis time allows easy control of hydrophilicity. The low scattering in the data shows that UV-assisted modification of PU films is a precise methodology by which a specific surface hydrophilicity for a particular application can be selected. The surface hydrophilicity obtained for PU UV-treated films depends mainly on a simple parameter: the photolysis time.

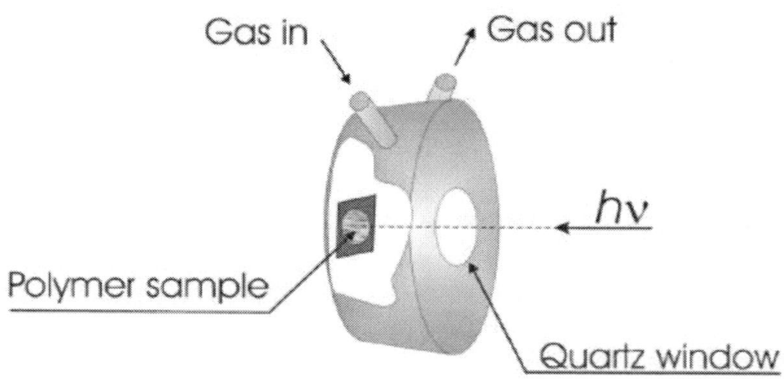

Figure 22. Photochemical reactor used for UV-assisted functionalization of PU films in the presence of a gas reactive atmosphere.

When UV irradiation of PU films was carried out in the presence of AA vapors, the WCA decreased from 85 to about 60° in a few minutes of treatment (see inset of Figure 23), reaching an approximately constant value of roughly 55° at longer irradiation times. These results contrasted with those using UV-assisted modification of PU in the presence of oxygen

(Figure 23) which led to a continuous decrease on the WCA when the photolysis time increased. Similar results were also obtained in plasma-induced graft co-polymerization of AA onto the PU surface [165] and in previous AA plasma treatment of PU films [166]. In this last study a plasma polymerization process was observed during treatment, forming a PAA like film on top of the PU film.

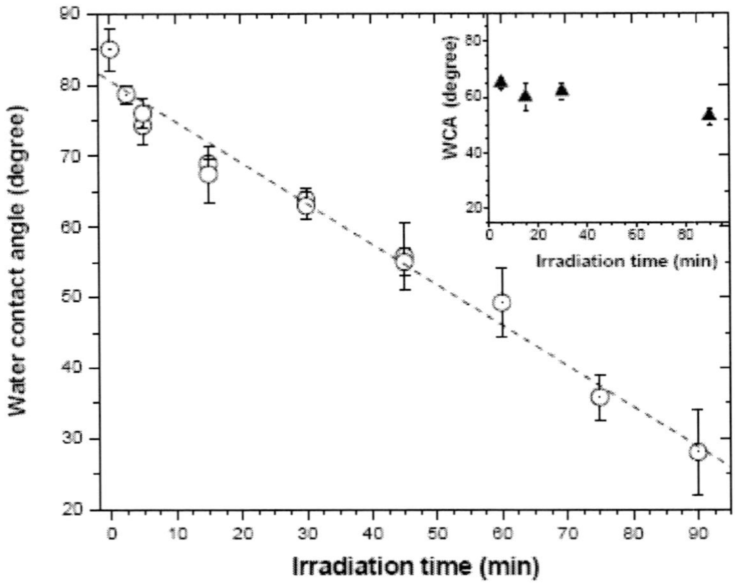

Figure 23. Water contact angle measurements of untreated and UV-modified PU films in the presence of oxygen or AA vapour (inset).

Experimental evidence obtained by water contact angle measurements, XPS and NEXAFS pointed out that there are important differences between UV-assisted treatments in the presence of oxygen or AA vapors. When oxygen was used at reactive atmosphere, the increase in hydrophilicity with the increase in photolysis time was attributed by XPS deconvolution of the C 1s to the relative increase of the COO and C=O contributions. UV-treatment in the presence of AA vapors also shows that the COO and C=O contributions increased with rise in photolysis time, but in this case the similarity between the final XPS spectra obtained and the corresponding XPS spectra of AA plasma modified PU films led the authors to a preliminary conclusion that a thin film, chemically resembling the PAA was formed during plasma treatment [90, 166] and also during UV-assisted treatment in the presence of AA vapors [17].

NEXAFS spectroscopy has been shown to be a powerful method for characterization of polymer materials (see section 3.2.2 in this chapter) [97, 98, 100, 167, 168] and it was chosen as an additional and very sensitive surface chemical probe. The sensitivity of the technique to the polymer and molecular structure has also been demonstrated [97, 98, 100]. In particular, polyurethanes and urethanes polymers have been carefully studied with high chemical sensitivity in the search for spectroscopic “fingerprints” that differentiate several formulations. Figure 24 presents the effects of UV irradiation time in the presence of AA vapors. When the photolysis time increased an intense signal at 288.3 eV appeared with increasing intensity with the rise of photolysis time. This signal shows, in agreement with XPS and WCA analyses, that a film with different chemical properties was formed with the increase in photolysis time.

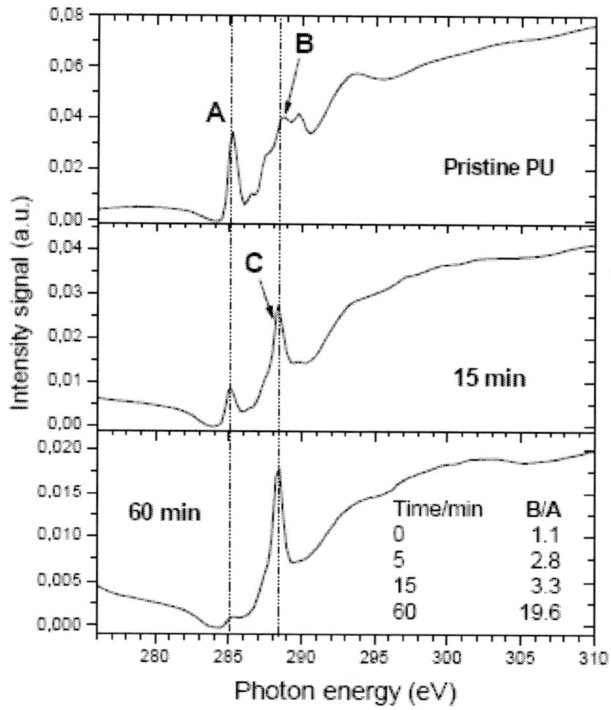

Figure 24. NEXAFS C *K*-edge spectra of UV-assisted polymerization of PAA films on PU at several irradiation times. A: C 1s $\rightarrow \pi^*_{C=C}$; B: C 1s $\rightarrow \pi^*_{C=O}$ and C: 1s $\rightarrow \pi^*_{ROC=O}$ from [98, 100, 105].

Simultaneously, the typical spectroscopic features of PU films slowly decreased until almost complete disappearance at 60 min of irradiation time. Plasma polymerized films prepared from AA were also studied by NEXAFS spectroscopy [105] as a function of plasma duty cycle. Swaraj *et al.* assigned the C 1s K-edge NEXAFS spectroscopic features to polymerized acrylic acid films with peaks at 284.4, 286.4, 288.3 and 295 eV corresponding to $\pi^*_{C=C}$, $\pi^*_{C=O}$, $\pi^*_{ROC=O}$, σ^*_{C-O} resonances excitations respectively. With the increase in duty cycle or with the plasma operating in a continuous mode, the transitions at 284.4 and 286.6 eV quickly disappeared and the spectrum showed only the $\pi^*_{ROC=O}$, σ^*_{C-O} resonances. This fact was explained as an increase in fragmentation of monomer molecules with the increase in effective power at higher duty cycle. The shape of the spectrum obtained by Swaraj *et al.*, at high duty cycle or in continuous modes resembles the NEXAFS spectrum showed in Figure 24-bottom. The position of the C 1s K-edge $\rightarrow \pi^*_{ROC=O}$, 288.3 eV [105] matched exactly the signal at 288.3 eV measured in Figure 24. In addition, the inset of Figure 24 shows a table where the changes in the signal ratio B/A, were measured form maximum peak high as a function of photolysis time. Clearly, when the irradiation time increased, the signal corresponding to a PAA film was dominant, leading to a change in the ratio B/A from ~ 1 to ~ 20 for a change in the illumination time from 0 to 60 min, respectively.

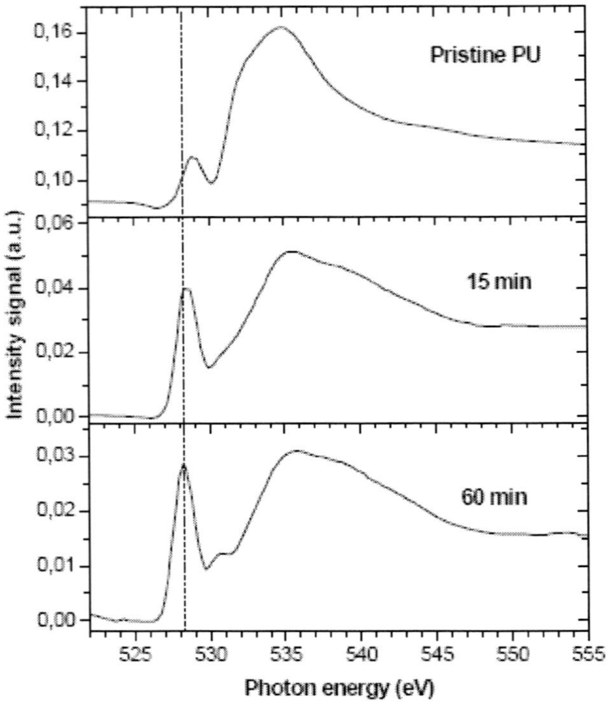

Figure 25. NEXAFS O *K*-edge spectra of UV-assisted polymerization of PAA films on PU at several irradiation times. The main absorptions correspond to O 1s → $\pi^*_{C=O}$ (~ 528.8 eV) and O 1s → σ^*_{C-C} (~ 535.3 eV). See section 3.2.2 and Figure 10 of this chapter for more details.

NEXAFS O *K*-edge spectra of UV-assisted polymerization of AA films on PU films were also recorded at several irradiation times and the results can be seen in Figure 25. The results obtained also showed a clear change in the NEXAFS spectra with the increase in the irradiation time. The transition O 1s → $\pi^*_{C=O}$ shows a shift of 0.85 eV from ~ 529.1 eV for untreated PU to ~528.25 eV for 60 min of treatment, which is more than three times larger than the beam line resolution of ~ 0.25 eV. This results confirms that a new C=O and COO functionalities are introduced into the PU surface after UV-assisted treatments.

Similar results obtained in surface modification of PU by UV-modification in the presence of reactive atmosphere were also obtained with polystyrene films [36]. Our recent results showed that grafting of polar functional groups and polymerization reactions of PAA initiated by UV irradiation without the presence of photoinitiators, led to analogous surface functionalization results in PU and PS than typical modifications obtained in plasma treatments of the same polymers. Figure 26 presents in qualitative terms XPS results that show a comparison between AA plasma treatment and UV-assisted surface modification in the presence of AA vapors of PU and PS films. The experimental conditions were very different (plasma and UV treatments); however, it was found a qualitative agreement between the two types of surface treatment methodologies. Consequently, the obtained NEXAFS, XPS, and WCA results showed that UV-assisted AA polymerization on PU and PS films is an efficient method to produce polymerization reactions and graft specific functional groups in a similar way as under plasma treatment without the need of a complex experimental set up.

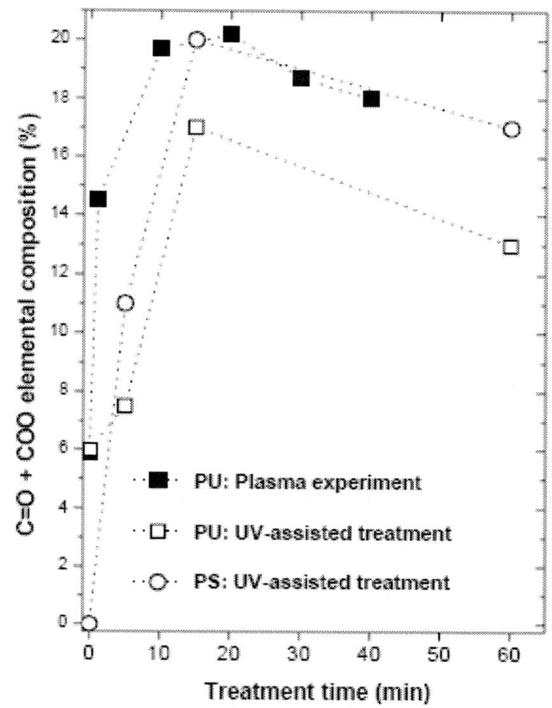

Figure 25. Dependence of the C=O plus COO elemental composition on treatment time in plasma and UV-assisted treatments of PU and PS films using AA vapours. Data obtained from C 1s XPS experiments [17, 36].

4.1.3. Vacuum-UV (VUV)-Surface Modification

The division of the electromagnetic radiation spectrum into ranges of UV, VUV, XUV or EUV (eXtreme or Extreme ultraviolet), soft X-rays and X-rays or hard X-rays is not trivial and there is no clear-cut distinction between their exact limits as pointed out by Stark and Smith [169]. Because air in standard temperature and pressure conditions absorbs radiation below about 200 nm, VUV radiation starts from this wavelength (λ < 200 nm). There is, however, more or less agreement among different authors that the range between 200 nm and 10 nm, reaching even 1 nm[27], should be called the VUV or EUV [170, 171]. For the purpose of the present chapter, VUV radiation will be referred according to the wide and accepted general definition of photons of wavelength below 200 nm up to reach the soft X-ray range.

Organic materials with multiple bonds or with atoms which have non-bonding electron pairs such as oxygen or nitrogen absorb light with a wavelength shorter than 200 nm very strongly [169]. This property limits the penetration depth of the photons to about 50–100 nm in the substrate (90% of the radiation is absorbed within a surface layer of this thickness). The very low penetration depth of the VUV radiation makes it very interesting for surface modification applications because the properties of the bulk material are preserved. Additionally, in many cases the experimental system construction is simple, the lamp has large emission area with high energy photons and different excitation wavelengths are available.

Surface treatment of polymers using plasma is a well know area of research in the academy and in the industrial arena and recently it has been recognized the important roll that VUV radiation may have in low pressure plasma treatments [129, 172]. The plasma group in UK obtained detailed results which indicated that under low pressure plasma conditions, VUV was the dominant component in producing free radical sites at the surface of polystyrene. Their conclusion was that the ionic plasma component has very little effect on the level of the resulting oxygen functionality at the treated polymer surface indicating that in this discharge regime, VUV (and possible excited neutral species) are responsible for the production of free radical sites. Those finding are very important because low-pressure r.f. plasmas were used to study VUV radiation since the early 1970s [19], but materials processing with VUV radiation has only recently received much attention [20-22, 24, 27, 173-176], mainly due to the developments of excimer lasers and of VUV sources based on dielectric barrier discharges, DBDs [150]. These highly efficient, intense VUV sources are used in numerous applications, for example surface cleaning of semiconductors, metallization, curing of paints, lacquers, adhesives, water purification, biomedical application and polymer surface activation/modification. VUV photochemistry has the advantage over its plasma counterpart that the surface sample can be chemical modified "selectively" via monochromatic radiation, due to more specific and selective reactions both on the solid surface and in the gas phase.

The most significant features of VUV radiation were well summarized by Skurat [27]:

a. The high photon energy of VUV radiation (6.2 eV at 200 nm up to 123.98 eV at 10 nm) is enough to break any chemical bond, going beyond to produce ionization processes with high efficiency. For example, absorption of photons with wavelengths lower that 124 nm (~10 eV) will be sufficient to ionize most organic molecules. From the energy point of view, the effect of VUV radiation would be similar to that of ionizing radiation (gamma-radiation or fast electrons) and very different to traditional UV photochemistry.
b. Electronic transitions are controlled by quantum mechanical selection rules and these selection rules are different for VUV radiation (optical transitions) and ionizing radiation. Optically allowed electronic singlet–singlet transitions are highly predominant in VUV irradiation of organic materials with very low probability of intersystem crossings (singlet–triplet transitions). On the other hand, secondary electrons with relative low energies ($\leq$ 100 eV) are responsible for the known chemical effects observed in the treatment of polymers and organic materials with ionizing radiation. In this case the optically allowed selection rules are not taken into account and the relative probabilities of triplet state formation are greatly enhanced. The reactivity of singlet and triplet excited states in a same material can be very different and the product distributions or final treatments obtained, in consequence, would not be the same. The optical allowed transition are a very important difference between the effects produced by VUV radiation and those of ionizing radiation and it can be used to try to obtain a more ''targeted'' surface chemistry via monochromatic radiation, due to more specific and selective reactions both on the solid surface and in the gas phase.
c. VUV radiation is characterized by large absorption coefficients and low penetration depth. For example, in polyethylene the extinction depth is 34 nm for VUV

excitation (λ = 147 nm) and 1.5 10^5 μm for ^{60}Co γ-rays. Therefore, a high density of energy is absorbed in a thin top region of the material when VUV is employed. This feature determines particular effects during VUV treatment that are not observed or they have low significance in radiolysis experiments [27].

d. Quantum yields of primary and final products of photochemical reactions, including polymers reactions, depend on the excitation wavelength. As the photon energy increases, the relative contributions of neutral triplet excited states increase compared to singlet excited states. At photon energies higher than about 10–12 eV, the product distributions and their energetic yields (*G* value defined as a number of product molecules formed during absorption of 100 eV) become similar to those for the case of radiolysis.

In the last ten years, there was a substantial increase in the use of VUV irradiation for surface treatment of polymeric materials, trying to compete in many cases with plasma technology. The main reasons for this augment are directly related with the VUV properties described above, in particular 4.3.1-*i* and *iii,* but also with recent technological advances which allowed the construction of VUV sources with different emission wavelengths together with important decreases in the production costs. Actually, it is now possible to select the wavelength of the VUV source, so that maximum absorption of photons and maximum photo-dissociation can be achieved, if the photochemical decomposition mechanisms and quantum yields are known in the VUV region. For example, Ar, Kr and Xe resonant lines in excimer sources emit nearly mono-chromatic VUV radiation at 126, 146 nm and 172 nm, respectively (see Figure 26). NH_3 and C_2H_4 have quite high absorption cross sections in the 100-220 nm region (absorption cross sections, σ in the range of ~10^{-17} cm^2 $molecule^{-1}$), while CO_2 presents a much lower σ, about 100 times lower than NH_3 and C_2H_4 (see Figure 26). Therefore, a judiciously selection of wavelength and lamp power has to be selected to achieve a particular target efficiently.

VUV spectroscopy of simple and more complex molecules is also a very active research field due to the existence of synchrotron radiation sources, in particular third generation sources that have been operating from late '90 are capable of providing photon beams with peak brilliance of at least 10^{18} photons/s/0.1%bw/(mm mrad)^2). Fourth generation SR sources will complement and increase the potential of older sources and it is expected that VUV spectroscopy research grows at even larger speeds. VUV spectroscopy is an exciting field to discover and exists excellent references that can be consulted [169, 182-184].

One of the first systematic studies on VUV irradiation of polymer substrates was carried out by Wilken, et al. [176]. Combining several experimental techniques, such us infrared reflection absorption spectroscopy (IRRAS), quartz crystal microbalance (QMB) measurements, and XPS together with chemical derivatization techniques, the VUV photolysis of polyethylene (PE), polypropylene (PP), and polystyrene (PS) was investigated. Different reaction degradation path ways under VUV irradiation were suggested by mass balance obtained from the quantification of the data. The results showed that although PE and PP behave similar, PP has a mass loss of about 8 times higher than PE. It was found that both polymers form double bonds and at extended treatment times they tend to crosslink. PS is rather stable due to the possibility of the energy dissipation by fluorescence. The authors concluded that VUV photolysis can play an important role in the low-pressure plasma treatment of polymers. Particles from the gas phase modified a surface region limited to a surface layer which is only some nanometers thick, while VUV radiation penetrates deeper. In this subsurface region only VUV photons can initiate reactions due to there are no particles from the gas phase involved there. The subsurface region finally determines how strong the outer surface with it desirable properties are bond to the bulk of the material. In addition, a high dosed degree of crosslinking can be very beneficial for the characteristics of the whole system. The data presented in this article was allowed to understand in more detail why the materials behave differently although they were exposed to the same plasma treatment conditions.

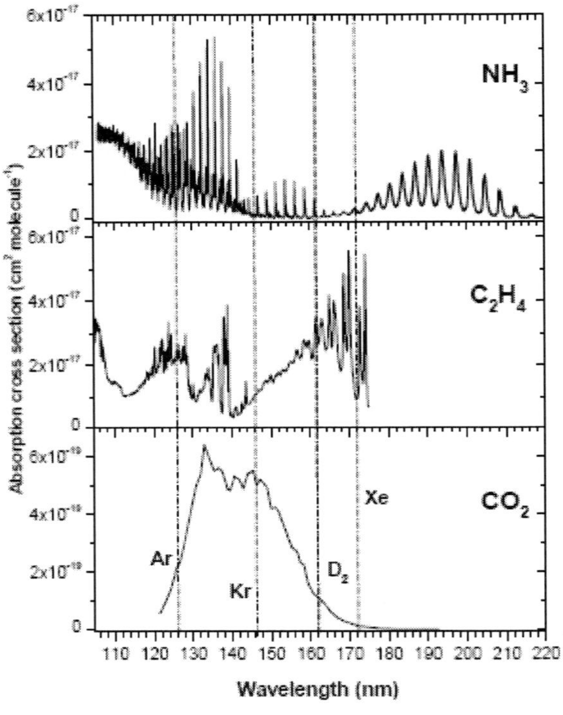

Figure 26. VUV absorption cross sections of gaseous NH_3[177-179], C_2H_4[180] and CO_2[181]. Typical main "monochromatic" emission lines of some excimer lamps are indicated in the figure (dash-dot lines).

Plasma treatments, as was highlighted in section 2.2 of this chapter, have a complex chemistry, difficult to predict and to control due to the several experimental parameters that are involved in a treatment. However, the selection of one energetic plasma component to carry out a particular treatment could allow, at least in theory, to achieve more specific and selective reactions obtaining better control of the overall process. This was the idea behind the work carried out in the last years by the group of Wertheimer in Canada [20, 22, 24, 26, 185, 186]. They have compared the chemical structures of nitrogen-rich organic thin films prepared by low-pressure plasma treatment and VUV-assisted deposition. They were interested in graft nitrogen containing functional groups at the surface because these nitrogen functionalities are particularly advantageous in bio-technology applications. For example, primary amines ($-NH_2$) apparently are the most effective groups. The authors claim that for the first time it was shown that polymerization of an organic ''monomer'' (C_2H_4) and its mixtures with ammonia (NH_3) [22] or nitrous oxide (N_2O) [20], designed to deposit nitrogen-rich and oxygen-rich thin film materials, can be achieved simply by VUV irradiation at reduced pressure. Additionally, the films formed by VUV irradiation were compared with plasma polymers deposited using glow-pressure r.f. glow discharges in similar gas ratio flow mixtures. It was showed that VUV photochemistry can potentially lead to better controlled surface reactions through the use of monochromatic radiation. Excimer rear gas sources emitting nearly mono-chromatic VUV radiation at 123.6 nm and 147 nm respectively were used. At these wavelengths the absorption cross sections values are high for C_2H_4 and NH_3 (see Figure 24) and N_2O [20]. Since, for example, the bond energy $D_0(NH_2–H) = 4.4$ eV, the absorption of energetic VUV photons at 147 nm ($h\nu = 8.43$ eV) leads to photo-dissociation of the NH_3 molecule, followed by a redistribution of the excess energy to the resulting fragments. In the 110–180 nm region, main primary reactions of NH_3 photolysis can be summarized as follow [187]:

$$NH_3 + h\nu \longrightarrow NH^* + H_2 \qquad 4.1.3.1$$

$$NH_3 + h\nu \longrightarrow NH_2^* + H \qquad 4.1.3.2$$

where NH^* and NH_2^* represent radicals in various excited states. Amino and imino radicals absorb only in the UV/VIS range and therefore, further photodissociation under VUV irradiation can be neglected.

Regarding the VUV photolysis of ethylene and nitrous oxide, the primary dissociative processes are the followings:

$$C_2H_4 + h\nu \longrightarrow C_2H_2 + H_2 \qquad 4.1.3.3$$

$$C_2H_4 + h\nu \longrightarrow [C_2H_3^* + H] \longrightarrow C_2H_2 + 2\,H \qquad 4.1.3.4$$

Further photodissociation of C_2H_2 is also possible, namely:

$$C_2H_4 + h\nu \longrightarrow C_2H_2^* + H_2 \qquad 4.1.3.5$$

$$C_2H_4 + h\nu \longrightarrow C_2H_2^* \longrightarrow C_2H + H \qquad 4.1.3.6$$

For N_2O the primary photochemical mechanism is:

$$N_2O + h\nu \longrightarrow N_2 + O_2 \qquad 4.1.3.7$$

$$N_2O + h\nu \longrightarrow NO + N^* \qquad 4.1.3.8$$

VUV and plasma treatments achieved similar values of nitrogen concentration at the surface, but the former produced a surface much richer in primary amines, with selectivity ($[NH_2]/[N]$) values up to 75% for a concentration ratio, R, of NH_3/C_2H_4 = 0.8. Table 2 summarizes the results comparing plasma and VUV treatments. The particular advantages in the used of VUV treatment lie in their combined high $[NH_2]/[N]$ and $[NH_2]/[C]$ values, so far very difficult to achieve by plasma chemistry [20, 22].

Table 2. Amine selectivities, $[NH_2]/[N]$ %, as function of the gas mixture ratio, R, for Nitrogen rich films deposited using a Kr VUV lamp and for low pressure plasma treatment [22]

	Polymerized films (%)		
$R = NH_3/C_2H_4$	VUV treatment	Plasma r.f. treatment	VUV/ Plasma r.f.
0	0	0	0
0.2	47	19	2.5
0.4	60	26	2.3
0.6	74	30	2.5
0.8	75	32	2.3
1.0	68	33	2.1

Last column of Table 2 shows the polymerization ratio between VUV and plasma treatment. In spite, the change in R from 0.2 to 1.0, the ratio in the polymerization reaction between the VUV and plasma treatment did not change in significant amounts. The authors did not comment about those results because there is an actual difficulty in the understanding of the complex plasma processes. Basically, the results obtained by the Canada group are indicating that much more effort has to be done in the research on basic processes of plasma and VUV photochemical reactions. In particular, it is very important to try to quantify the VUV component within a plasma process.

Lopez Gejo, et al. [139] have reported results on VUV irradiation of polystyrene (PS) films in the presence of oxygen using a pulse Xe_2^* excimer radiation source ($\lambda = 172 \pm 12$ nm). The results showed that the initiated oxidation was very efficient introducing microstructures on PS surfaces. VUV photochemistry produced not only oxidative functionalized surfaces, but generated also morphological changes. If the lamp is set at a short distance from the polymer in normal atmosphere, oxidative fragmentation of the polymer takes place permitting photolithographic pattern to be printed on the substrate surface. Controlling the distance between the lamp and the substrate surface, and the intensity of the lamp (number of pulses) hydroxyl and carbonyl groups of appropriated concentrations can be introduced by VUV irradiation (oxidative functionalization). Evidence was given through different experimental conditions that two different reaction pathways occur: introduction of OH- and CO-groups is mainly due to the reaction of reactive oxygen species (hydroxyl radicals, atomic oxygen, and ozone) produced in the gas phase between the VUV-radiation source and the substrate. On the other hand, VUV-electronic excitation of the polymer surface leads to C-C bond breaking, which subsequent yield peroxyl radicals by trapping of the C-centered radicals by molecular oxygen. This last process triggers a complex oxidation manifold reactions leading finally to mineralization of the organic fragments. The authors concluded that ozonization of the PS leads to oxidative functionalization of the polymer surface but cannot initiate the fragmentation of the polymer backbone.

With the objective to understand the main interactions involved in plasma-polymer process such us etching, deposition, and surface modification, Bruce, et al. [174] have carried out a detailed comparison of the etch behavior of polymers in Ar plasma discharges with low and high energy ions by changing the substrate bias. In addition, the authors compared the treatment process at different substrate temperature conditions, and finally compared fully plasma-exposed conditions and vacuum ultraviolet VUV-only conditions by employing a magnesium fluoride window to prevent ion bombardment in the VUV-only case. The study revealed that ions, VUV radiation, and temperature all had significant impact on the etch behavior of polymers. It was shown that a dehydrogenated, amorphous carbon layer is created in polymers with thickness determined by the ion penetration depth. The thickness of this layer is stabilized after the initial formation state and afterwards its properties remain constant. The authors used Molecular Dynamic (MD) simulations to reproduce the plasma experiments with a self-bias of −100 V, so that ions impact the surface with energies close to those simulated by MD. Using the damage layer thickness found in MD simulation, the experimental data are fitted by varying the refractive index and extinction coefficient of the damaged surface obtaining results of density and hydrogen constant very close to those simulated by MD. The dependence of different polymer structure on etches and surface roughness was also compared. Taking into account the modification carried on in the VUV-only experiments, the experimental data fitted well, implying the importance of VUV radiation for understanding the processing of the materials. Some polymers with higher content of ester groups showed much greater etch rates at the beginning of the treatments, which corresponds with greater thickness removed to establish the damage layer. This is because to make a carbon rich layer, more material needs to be removed. The polymers that had the greatest amount of removed thickness during densification and damage layer creation showed the highest amount of roughness. The authors show that the surface roughness in polymers exposed to plasma depends on several factors and cannot be described by only physical sputtering. Both ions and VUV radiation must be considered and can have substantial effect on polymer modification. Also the polymer structure is especially important to take into account when a particular treatment condition has to be used.

Very recent work in the UFRGS group on surface modification of polypropylene (PP) by VUV irradiation in air atmosphere has demonstrated important properties of the VUV irradiation in polymer film treatments. PP is widely used in many sectors of industry, but some of its properties still need to be improved, such as heat resistance and chemical resistance. Crosslinking provides an important method to improve the thermal and chemical resistance of polyolefins. Common crosslinking methods employ to crosslink polyolefins are silane crosslinking, peroxide crosslinking, and radiation crosslinking. Another problem associated with widely available polymers is their surface properties. For example, PP is a hydrophobic polymer with a water contact angle between about 98 and 110° depending on the roughness of the film. There is an increasing interest to improve the surface properties of commodity polymers, such us PP increasing the range of potential applications for these materials. For certain applications the required treatment need not be specific, for example, to overcome the adhesion problems associated with hydrocarbon polymers. In other cases, highly specific treatments are required, for example, in the modification of polymer surface properties for potential use as biomaterials. The experiments developed at UFRGS were carried out using a "monochromatic" Deuterium lamp emitting mainly at 162 nm. Our preliminary results showed that by using VUV treatment in the presence of air atmosphere or a controlled oxygen atmosphere permanent surface changes can be obtained by VUV irradiation[188]. Thin PP films prepared by spin-coating technique on stain steel substrates have a water contact angle of about 110°. After 60 minutes of VUV irradiation in air atmosphere the water contact angle decreased to about 60°, reaching a constant value of about 55° after 90 minutes of irradiation time. An aging study showed that PP surfaces remained with almost the same water contact angle after treatment for several weeks (see Figure 27).

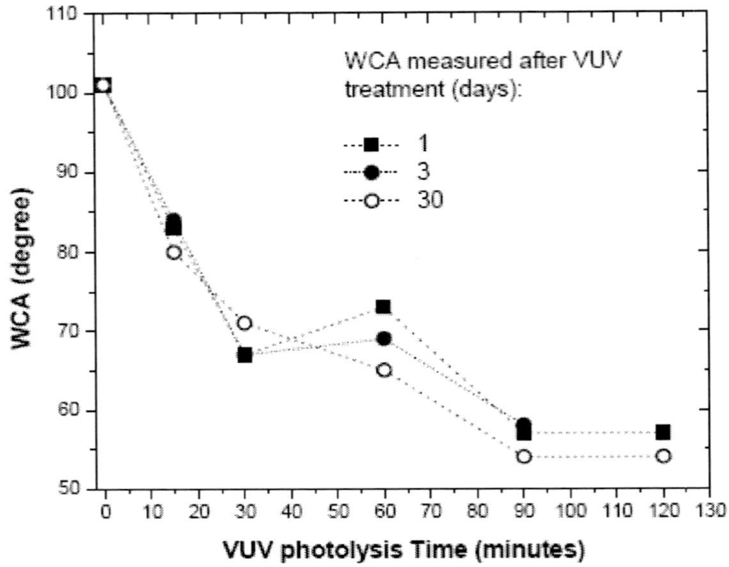

Figure 27. Aging study of PP films treated by VUV irradiation at several photolysis times in air atmosphere.

Our results confirmed that permanent surface changes can be obtained by VUV irradiation of PP films. This is an important result for further processing and industrial applications of polymer films, because the treated films can be stored or transported far from the treatment place without worried about surface hydrophobic recovery. Aging is a widespread phenomenon that often has not been recognized, particularly in some of the earlier studies on the use of plasma-fabricated surfaces for bio-interfacial applications, and can markedly alter the surface chemistry. Plasma treatment on many polymer substrates suffers from aging, with surface adaptation leading to the movement of surface modification effects into the polymer [47]. This effect was particularly shown in films of pure polyolefins (low density polyethylene and biaxially oriented polypropylene, BOPP) that were surface-modified by two different methods, namely VUV irradiation by using a Kr resonant lamp and atmospheric pressure glow discharge plasma treatment [189]. The authors show that the hydrophobic recovery after the treatments is due to a motion of polymer chains and chain segments, which governs an apparent "loss" of functional groups, within the first monolayers of the surface. In these sense, comparing BOPP samples modified by both techniques, it is shown that aging can be reduced by cross-linking near the surface, which was illustrated by depth-sensing nano-indentation experiments. In our experiments, important cross-linking near the surface may be responsible for the long term stability of the treated PP surface. Work is been carried out in our laboratory at the present time to confirm this hypotheses which was highlighted by the preliminary findings of Wilken, et al. [176].

As a final example on VUV polymer surface processing, a recent study on VUV photodegradation and micropatterning of aminosilane films using 172 nm light emission from Xe_2^* excimers source was used to induce strong surface modification effects on polymeric materials in the top 100 nm layer [173]. In this study a previous series of quantum chemical calculations were carried out in order to identify suitable monomers for this VUV-based process and the photodegradation mechanism of different organosilanes of the general structure $R–CH_2–Si(OCH_3)_3$ was elucidated. The photodegradation of 3-amino propyltrimethoxysilane films by the use of a 172 nm excimer lamp under different irradiation conditions was described and completed by micropatterning experiments. It has been shown that by a Xe_2^*-excimer lamp irradiation the degradation and micropatterning of 3-aminopropyltrimethoxysilane-based films proceeded within acceptable times and by a manageable technical complexity. It was found that the Si plus O concentration at the surface increased from 50 at.% to 88 at.% during irradiation. The presence between 1000 and 5000 ppm of oxygen was found to promote the transformation process to an inorganic-like surface. The ratio of the inorganic over the organic fraction of the coating was expressed as the (Si+O)/(C+N) ratio. The study of the dependence of the (Si+O)/(C+N) ratio on the VUV-treatment time showed that up to an oxygen concentration of 1000 ppm this ratio increased with the amount of oxygen in the irradiation zone and decreased when the oxygen concentrations of 5000 ppm were used. Additional data showed that the modification processes on the top surface proceed in fact comparably fast. Main changes produced in the surface composition of the films were found during the first 2 min. An irradiation time of more than 2 min had no further influence on the surface energy and irrespective of the oxygen concentration, the surface energy increased from 46 to 68 N/m, affecting mostly the polar part, while the dispersive part did not contribute.

4.1.4. Laser Surface Modification

Pulsed UV and VUV-laser sources have become commercially available and new surface polymer applications are being explored in many fields in the last decades. Pulsed laser irradiation can be significantly more effective in enhancement of hydrophilicity, wetability, and gas permeability in many applications. Laser based surface treatment results in combined surface modification of topography and chemical changes because laser illumination induced thermal and chemical processes which will change the micromechanical and chemical properties of the material. The advantage of the application of laser irradiation is the available sub-micrometer resolution and the possibility to control the contribution of topographical and chemical effects by changing the experimental conditions. Pulsed laser beams can also be used for surface modification of metal deposition on polymer surfaces like PTFE, and others, which can be exposed in air, reactive gases, or aqueous solutions depending on the final applications.

The use of laser techniques, such us rapid prototyping, rapid tooling and rapid manufacturing, gain more and more significance in modern manufacturing of different industrial prototypes [190, 191]. Rapid prototyping is a big advantage during the design stage of these elements, since flow characteristics of complex structures are difficult to model and are dependent on dimensions. There is also a growing interest in the development of a number of alternative serial fabrication techniques based on the removal of materials and adapted to prototyping polymeric devices, such as micro-milling and laser micromachining, because these direct fabrication techniques do not require access to clean-room facilities. Also there is no need for using masks, like photolithographic techniques, or molds, like the transforming techniques, which adds an additional step to the microchannel fabrication process and reduces flexibility because minor modifications to the original design require new masks or molds. Micromilling by using laser sources is a well-established technique, efficient, and flexible which is used in many areas of industrial production.

Lasers are unique energy sources characterized by their spectral purity, spatial and temporal coherence, and high peak intensity. Although laser prices are usually much more expensive than traditional tools, their advantages have allowed developing applications such in laser micromachining that is a field growing rapidly over the last decades. All laser micromachining techniques use the process of laser ablation. Laser ablation usually relies on the strong absorption of laser photons by the sample material, which means that the wavelength of the laser has to be chosen carefully for maximum absorption. New technological advances in laser techniques has overcome this approach by using ultrafast laser pulses where ablation takes place as a result of non-resonant process by multi-phonon absorption at high peak intensities. When ultra fast laser sources are used even materials normally transparent to the laser wavelength can be machined.

The ablation mechanism is typical for laser treatment techniques and it is a complex combination of photochemical and photothermal processes which is dependent on the laser characteristics and materials properties. For example, depending on the laser wavelength the ablation process can be relatively "cold" or on the contrary "hot". When UV laser radiation is used, UV photons have enough energy for direct bond breaking in an organic material strongly absorbing at the wavelength of the laser emission. This strong photon absorption leads to ablative photodecomposition, a relatively 'cold' process with minimum thermal damage to the substrate. In contrast, if long wavelength laser radiation is employed the radiation mainly initiate thermal reactions that will first melt and then decompose the material, leaving a void in the work-piece. An interesting article published by Yingling, et al.[192] have investigated those effects by molecular dynamics simulations. In order to investigate pure photoablation of organic solids, a simulation containing only photochemical excitations was performed. The results of this pure photochemical system were compared to the results from the system containing both photochemical and thermal processes. The molecular dynamics simulations of a system containing only photochemical processes showed the ejection of big molecular clusters from the surface that were originated from the propagation of a strong pressure wave and a sharp enthalpy rise. The analysis of the clusters internal energies showed that clusters were colder and denser than clusters in the system where photochemical and thermal processes are present. The removal of damaged material through big cluster ejection can be attributed to experimental observations of low heat damage of material by far-UV laser irradiation. The simulations provided a unique perspective on the laser ablation phenomena at the molecular level been consistent with experimental observations of low heat damage of material.

Laser pulse duration also plays a critical role in the laser–material interaction affecting the quality of the micro-features created and the material removal rate. Absorption mechanisms in ultra-short laser pulses (subpicosecond and femtosecond duration) are different from those involved when longer pulses are used, because the energy is deposited too fast than thermalization processes are slower than absorption rate of photons. The potential of ultra-short laser pulses for material processing is based primarily on the possibility to localize energy deposition to dimensions smaller than the diffraction limit of the focusing optics on the surface or in the bulk of materials through nonlinear absorption mechanisms, and to reduce residual damage by minimizing thermal effects. Laser technological advances have been impressive in the last two decades with new uses and applications appearing in increasing amounts. In this section, a few representative examples will be given to show the potential of the use of laser technology for surface polymer modification.

While micro-devices such as sensors or actuators based on glass, quartz or silicon are well established commercial products in microsystem technology, most applications demand devices which can be used as disposables in order to eliminate risks of sample contamination. Polymer materials and their fabrication methods have proved to be the solution to this challenge. In particular polymer based, micro-fluidic systems have been recently object of intensive research [193] and also they were introduced to the market. Mycrosistems produced by UV-laser-assisted processes are of particular interest for applications in microfluidics, bio-analytics, bioreactors and micro-optics. Pfleging, et al. [194] has recently investigated the modification of polystyrene surfaces for cell culture applications by laser processing using excimer laser radiation sources at 193 nm (pulse length 20 ns) and 248 nm (pulse length 4–6 ns). UV-laser assisted ablation and modification processes were studied in three different process regimes with respect to the laser influences: below the ablation threshold, ε_t, in the range of ε_t and significantly above ε_t. Each laser regime produces a particular surface topography modification. Above ε_t three-dimensional shapes with a very small surface roughness can be realized. In the range of ε_t the surface roughness significantly increases and cone formation can frequently be observed. Below ε_t in general no significant change in topography can be determined. However, for a large laser pulse number ''ablation'' or formation of nanopores caused by polymer degradation can occur below the threshold ε_t [194]. Ex situ XPS measurements of PS surfaces after laser treatment showed that the highest amount of oxygen compounds at the surface were detected after laser treatment at 5 mJ/cm^2 in He or O_2 processing gas atmosphere. The results suggested that oxidization occurs mainly after laser treatment, when the sample is handled ex situ in air. Radical laser formation is the main mechanism to generate radicals in PS surfaces due to breaking of phenyl rings or cracking of bonds of the main polymer chain or of side chains. Laser processing allowed the generation of patterns on nanometer and micrometer scale. The patterning also encompassed ablation of the PS surface with production of two- and three-dimensional shapes in polymer surfaces. It was demonstrated that the adhesion of cell clusters, single-cells as well as the formation of cell networks on polystyrene surfaces could be controlled with high accuracy. By using other ambient conditions than He or O_2 during laser processing, different types of chemical groups (e.g. –COOH, –NH_2) could be patterned onto polymer surfaces containing features with two or three dimensions. Besides the chemical modifications produced at the polymer surface by UV laser irradiation, also a change in topography on nm- and μm-scale could be used for a change of cell adhesion and wetability. It is important to highlight that in all the laser processes investigated there was no need for a vacuum environment and the methodology used enabled also the use of liquid media as possible reactant.

PTFE absorbs strongly in the VUV range and in particular 157 nm radiation can be easily obtained by using a F_2 excimer laser. For this reason several investigators have exploited 157 nm excimer lasers to produce high quality structures on PTFE or other polymers by direct ablation, with little thermal damage [23, 25, 30, 195]. For example, a recent work reported the time- and mass-resolved measurements on neutral particles emitted from PTFE during exposure to 157 nm laser radiation at fluencies high enough to observe etching of the polymer [196]. The results showed intense emissions of $(CF_2)_n$ fragments. The authors note that the great majority of material removed under the conditions of their work appears to be in the form of fragments with masses well above 300 amu. Due to the mass limit of the quadruple mass spectrometer used of 300 amu, their affirmation was quite speculative and could not been confirmed. Irradiation of PTFE with 248 nm laser radiation contrasted with 157 nm laser excitation. Time-of-flight signals measured at 100 amu (C_2F_4 monomer) during 248 nm and 157 nm irradiation were fitted to the Maxwell–Boltzmann velocity distribution for particle effusion (appropriate for thermal desorption from a hot surface). The Maxwell–Boltzmann distribution provides a reasonable fit to the entire time-of-flight curve acquired at 248 nm and yields a plausible surface temperature of 1180 K. On the contrary, it was observed that the leading edge at the 248 nm signal was much slower than the leading edge at 157 nm and the fit to the leading edge of the signal at 157 nm yields an implausibly high temperature of 4200 K. These results showed that the high kinetic energies of the faster fragments are consistent with photochemical emission. In addition, the observation of intense emissions at masses other than the monomer rule out thermally activated unzipping of the polymer chain as an important source of material removal. The complex ablassion mechanism also included polymerization reactions initiated by ions and radicals that may efficiently incorporate these larger fragments into deposited films and help account for the high quality of PTFE films produced by laser ablation deposition at 157 nm.

The culture of cells on a flat substrate is required by cell biologists to allow three-dimensional (3D) observation of complex large group of cells. Transparent substrates for individual cell analysis are indeed a very combining substrate. In general, a glass slide or polymers are often used as the substrate, due to their high transparency and high chemical stability and fabrication of microwells on a glass or polymer substrate is applied to control cell attachments. These methods are not suitable for dynamic cell analysis since cell shapes are mainly restricted to the micropatterned wells on the substrate. Additionally, refractive indexes of most transparent substrates are larger than 1.4, which is disadvantage for cultured on the substrate due to the refractive index mismatch between the substrate and the culture medium. Recently, Hanada, et al. [197] modified the surface of an amorphous flour-polymer with a refractive index of 1.34, almost comparable to that of culture medium as well as water, using a F_2 laser (157 nm wavelength, 20 ns pulse width, 30 Hz repetition rate). The F_2 laser was chosen due to the strong absorption of the flour-polymer to 157 nm radiation. The objective was to modify the pristine hydrophobic surface of the flour-polymer to hydrophilic by laser modification allowing the attachment of the cells and simultaneous 3D observation of cultured cells. The water contact angle of untreated polymer of 110° decreased to 73° after F_2 laser irradiation in dry nitrogen gas ambient obtaining an optimum wetability when 300–500 laser pulse were fired. After those amount of pulsed there was found a partial recovery of the hydrophobicity. AFM images of untreated and modified polymer surfaces after 500 and 1500 laser pulse irradiation showed that the average roughness increased from about 2.5 nm with 300–500 laser pulses to 6.06 nm at 1500 laser pulse irradiation. The increase in the average roughness was ascribed to laser induced ablation or desorption, due to incubation effect of increased laser absorption site after previous several hundred laser irradiations. The authors speculate that monotonically improvement of wetability up to 500 laser pulses is due to photochemical reactions initiated by the F_2 laser irradiation. Using F_2 laser irradiation it was possible to increase the wetability of the flour-polymer without deterioration of its original optical properties. Cells were selectively cultured in the treated substrates with 3D observation cultured cells obtaining clear images of the cells from the side view, due to almost the same refractive index of polymer as that of culture medium. The authors concluded that surface treatment of a flour-polymer by F_2 laser irradiation would be attractive as a microchip platform for 3D microscopic observation of adherent cells.

Recently, first results on surface modification of polyethylene terephthalate (PET) were published using extreme ultraviolet (EUV) radiation. A 10-Hz laser-plasma source emitting radiation from VUV to soft X-rays [170] was used. The source is based on the double-stream gas-puff target irradiated with the 3 ns/0.8 J Nd:YAG laser pulse. The target was created by pulsed injection of a krypton jet into a hollow stream of helium (Kr/He target) by employing an electromagnetic valve system equipped with a double nozzle set-up. The effective focused radiation of Kr or Xe plasmas within the wavelength range of 9–70 nm has higher intensity and a relatively narrow wavelength range centered at 10 ± 1 nm. Different kinds of microstructures were created in the near-surface layer as a result of EUV irradiation of the polymer foil. When a few EUV pulses irradiated a PET sample located in the focal plane the formation of some rippled microstructure was observed. It was shown that the shape of the ripples changes with consecutive EUV pulses and after irradiation with 25 pulses they form a microstructure with a preferred direction. A wall-like structure was created uniformly in the area of irradiation with diameter of approximately 2.2 mm. Irradiation with 200 pulses or more leads to further changes in the surface microstructure. Basically two clear zones were created with significantly different structures. A circular central zone, with diameter of approximately 1.4 mm was covered with a nap-type microstructure and sparsely dispersed cones. The authors speculated that their creation is probably connected with the shielding effect of some surface or volume impurities or small parts of the polymer sample locally modified by the EUV radiation. This might lead to a higher ablation threshold than that for normal PET as it was suggested for UV laser ablation [170]. The second zone has a form of a ring surrounding the first zone and there was lower material ablation in this region than the center. A densely packed conical structure was typical from this second zone. It can be then assumed that the EUV fluence at the border corresponds to the ablation threshold. XPS measurements showed a change of the relative concentration of carbon and oxygen connected with decomposition of the polymer chains as it was expected. Total concentration of oxygen decreased from 27.94% in untreated PET foil to 19.77% in irradiated PET. The changes were connected with bond breaking and preferential removing of oxygen. The O 1s de-convoluted signal was almost unaffected after irradiation suggesting that oxygen is being removed as a whole O–C=O functional group.

4.1.5. Synchrotron Radiation Surface Modification

Synchrotron radiation (SR) is a very powerful source that has not been explored yet for surface functionalization of solid materials. The extraordinary properties of synchrotron radiation, in particular in the new generation sources, and its rapidly increasing availability throughout the world are having a profound impact on a broad range of scientific and technological disciplines including polymer science. The most important properties of synchrotron radiation from an experimental point of view are the total high power radiated, the highly focused angular distribution, the wide spectral distribution, and that the radiation is highly polarized. Other very important properties of the radiation depend on the design of the particular machine. These include the pulsed time structure and the optical properties of the source point (i.e. electron beam size and divergence convoluted with the angular distribution of the emitted radiation).

Remarkably, a few reports have appeared in the literature on polymer surface functionalization by SR. One exception is the work carried out in the Graduate School of Science in the University of Hyogo, Japan. During the last years, the research group has focused its work on the study of surface modification of poly(tetrafluoroethylene) (PTFE) by SR exposure under several experimental conditions [32-34, 198-201]. PTFE has wide industrial applications due to its several excellent properties, such us thermal and chemical stabilities, low surface energy and high resistivity. However, the PTFE surface hydrophobicity has restricted the application of fluorocarbon polymers in several other fields, such us microfabrication, painting and adhesion. The Japan group used white radiation, non monochromatic SR light from IR to soft X-ray radiation, to modify the PTFE surface while varying the substrate temperature from room temperature to 200° C. The water contact angle decreased from 96° to 72° by the irradiation at room temperature, while the contact angle increased up to 143° when the substrate temperature was 200° C. The increase of the hydrophobicity was ascribable to the CF_2 components on the PTFE surface. XPS results showed that the amount of CF_2 components on the PTFE surface decreased from 90.1% to 0.9% and that of C–C component increased from 4% to 79% by SR exposure at room temperature. In addition, microstructures formed at the PTFE surface during SR irradiation explain the increase of the contact angle at higher temperatures. The results showed that wetability of a PTFE surface can be controlled from hydrophilic to hydrophobic by SR exposure.

However the above series of studies, only our UFRGS group has started a series of experiments on selective surface functionalization using monochromatic SR. The aim of those studies was to control the selectivity of the functionality that is bonded to the polymer surface. The hypothesis was that by using a highly monochromatic radiation, a selective functionalization of the polymer surface can be obtained. A simple polymer such as Polystyrene (PS) was chosen because new functional groups containing oxygen, for example, would be easy detected by surface sensitive techniques. By using oxygen as reactive atmosphere, we were able to find experimental conditions to functionalize selectively the polymer surface introducing high concentrations of COO, C=O and C-O groups at the most top monolayer of the films [35, 36]. SR experiments were carried out at the Brazilian Synchrotron Light Source (LNLS), Campinas, Brazil. The SGM (Spherical Grating Monochromator) beam line, for VUV and Soft X-ray Spectroscopy (250 - 1000 eV) which gives a spectral resolution ($E/\Delta E$) better than 2000, was used as monochromatic photon source. SR beam was defocused ~5 mm in diameter to avoid sample charging throughout the experiments [35]. When SR was used to functionalize the polymer surface, specific transitions were selected and during a fix period of time of 15 minutes the polymer was irradiated under UHV conditions. After irradiation, pure oxygen at atmospheric pressure was introduced into the UHV chamber for 30 min to neutralize the radicals remained at the PS surface. Figure 28 shows a scheme of the experimental set up used. XPS and NEXAFS spectroscopy were obtained before and after irradiation to study the changes produced by the treatments.

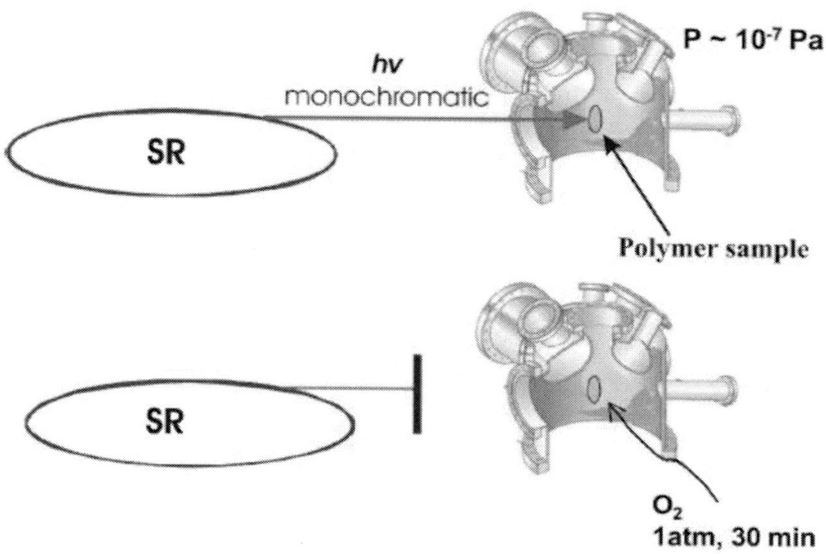

Figure 28. Schematic set up of SR monochromatic functionalization of polystyrene thin films. During 15 minutes the polymer was irradiated under UHV conditions (top of the Figure). After irradiation, pure oxygen at atmospheric pressure was introduced into the UHV chamber for 30 min without irradiation to neutralize the radicals remained at the PS surface (bottom of the Figure).

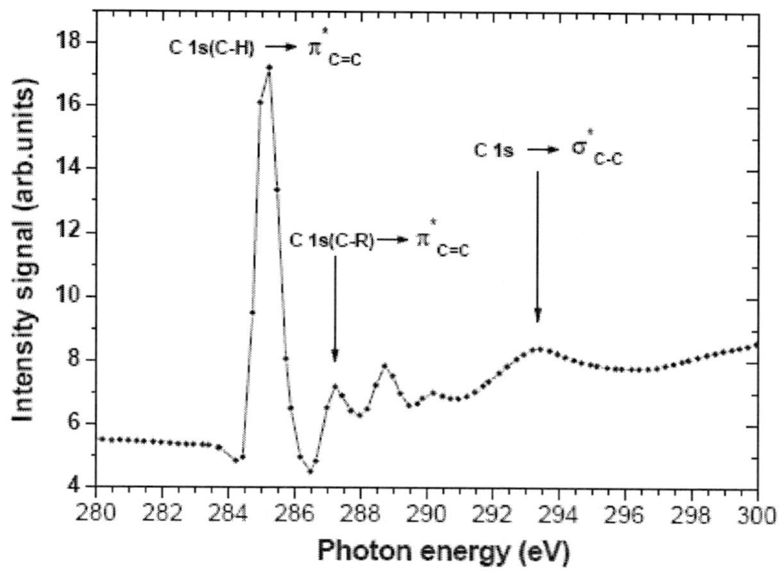

Figure 29. Carbon *K*-edge NEXAFS spectrum of untreated poly-(styrene) showing the main transitions selected for selective surface functionalization [35].

By selecting specific *K*-shell excitation energies, it was shown that a high efficient surface functionalization can be obtained. After the treatments, several oxygen functionalities were detected by XPS and NEXAFS spectroscopy. In this study specific transitions were chosen (see Figure 29). The results showed that the surface functionalization depended strongly on the excitation energy. For example, by excitation of C 1s → σ^*_{C-C} transitions at 293.5 eV an extremely high efficient oxidation process of the most top monolayers of PS films was obtained (see Figure 30). Angle resolved XPS measurements revealed that the functionalization of the PS films occurred on uppermost monolayer of the polymer surface. The COO and C=O elemental contributions in the C 1s envelope at the PS surface was higher than 70%. On the contrary, other excitation energies did not introduce COO functionalities efficiently.

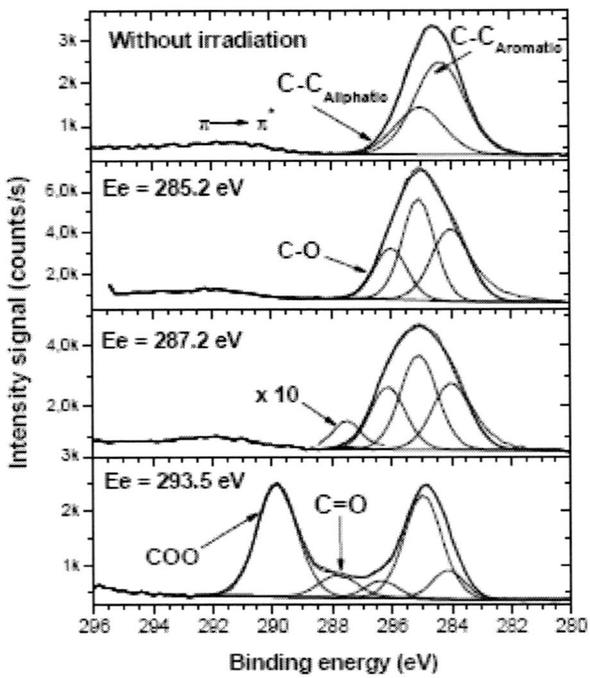

Figure 30. High-resolution XPS spectra of the C 1s envelope of PS without treatment and after SR functionalization at the selected excitation energies shown in Figure 29.

Because in NEXAFS spectroscopy the C 1s → π^* phenyl ring resonance is one of the most prominent spectral features, its changes can be easily monitored during a surface modification treatment. Figure 31 shows the carbon *K*-edge NEXAFS spectra of PS before and after selective SR treatment at 293.5 eV. As can be seen in Figure 31, important changes in the C 1s NEXAFS spectra were found after SR treatment. Figure 31 shows the carbon *K*-edge NEXAFS spectra of PS before and after selective SR treatment at 293.5 eV. The C 1s → $\pi^*_{C=C}$ signal intensity strongly decreased after irradiation indicating a lower contribution of this transition in the new modified polymer surface. The C 1s → σ^*_{C-C} primary excitation affected mainly the phenyl polymer structure. This result confirmed the XPS data that showed a decrease in the aromatic contribution of the C 1s peak at the three excitation energies studied. In addition, it was necessary to add a new peak at ~ 288.5 eV to get a satisfactory fitting in the treated spectrum.

The new signal aroused from carbon with a double bond to oxygen, $\pi^*_{C=O}$, which appears located at energies between 286.5 and 289 eV [104, 202] depending on the polymer structure.

Additionally, the A/B ratio of the peaks labels in Figure 31-top decreased from 4.7 (untreated) to 2.0 (treated) when the excitation energy was tuned to 293.5 eV showing the effect of the surface functionalization process.

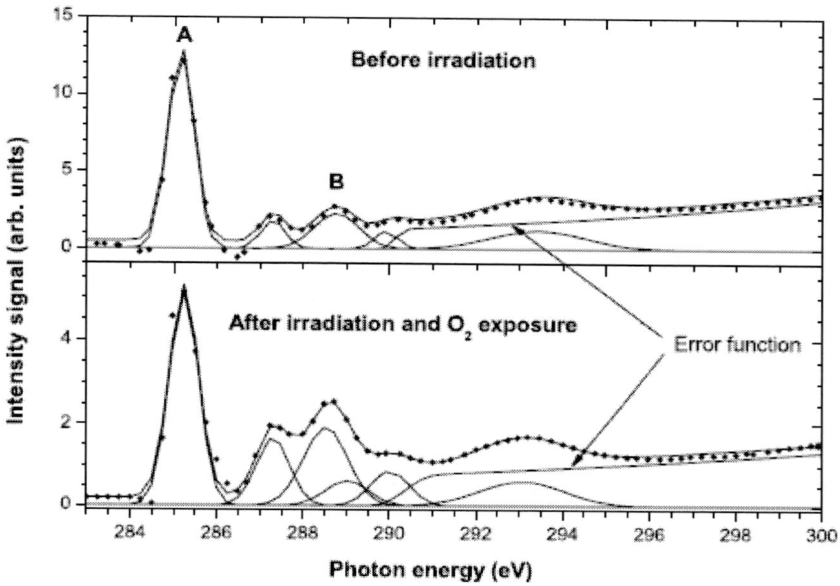

Figure 31. Carbon *K*-edge NEXAFS spectra of untreated (top) and SR treated (bottom) PS films. Ratio A/B: see text for details.

Oxygen *K*-edge NEXAFS spectra clearly revealed the presence of oxygen functionalities in the SR-treated polymer films (see Figure 32). Before irradiation no oxygen signal was detected. However, after treatment two signals corresponding to O 1s $\rightarrow \pi^*_{C=O}$ and O 1s $\rightarrow \sigma^*$ excitations [104, 203] can be seen in Figure 32. Because PS does not has any oxygen contribution, the oxygen *K*-edge signal could be detected with high sensitivity over the background signal demonstrating the presence of new functionalities on the PS surface.

The results were interpreted taking into account previous photon stimulated positive-ions desorption studies of polymers [204-206]. It was shown that core electron excitation to an anti-bonding σ^* state, led to a much more effective bond scission at the site of excitation, compared to π^*. The fragmentation and desorption of some polymer groups was taken as an indication of the higher efficiency of ion production by the σ^* excitation. Therefore, excitation to σ^* states may leave at the surface radical species with probability higher than π^* excitations. Those radicals can remain in UHV conditions to be neutralized efficiently, for example, with oxygen introduction in a last step of the process.

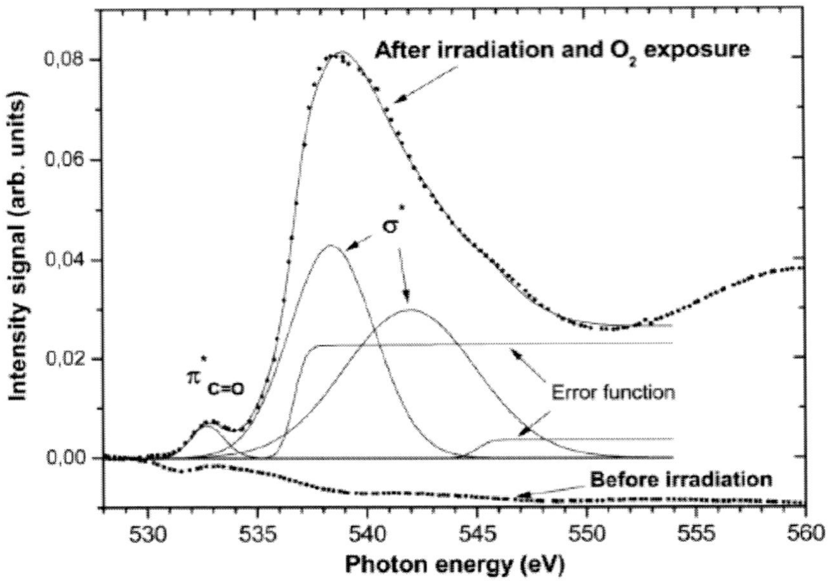

Figure 32. Oxygen *K*-edge NEXAFS spectra of untreated and SR treated PS films.

Conclusion

In this chapter, it was presented an overview of the current state-of the- art in polymer surface modification using electromagnetic radiation as driven technology. The examples described here were mainly limited to the literature published in the last decade. The academic interest in this area is motivated by two reasons: the first one is the advances in the understanding and manipulation of nano and micro structures designs at surfaces toghether with the chemistry involved at the molecular level. The second is associated with the unique potential associated with the use of light to transform and pattern surfaces. The extensive base of knowledge of photochemistry accumulated in the last half century, in particular the advances in surface photochemistry, offers exciting potential for modification of surfaces with control of both the density and nature of surface chemical groups. The availability of new technologies and fabrication infrastructure for both academic and industrial project has played an important role in the rapid growth of the field. The capacity to obtain a detailed chemical and morphological characterization of the treated polymeric surfaces by high surface sensitive techniques has allowed the development of new technological approaches specially using VUV, laser and SR excitation sources.

Also, we addressed the principal factors affecting electromagnetic radiation-induced surface modification of polymer materials. Dry and wet methods of grafting were presented describing a wide number of applications.

Special attention was devoted to new methodologies of radiation driven modification of surfaces, such us the emerging field of VUV-photochemistry that can potentially lead to better-controlled overall reactions through the use of monochromatic radiation. In this sense, it will be interesting to monitor future tends in pushing down the limits of chipmaking technology. VUV (EUV) lasers are promising sources to overcome the wavelentght limit of traditional laser sources and decrease the size of the semiconductor chips below 32 and even 16 nm.

Examples were presented in which UV and VUV processing of polymer surfaces reached similar or better results than plasma treatments in quali and quantitative terms. Synchrotron radiation treatment of polymer presents unique possibilities for basic academic research due to the characteristics and tuneability of this source. It is expected a development of new strategies and applications in selective surface functionalization of polymers in the near future using this powerful excitation source due to new generation of synchrotron radiation facilities planned and in construction around the world. New laser technologies in the femtosecond range will be also available in the near future at lower prices, which will increase the use and application of laser excitation sources.

A great challenge of materials science and technology in the field of surface functionalization of polymers is to control the grafting of specific functional groups at the surface to tune the response of the material to specific stimulos, especially in bioapplications.

Selective surface functionalization as it was shown in particularly examples using VUV or synchrotron radiation sources would allow obtaining a selected chemistry at the molecular level to target a specific final product. This frontier will gather much interest from both scientists and engineers due to its potential applications.

ACKNOWLEDGMENTS

The author work included in the present chapter was partially supported by the Conselho Nacional de Desenvolvimento Científico e Tecnológico (CNPq), Brazil and LNLS – National Synchrotron Light Laboratory, Brazil. The author would also like to acknowledge the technical assistance of the Accelerator Group, especially the VUV and Soft X-ray Spectroscopy Group.

REFERENCES

[1] Stepanov, A. L. *Technical Physics* 2004, 49, 143-153.

[2] Moran-Mirabal, J. M.; Slinker, J. D.; DeFranco, J. A.; Verbridge, S. S.; Ilic, R.; Flores-Torres, S.; Abruna, H.; Malliaras, G. G.; Craighead, H. G. *Nano Letters* 2007, 7, 458-463.

[3] Snaith, H. J.; Moule, A. J.; Klein, C.; Meerholz, K.; Friend, R. H.; Gratzel, M. *Nano Letters* 2007, 7, 3372-3376.

[4] Browne, W. R. *Coordination Chemistry Reviews* 2008, 252, 2470-2479.

[5] Browne, W. R.; Feringa, B. L. *Annu. Rev. Phys. Chem.* 2009, 60, 407-428.

[6] Wesenhagen, P.; Areephong, J.; Fernandez Landaluce, T.; Heureux, N.; Katsonis, N.; Hjelm, J.; Rudolf, P.; Browne, W. R.; Feringa, B. L. *Langmuir* 2008, 24, 6334-6342.

[7] Grassie, N.; Scott, G. Polymer Degradation and Stabilization; Cambridge University Press: New York, 1985.

[8] Guillet, J. Polymer Photophysics and Photochemistry; Cambridge University Press: Cambridge, 1985.

[9] Calvert, J. G.; Pitts, J. N. Photochemistry John Wiley: New York, 1966, 899.
[10] Malanowski, P.; Huijser, S.; van Benthem, R.; van der Ven, L. G. J.; Laven, J.; de With, G. *Polymer Degradation and Stability* 2009, 94, 2086-2094.
[11] Neumann, M. G.; Schmitt, C. C.; Goi, B. E. *Journal of Applied Polymer Science* 2010, 115, 1283-1288.
[12] Tyler, D. R. *Coordination Chemistry Reviews* 2003, 246, 291-303.
[13] Spanring, J.; Buchgraber, C.; Ebel, M. F.; Svagera, R.; Kern, W. *Macromol. Chem. Phys.* 2005, 206, 2248-2256.
[14] Buchgraber, C.; Spanring, J.; Kern, W.; Pogantsch, A. *Macromol. Chem. Phys.* 2005, 206, 2362-2372.
[15] Spanring, J.; Buchgraber, C.; Ebel, M. F.; Svagera, R.; Kern, W. *Polymer* 2006, 47, 156-165.
[16] Deng, J. P.; Wang, L. F.; Liu, L. Y.; Yang, W. T. *Progress in Polymer Science* 2009, 34, 156-193.
[17] Weibel, D. E.; Michels, A. F.; Horowitz, F.; Cavalheiro, R. D. S.; Mota, G. V. d. S. *Thin Solid Films* 2009, 517, 5489–5495.
[18] Daniloska, V.; Blazevska-Gilev, J.; Dimova, V.; Fajgar, R.; Tomovska, R. *Appl. Surf. Sci.* 2010, 256, 2276-2283.
[19] Hudis, M. *J. Appl. Pol. Sci.* 1972, 16, 2397-and.
[20] Ruiz, J. C.; Girard-Lauriault, P. L.; Truica-Marasescu, F.; Wertheimer, M. R. *Radiation Physics and Chemistry* 2010, 79, 310-314.
[21] Kim, Y. J.; Taniguchi, Y.; Murase, K.; Taguchi, Y.; Sugimura, H. *Appl. Surf. Sci.* 2009, 255, 3648-3654.
[22] Truica-Marasescu, F.; Wertheimer, M. R. *Macromol. Chem. Phys.* 2008, 209, 1043-1049.
[23] Sarantopoulou, E.; Kovac, J.; Kollia, Z.; Raptis, I.; Kobe, S.; Cefalas, A. C. *Surf. Interf. Anal.* 2008, 40, 400-403.
[24] Truica-Marasescu, F.; Pham, S.; Wertheimer, M. R. Nucl. Instrum. *Methods Phys. Res. Sect.* B 2007, 265, 31-36.
[25] Raptis, I.; Kovac, J.; Chatzichhstidi, M.; Sarantopoulou, E.; Kollia, Z.; Kobe, S.; Cefalas, A. C. *Journal of Laser Micro Nanoengineering* 2007, 2, 200-205.
[26] Truica-Marasescu, F. E.; Wertheimer, M. R. *Macromol. Chem. Phys.* 2005, 206, 744-757.
[27] Skurat, V. *Nucl. Instrum. Methods Phys. Res. Sect. B* 2003, 208, 27-34.
[28] Wayne, R. P. Chemistry of Atmospheres; Oxford University Press: Oxford, 1992.
[29] Kitsara, M.; Goustouridis, D.; Chatzandroulis, S.; Beltsios, K.; Raptis, L. *Microelectronic Engineering* 2006, 83, 1192-1196.
[30] Sarantopoulou, E.; Kollia, Z.; Cefalas, A. C.; Manoli, K.; Sanopoulou, M.; Goustouridis, D.; Chatzandroulis, S.; Raptis, I. *Appl. Surf. Sci.* 2008, 254, 1710-1719.
[31] Clough, R. L. *Nucl. Instrum. Methods Phys. Res. Sect.* B 2001, 185, 8-33.
[32] Kanda, K.; Kato, Y.; Idetai, T.; Haruyama, Y.; Ishigaki, H.; Matsui, S. Japanese Journal of Applied Physics Part 1-Regular Papers Brief Communications and Review Papers 2005, 44, 3242-3244.
[33] Kanda, K.; Ideta, T.; Haruyama, Y.; Ishigaki, H.; Matsui, S. J. *J. Appl. Phys. Part* 1 2003, 42, 3983-3985.

[34] Kato, Y.; Kanda, K.; Haruyama, Y.; Matsui, S. Japanese Journal of Applied Physics Part 1-Regular Papers Brief Communications and Review Papers 2006, 45, 8839-8841.

[35] Weibel, D. E.; Kessler, F.; da Silva Mota, G. V. *Polymer Chemistry* 2010, 1, 645–649.

[36] Weibel, D. E. *Composite Interfaces* 2010, 17, 127-136.

[37] Liu, X. H.; Ma, P. X. *Annals of Biomedical Engineering* 2004, 32, 477-486.

[38] Schaubroeck, D.; De Baets, J.; Desmet, T.; Dubruel, P.; Schacht, E.; Van Vaeck, L.; Van Calster, A. *Appl. Surf. Sci.* 2010, 256, 6269-6278.

[39] Yaghoubi, H.; Taghavinia, N.; Alamdari, E. K. *Surface and Coatings Technology* 2010, 204, 1562-1568.

[40] Kong, J. S.; Lee, D. J.; Kim, H. D. *J. Appl. Pol. Sci.* 2001, 82, 1677-1690.

[41] Tao, G. L.; Gong, A. J.; Lu, J. J.; Sue, H. J.; Bergbreiter, D. E. *Macromolecules* 2001, 34, 7672-7679.

[42] Glodek, J.; Milka, P.; Krest, I.; Keusgen, M. *Sensors and Actuators B: Chemical* 2002, 83, 82-89.

[43] Goddard, J. M.; Hotchkiss, J. H. *Progress in Polymer Science* 2007, 32, 698-725.

[44] Chan, C. M.; Ko, T. M.; Hiraoka, H. *Surf. Sci. Rep.* 1996, 24, 3-54.

[45] Denes, F. S.; Manolache, S. *Progress in Polymer Science* 2004, 29, 815-885.

[46] Fisher, E. R. *PLASMA PROCESSES AND POLYMERS* 2004, 1, 13-27

[47] Siow, K. S.; Britcher, L.; Kumar, S.; Griesser, H. *J. Plasma Processes and Polymers* 2006, 3, 392-418.

[48] Boswell, R. W.; Henry, D. *Applied Physics Letters* 1985, 47, 1095-1097.

[49] Takahashi, K.; Hori, M.; Goto, T. *Japanese Journal of Applied Physics Part 2-Letters* 1993, 32, L1088-L1091.

[50] Samukawa, S.; Furuoya, S. *Applied Physics Letters* 1993, 63, 2044-2046.

[51] Anders, A. *Surf. Coat. Technol.* 2002, 156, 3-12.

[52] Chu, P. K. *Plasma Physics and Controlled Fusion* 2003, 45, 555-570.

[53] Rangel, E. C.; Bento, W. C. A.; Kayama, M. E.; Schreiner, W. H.; Cruz, N. C. *Surf. Interf. Anal.* 2003, 35, 179-183.

[54] Kostov, K. G.; Ueda, M.; Tan, I. H.; Leite, N. F.; Beloto, A. F.; Gomes, G. F. *Surf. Coat. Technol.* 2004, 186, 287-290.

[55] Shah, J. J.; Geist, J.; Locascio, L. E.; Gaitan, M.; Rao, M. V.; Vreeland, W. N. *Electrophoresis* 2006, 27, 3788-3796.

[56] Najafi, E.; Kim, J. Y.; Han, S. H.; Shin, K. *Colloids and Surfaces a-Physicochemical and Engineering Aspects* 2006, 284, 373-378.

[57] Bhurke, A. S.; Askeland, P. A.; Drzal, L. T. *Journal of Adhesion* 2007, 83, 43-66.

[58] Han, S. J.; Kim, J. H.; Kim, J. W.; Min, C. K.; Hong, S. H.; Kim, D. H.; Baek, K. H.; Kim, G. H.; Do, L. M.; Park, Y. *J.Appl.Phys.* 2008, 104.

[59] Ozcam, A. E.; Efimenko, K.; Jaye, C.; Spontak, R. J.; Fischer, D. A.; Genzer, J. *J. Electron Spectrosc. Relat. Phenom.* 2009, 172, 95-103.

[60] Sham, M. L.; Li, J.; Ma, P. C.; Kim, J. K. *Journal of Composite Materials* 2009, 43, 1537-1564.

[61] Hu, T.; Zhang, F. J.; Xu, Z.; Zhao, S. L.; Yue, X.; Yuan, G. C. *Synthetic Metals* 2009, 159, 754-756.

[62] Ammar, M. R.; Legeay, G.; Bulou, A.; Bardeau, J. F. *Surf. Coat. Technol.* 2009, 203, 2202-2206.

[63] Golczak, S.; Kanciurzewska, A.; Langer, J. J.; Fahlman, M. *Polymer Degradation and Stability* 2009, 94, 350-354.
[64] Teare, D. O. H.; Emmison, N.; Ton-That, C.; Bradley, R. H. *Langmuir* 2000, 16, 2818-2824.
[65] Teare, D. O. H.; Ton-That, C.; Bradley, R. H. *Surf. Interf. Anal.* 2000, 29, 276-283.
[66] Lubarsky, G. V.; Davidson, M. R.; Bradley, R. H. *Surf. Sci.* 2004, 558, 135-144.
[67] Murakami, T. N.; Fukushima, Y.; Hirano, Y.; Tokuoka, Y.; Takahashi, M.; Kawashima, N. *Appl. Surf. Sci.* 2005, 249, 425-432.
[68] Ertl, G.; Kuppers, J. Low Energy Electrons and Surface Chemistry V C H Publishers, Incorporated: Weimheim, 1985, 374.
[69] Chan, C.-M. Polymer surface modification and characterization; Hanser Publishers: Cincinnati, 1993, 285.
[70] Briggs, D.; Seach, M. P. Practical Surface Analysis. Volume 1. Auger and X-ray Photoelectron Spectroscopy; John Wiley and Sons, Chichester, England, 1996.
[71] Benninghoven, A.; Rudenauer, F. G.; Werner, H. W. Secondary Ion Mass Spectrometry: Basic Concepts, Instrumental Aspects, Applications and Trends (Chemical Analysis); John Wiley and Sons New York, 1987, 1264.
[72] Vickerman, J. C.; Briggs, D. ToF-SIMS: Surface Analysis by Mass Spectrometry; IMP Publications and SurfaceSpectra Limited: Manchester and Chichester, 2001, 1-789.
[73] Neinhuis, C.; Barthlott, W. *Annals of Botany* 1997, 79, 667-677.
[74] Sun, T. L.; Feng, L.; Gao, X. F.; Jiang, L. *Accounts Chem. Res.* 2005, 38, 644-652.
[75] Lee, W.; Jin, M. K.; Yoo, W. C.; Lee, J. K. *Langmuir* 2004, 20, 7665-7669.
[76] Gao, X. F.; Jiang, L. *Nature* 2004, 432, 36-36.
[77] Swain, P. S.; Lipowsky, R. *Langmuir* 1998, 14, 6772-6780.
[78] Werner, O.; Wagsberg, L.; Lindstrom, T. *Langmuir* 2005, 21, 12235-12243.
[79] Nishino, T.; Meguro, M.; Nakamae, K.; Matsushita, M.; Ueda, Y. *Langmuir* 1999, 15, 4321-4323.
[80] Leger, L.; Joanny, J. F. *Rep. Prog. Phys.* 1992, 55, 431-486.
[81] Barbieri, L.; Wagner, E.; Hoffmann, P. *Langmuir* 2007, 23, 1723-1734.
[82] Li, X.-M.; Reinhoudt, D.; Crego-Calama, M. *Chem. Soc. Rev* 2007, 36, 1350–1368.
[83] Kwon, Y.; Patankar, N.; Choi, J.; Lee, J. *Langmuir* 2009, 25, 6129-6136.
[84] Balaur, E.; Macak, J. M.; Tsuchiya, H.; Schmuki, P. *Journal of Materials Chemistry* 2005, 15, 4488-4491.
[85] Bico, J.; Marzolin, C.; Quere, D. *Europhysics Letters* 1999, 47, 220-226.
[86] Shibuichi, S.; Onda, T.; Satoh, N.; Tsujii, K. *Journal of Physical Chemistry* 1996, 100, 19512-19517.
[87] Wenzel, R. N. *Ind. Eng. Chem.* 1936, 28, 988.
[88] Cassie, A. B. D.; Baxter, S. *Trans. Faraday Soc.* 1944, 40, 546.
[89] Stöhr, J. NEXAFS Spectroscopy Springer-Verlag: Berlin Heidelberg New York, 2003, 403.
[90] Weibel, D. E.; Vilani, C.; Habert, A. C.; Achete, C. A. *Journal of Membrane Science* 2007, 293, 124-132.
[91] Haidopoulos, M.; Horgnies, M.; Mirabella, F.; Pireaux, J. *J. Plasma Processes and Polymers* 2008, 5, 67-75.
[92] Cumpson, P. J. *J. Electron Spectrosc. Relat. Phenom.* 1995, 73, 25-52.
[93] Ueda, K.; Itoh, Y. *Surf. Sci.* 2006, 600, 1684-1688.

[94] Baer, D. R.; Engelhard, M. H.; Lea, A. S. *Surf. Sci. Spectra* 2003, 10, 47-56.
[95] Unger, W. E. S.; Lippitz, A.; Woll, C.; Heckmann, W. Fresenius *J. Anal. Chem.* 1997, 358, 89-92.
[96] Otero, E.; Shipman, P. O.; Abd-El-Aziz, A. S.; Urquhart, S. G. *Macromolecules* 2008, 41, 9532-9541.
[97] Cooney, R. R.; Urquhart, S. G. *J. Phys. Chem. B* 2004, 108, 18185-18191.
[98] Urquhart, S. G.; Hitchcock, A. P.; Smith, A. P.; Ade, H. W.; Lidy, W.; Rightor, E. G.; Mitchell, G. E. *J. Electron Spectrosc. Relat. Phenom.* 1999, 100, 119-135.
[99] Cavalleri, A.; Rini, M.; Chong, H. H. W.; Fourmaux, S.; Glover, T. E.; Heimann, P. A.; Kieffer, J. C.; Schoenlein, R. W. *Physical Review Letters* 2005, 95, 067405.
[100] Urquhart, S. G.; Smith, A. P.; Ade, H. W.; Hitchcock, A. P.; Rightor, E. G.; Lidy, W. *J. Phys. Chem. B* 1999, 103, 4603-4610.
[101] Endo, O.; Sumii, R.; Amemiya, K.; Osaki, H. In Photon Factory Activity Report 2008 #26 Part B: Tsukuba, 2008, p Pa
[102] Hofmann, S. *Appl. Surf. Sci.* 2005, 241, 113-121.
[103] Outka, D. A.; Stohr, J. *J. Chem. Phys.* 1988, 88, 3539-3554.
[104] Klein, R. J.; Fischer, D. A.; Lenhart, J. L. *Langmuir* 2008, 24, 8187-8197.
[105] Swaraj, S.; Oran, U.; Lippitz, A.; Friedrich, J. F.; Unger, W. E. S. *Surf. Coat. Technol.* 2005, 200, 494-497.
[106] Stohr, J.; Jaeger, R. *Physical Review* B 1982, 26, 4111-4131.
[107] Nagayama, K.; Sei, M.; Mitsumoto, R.; Ito, E.; Araki, T.; Ishii, H.; Ouchi, Y.; Seki, K.; Kondo, K. *J. Electron Spectrosc. Relat. Phenom.* 1996, 78, 375-378.
[108] Gamble, L. J.; Ravel, B.; Fischer, D. A.; Castner, D. G. *Langmuir* 2002, 18, 2183-2189.
[109] Chua, L. L.; Dipankar, M.; Sivaramakrishnan, S.; Gao, X. Y.; Qi, D. C.; Wee, A. T. S.; Ho, P. K. H. *Langmuir* 2006, 22, 8587-8594.
[110] Vickerman, J. C.; Briggs, D.; Henderson, A. The Static SIMS Library; SurfaceSpectra Ltd: Manchester, 2002.
[111] Van Vaeck, L.; Adriaens, A.; Gijbels, R. *Mass Spectrometry Reviews* 1999, 18, 1-47.
[112] Adriaens, A.; Van Vaeck, L.; Adams, F. *Mass Spectrometry Reviews* 1999, 18, 48-81.
[113] McDonnell, L. A.; Heeren, R. M. A. *Mass Spectrometry Reviews* 2007, 26, 606-643.
[114] Vickerman, J. C. *Surf. Sci.* 2009, 603, 1926-1936.
[115] Winograd, N.; Garrison, B. *J. In Annual Review of Physical Chemistry*, Vol 61, 2010, p 305-322.
[116] Weibel, D.; Wong, S.; Lockyer, N.; Blenkinsopp, P.; Hill, R.; Vickerman, *J. C. Anal. Chem.* 2003, 75, 1754-1764.
[117] Winograd, N.; Postawa, Z.; Cheng, J.; Szakal, C.; Kozole, J.; Garrison, B. *J. Appl. Surf. Sci.* 2006, 252, 6836-6843.
[118] Delcorte, A. Appl. Surf. Sci. 2008, 255, 954-958.
[119] Wucher, A.; Cheng, J.; Zheng, L. L.; Willingham, D.; Winograd, N. *Appl. Surf. Sci.* 2008, 255, 984-986.
[120] Weibel, D. E.; Lockyer, N.; Vickerman, J. C. *Applied Surface Science* 2004, 231-232, 146-152.
[121] Wiley, W. C.; McLaren, I. H. Rev. *Sci. Instrum.* 1955, 26, 1150-1157.
[122] Eccles, A. J.; Steele, T. A. *Int. J. Adhes. Adhes.* 2001, 21, 281-286.
[123] Delcorte, A.; Bertrand, P. *Surf. Sci.* 1998, 413, 97-124.
[124] Bertrand, P.; Delcorte, A.; Garrison, B. *J. Appl. Surf. Sci.* 2003, 203, 160-165.

[125] Briggs, D. Surface Analysis of Polymers by XPS and Static SIMS Cambridge Univ Pr: Cambridge, 2005, 198.
[126] Chilkoti, A.; Ratner, B. D.; Briggs, D. *Chem. Mater.* 1991, 3, 51-61.
[127] Voronin, S. A.; Zelzer, M.; Fotea, C.; Alexander, M. R.; Bradley, J. W. *J. Phys. Chem. B* 2007, 111, 3419-3429.
[128] Voronin, S. A.; Alexander, M. R.; Bradley, *J. W. Surf. Coat. Technol.* 2006, 201, 768-775.
[129] Dhayal, M.; Alexander, M. R.; Bradley, J. W. *Appl. Surf. Sci.* 2006, 252, 7957-7963.
[130] Thierry, B.; Jasieniak, M.; de Smet, L.; Vasilev, K.; Griesser, H. J. *Langmuir* 2008, 24, 10187-10195.
[131] Winograd, N. *Anal. Chem.* 2005, 77, 142A-149A.
[132] Sostarecz, A. G.; McQuaw, C. M.; Wucher, A.; Winograd, N. *Anal. Chem.* 2004, 76, 6651-6658.
[133] Wong, S. C. C.; Hill, R.; Blenkinsopp, P.; Lockyer, N. P.; Weibel, D. E.; Vickerman, J. C. *Applied Surface Science* 2003, 203, 219-222.
[134] Czerwinski, B.; Samson, R.; Garrison, B. J.; Winograd, N.; Postawa, Z. *Vacuum* 2006, 81, 167-173.
[135] Fletcher, J. S.; Lockyer, N. P.; Vaidyanathan, S.; Vickerman, J. C. *Anal. Chem.* 2007, 79, 2199-2206.
[136] Wucher, A.; Cheng, J.; Zheng, L.; Winograd, N. *Anal. Bioanal. Chem.* 2009, 393, 1835-1842.
[137] Hind, A. R.; Bhargava, S. K.; McKinnon, A. *Advances in Colloid and Interface Science* 2001, 93, 91-114.
[138] Liu, P. *Applied Clay Science* 2007, 38, 64-76.
[139] Lopez-Gejo, J.; Manoj, N.; Sumalekshmy, S.; Glieman, H.; Schimmel, T.; Worner, M.; Braun, A. M. *Photochem. Photobiol. Sci.* 2006, 5, 948 - 954.
[140] Rajajeyaganthan, R.; Kessler, F.; Leal, P. H. d. M.; Kühn, S.; Weibel, D. E. Macromolecular Symposium 2010, In press.
[141] Casserly, T. B.; Gleason, K. K. *Chemical Vapor Deposition* 2006, 12, 59-66.
[142] Jalili, N.; Laxminarayana, K. *Mechatronics* 2004, 14, 907-945.
[143] Zheludkevich, M. L.; Serra, R.; Grundmeier, G.; Yang, L. H.; Ferreira, M. G. S. *Surf. Coat. Technol.* 2006, 200, 4040-4049.
[144] Muir, B. W.; McArthur, S. L.; Thissen, H.; Simon, G. P.; Griesser, H. J.; Castner, D. G. *Surf. Interf. Anal.* 2006, 38, 1186-1197.
[145] Grythe, K. F.; Hansen, F. K. *Langmuir* 2006, 22, 6109-6124.
[146] Ensikat, H. J.; Schulte, A. J.; Koch, K.; Barthlott, W. *Langmuir* 2009, 25, 13077-13083.
[147] Bozzola, J. J.; Russell, L. D. Electron Microscopy; Jones and Bartlett Publlishers 1999, 688 pp.
[148] Park, E. J.; Carroll, G. T.; Turro, N. J.; Koberstein, J. T. *Soft Matter* 2009, 5, 36-50.
[149] Zhao, B.; Brittain, W. J. *Progress in Polymer Science* 2000, 25, 677-710.
[150] Kogelschatz, U. *Plasma Chemistry and Plasma Processing* 2003, 23, 1-46.
[151] Allen, N. S.; Rabek, J. F. New trends in the photochemistry of polymers Elsevier Applied Science Pub.: London, 1985, 321.
[152] Guillet, J. Polymer photophysics and photochemistry : an introduction to the study of photoprocesses in macromolecules Cambridge University Press: Cambridge, 1985, 391.
[153] Ghiggino, K. P.; Smith, T. A. *Progress in Reaction Kinetics* 1993, 18, 375-436.

[154] Allen, N. S. Photopolymerisation and photoimaging : science and technology Elsevier: London, 1989, 310.

[155] Turro, J. N. Modern Molecular Photochemistry; The Benjamin/Cummings Publilshing Company, Inc.: California, 1978, 628.

[156] Millan, M. D.; Locklin, J.; Fulghum, T.; Baba, A.; Advincula, R. C. *Polymer* 2005, 46, 5556-5568.

[157] Costamagna, V.; Strumia, M.; Lopez-Gonzalez, M.; Riande, E. *J Polymer Sci. B Polymer Phys.* 2007, 45, 2421-2431.

[158] Hu, M. X.; Yang, Q.; Xu, Z. K. *J. Membr. Sci.* 2006, 285, 196-205.

[159] Asano, M.; Chen, J.; Maekawa, Y.; Sakamura, T.; Kubota, H.; Yoshida, M. *J. Polym. Sci. Pol. Chem.* 2007, 45, 2624-2637.

[160] Costamagna, V.; Strumia, M.; Lopez-Gonzalez, M.; Riande, E. *J Polymer Sci. B Polymer Phys.* 2006, 44, 2828-2840.

[161] Wang, H. L.; Brown, H. R. J. Polym. *Sci. Pol. Chem.* 2004, 42, 253-262.

[162] Colavita, P. E.; Sun, B.; Tse, K. Y.; Hamers, R. J. *J. Vac. Sci. Technol. A* 2008, 26, 925-931.

[163] Buchgraber, C.; Svagera, R.; Ebel, M. F.; Schrottner, H.; Kern, W. *Macromol. Chem. Phys.* 2007, 208, 1159-1167.

[164] Bormashenko, E.; Pogreb, R.; Whyman, G.; Bormashenko, Y.; Jager, R.; Stein, T.; Schechter, A.; Aurbach, D. *Langmuir* 2008, 24, 5977-5980.

[165] Choi, H. S.; Kim, Y. S.; Zhang, Y.; Tang, S.; Myung, S. W.; Shin, B. C. *Surf. Coat. Technol.* 2004, 182, 55-64.

[166] Vilani, C.; Weibel, D. E.; Zamora, R. R. M.; Habert, A. C.; Achete, C. A. *Appl. Surf. Sci.* 2007, 254, 131–134.

[167] Ade, H.; Smith, A. P.; Zhang, H.; Zhuang, G. R.; Kirz, J.; Rightor, E.; Hitchcock, A. *J. Electron Spectrosc. Relat. Phenom.* 1997, 84, 53-72.

[168] Coffey, T.; Urquhart, S. G.; Ade, H. *J. Electron Spectrosc. Relat. Phenom.* 2002, 122, 65-78.

[169] Stark, G.; Smith, P. L. In Atomic, Molecular and Optical Physics Handbook; G. W. F. Drake, Ed.; AIP Press: New York, 2006, p 642-652.

[170] Bartnik, A.; Fiedorowicz, H.; Jarocki, R.; Kostecki, J.; Szczurek, M.; Bilinski, A.; Chernyayeva, O.; Sobczak, J. W. *Applied Physics a-Materials Science and Processing* 2010, 99, 831-836.

[171] Tagawa, M.; Yokota, K.; Kishida, K.; Okamoto, A.; Ishizawa, J.; Minton, T. K. *High Performance Polymers* 2010, 22, 213-224.

[172] Dhayal, M.; Parry, K. L.; Short, R. A.; Bradley, J. W. *J. Phys. Chem.* B 2004, 108, 14000-14004.

[173] Elsner, C.; Naumov, S.; Zajadacz, J.; Buchmeiser, M. R. *Thin Solid Films* 2009, 517, 6772-6776.

[174] Bruce, R. L.; Engelmann, S.; Lin, T.; Kwon, T.; Phaneuf, R. J.; Oehrlein, G. S.; Long, B. K.; Willson, C. G.; Vegh, J. J.; Nest, D.; Graves, D. B.; Alizadeh, A. *J.Vac.Sci.and Techonol.B* 2009, 27, 1142-1155.

[175] Lopez-Gejo, J.; Gliemann, H.; Schimmel, T.; Braun, A. M. *Photochem. Photobiol.* 2005, 81, 777-782.

[176] Wilken, R.; Hollander, A.; Behnisch, *J. Plasmas and Polymers* 2002, 7, 19-39.

[177] Wu, Y. J.; Lu, H. C.; Chen, H. K.; Cheng, B. M.; Lee, Y. P.; Lee, L. C. *Journal of Chemical Physics* 2007, 127.

[178] Cheng, B. M.; Lu, H. C.; Chen, H. K.; Bahou, M.; Lee, Y. P.; Mebel, A. M.; Lee, L. C.; Liang, M. C.; Yung, Y. L. *Astrophysical Journal* 2006, 647, 1535-1542.

[179] Chen, F. Z.; Judge, D. L.; Wu, C. Y. R.; Caldwell, *J. Planetary and Space Science* 1999, 47, 261-266.

[180] Liu, H. C.; Chen, H. K.; Cheng, B. M. *Analytical Chemistry* 2004, 76, 5965-5967.

[181] Parkinson, W. H.; Rufus, J.; Yoshino, K. *Chemical Physics* 2003, 290, 251-256.

[182] Dattoli, G.; Renieri, A. Nuclear Instruments and Methods in Physics Research Section a-Accelerators Spectrometers Detectors and Associated Equipment 2003, 507, 464-469.

[183] Leach, S.; Schwell, M.; Jochims, H. W.; Baumgartel, H. *Chemical Physics* 2006, 321, 171-182.

[184] Eden, S.; Limao-Vieira, P.; Hoffmann, S. V.; Mason, N. *J. Chemical Physics* 2006, 323, 313-333.

[185] Nelea, V.; Vasilets, V. N.; Skurat, V. E.; Truica-Marasescu, F.; Wertheimer, M. R. *Plasma Processes and Polymers* 2010, 7, 431-444.

[186] Truica-Marasescu, F.; Guimond, S.; Wertheimer, M. R. *Nucl. Instrum. Methods Phys. Res. Sect. B* 2003, 208, 294-299.

[187] Okabe, H. Photochemistry of Small Molecules; John Wiley and Sons: New York 1978, 431.

[188] Rajajeyaganthan, R.`; Kessler, F. and Weibel, D. E.`; manuscript in preparation.

[189] Truica-Marasescu, F.; Guimond, S.; Jedrzejowski, P.; Wertheimer, M. R. *Nucl. Instrum. Methods Phys. Res. Sect. B* 2005, 236, 117-122.

[190] Malek, C. G. K. *Anal. Bioanal. Chem.* 2006, 385, 1362-1369.

[191] Malek, C. G. K. *Anal. Bioanal. Chem.* 2006, 385, 1351-1361.

[192] Yingling, Y. G.; Garrison, B. *J. Nucl. Instrum. Methods Phys. Res. Sect. B* 2003, 202, 188-194.

[193] Becker, H.; Gartner, C. *Anal. Bioanal. Chem.* 2008, 390, 89-111.

[194] Pfleging, W.; Bruns, M.; Welle, A.; Wilson, S. *Appl. Surf. Sci.* 2007, 253, 9177-9184.

[195] Jasik, J.; Heitz, J.; Pedarnig, J. D.; Veis, P. *Spectroc. Acta Pt. B-Atom. Spectr.* 2009, 64, 1128-1134.

[196] John, S. R.; Langford, S. C.; Dickinson, J. T. *Applied Physics a-Materials Science and Processing* 2008, 92, 981-985.

[197] Hanada, Y.; Sugioka, K.; Kawano, H.; Tsuchimoto, T.; Miyamoto, I.; Miyawaki, A.; Midorikawa, K. *Appl. Surf. Sci.* 2009, 255, 9885-9888.

[198] Ukita, Y.; Kishihara, M.; Kanda, K.; Matsui, S.; Mochiji, K.; Utsumi, Y. *Japanese Journal of Applied Physics* 2008, 47, 337-341.

[199] Ukita, Y.; Kanda, K.; Matsui, S.; Kishihara, M.; Utsumi, Y. Microsystem Technologies-Micro-and Nanosystems-Information Storage and Processing Systems 2008, 14, 1567-1572.

[200] Kato, Y.; Kanda, K.; Haruyama, Y.; Matsui, S. *Radiation Physics and Chemistry* 2006, 75, 2049-2053.

[201] Kato, Y.; Kanda, K.; Haruyama, Y.; Matsui, S. *Japanese Journal of Applied Physics* Part 1-Regular Papers Brief Communications and Review Papers 2006, 45, 7978-7979.

[202] Dhez, O.; Ade, H.; Urquhart, S. G. *J. Electron Spectrosc. Relat. Phenom.* 2003, 128, 85-96.

[203] Koprinarov, I.; Lippitz, A.; Friedrich, J. F.; Unger, W. E. S.; Woll, C. *Polymer* 1998, 39, 3001-3009.
[204] Tanaka, K.; Tinone, M. C. K.; Ikeura, H.; Sekiguchi, T.; Sekitani, T. *Rev. Sci. Instrum.* 1995, 66, 1474-1476.
[205] Sekiguchi, H. I.; Sekiguchi, T.; Tanaka, K. *Phys. Rev. B* 1996, 53, 12655-12658.
[206] Ueno, N.; Tanaka, K. *Jpn. J. Appl. Phys. Part 1* 1997, 36, 7605-7610.

In: Polymer Films: Properties, Performance and Applications ISBN 978-1-61209-201-0
Editors: Stefan A. Romano and George P. Somners

Chapter 3

Surface Modification of Cation Exchange Membranes by Incorporation of Cationic Polymer Films for Improvement of Proton Selectivity

Xuan Tuan Le**, Pascal Viel and Serge Palacin
Laboratory of Chemistry of Surfaces and Interfaces,
CEA Saclay, IRAMIS /SPCSI, F,
Gif-sur-Yvette Cedex, France

Abstract

One of the most important applications of electrodialysis using ion exchange membranes is the recovery of strong acids from effluents of hydrometallurgy or electroplating industries. Rinse streams of these processes are generally concentrated equally in multivalent metal ions. All these different industrial applications of electrodialysis, in order to be feasible, require the use of some specific cation exchange membranes which are preferentially selective to monovalent cations. It was recognized that depositing a thin polyamine layer on the membrane surface which becomes positively charged in contact with a mixed solution of acid and metallic multivalent salts, thus creates an electrical repulsion barrier towards the highly charged and large hydrated ions is a well-known technique for enhancement of proton selectivity. This chapter describes the current state of the art in our understanding of surface modification routes involving plasma process, electrochemical and chemical methods to obtain the proton permselective membranes. We also report our efficient and one-step chemical process through the diazonium chemistry which can be considered within the field and provide some directions for future endeavours. Evidently, proton selectivity of the modified membranes as well as the important parameters with respect to the electrodialytic efficiency such as electrical resistance, ion exchange capacity, and long-term stability will be discussed herein.

* Fax: +33 1 69 08 64 62, E-mail: xuan.tuan.le@ulb.ac.be ; pascal.viel@cea.fr ; serge.palacin@cea.fr

Keywords: monovalent selectivity, surface modification, plasma process, electrochemical and chemical methods, electrodialysis, durability, membrane conductivity.

1. Ion Exchange Membranes and Electrodialysis

Ion exchange membranes, by common definition, are insoluble solid materials which carry exchangeable cations or anions. These ions can be exchanged for a stoichiometrically equivalent amount of other ions of the same sign when the membrane is in contact with an electrolyte solution [1]. There are two different types of ion exchange membranes: 1) cation exchange membranes (CEM) which contain negatively charged groups fixed to the polymer matrix and 2) anion exchange membranes (AEM) which contain positively charged groups fixed to the polymer matrix. In a cation exchange membrane, the fixed negative charges are in electrical equilibrium with mobile cations in the interstices of the polymer as indicated in Figure 1 which shows schematically the matrix of a CEM with fixed anions and mobile cations; the latter are referred to as counter-ions. Mobile anions are referred to as co-ions. They are more or less completely excluded from the polymer matrix (Donnan exclusion) because of their electrical charge which is identical to that of the fixed ionic groups. Due to the exclusion of the co-ions, a cation exchange membrane merely permits transfer of cations while an anion exchange membrane is permeable to anions only [2].

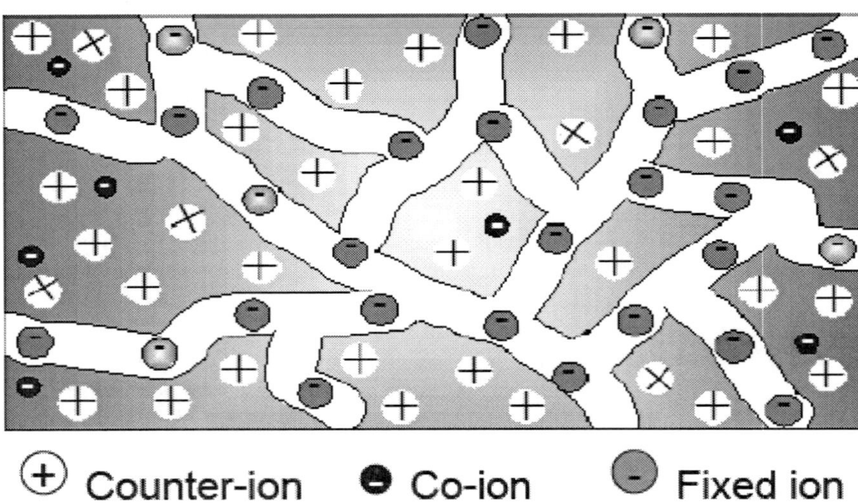

Figure 1. Schematic drawing illustrating the structure of a cation exchange membrane, after Ref.[2].

When cation and anion exchange membranes are alternatively placed between a cathode and an anode (Figure 2a), the ionic species are transported through the electro-membranes thanks to the action of electrical current. The cations which are able to pass the cation exchange membrane are blocked by the anion exchange membranes and reciprocally. This results an increasing concentration of HX acid in the compartments of the same parity (concentrate) and a decrease in the adjacent ones (diluate). The recovery of strong acids from effluents of hydrometallurgy or electroplating industries whose principle is presented in

Figure 2a is one of the most important applications of electrodialysis using ion exchange membranes. In an industrial-size electrodialysis stack 100 to 200 cell pairs are arranged between the electrodes. Various spacers and stack constructions such as the so-called sheet flow or tortuous path flow stack design are used in practical applications. Details concerning the components in an electrodialysis stack are found in Figure 2b.

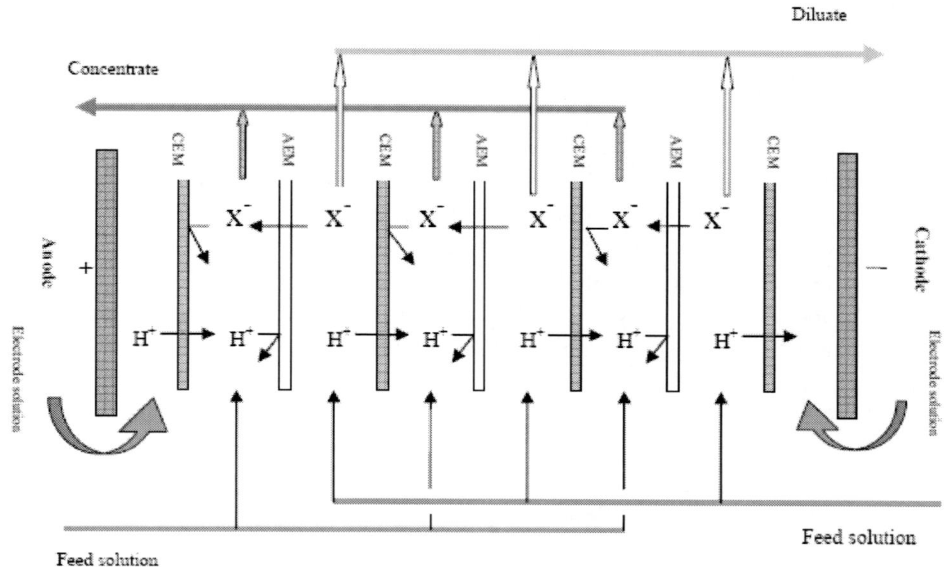

Figure 2a. Principle of electrodialysis for recovery of HX acid (three compartments system made from 4 cation exchange membranes and 3 anion exchange membranes).

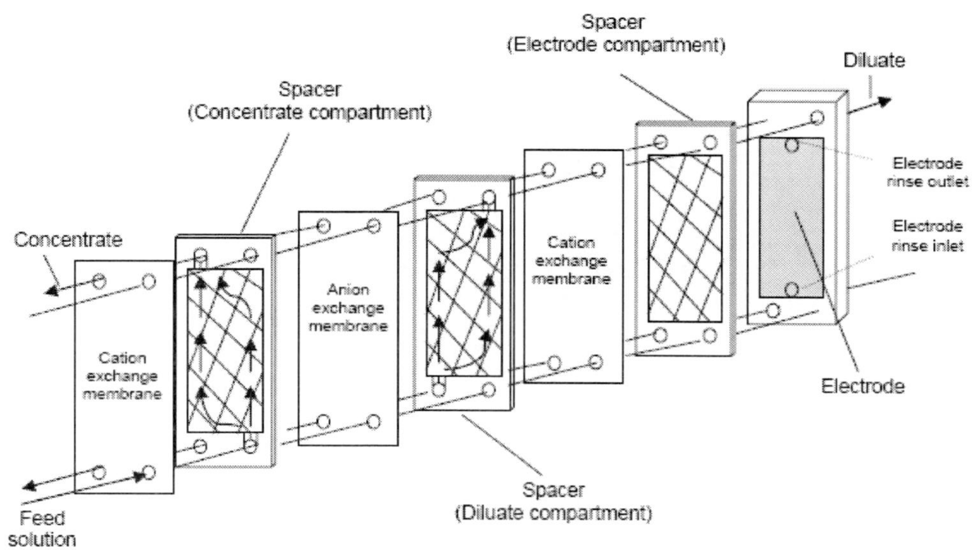

Figure 2b. Components in an electrodialysis stack.

However, rinse streams of metal finishing processes are usually concentrated in bivalent metal ions and frequently in strong acids. Disposals of these streams pose therefore important pollution problems. Some specific metal recovery methods are thus required to process these metal-laden rinse solutions and the use of electrodialysis is often considered as an appropriate technique. The application of electrodialysis may also enable to control and improve the quality of electroplating baths. The benefit of this methodology, with respect to the composition of these baths, is to reduce the need for their frequent replenishment. The recovery of acid from spent pickling baths by electrodialysis constitutes another procedure of considerable interest for the treatment of waste effluents. All these different industrial applications of electrodialysis, in order to be feasible, require the use of some specific cation exchanges membranes which are preferentially selective to monovalent cations [3].

It was reported that simple immersion of the cation exchange membranes in a surface active agent solution or in a cationic polyelectrolyte solution which are able to ion-exchange and/or adsorb on the surface leads to a monovalent permselective membrane [4-11]. Some cationic group should be converted to polysalt by reaction with sulfonic groups on the membrane surface and other cationic groups should be effective as a barrier for permeation of multivalent cation because of its higher electrostatic repulsion in comparison with that of the monovalent cations (Figure 3).

Similarly, depositing a thin polyamine layer (polyethyleneimine, polyanilin, polypyrrole,...) on the membrane surface which becomes positively charged in contact with a mixed solution of acid and metallic multivalent salts, thus creates an electrical repulsion barrier towards the highly charged ions is a well-known technique for enhancement of proton selectivity.

Because the traditional immersion techniques did not provide sufficient stable modified layers on the membrane surfaces, various alternative possibilities have been described during the past 20 years. This chapter covers the most significant studies on surface modification routes involving plasma process, electrochemical and chemical methods to obtain proton permselective membranes.

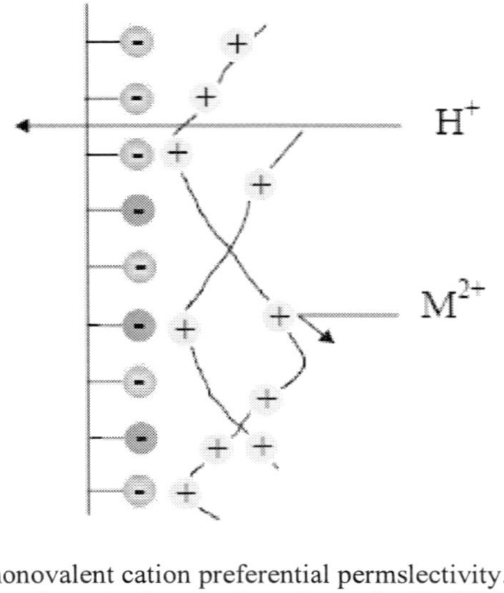

Figure 3. Mechanism of the monovalent cation preferential permslectivity. CEM modified by adsorption and/or ion exchange of a cationic polyelectrolyte, after Ref [11].

In this case, it is important to note that all surface modification techniques have to overcome a difficult problem: the preparation of excellent ion exchange membranes with both small increase in electrical resistance and high preferential permselectivity for proton ions. Besides discussing the proton selectivity of the modified membranes, we therefore focus on the important parameters that describe the actual electrodialytic efficiency such as electrical resistance, ion exchange capacity, and long-term stability to bring out a comparison between the mentioned methods.

2. Surface Modification by Plasma Proces

One of the first works dealing with using plasma technique for surface modification of cation exchange membrane in order to improve the proton selectivity was reported by Ogumi et al.[12-14] in the years of 1990. Plasma modification is a useful method for forming a thin uniform layer which tightly adheres to various substrates. Plasma polymers are known to be generally cross-linked and to have high chemical and thermal stability. Otherwise, by controlling the thickness of the modified layer, it is expected to keep the resistance of the plasma modified membrane relatively low. For these reasons, plasma polymerization was selected for forming a thin layer of anion exchanger on the surface of Nafion 117 membrane by using 4-vinyl pyridine as monomer. The resulting plasma polymer layers were found to be pinhole-free and of uniform thickness (~ 0.2 μm) which is indeed relatively thin compared to the thickness of the Nafion 117 (about 200 μm). The selective proton transport model is schematically described in Figure 4, using the Fe^{2+}-H^{+}-Cl^{-} system.

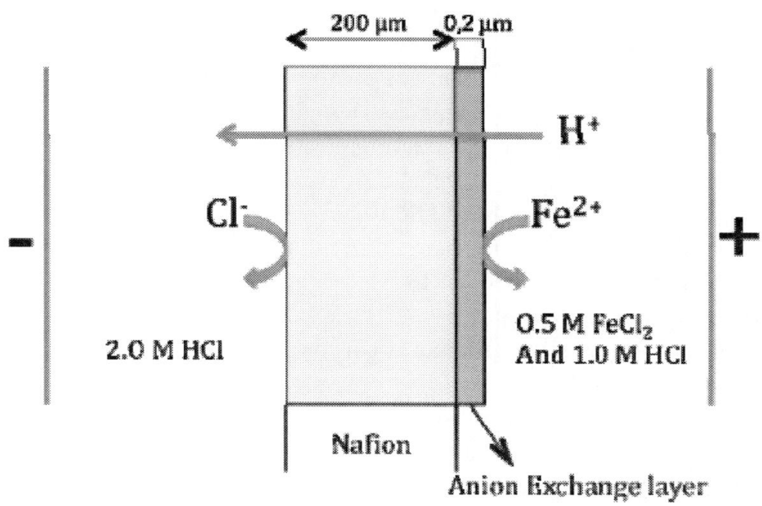

Figure 4. Principle of the enhancement of H^{+} ion permselectivity through a cation exchange membrane, after Ref.[13].

When a potential field is applied as shown in the figure, Cl^{-} cannot be transported through the membrane due to the Donnan exclusion of the cation exchange membrane. The transport of H^{+} and Fe^{2+} ions is also suppressed due to the repulsion by the fixed cation groups in the thin modified layer. However, when the ion exchange capacity of that anion exchange layer is negligible with respect to that of the cation exchange membrane, some cations may be transported through the thin layer. Since the electrostatic repulsion for protons from the fixed cations is weaker than that for Fe^{2+}, protons can transport more easily.

A total charge of 100C was passed at a constant current of 10 mA (12.8 mA/cm^2) through the system. After the electrolysis, the total amount of Fe^{2+} in the catholyte was measured by absorption spectrometry to evaluate the amount of Fe^{2+} transported through the plasma-modified Nafion. Values of t_{Fe} were calculated from the amount of Fe^{2+} transported and the total charge passed during the electrolysis. As expected, when the Nation was plasma modified in the in-glow region, the resulting plasma-modified membranes exhibited very low values of Fe^{2+} ion transference number ($t_{Fe} \sim 0.0003$); however, the membrane resistances were unacceptably high (30≈40 Ωcm^2). These values are about ten times higher than area resistances of the Nafion 117 which normally fall within a quite small range of a few Ωcm^2. The observed behaviour can be explained by noting that the substrate was set in the in-glow region. Since both the substrate and the polymer surface were subjected to repeated attack by the activated species in the in-glow region, it can be inferred that functional groups in the Nation surface and in the deposited polymer layer were destroyed and that the plasma polymer was highly cross-linked, thereby increasing the resistance of the membrane. The considerable increase in membrane resistance brought about by oxygen sputtering and plasma modification is, unfortunately, a problem that still remains to be solved.

Research efforts had therefore been focused on the resistance of plasma-modified membranes [15,16]. Indeed, by controlling the parameters of the plasma enhanced chemical vapor deposition (PECVD) such as: monomer nature and flux, input power... one could vary the structure and chemical composition of plasma polymers to a large extent and so adjust them according to the required selective properties [16]. The authors succeeded in depositing an ultra-thin layer of polyethyleneimine-like on the surface of the Nafion® and sulfonated polyimide membranes by introducing a gas mixture of ethylene and ammonia into the PECVD reactor. Despite its very low thickness (less than 10 nm), this layer strongly acted on the transport properties of the membranes during electrodialysis process for recovery of acids from effluents containing copper salts. Depending on the PECVD conditions performed, $t_{Cu}{}^{2+}$ decreases from 13% to 74% for the Nafion® membrane and from 18% to 59% for the sulfonated polyimide membrane. In particular, although the conductivities of the two modified membranes were decreased up to 62 % and 76 % respectively, they still remain satisfactory for an electrodialysis process. It was also reported in this work that PECVD technique appeared to be more efficient than the conventional electrodeposition of a positively charged polyelectrolyte.

It has clearly been shown that plasma modified membranes are a good candidate for electrodialysis processes. However, it should be kept in mind that plasma modification requires a complex instrumental set-up consisting of a tubular glass reactor equipped 2 plate electrodes, at least 2 gas inlets, a vacuum pump, a pressure gauge, a power supply... as shown in Figure 5. This technique has therefore no practical advantage when the sample dimensions or forms make an approach difficult or when the instruments are not available on the other hand. We will focus next on the conventional electrodeposition method, in which the experimental procedures are relatively simple.

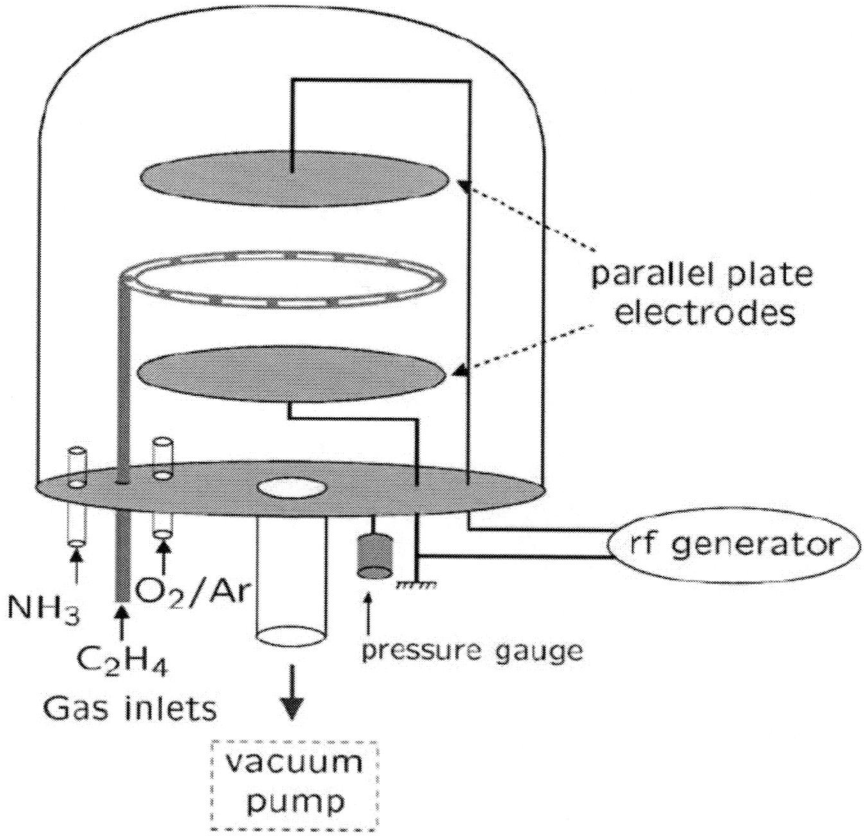

Figure 5. Schematic of rf plasma deposition system, reproduced from Ref.[16] with permission of Elsevier.

3. Electrodeposition of Cationic Polyelectrolytes

Among the modification techniques, the simplest consists in immersing the CEM in an aqueous polyelectrolyte solution bearing positive fixed charges in order to create an anion-exchange layer on the surface of the CEM. However, the detachment of the polyelectrolyte layers was observed in certain cases during electrodialysis processes which function for a long-time. Moreover, the polyelectrolyte layers formed through the immersion method did not cover completely the membrane surface. It would be desirable to increase the charge density of the polyelectrolyte layer [8]. Electrodeposition of ionic macromolecules is frequently used for coating metal surfaces, that is, a macroanion or macrocation is deposited by an aqueous dispersion that contains base or acid and solubilizer on the anode and cathode, respectively. Similarly, a thin compact cationic polyelectrolyte layer could be formed on the surface of the cation exchange membrane, which is placed between the cathode and anode, by applying direct current through the solution containing a cationic polyelectrolyte. Figure 6 presents a simple two-compartment cell used for electrodeposition procedure. A cation exchange membrane was placed in the two-compartment cell of the same volume (120 cm^3) with Ag/AgCl electrodes, the anode compartment was filled with a solution that contained a sodium chloride solution and the cationic polyelectrolyte of a definite concentration while the cathode compartment was filled with a sodium chloride solution only. As soon as the cells were filled the direct current was applied at high-current density, 10 or 20 mA/cm^2, for a definite period and stirred at 1500 ±100 rpm. Due to the high molecular weight, the macroion cannot cross the membrane; the cationic polyelectrolyte was thus deposited on the desalting side of the membrane surface [8].

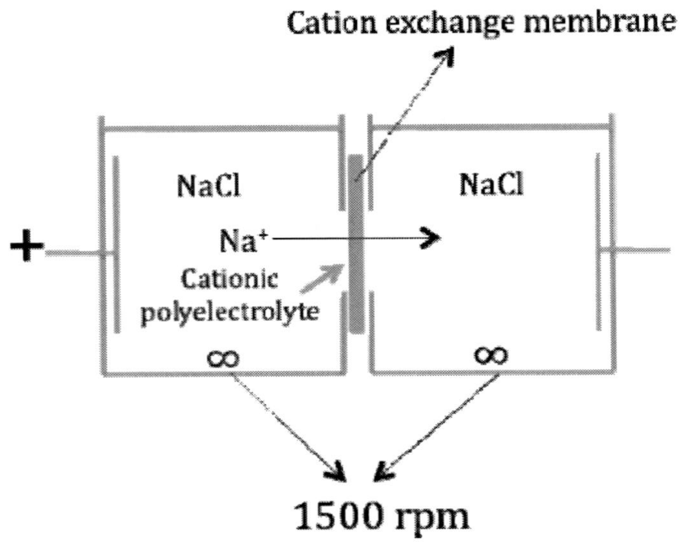

Figure 6. Schematic diagram of the two-compartment cell used for electrodeposition of cationic polyelectrolyte on the membrane surface.

Due to the simplicity, electrodeposition on the cation exchange membranes performed directly in an electrodialysis cell, firstly proposed by Sata and Mizutani groups through the name "current-passing method", has been tackled by many authors [8,9,17-20]. On the basis of this method, Sistat et al.[17] reported that polyethyleneimine (PEI) could be in situ electrodeposited on the surface of the membrane in a pre-industrial pilot working under the process conditions. In the case of a decrease in selectivity, the advantage of this process lays in the possibility of renewing the modification in situ, by regenerating the surface PEI film without having to remove the membrane from the stack. The membranes modified by a conducting polymer (polyethyleneimine, polyaniline, polypyrrole) were mainly used for recovery of sulphuric acid from effluents containing one or several metallic salts, or for separation of sodium ions from trivalent chromium in leather tanning wastewater [17,19] with very high electrodialysis current efficiency (usually higher than 90 %). Here, an important point to be underlined is that this method is not limited to grafting of the mentioned polyamines but has been extended to other kinds of cationic polyelectrolytes such as poly(4-vinyl-N-methylpyridinium-iodide) [9], quaternized chitosan [20]... The investigations allowed enlarging applications of modified cation exchange membranes to the field of sea water concentration. As discussed above, PEI is an effective modifier. However, its weak basicity is a serious drawback. The pH of sea water is about 8.3. In order to make PEI effective in the sea water concentration process, i.e. become positively charged through the protonation reaction, the pH of working solution must be adjusted to 7 or lower. Clearly, the pH adjustment of a large amount of sea water is disadvantageous in practice.

In comparison with the immersion method, current-passing technique is clearly more effective. Such superior effectiveness can be explained as follows: the polymer in the bulk solution is forced more rapidly to interact with the membrane surface thanks to the electrical field and probably forms a thin layer with a tighter structure. The higher compactness obtained by electrodeposition enables to prevent an increase in electrical resistance, as that formed by the immersion method [8,9]. More importantly, suitable conditions like pH of working solution, applied current density, electrodeposition time can be controlled in order to obtain membranes with better selectivity and low electrical resistance. In general, when the amount of deposited PEI increased, the electrical resistance of the membrane increased and transport number of zinc decreased [18].

Another possibility using electrochemistry to modify cation exchange membranes with polyaniline (PANI) for improvement in permselectivity has recently been reported [21,22]. In these works, cation exchange membranes coated on a conducting substrate plays a role of working electrode in an electrochemical cell composed of three electrodes (working, counter and reference electrodes). The electropolymerization was performed by repeated cyclic voltammetry. For example, the growth of PANI on the membrane surface could be achieved by repeated linear potential scanning between −200 mV and 1050 mV at a scan rate of 100 mV s^{-1}. The electrosynthesis was accomplished in a solution containing 0.1 M aniline and 1.0 M H_2SO_4. As can be seen in figure 7, the cyclic voltammetry curves are characteristic of the electropolymerization of an electroactive conjugated film. Three anodic peaks located at +547 mV, +683 mV and +796 mV, as well as their cathodic counterparts are observed, for which the current intensity increases notably with each further scan. The redox processes located at +547 mV and +683 mV are attributed to the partial and full oxidation of PANI, respectively, while the peak at +796 is mostly associated with the degradative hydrolysis products of PANI.

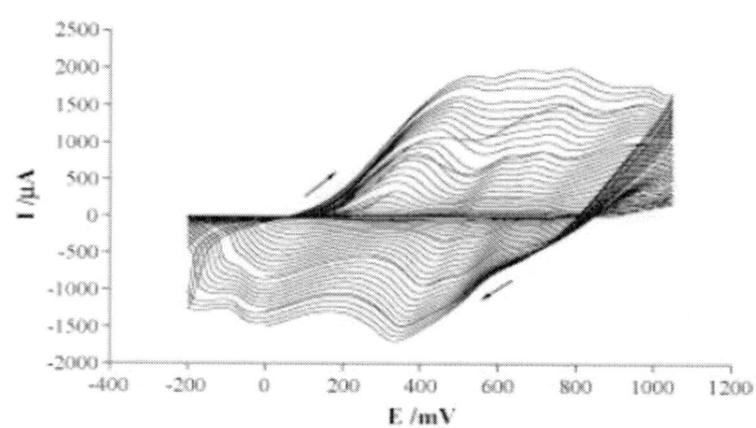

Figure 7. Electropolymerization of aniline performed by repetitive scans between −200 mV and 1050 mV on an ion exchange membrane adhered to carbon paste electrode in H_2SO_4 1.0 M. Potential scan rate: 100 mV s^{-1}; reproduced from Ref. [22] with permission of Elsevier.

This method allowed the control of characteristics of the obtained film, such as thickness and morphology. Even though PANI was really grafted on the membrane surfaces as evidenced from electrochemical characteristics and there was at least 20 times reduction in the transference number of Zn^{2+} through the modified membranes, discussion on the ion exchange capacity, thickness and electrical resistance of the obtained membranes is insufficient. Sivaraman et al.[21] reported that the thickness of PANI coating was approximately of 35 μm, which is too large compared to that of the pristine membrane (≈ 80 μm). In such situation, the modified membrane becomes a true "bipolar membrane" in contact with acidic media, the ion exchange capacity (IEC) obtained with their experiments is due to the combined IEC, this value thus do not represent correctly the trend at which PANI deposited on the membrane. The influence of IEC, thickness of the modified layer on the membrane resistance will be further discussed in the following sections of this chapter.

With respect to the electrochemical techniques, it is worth noting that the relationship between the molecular structure of the cationic polyelectrolytes and the durability of the preferential permselectivity remains unclear [9]. The polyelectrolytes can be primarily maintained at the surface of the membrane by electrostatic interaction or by adsorption. Besides, modification by electrodeposition leads to a penetration of the polyelectrolytes inside the membrane pores, contrary to modification by immersion [18, 21]. That phenomenon may be the main reason why long-term stability of the polymers and the resulting membranes were insufficient. Thus, studies on the stable chemical bonding between the cationic charged layer and the cation exchange membranes are really important.

4. Chemical Modification of Membrane Surfaces

The first method in the field developed by Sata's group [23] is to form acid-amide bonding between a cationic polyelectrolyte having primary or secondary amines groups, such as polyethyleneimine, and a membranous copolymer bearing sulfonyl chloride groups. Chlorosulfonyl groups were introduced as reactive groups into the membrane composed of a styrene-divinylbenzene copolymer and poly (vinyl chloride) fabric (i.e. a base membrane without any ion exchange sites). The obtained membrane was then reacted with various amino compounds. Suitable reaction conditions of the copolymer membrane with chlorosulfonic acid were examined in detail. After the chlorosulfonated membrane had been immersed in an aqueous solution of amino compounds to form a polyamino layer through the acid amide bonding, mainly on the membrane surface, the membrane was immersed in aqueous sodium hydroxide solution to hydrolyze unreacted chlorosulfonyl groups to sodium sulfonate groups. The amino compounds used were p-aminosalicylic acid, L-tyrosine and polyamino compounds such as triethylenetetramine, polyethyleneimines etc. Formation of the polyamino compound layer on the surface of the cation exchange membrane by acid amide bonding conferred stable monovalent cation permselectivity on the membrane in long term continuous electrodialysis (more than six months). Clearly, the "acid-amide bonding" method is a promising technique because some resulting cation exchange membranes provided a long service time concerning monovalent cation permselectivity. In this case it should be noted that sulfonyl chloride groups are introduced in a copolymer of styrene-divinylbenzene using chlorosulfonic acid solution. Since the reaction of the chlorosulfonic acid with the membranous copolymer is severe, the polymer matrix of the membrane might partially decompose. Further, after the reactions, removal of chlorosulfonic acid adhered in and on the membrane is troublesome [24]. The method had therefore been improved as follows: instead of starting with a styrene-divinylbenzene copolymer, p-styrenesulfonyl chloride was firstly synthesized. After that, copolymerization performed with p-styrenesulfonyl chloride and divinylbenzene monomers was realized to obtain a styrenesulfonyl chloride– diviylbenzene copolymer. The resulting membranous copolymer was then used to form a cationic polyelectrolyte layer on the surface by acid-amide bonding between the sulfonyl chloride groups and amino compounds (ethylenediamine, tetraethylenepentamine, polyethyleneimine). Finally, the remaining sulfonyl chloride groups in the inner part of the membranous copolymer were hydrolyzed to sulfonic acid groups [24, 25].

In 1996, Hayashita et al.[26] reported that a Nafion sulfonyl chloride membrane was prepared by reaction of the perfluorinated ion-exchange Nafion membrane with a phosphorus pentachloride/phosphorus oxychloride mixture. This modified membrane was then reacted with an amine to form a sulfonamide bond at the membrane surface. The proton-coupled competitive transport of alkali metal cations was found to be strongly affected by the modification procedure. Following these works, Belanger and coauthors [27,28] proposed a new three-step chemical process by formation of a sulfonamide bond in order to graft ammonium groups on the surface of the CMX commercial cation exchange membranes (from Tokuyama Soda Inc.). The modification procedure involved the chlorosulfonation of the sulfonate groups of the base membrane with thionyl chloride, followed by a reaction with a diamine to yield a sulfonamide bond and a terminal amine (Figure 8). The latter could be quaternized by reaction with methyl iodide or protonated by soaking in acidic media.

Notwithstanding the stable chemical bonding, decrease in ion exchange capacity of the modified membranes was observed in several cases [24,25,27]. As well-known, ion exchange capacity of electromembranes is an important parameter because this amount affects the water content, hence membrane conductivity and permselectivity. A highly ion exchange capacity membrane has the tendency to dilute itself and its resulting swelling is more pronounced than the ones with lower capacity [29]. Furthermore, modification procedure consists of at least two steps and was realized in the highly stringent conditions (using organic solvents, optimization of reactive ratio as well as molecular weight of polyamines, heating condition for hydrolyzed reaction...).

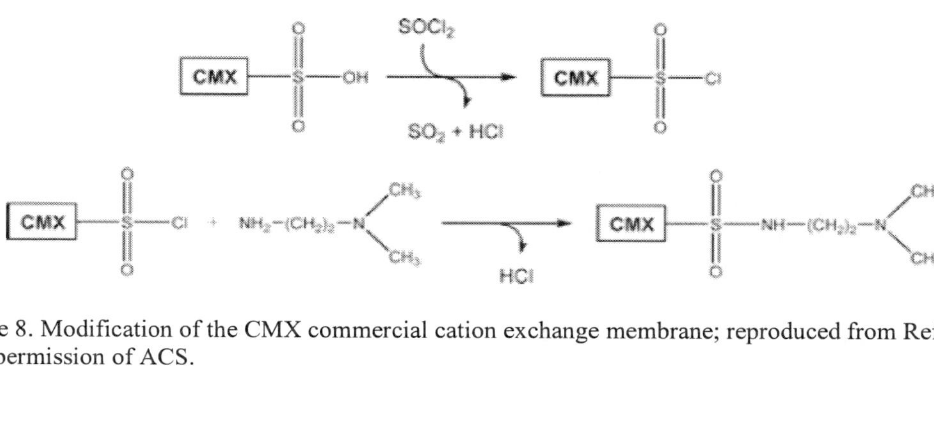

Figure 8. Modification of the CMX commercial cation exchange membrane; reproduced from Ref. [27] with permission of ACS.

Beyond the methods mentioned above, chemical polymerization of aniline, pyrrole on the membrane surface can be given here as another representative example [30-37]. It was reported that a polypyrrole layer formed on a cation exchange membrane by contacting a ferric ion form membrane with an aqueous pyrrole solution decreased the transport number of alkaline earth cations relative to sodium ions [30]. This is due to the synergistic effect of the sieving of alkaline earth cations by a tight, rigid polypyrrole layer and the difference in the electrostatic repulsion force between divalent cations or the sodium ions and the cationic charge in the polypyrrole layer. In this case, polypyrrole gradually penetrates into the membrane matrix with increasing polymerization time because an oxidizing agent, ferric ions, exists in the membrane matrix, causing an increase in the electrical resistance of the resultant membranes. On contrary, aniline is easily polymerized by peroxodisulfate ions. If the cation exchange membrane contacts an anionic oxidizing agent after aniline was ion-exchanged with a membrane previously, a thin layer of polyaniline is expected to be formed on the membrane surface [31].

Tan et al.[32-34] continued this direction and improved considerably this method by investigating the effects of several parameters on the chemical polymerization of aniline (the choice of oxidant and its concentration; and the pH of the reaction medium). Their results demonstrated that it is possible to control the polymerization site of aniline which in turn affects the membrane selectivity properties which was achieved by using different oxidants of opposite charges and a single-step reaction where both the monomer and the oxidant diffuse through the membrane at the same time from each side of the membrane. Indeed, a negatively charged oxidant, $S_2O_8^{2-}$, led to a very thin but homogeneous layer at the surface of the CMX membrane, whereas a positively charged oxidant, Fe^{3+}, led to the formation of PANI within the CMX membrane. Moreover, it was also demonstrated that to block the transport of Zn^{2+} while maintaining the transport of H^+ through the membrane during electrodialysis, only a very thin but homogeneous layer of PANI at the surface of the CMX membrane without increasing significantly the resistance to proton conductivity was required. The presence of PANI solely within the membrane does not seem to block the transport of Zn^{2+}. Otherwise, a three layer model for the most permselective CMX-PANI composite membrane was also proposed based on XPS and IEC data: the outer layer would be constituted of positively charged PANI doped with chloride ions, the intermediate layer of positively charged PANI compensated by the fixed sulfonate groups of the CMX, and the last layer of the CMX bulk membrane (Figure 9). Some other works reported by Shahi's group [35-37] contributed to confirm that chemical polymerization of aniline is a promising technique for separation processes.

It should be remarked that for the field of membrane surface modification, aside from the improvement of monovalent cation permselectivity and the long performance durability, increase in resistance of the modified membranes has to be taken into attention [24]. From the literature [18,24,25], it is reasonable to explain the increase in the electrical resistance by the additional thickness of the membrane obtained after modification. On the other hand, it was reported that the presence of the modified layer on the surface leads to a decrease in the membrane ion exchange capacity [32], where a contraction of the membrane thickness by approximately 10 μm was observed after modification and a decrease in IEC from 1.48 meq./g to a value of 1.41 meq./g for the pristine and modified CMX cation exchange membranes, respectively. Nagarale et al.[35] also reported that the IEC values decrease with polymerization time for the PANI modified cation exchange membranes confirms the compensation of the sulfonate functional groups of the membranes. More effective modification processes which particularly provide the covalently grafted polymer films on the membrane surfaces but also do not affect significantly the modified membrane resistance are always required and still an opened research subject.

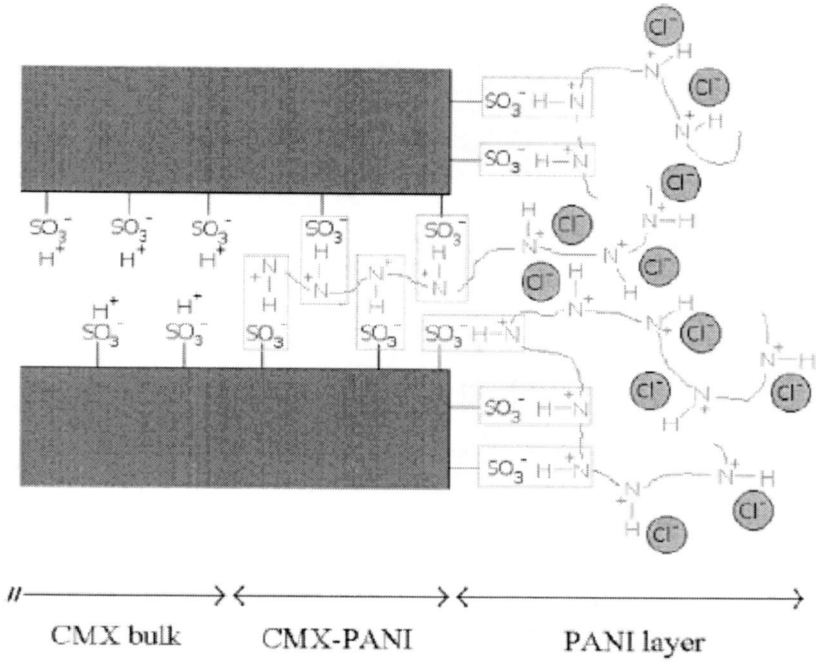

Figure 9. Idealized schematic representing the three different layers found on a CMX-PANI composite membrane. This is a representation for the membrane showing the best permselectivity for their experimental conditions; reproduced from ref. [33] with permission of ACS.

5. Covalent Grafting of Polyaniline-Like Layer Through the Diazonium Induced Anchoring Process

We have recently proposed a new and one-step chemical process, the Diazonium Induced Anchoring Process (DIAP[1]), which provides covalently "polyaniline-like" thin film on the surface of Selemion CMV cation exchange membranes [38]. The following text will describe how that new process may be useful for improvement of proton selectivity. Actually, it has been proved in our previous works that a thin "polyaniline-like" film can be grafted covalently on any surface, either conducting or insulating, through the Diazonium Induced Anchoring Process [39-41]. This method can be described as a spontaneous grafting based on the aryldiazonium salt chemistry using additional reducing agents (iron powder, hypophosphorous acid…) without any electrochemical assistance. Up to now, the use of aryldiazonium salt chemistry for covalent attachment of organic layer developed firstly by

[1] licensed under the name GraftFast™

Pinson's group [42], is indeed quite common since diazonium salts are easily and rapidly prepared in one step from a wide range of anilines [43,44]. In this case, an important point to be underlined is that a very thin aryl films close to 20 nm in thickness have been grafted on gold substrates through our published DIAP [40]. Furthermore, Combellas et al.[45] have reported that when grafting takes place spontaneously (chemically) by simple dipping of the surface into a solution of the diazonium salt, the parameters controlling the thickness of the layer are the concentration of the solution and the immersion time. It is thus interesting to test our DIAP method on commercial cation exchange membranes in order to provide a grafted "polyaniline-like" film on the surface.

The grafting of "polyaniline-like" film on the CMV membranes was carried out at open air and room temperature as follows: first, p-phenylenediamine was solubilized as a 0.1M solution in water at pH = 0.3 (HCl 0.5 M). To generate the mono-aryldiazonium salt, one equivalent of sodium nitrite was used; i.e. 5 mL of sodium nitrite (0.1M) was added dropwise to the same volume of the p-phenylenediamine solution. After homogenizing the solution, 0.25 g of iron powder was added to reduce the diazonium salt in solution. Once the solution bubbled because of dihydrogen and dinitrogen evolution, we introduced the membrane samples (3×3 cm). The samples were taken out from the solution after 90 minutes reaction time and carefully rinsed with HCl, NaCl and deionized water in an ultrasonic bath for 120 minutes. According to this synthesis process, the mechanism for grafting a polyaniline-like layer on the membrane surface can be summarized as presented in Figure 10 [38,41]: (a) the addition of one equivalent of $NaNO_2$ leads to the conversion of one of the two amine groups to the corresponding aryldiazonium salt.

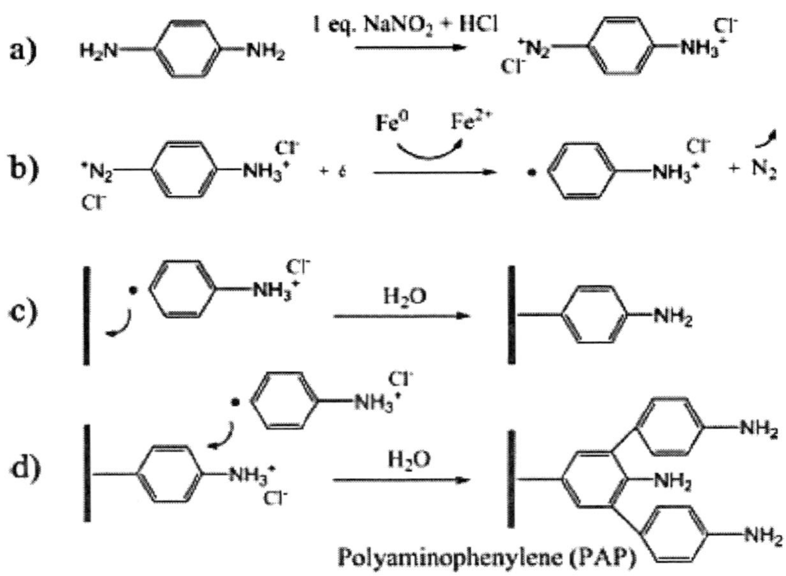

Figure 10. Chemical route to produce grafted polyaniline-like layer: (a) Diazotization of the *p*-phenylenediamine in solution, (b) Reduction of the aryldiazonium group through redox mechanism with iron particles, (c) Grafting of the resulting aryl radicals onto the surface, (d) Growing of PAP multilayer; reproduced from Ref. [38] with permission of RSC.

It is important to point out that the acidic medium protects the remaining amine group from reacting with newly formed diazonium salts, which is likely to improve both the overall yield of grafting and the covering ratio; (b) chemical reduction of aryl diazonium salts and protons at iron surface leading to the formation of the corresponding radical species; (c) formation of the grafted polyaniline-like layer; (d) growth of the grafted polyaniline-like layer. Note that the membrane was first rinsed by 0.5 M HCl solution after the modification process in order to remove the iron powder which may remain on the membrane surface. Then, it was conditioned in 1.0 M NaCl solution to replace the hydrogen counter-ions inside the membrane by sodium ions. The modified membrane was finally rinsed by deionized water to desorb the excess NaCl.

The presence of the "polyaniline-like" film on the surface of the membranes was confirmed by means of SEM, IR and XPS techniques [38]. It is important to remark that the resistance of our obtained polymer toward ultrasonic cleaning in various solvents suggests the formation of a covalent bond between the substrate and the polymer layer. As mentioned in the literature, these covalently bound layers can only be removed by mechanical polishing or by scratching with an AFM tip [43]. This resistance to aggressive reagents, to many organic solvents, to temperature clearly excludes a weak bonding such as hydrogen bonds or Van-der-Waals interactions [43]. Obviously, the proton selectivity of the modified membrane has been performed with electrodialysis test in $NiSO_4$ + H_2SO_4 mixed solution [38]. The transport number of cation Ni2+ relative to hydrogen ions values equal to 0.056 and 0.006 respectively for the bare and modified membranes which shows the increase in selectivity towards proton ions. Such result is comparable to those obtained with the other proton selective cation exchange membranes modified by the conventional modification processes [46,47].

It is clear that the modification layer does not only improve the monovalent selectivity but also leads to a loss in membrane conductivity, then affects the electrodialysis efficiency as already mentioned in this work. It is thus interesting to study the electrical resistance of our modified membranes. We remark that the resistance increases very slightly from 0.86 to 0.93 Ωcm^2 after the modification (about 8 %) when the membranes are in contact with 0.25 M H_2SO_4. The rather small increase in resistance observed in our case can be explained by the very thin polyaniline-like layer grafted on the studied membrane. As the Cl 2p signal of PVC backbone is observed in the XPS spectra of the CMS membrane and the modified CMV membrane [38], one can deduce that the thickness of the polyethyleneimine layer on the CMS membrane surface is in the same order with that of the modified polyaniline-like layer in this work. More importantly, taking into account the depth sensitivity of the XPS technique, we can conclude that the thickness of the modified layers in both our CMV membrane and the CMS commercial membrane does not exceed 20 nm. Such observation is in a good agreement with the study on the ion exchange capacity. As expected, the thin surface layer did not affect the ion exchange capacity of the modified membrane which was found to be the same as that of the bare membrane (2.11 and 2.13 meq./g respectively). Moreover, the value of 0.93 Ωcm^2 obtained for the modified membrane can be compared to that of the Neosepta CMS commercial cation exchange membrane published in our previous work [3] (0.96 Ωcm^2). All membrane resistances were performed with the same experimental conditions. Note that the CMS membrane which was frequently used for the treatment of metal finishing waste water has the same PS-DVB structure as the CMV membrane and its surface is modified by a polyethyleneimine layer via the electrodeposition method in order to improve the selectivity for the monovalent cations [3,38,47,48]. Besides, the absorption properties of both membranes were also studied at equilibrium with a mixed H_2SO_4 + $NiSO_4$ solution. Application of conductivity measurements to these systems has led to determine the ratio of H^+ mobility relative to that of Ni^{2+} ion in the membranes ($\overline{u}_{H^+} / \overline{u}_{Ni^{2+}}$) [38]. The value of 40.2 obtained with the modified membrane is compared to the ratios of the hydrogen ion mobility to the mobility of nickel and other bivalent cations in the CMS monovalent cation permselective membrane reported in the literature: 39.8, 37.9 and 38.7 for Ni^{2+}, Cu^{2+} and Zn^{2+} respectively [3,48,49].

CONCLUSION

This work summarized the state-of-art in our understanding of surface modification techniques for enhancement of proton selectivity of cation exchange membranes. Significant investigations concerning plasma process as well as electrochemical and chemical methods to graft a polyamine layer on the membrane surfaces have been reported. We mainly focused on application of electrodialysis using the proton selective membranes for treatment of strong acids from effluents of hydrometallurgy or electroplating industries which are generally concentrated equally in multivalent metal ions. In such situation, the positively charged surface modification layer plays a role of repulsion barrier towards the multivalent cations and thus permits the transfer of proton ions only, as recognized in many published works in the field. This point, however, remains unclear in certain cases and is of potential interests from the scientific point of view. It was observed that the ionic compositions of both the pristine and polyethyleneimine modified membranes were almost the same when the membrane had been in contact with the mixed solution containing monovalent and bivalent cations at equilibrium state while they completely differed under electrodialysis condition (a current was applied through the membranes) [10]. The phenomena proved that multivalent cations were rejected by the cationic charged layer on the membrane surface in the course of electrodialysis. According to Belanger and coauthors, the blocking action of the polyaniline layer seemed to be more related to the presence of electrostatic repulsion than to the presence of a physical layer [34]. In the work of Chapotot et al. [46,50], as the authors did not observe any significant change in the ion exchange isotherm after the surface modification while the apparent rate constant of the entry step of Ca^{2+} into the membrane was substantially changed, one could conclude that the origin of the active mechanism of the modified polyethyleneimine layer on the CRA membrane surface is kinetic and not thermodynamic. In contrary, Siali and Gavach [48] reported that the presence of a layer of positively charged ionic groups on the surface of the CMS commercial membrane had no significant influence on permselectivity. The high interaction of the cupric ion with the membrane material caused a relative decrease of the bivalent ion activity coefficient with respect to the proton activity coefficient as compared to the values in external solution. More recently, it was reported that the transport flux depends on both the composition of the Nafion membrane resulting from the solution-membrane exchange equilibrium and the mobility of the counter-ions within the membrane [51]. In our case where the CMV membrane was modified by a thin layer of "polyaniline-like" layer [38], the equilibrium exchange isotherm is slightly affected by the modified polyaniline-like layer and the ratios $\overline{u}_{H^+} / \overline{u}_{Ni^{2+}}$ are considerably different in the two membranes. Taking into account those facts, it would be very interesting to extend our investigation to kinetic aspects, as realized for the other membranes mentioned above. This would allow a better understanding of the membrane-solution interphase as a function of surface modification.

Kinetic and thermodynamic issues are however not the purpose of the present work, which aims only at technical aspects to provide an effective surface modified layer for proton selective membranes, or more generally, monovalent cation exchange membranes used in electrodialysis. As clearly stated in the introduction, the two key parameters of electrodialytic efficiency (membrane resistance and ion exchange capacity) have been taken into account in this paper. It is worth to remark that we mainly focused our attention on the decrease in the membrane conductivity after the modification. Influence of the surface layer on the membrane resistance as well as on the electrotransport of the proton ion near and through the modified cation exchange membranes was discussed in detail in the literature [3,10,28,52,53]. Investigation on the transport properties of the perfluorinated composite membranes modified by polyaniline reported by Berezina et al.[53] can be given here as an representative example where conductivity, diffusion and electroosmotic permeability, proton permselectivity – as well as porosimetry and polarization behaviours have been investigated as functions of aniline polymerization parameters and acid concentration. The fibrous-cluster model of a composite membrane has been proposed for the estimation of transport and structural parameters, taking into consideration different mechanism of charge transfer in structural fragments of the composite. The atomic force microscopy images and curves of water distribution on the effective pore radii in the composite membranes testify to a morphological transition from the nano- to the microsize of polyaniline intercalations with increasing the aniline polymerization time. This effect is confirmed by the analysis of two-phase model transport and structural parameters. By taking into account this two-phase model [54] towards the CMV and CMS commercial cation exchange membranes in contact with a mixed solution H_2SO_4–$NiSO_4$, our previous study also left open fundamental questions about the molecular mechanisms responsible for the ion selective transport inside polymer electromembranes [3].

While most of the discussion in this work has concentrated on electrodialysis applications for water treatment where membrane conductivity and long-term stability are critical, these types of modified membrane may serve for many important areas. Actually, the range of applications for ion exchange membranes would be extended if such membranes had a higher selectivity among counterions. For example, the performance of a redox flow battery is strongly dependent on the separator permselectivity among ions of the same polarity. Thus, the development of the redox flow battery would be greatly aided by the availability of a cation exchange membrane having a high proton permselectivity [13,55]. On the other hand, modification of Nafion membranes by incorporation of cationic polymers was equally proved efficient for development of fuel cell technologies [56,57]. Additionally, it has recently been reported that ties between ion-conducting polymer electrolytes and water treatment membranes are made to illustrate that the underlying mechanisms that control ion transport in fuel cell membranes may also be harnessed to catalyze the development of new membrane materials for water purification [58]. Finally, we hope that this work will provide practical information for choosing a surface modification method compatible with real working conditions.

REFERENCES

A. Chapotot, G. Pourcelly, C. Gavach, F. Lebon. Electrotransport of proton and divalent cations through modified cation-exchange membranes. *J. Electroanal. Chem.* 386 (1995) 25.

A. Chapotot, G. Pourcelly, C. Gavach, Transport competition between monovalent and divalent cations through cation-exchange membranes. Exchange isotherms and kinetic concepts, *J. Membr. Sci.* 96 (1994) 167.

A.M. Rojas, Y.O. Maldonado, L.M.T. Rodríguez, An easy method to modify the exchange membranes of electrodialysis with electrosynthetized polyaniline, *J. Membr. Sci.* 300 (2007) 2.

A.T. Cherif, C. Gavach, J. Molenat, A. Elmidaoui, Transport and separation of Ag^{+} and Zn^{2+} by donnan dialysis through a monovalent cation selective membrane, *Talanta* 46 (1998) 1605.

B. Tian, C.W. Yan, F.H. Wang, Modification and evaluation of membranes for vanadium redox battery applications, *J. Appl. Electrochem.* 34 (2004) 1205.

C. Combellas, D. Jiang, F. Kanoufi, J. Pinson, F. Podvorica, Steric Effects in the reaction of aryl radicals on surfaces, *Langmuir* 25 (2009) 286.

C. Vallois, P. Sistat, S. Roualdès, G. Pourcelly, Separation of H^{+}/Cu^{2+} cations by electrodialysis using modified proton conducting membranes, *J. Membr. Sci.* 216 (2003) 13.

D.H. Kang, D. Kim, Modification of Nafion membranes by incorporation of cationic polymers for reduction of methanol permeability, *Korean J. Chem. Eng.* 24 (2007) 1101.

F. Helfferich, Ion exchange, McGraw-Hill Book Company, New York, 1962.

French patent: FR 06 55653, US 11/686,076 (14/03/2007).

G. Chamoulaud, D. Belanger, Chemical Modification of the Surface of a Sulfonated Membrane by Formation of a Sulfonamide Bond, *Langmuir* 20 (2004) 4989.

G. Chamoulaud, D. Belanger, Modification of ion-exchange membrane used for separation of protons and metallic cations and characterization of the membrane by current–voltage curves, *J. Colloid Interface Sci.* 281 (2005) 179.

G.S. Gohil, R.K. Nagarale, V.V. Binsu, V.K. Shahi, Preparation and characterization of monovalent cation selective sulfonated poly(ether ether ketone) and poly(ether sulfone) composite membranes, *J. Colloid Interface Sci.* 298 (2006) 845.

G.S. Gohil, V.V. Binsu, V.K. Shahi, Preparation and characterization of mono-valent ion selective polypyrrole composite ion-exchange membranes, *J. Membr. Sci.* 280 (2006) 210.

H. Strathmann, Ion-exchange membrane separation processes, 9, Elsevier Publishers, 2004.

J. Lambert, M. Avila-Rodriguez, G. Durand, M. Rakib, Separation of sodium ions from trivalent chromium by electrodialysis using monovalent cation selective membranes, *J. Membr. Sci.* 280 (2006) 219.

J. Pinson, F. Podvorica, Attachment of organic layers to conductive or semiconductive surfaces by reduction of diazonium salts, *Chem. Soc. Rev.* 34 (2005) 429439.

K. Shimasaki, H. Ihara, Y. Mizutani, Modification of cation exchange membrane by grafted poly(4-vinyl-N-methylpyridinium-iodide), *J. App. Polym. Sci.* 34 (1987) 1093.

K. Takata, H. Ihara, T. Sata, Modification of transport properties of ion exchange membranes. XIV. Effect of molecular weight of polyethyleneimine bonded to the surface of cation exchange membranes by acid–amide bonding on electrochemical properties of the membranes, *J. Membr. Sci.* 179 (2000) 101.

K. Takata, Y. Yamamoto, T. Sata, Modification of transport properties of ion exchange membranes. XIII. Surface-modified cation exchange membranes prepared from polyamines and a copolymer membrane containing p-styrenesulfonyl chloride, *Die Angewandte Makromolekulare Chemie* 236 (1996) 67.

M. Amara, H. Kerdjoudj, Modification of cation-exchange membrane properties by electro-adsorption of polyethyleneimine, *Dessalination* 155 (2003) 79.

M. Delamar, R. Hitmi, J. Pinson, J.M. Saveant, Covalent modification of carbon surfaces by grafting of functionalized aryl radicals produced from electrochemical reduction of diazonium salts, *J. Am. Chem. Soc.* 114 (1992) 5883.

M. Joselevich, F. Williams, F.J. Synthesis and characterization of diazonium functionalized nanoparticles for deposition on metal surfaces, *Langmuir* 2008, 24, 11711.

M. Siali and C. Gavach. Transport competition between proton and cupric ion through a cation-exchange membrane: I. Equilibrium properties of the system: membrane-$CuSO_4$+H_2SO_4 solution. *J. Membr. Sci.* 71 (1992) 181.

M. Taky, G. Pourcelly, A. Elmidaoui. Transport properties of a commercial cation-exchange membrane in contact with divalent cations or proton-divalent cation solutions during electrodialysis, *Hydrometallurgy* 43 (1996) 63.

M.A. Hickner, Ion-containing polymers: new energy and clean water, *Materialstoday* 13 (2010) 34.

N.P. Berezina, N.A. Kononenko, A.A.-R. Sytcheva, N.V. Loza, S.A. Shkirskaya, N. Hegman, A. Pungor, Perfluorinated nanocomposite membranes modified by polyaniline: Electrotransport phenomena and morphology, *Electrochimica Acta* 54 (2009) 2342.

N.P. Berezina, N.A. Kononenko, O.A. Dyomina and N.P. Gnusin, Characterization of ion-exchange membrane materials: properties vs structure, *Adv. Colloid Interface Sci.* 139 (2008) 3.

P. Sistat, G. Pourcelly, C. Gavach, N. Turcotte, M. Boucher, Electrodialysis of acid effluents containing metallic divalent salts: recovery of acid with a cation-exchange membrane modified in situ, *J. App. Electrochem.* 27 (1997) 65.

P. Sivaraman, J.G. Chavan, A.P. Thakur, V.R. Hande, A.B. Samui, Electrochemical modification of cation exchange membrane with polyaniline for improvement in permselectivity, *Electrochimica Acta* 52 (2007) 5046.

P. Viel, X.T. Le, V. Huc, J. Bar, A. Benedetto, A. Le Goff, A. Filoramo, D. Alamarguy, S. Noël, L. Baraton, S. Palacin, Covalent grafting onto self-adhesive surfaces based on aryldiazonium salt seed layers, *J. Mater. Chem.* 18 (2008) 5913.

R.K. Nagarale, G.S. Gohil, V.K. Shahi, G.S. Trivedi, R. Rangarajan, Preparation and electrochemical characterization of cation- and anion-exchange/polyaniline composite membranes, *J. Colloid Interface Sci.* 277 (2004) 162.

S. Logette, C. Eysseric, G. Pourcelly, A. Lindheimer, C. Gavach, Selective permeability of a perfluorosulphonic membrane to different valency cations. Ion-exchange isotherms and linetic aspects, *J. Membr. Sci.* 144 (1998) 259.

S. Tan, A. Laforgue, D. Belanger, Characterization of a Cation-Exchange/Polyaniline Composite Membrane, *Langmuir* 19 (2003) 744.

S. Tan, D. Belanger, Characterization and Transport Properties of Nafion/Polyaniline Composite Membranes, *J. Phys. Chem. B* 109 (2005) 23480.

S. Tan, J.H. Tieu, D. Belanger, Chemical Polymerization of Aniline on a Poly(styrene sulfonic acid) Membrane: Controlling the Polymerization Site Using Different Oxidants, *J. Phys. Chem. B* 109 (2005) 14085.

S. Tan, V. Viau, D. Cugnod, D. Belanger, Chemical Modification of a Sulfonated Membrane with a Cationic Polyaniline Layer to Improve its Permselectivity, *Electrochemical and Solid-State Letters*, 5 (2002) E55.

T. Hayashita, J.C. Lee, R.A. Bartsch, Transport of alkali metal cations through monoazacrown ether-modified Nafion^{TM} 117 membrane, *J. Membr. Sci.* 116 (1996) 243.

T. Sata, Modificatien of properties of ion exchange memhranes. III. Interaction between ion exchange membranes and surface active agents, *Colloid and Poiymer Sci.* 256 (1978) 62.

T. Sata, Modificatien of properties of ion exchange memhranes. IV. Change of transport properties of cation exchange membranes by various polyelectrolytes, *J. Polym. Sci.: Polym. Chem. Ed.* 16 (1978) 1063.

T. Sata, R. Izuo, K. Takata, Modification of the transport properties of ion exchange membranes. IX. Layer formation on a cation exchange membrane by acid-amide bonding, and transport properties of the resulting membrane, *J. Membr. Sci.* 45 (1989) 197.

T. Sata, R. Izuo, Modificatien of properties of ion exchange memhranes. V. Structure of cationic polyelectrolyte on the surface of cation exchange membranes, *Colloid and Poiymer Sci.* 256 (1978) 757.

T. Sata, R. Izuo, Y. Mizutani, R. Yamane, Transport properties of ion-exchange membranes in the presence of surface active agents, *J. Colloid Interface Sci.* 40 (1972) 317.

T. Sata, T. Funakoshi, K. Akai, Preparation and Transport Properties of Composite Membranes Composed of Cation Exchange Membranes and Polypyrrole, *Macromolecules* 29 (1996) 4029.

T. Sata, Y. Ishii, K. Kawamura, K. Matsusaki, Composite Membranes Prepared from Cation Exchange Membranes and Polyaniline and Their Transport Properties in Electrodialysis, *J. Electrochem. Soc.* 146 (1999) 585.

T. Sata, Y. Mizutani, Modificatien of properties of ion exchange membranes. VI. Electrodialytic transport properties of cation exchange membranes with a electrodeposition layer of cationic polyelectrolytes, *J. Polym. Sci.: Polym. Chem. Ed.* 17 (1979) 1199.

T. Sata, Y. Mizutani, Modificatien of properties of ion exchange membranes. VII. Ionic composition in cation exchange membranes with and without a cationic polyelectrolyte layer at equilibrium and during electrodialysis, *J. Membr. Sci.* 45 (1989) 209.

V. Compan, E. Riande, F.J. Fernandez-Carretero, N.P. Berezina, A.A.-R. Sytcheva, Influence of polyaniline intercalations on the conductivity and permselectivity of perfluorinated cation-exchange membranes, *J. Membr. Sci.* 318 (2008) 255.

V. Mevellec, S. Roussel, L. Tessier, J. Chancolon, M. Mayne-L'Hermite, G. Deniau, P. Viel, S. Palacin, Grafting polymers on surfaces: a new powerful and versatile diazonium salt-based one-step process in aqueous media, *Chem. Mater.* 19 (2007) 6323.

X.T. Le, C. Buess-Herman, H.D. Hurwitz, Absorption equilibrium and permselectivity of cation exchange membranes in sulfuric acid, sodium chloride and nickel sulfate media, *J. Membr. Sci.* 323 (2008) 288.

X.T. Le, P. Viel, P. Jegou, A. Garcia, T. Berthelot, T.H. Bui, S. Palacin, Diazonium-induced anchoring process: an application to improve the monovalent selectivity of cation exchange membranes, *J. Mater. Chem.* 20 (2010) 3750.

X.T. Le, Permselectivity and microstructure of anion exchange membranes, *J. Colloid Interface Sci.* 325 (2008) 215.

Y. Hu, M. Wang, D. Wang, X. Gao, C. Gao, Feasibility study on surface modification of cation exchange membranes by quaternized chitosan for improving its selectivity, *J. Membr. Sci.* 319 (2008) 5.

Y. Mizutani, Ion exchange membranes with preferential permselectivity for monovalent ions, *J. Membr. Sci.* 54 (1990) 233 and Refs. therein.

Z. Ogumi, Y. Uchimoto, M. Tsujikawa, K. Yasuda and Z.I. Takehara, Modification of ion-exchange membrane surface by plasma process. Part 2. Monovalent cation permselective membrane from perfluorosulfonate cation exchange membrane. *J. Membr. Sci.* **54** (1990) 163.

Z. Ogumi, Y. Uchimoto, M. Tsujikawa, K. Yasuda, Z.I. Takehara, Modification of ion-exchange membrane surface by plasma process. Part 3. Interfacial resistance of monovalent cation permselective membrane from Nafion®, Bull. *Chem. Soc. Jpn.* ***63*** (1990), 2150.

Z. Ogumi, Y. Uchimoto, M. Tsujikawa, Z. Takehara, Enhancement of proton selectivity of cation exchange membrane by plasma modification, *J. Electrochem. Soc.* 136 (1989) 1247.

Z. Ogumi, Y. Uchimoto, M. Tsujikawa, Z. Takehara, F.R. Foulkes, Modification of ion exchange membrane surface by plasma process. 1. H^+ Ion Perm-Selective Membrane from Nation for Redox-Flow Battery, *J. Electrochem. Soc.* 137 (1990) 1430.

In: Polymer Films: Properties, Performance and Applications ISBN 978-1-61209-201-0
Editors: Stefan A. Romano and George P. Somners

Chapter 4

MODIFIED PERFLUORINATED POLYMER FILMS AS HIGH-PERFORMANCE, SOLUTION-DEPOSITED BARRIER- AND DIELECTRIC LAYERS FOR ORGANIC ELECTRONIC DEVICES

Jimmy Granstrom
George W. Woodruff School of Mechanical Engineering
Georgia Institute of Technology, Atlanta, GA, US

ABSTRACT

This short communication provides a review on Cytop™, a commercial perfluorinated polymer, as a barrier- and dielectric layer. It also describes ideas for future work on optimizing Cytop™ for aforementioned applications. Solution-deposited films of Cytop™ have shown great promise as dielectric films and/or encapsulation barrier layers for organic field-effect transistors (OFETs) and/or other organic electronic devices, e.g. organic light-emitting diodes (OLEDs) and organic photovoltaic (OPV) devices. OLEDs encapsulated with Cytop™ exhibited up to five times longer continuous operation under identical environmental and driving conditions compared with devices that were not encapsulated with Cytop™. A general method has also been developed to make thin and smooth films of Cytop™ encapsulation barrier layers against water and oxygen permeation. Based on a phenomenological analysis, it was found that disturbances in flow due to Rayleigh-Benard-Marangoni instability during drying of spin-cast perfluorinated polymer films caused high surface roughness and the formation of "pinholes". Casting films from solutions with higher polymer concentration and from solvents with higher viscosity suppressed the instability and significantly reduced the roughness. Suppression of the instability resulted in improved barrier properties as indicated by the calcium thin film optical transmission test. Spin-coated films of Cytop exhibit good gate insulating properties and also provide excellent OFET operational stability. However, these devices operate at large voltages since a dielectric thickness of ~ 450-600 nm is required for stable device operation. A cross-linked Cytop (C-Cytop) formulation deposited by spin-coating enabled the fabrication of OFETs with excellent performance using C-Cytop dielectric films as thin as 50 nm. The use of anti-plasticizing additives, possibly in combination with cross-linkers, may further improve the performance of dielectric films and/or encapsulation barrier layers for OFETs and/or OLEDs and OPVs.

Cytop™ is a hydrophobic, perfluorinated polymer which has been used for a wide range of applications, ranging from anti-reflective coatings to optical filters for plasma display panels [1]. Cytop™ has also been used as a barrier layer for organic light-emitting devices [2] and as a dielectric layer for organic field-effect transistors (OFETs) [3]. The use of Cytop™ for these applications can be related to its water-repelling properties and the solvents used to dissolve Cytop™ - these solvents are orthogonal to the underlying device layers, meaning that the solvents used to dissolve Cytop™ do not dissolve the underlying layers during deposition and film formation.

Cytop™ was used to encapsulate light-emitting field-effect transistors (LEFETs) and organic light-emitting diodes (OLEDs) [2]. LEFETs were functional after immersion into H_2O for two minutes, while the operational lifetimes of OLEDs were improved five times for OLEDs encapsulated with Cytop™ compared with identical devices which were not encapsulated with Cytop™.

As already mentioned, Cytop™ has been used as a gate dielectric for OFETs [3] and for stable complementary inverters [4]. It has already been demonstrated that a solution deposited titanium sub-oxide (TiO_x) barrier layer can significantly extend the lifetimes of organic field-effect transistors [5]. The mobility of a P3HT-based OFET device had decreased by an order of magnitude after 500 hours in air, compared with negligible reduction for an equivalent P3HT-based OFET device with TiO_x. It might therefore be interesting to investigate the use of Cytop™ as a top barrier as well as a polymer layer in a Cytop™/TiO_x multilayer encapsulation for OFETs.

It has also recently been demonstrated that top-gate OFETs utilizing PCDTBT, a poly(2,7-carbazole) bulk heterojunction polymer derivative which is stable at temperatures exceeding 200° C in nitrogen, can be fabricated with a gate dielectric consisting of a polymer cross-linked at 250° C (without degrading the underlying PCDTBT layer) [6]. Possible future work may involve fabrication of top-gate OFETs with PCDTBT and Cytop™ dielectric layers cured at 200° according to conditions previously described in the literature. [7].

The performance of thin Cytop™ films were optimized by utilizing instability suppression with a high-viscosity co-solvent [8] in order to reduce the cost/performance ratio of these films. For lowest cost, one would naturally like to obtain the thinnest possible Cytop™ films with retained barrier properties. The thickness of polymer films made from spin casting can easily be decreased by reducing the concentration of polymer in the solution used for spin-casting. However, Cytop™ films become rougher as they become thinner, which severely compromises their barrier properties. It was found that the increased roughness results from a combination of Rayleigh-Benard and Benard-Marangoni instabilities. By suppressing the flow instability, it was possible to make Cytop™ films as thin as 150 nm that functioned as effective barrier layers.

Cytop™ has also been used in multilayer barrier films comprising nitrogen spacers, both as a barrier layer [9] and as an anti-reflective coating [10]. It is also anticipated that Cytop™ may fill an important role in Russian Doll encapsulation architectures [11]. A barrier matrix could conceptually consist of a sorption/diffusion/desorption multi-layer barrier, where the diffusion barrier is a free-standing barrier film with a low free volume, while the sorption barrier is a hydrophobic material, e.g. Cytop™, and the desorption barrier is a permeant-binding hydrophilic molecule, e.g. the previously mentioned TiO_x material.

Nanoparticles may be used to improve both the barrier- and dielectric properties of Cytop™. Well dispersed nanoparticles appear as additives to small molecule diffusion. As a consequence, the tortuosity of the diffusion pathway increases resulting in a slower diffusion process. High-performance organic transistors using solution-processed nanoparticle-filled high-κ polymer gate insulators have been demonstrated for PVA:$BaTiO_3$ composites [12]. The nanoparticles in Cytop™ composites may segregate in the polymer matrix, possibly resulting in high k nanoparticles at the gate/dielectric interface and the low k polymer at the dielectric/semiconductor interface.

Cross-linking has been used to improve the dielectric properties of Cytop™ [13]. The cross-linker 1,6-bis(trichlorosilyl)hexane (C_6-Si) was dispersed into Cytop™ to obtain cross-linked Cytop (C-Cytop) formulations. Films were deposited by spin-coating to enable the fabrication of OFETs with excellent performance for C-Cytop dielectric films as thin as 50 nm. Both the very high device yield (~ 100%) for thin films (< 100 nm) and a highly reduced sensitivity to the underlying semiconductor film morphology were typical characteristics for these C-Cytop gate dielectric films.

It is desirable to reduce the free volume of Cytop™ to possibly improve both the barrier- and dielectric properties of the material. One approach to lower the free volume of Cytop™ is to introduce small molecules into the polymer matrix [14]. Certain low molar mass additives incorporated in glassy polymers induce a mobility restriction of the polymeric chains, which is the opposite effect of adding plasticizers to a polymer matrix. This behavior is called "antiplasticization", and is somehow related to a loss in the fractional free volume (FFV) and subsequent reduction in molecular mobility. Usually, the antiplasticization phenomenon also leads to a reduction in the gas and vapor permeability due to a loss of the FFV resulting in a lower diffusion coefficient. The magnitude of these effects depend on some charachteristics of the additives, such as size, shape and stiffness [15].

Small molecules also affect the viscosity of polymer solutions. As already discussed, polymer solutions with higher viscosities suppress the Marangoni instability resulting in smoother spin-cast films. Depending on the compatibility between the polymer and the small molecule, additives may either increase or reduce polymer solution viscosity. Thus, a trade-off might occur with antiplasticization resulting in a reduction in permeability through reduced FFV but also a reduction in polymer solution viscosity contributing to the on-set of the Marangoni instability. The opposite situation could also occur – that plasticization results in increased permeability through increased FFV but also an increase in polymer solution viscosity resulting in an on-set of the Marangoni instability for thinner Cytop™ polymer films.

It might be interesting to investigate whether it is possible to combine two additives, e.g. cross-linkers and anti-plasticizing additives or nanoparticles, to maximize the free volume reduction of Cytop™. It might be possible that different kinds of additives fill up different parts of the polymer free volume, e.g. between chains and around chain ends, which could help to maximize the free volume reduction.

References

[1] K. Yamamoto, G. Ogawa, *Journ. of Fluor Chem.* 2005, Vol. 126, 1403.

[2] J. Granstrom, J. S. Swensen, J. S. Moon, G. Rowell, J. Yuen, A. J. Heeger, *Appl. Phys. Lett.* 2008, Vol. 93, 193304.

[3] W. Kalb, T. Mathis, S. Haas, A. F. Stassen, B. Batlogg, *Appl. Phys. Lett.* 2007, Vol. 90, 092104.

[4] M. P. Walser, W. L. Kalb, T. Mathis, T. J. Brenner, B. Batlogg, *Appl. Phys. Lett.* 2009, Vol. 94, 053303.

[5] S. Cho, K. Lee, A. J. Heeger, *Adv. Mater* 2009, Vol. 21, 1.

[6] S. Cho, J. H. Seo, S. H. Park, S. Beaupre, M. Leclerc, A. J. Heeger, *Adv. Mater* 2010, Vol. 22, 1253.

[7] J. Smith, R. Hamilton, I. McCulloch, M. Heney, J. E. Anthony, D. C. Bradley, T. D. Anthopoulos, *Synth. Met.* 2009, Vol. 159, 2365.

[8] J. Granstrom, A. Roy, G. Rowell, J. S. Moon, E. Jerkunica, A. J. Heeger, *Thin Solid Films* 2010, Vol. 518, 3767.

[9] J. Granstrom, M. Villet, T. Chatterjee, J. A. Gerbec, E. Jerkunica, A. Roy, *Appl. Phys. Lett.*, 2009, Vol. 95, 093306.

[10] J. Granstrom, M. Gällstedt, O. Arias, T. Blomfeldt, T. Chatterjee, Y. Kim, H. Kim, S.-W. Cho, M. Hedenqvist, S. Graham, Under review for *Thin Solid Films*.

[11] J. Granstrom, M. Gällstedt, M. Villet, J. S. Moon, T. Chatterjee, *Thin Solid Films* 2010, Vol. 518, 5282.

[12] R. Schroeder, L. A. Majewski, M. Grell, *Adv. Mater* 2005, Vol. 17, 1535.

[13] X. Cheng, M. Caironi, Y.-Y. Noh, J. Wang, C. Newman, H. Yan, A. Facchetti, H. Sirringhaus, *Chem Mater* 2010, Vol. 22, 1559.

[14] E. C. Lugert, T. P. Lodge, P. Buhlmann, *Jour. of Pol. Sci. Part B: Polymer Physics* 2008, Vol. 46, 516.

[15] S. E. Vidotti, A. C. Chinellato, L. A. Pessan, *Jour. of Appl. Polym. Science* 2007, Vol. 103, 2627.

In: Polymer Films: Properties, Performance and Applications ISBN 978-1-61209-201-0
Editors: Stefan A. Romano and George P. Somners

Chapter 5

STABILITY OF SELF-ASSEMBLED POLYELECTROLYTES

Jérôme Dejeu*
University of Franche-Comté, FEMTO-ST Institute

Abstract

We discussed the influence of post-treatment rinsing after the formation of self-assembled polyelectrolyte films made with the different polymers pairs: PAH/PSS, P4VP/PMA, PMA/QPVP. The stability of the film was studied using optical fixed-angle laser reflectometry to measure the release of polymeric material, AFM experiments to reveal the change of morphology and thickness, ellipsometry, wetting characteristics, confocal laser scanning microscopy. We found that the polymer films were stable upon rinsing when the pH and the ionic strength were the sames in the solution as those used in the buildup. The films released most of the polymeric material when rinsed with pH values where the polymer charge decreases, but a layer remained that corresponded to a monolayer directly bound with the silica surface. Films containing at least four bilayers were stable upon rinsing at pH values where the interaction charge increases, but the stability of thinner films depended on the type of the last polymer deposited. They were stable in the case of polymer diffused on the film as an outermost deposit, but they released a large part of their material in the case of polymer adsorbed on the substrate and on opposite polymer groups pointing out of the macromolecular complexes. The stability results were determined using a simple model of the step-by-step assembly of the polymer film described formerly.

PACS 68.55.-a, 68.55.A-, 68.37.Ps, 68.37.-d.

Keywords: Stability, self-assembled, polyelectrolytes, laser reflectometry, AFM, colloids.

1. Introduction

Self-assembled polymer films (also called multilayer films) have received much attention since first described by Decher et al. in the early 1990s [1]. They are easily obtained

*E-mail address: jerome.dejeu@univ-fcomte.fr

by the successive and alternate adsorption of two polyelectrolytes of opposite charges. In some cases, they are constructed with polymers bound via hydrogen bonds or a combination of electrostatic and hydrogen bonds [2, 3]. The plethora of recent papers and reviews addressing the buildup and the properties of these films has contributed much to our knowledge of the field [4]. Given the broad range of materials and applications potentially available, there is every reason to pursue the study of macromolecular polymer assembly [5–7]. Moreover, the layer-by-layer assembly process facilitates the elaboration of mixed or complex nanostructured materials containing small molecules, ions, proteins, DNA, or other particles [8–10]. One of the most encouraging perspectives of polymer self-assembly is the design of membranes for the controlled release of medicinal or other active molecules by tuning the permeation properties through the appropriate choice of the polymers and (or) the buildup procedure [11–15].

Although much has been learned and debated regarding the main factors affecting the construction of polymer films using the layer-by-layer process, less attention has been given to their stability. It is, however, of crucial importance to address this aspect of the elaborated materials in view of their use and the development of new applications. Several studies have shown that polyelectrolyte films are sensitive to the modification of the environmental conditions, with respect to the conditions of the buildup of the film. This is the case with a change in pH when one polymer is a weak base or a weak acid [16], a change in salt content [13], or a change in temperature [17]. Since the layer-by-layer process relies mostly on electrostatic binding between two oppositely charged macromolecules, one would expect the physicochemical parameters affecting electrostatic interaction forces to alter both the binding of the film to the surface and the cohesion of the film. However, the study of film stability is complicated by the largely irreversible character of interactions involving macromolecules, particularly in the case of the adsorption and complexation of polyelectrolytes, which are the two principal reactions involved in the growth of self-assembled polymer films.

Several papers [16, 18–20] have investigated the role of pH in the buildup of weak polyelectrolyte films, but few have described the effects of a change in pH after film buildup. The pH-induced alterations of polyallylamine hydrochloride/polystyrensulfonate (PAH/PSS) polymer films include a change in morphology [21], thickness [13], wettability [22], a release of part of the materials [23], and sometimes a complete destruction [24]. Of course, the behavior depends on the polymers used but also on the construction process, film thickness, and sometimes the type of molecule last deposited [25] or the deposition of a preliminary layer of polymers [26]. The destruction of the multilayer films has been observed at low pH values for the polyelectrolyte pairs poly(methacrylic acid) (PMAA)-poly(4-vinyl pyrinide) quaternized (QP4VP) [27] and in acid and basic medium for PMAA/P4VP [28] and PMAA/PAH [20, 24]. However, an increase in the molecular weight of the polyelectrolytes has been shown to prevent the destruction of the film but not the swelling [24].

The stability of multilayers is of great interest in the study of encapsulated material release.29,30 Several studies have been devoted to this property by Sukhorukov et al. [29–31], who made capsules in which the porosity of the membrane depended on the pH of the rinsing solutions. Upon change in pH, the polyelectrolyte membrane switched back and forth between an open and a closed state for the release of Dextran molecules. In the simple adsorption of a polymer layer, it has been shown that the adsorbed amount is not the same at a given pH whether the adsorption was made directly at that pH or via the release of an excess of polymer adsorbed at another pH where the molecules were weakly charged [32]. The literature contains a large number of examples showing that polymer adsorption is either not or only partially reversible upon dilution of the bulk solution, change of salt content, change of pH, etc [23,33,34]. The stability of single polymer layers remains unclear; that of multistep deposited films is even more ambiguous.

The present chapter addresses the stability of self-assembled polymer films, predominantly, formed with the weak base poly(allylamine hydrochloride) (PAH) and the strong acid poly(styrene sulfonate) (PSS) on silica. After construction at pH 9 in 10^{-2} M NaCl solutions, the films containing a variable number of polymer layers were immersed in polymer-free solutions at different pH values. Alteration or destruction of the films was monitored using the evolution of a laser reflectometric signal in time. In addition, some samples were examined with AFM to determine the morphological changes. These results was compared with the literature and general conclusions will be emited.

2. Laser Reflectometry

The polymer films were grown in the reflectometric cell described in detail previously [35]. Briefly, the reflectometer was equipped with an impinging jet cell into which the solution was injected through a cylindrical channel drilled into the prism. This geometry led to the formation of a stagnation point in the center of the cell to which the molecules are transported by convective diffusion. The polarized laser beam (He-Ne, 632.8 nm) reflected by the adsorbing substrate in the cell was decomposed by a beamsplitter in the parallel (p) and perpendicular (s) components with respect to the plane of incidence. The respective intensities (I_p and I_s) were measured with photodiodes, and the reflectometric signal S was calculated from:

$$S = \frac{I_p}{I_s} \tag{1}$$

Practically, the output was expressed as the ratio of $(S-S_0)/S_0$), which is usually related to the surface excess as follows [36]:

$$\frac{S - S_0}{S_0} = \frac{\Delta S}{S_0} = f(\Gamma) \tag{2}$$

where Γ is the surface excess (mg.m^{-2}). It is a sensitivity coefficient that depends on several experimental parameters [36]: the thickness and refractive index of the substrate, the refractive index increment of the polymer solution (dn/dc), and the refractive index of the adsorbed layer (n_{ads}).

In the cases of polyelectrolyte monolayer adsorption and after rinsing a self-assembled film at pH 12, the reflectometric output could be converted into the deposited weight using the relation [36]:

$$\Gamma = \frac{1}{2a}\left(-b+\sqrt{b^2+4a\left(\frac{\Delta S}{S_0}-c\right)}\right) \tag{3}$$

where a, b, and c were determined for each pH after modeling of the interface.

Reflectometric measurements started with the injection of a pH 9 polyelectrolyte-free solution containing NaCl 10^{-2} M for 5 min to record S_0. The polyelectrolyte solutions were then injected alternately for 10 min to build-up the PAH/PSS films. No rinsing was done between adsorption steps. Since silicon oxide is charged negatively [37], the PAH solution was injected first. At the end of the build-up, polymer-free solutions were circulated at different pH values but with the same salt concentration (NaCl 10^{-2} M). The rinsing steps lasted the time needed to reach a stationary value of the output.

3. Growth Mechanism of Self-assembled Polyelectrolyte [10,38]

Before to study the stability of the multilayer film, it is important to resume the multilayer build-up. There were two stages in the initial steps of film growth: the earlier one, during which the surface is active via the adsorption of PAH, and the later one, during which growth relied solely on interactions between the two polymers. These two stages of the build-up are shown schematically in Figure 1 and will be outlined briefly next.

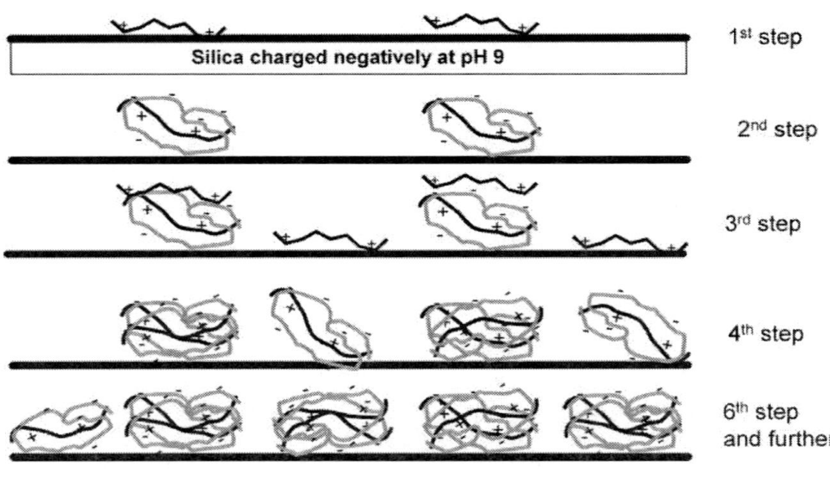

Figure 1. Sketch of the growth mechanism of the step-by-step self-assembly of two polyelectrolytes of opposite charge. Step 1: electrostatically limited adsorption of the positive polymer (PAH on silica). Step 2: formation of polymer complexes by adsorption of the negative polymer (PSS) on the first deposited PAH. Step 3: adsorption of the positive polymer on the polymer complexes PAH/PSS and on uncovered parts of silica. Step 4: adsorption of PSS on preformed PSS/PAH aggregates and formation of new pAH/PSS aggregates with isolated PAH molecules, etc. Step 6: the surface is completely covered with polymer aggregates, and further adsorption of the polyelectrolytes relies only on polycation-polyanion interactions.

In the first stage, the adsorption of PAH is limited electrostatically, slightly above the neutralization of the surface charge of silica [10, 39]. At this point, the effective surface potential (streaming or zeta potential) is about 40 mV [10]. An important feature at this stage is that a significant part of the surface remains polymer free (step 1 in Figure 1). In the second stage, PSS adsorbs only on the formerly deposited PAH macromolecules. It is also limited electrostatically, up to a surface potential near -40 mV. Detailed investigations have shown PSS to bind with the free protonated groups of PAH after displacement of the Cl- counter-ions. It also binds with a large amount of newly ionized amine groups thanks to the additional protonation due to the electrostatic screening of PSS. At pH 9, we found that the binding with PSS increases the ionization of PAH to about 30 % [10]. Finally, as suggested by Hoogeven et al. [23], we presume that many of the bonds formed between PAH and silica surface in the first step break upon adsorption of PSS and free silanol groups to make new bonds in ensuing steps. Once bound, the two macromolecules form hydrophobic entities, much like they do in solution through a coacervation process [4]. A few bonds between PAH and silica keep the complexes attached to the substrate.

In the third adsorption step, PAH binds both with unbound silanol groups and with sulfate groups pointing out of the macromolecular complexes. In the fourth step, again PSS may only bind with PAH, enlarging existing PAH-covered aggregates and forming new aggregates with the isolated PAH molecules deposited in the third step. The process continues until the entire silica surface is covered with macromolecular aggregates (step 6 in Figure 1). In subsequent stages, which constitute the second growth regime, the two polymers adsorb only via polymer/polymer interactions (i.e., the surface no longer contributes to the adsorption process). The transition step between the two growth stages depends on the many parameters influencing the polymer-surface and polymer-polymer interactions. In the case of PAH/PSS in 10^{-2} M NaCl solutions, the transition took place around step number 5 [38]; in the case of Madquat/PAA polymers, we reported a transition between steps 4 and 6, depending on the pH and ionic strength [40]. For the PSS-poly(methyl vinylpyridinium) (PVP^+) pair, Hoogeven et al. [23] found a transition beyond the third layer on TiO_2 and the sixth layer on SiO_2. AFM images confirmed the mechanism described previously: aggregates show up as small islands that vary in size and number at each adsorption step [38]. The mechanism described previously was well-illustrated by Picart et al. in the build-up of films with poly(L-lysine) and hyaluronic acid on glass [41].

4. Polymer-Surface Interaction

4.1. Stability of the First Layer

4.1.1. Polyalylamine (hydrochloride) (PAH)

The adsorption of a single layer of PAH on silica was measured in the reflectometric cell at different pH values from a polymer solution at a concentration of 10 $mg.L^{-1}$ and NaCl 10^{-2} M. The film was further rinsed with a PAH-free solution at the same pH and ionic strength, which did not induce any change in the output. The variation of the adsorption with pH reported in Table 1 revealed a maximum at pH 11, which is to be expected since the driving force for adsorption is mostly electrostatic. At low pH values (pH 4), the surface charge density of silica is very low and that of PAH is high; therefore, the adsorption is significantly lower than at pH 10 where reverse conditions apply. The lower adsorption of PAH at pH 11 and 12 can be explained by its lower charge density and lower screening capacity when compared to the small counter-ions Na+, which adsorb preferentially. The same adsorption trend was found for the weak polybase poly(vinyl imidazole) (PVI) on silica, with maximum adsorption at about pH 8 [42]. Note that it is difficult to compute the charge balance between adsorbate and substrate after completion of adsorption with precision since the known charge density of each isolated compound (PAH and SiO_2) increases significantly upon interaction [10].

Table 1. Adsorption of PAH on silica wafer at different pH values (NaCl 10^{-2} M): reflectometric measurements

pH	4	5	9	10	11	12
Γ (mg.m^{-2})	0.10	0.13	0.47	0.51	0.58	0.11

A single layer of PAH was adsorbed on silica at pH 9, and then a polymer-free solution at different pH values was circulated in the cell. Figure 2 shows that the polymer layer was stable upon rinsing with a solution at pH 9 (the build-up pH) or above. Conversely, the polymer desorbed partially in more acidic solutions: more lower is the pH, more polymer is desorbed (Table 2). As stated previously, unless there is a significant contribution of specific interactions (i.e., non electrostatic binding forces), polyelectrolytes adsorb up to slightly above the neutralization of the surface charge. Thus, in these experiments, we found that the desorption of PAH did indeed follow from the decrease in the silica charge density. Table 2 gives the remaining amount of polymer after rinsing: the polymer weight was calculated using the relation (equation 3) and the appropriate values of the parameters a-c [36].

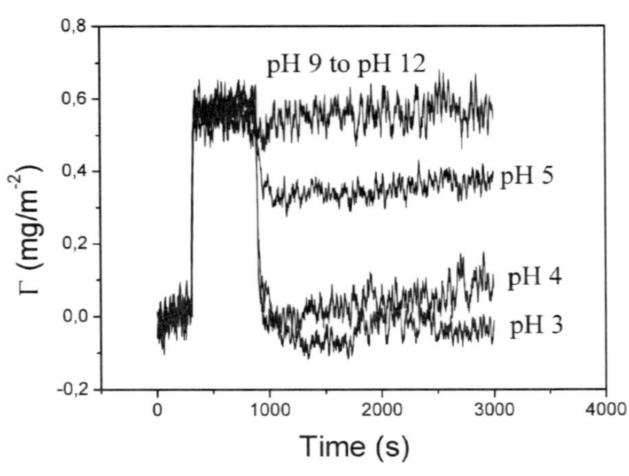

Figure 2. Reflectometric signal induced by rinsing with solutions at different pH values of a single PAH layer adsorbed at pH 9 (NaCl 10^{-2} M in all measurements). The different traces correspond to the repetition of the experiments.

Table 2. Remaining amount of polymer after rinsing a single PAH layer formed in NaCl 10^{-2} M solution at pH 9

Rinsing pH	3	4	5	9	12
remaining film wt ($mg.m^{-2}$)	0	0.1	0.35	0.53	0.53

The stability of the film when rinsing with solutions at higher pH values behaves similarly to PVI on silica [34]. Since the adsorbed amount is generally limited kinetically by an electrostatic barrier and not thermodynamically, it is process dependent. In many cases, as here, adsorption at a given pH is higher when it is obtained by the release of polymer adsorbed at a more favorable pH value than when it is adsorbed directly from the solution at that pH. This phenomenon will be discussed further in the last section. Hoogeveen et al. [23] observed also a reflectometric signal decreasing when they rinsed the Poly((dimethylamino)ethyl methacrylate) (PAMA) film at pH 4 after adsorption at pH 8 on $Ti0_2$. This decreasing are explained as desorption of the polymer. The silica and titane charge density are quite similar.

With this knowledge of the build-up mecanism exposed previously, the phenomena observed is easily understandable. Since lowering the pH decreased the surface charge density, part of the PAH molecules desorbed, leaving nearly the required amount for charge neutralization. When rinsing at pH 3, all molecules deposited at pH 9 left the surface since it was uncharged at this pH value. At pH 4 and 5, the remaining amount of PAH was 0.1 and 0.35 $mg.m^{-2}$, respectively (accuracy 5%). However, although the isolated PAH molecules were not charged, they remained adsorbed after rinsing at pH 12. There is an interesting point to mention here: in the case of direct adsorption of PAH at pH 12, the amount adsorbed was 0.11 $mg.m^{-2}$ (Table 2), which is significantly lower than the amount remaining on the film made at pH 9 after rinsing at pH 12 (0.5 $mg.m^{-2}$). The same phenomenon has been described for the adsorption on silica of another weak polybase: PVI [34]. We found that PVI adsorbed about 40% less from a solution at pH 10 than via the release at pH 10 of the excess of polymer adsorbed at pH 8. This phenomenon was explained by a kinetic phenomenon associated with the time required to detach and rearrange the entangled chains of the polymer. There is much evidence from the literature that polyelectrolyte adsorption is controlled kinetically, via the formation of an electrostatic barrier between the deposited molecules and those coming from the solution [33, 43]. One consequence is the lack of reversibility and a large dependence of the polymer uptake on the experimental protocol, as we see here. Instead, when the pH was switched from the build-up at pH 9 to pH 4 and 3, the release of PAH was immediate because the charge density of the PAH molecules increased strongly, which created an expelling force by electrostatic repulsion.

4.1.2. Poly(ethylene imine)(PEI)

The stability analyse was performed in the same condition than the PAH layer (figure 3).

The PEI layer is stable between pH 7.5 and 10. Except these pH, desorption of the cationic polymer is recorded. To compare both polycations, PAH and PEI, their reflecto-

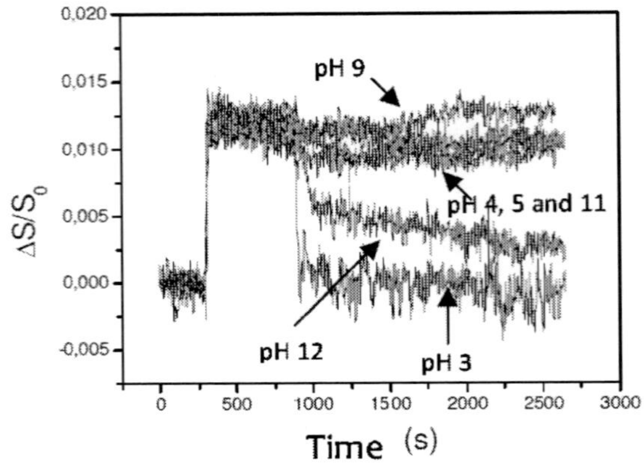

Figure 3. Reflectometric signal induced by rinsing with solutions at different pH values of a single PEI layer adsorbed at pH 9 (NaCl 10^{-2} M in all measurements)

metric signals variations according to the rinsing pH are collected in the table 3.

Table 3. Relative variation of the reflectometric signal (in percent) versus the rinsing at Different pH Values of PAH or PEI films built at pH 9 (all experiments in NaCl 10^{-2} M solution)

Rinsing pH	PAH Charge	PAH desoprtion	PEI charge	PEI desorption
3	100 %	100 %	89 %	100 %
4	100 %	100 %	79 %	22 %
5	98 %	34 %	70.5 %	22 %
7.5	76 %	0 %	47 %	0 %
9	43.5 %	0 %	24.6 %	0 %
10	23 %	0 %	12.3 %	0 %
11	9.5 %	0 %	3.4 %	18 %
12	0 %	0 %	0 %	100 %

A variation of stability is observed between both polymers in the more acid or basic pH with regard to the pH of adsorption. In pH acid, the PEI is more stable than the PAH, while in the basic domain it is the opposite which occurs. This difference of behavior towards the solution of rinsing explains by a difference of protonation of the PEI with regard to PAH in a given pH. In the basic pH, the null PEI charge is more quickly affected and involve a polymer desorption more important. On the other hand its charge is less important in the acid pH, so its desorption is weaker than PAH. In pH 3, it is the silica null charge which involves a total desorption of the polyelectrolytes.

4.2. Stability of Multilayer Film

The multilayer film study in this part are the multilayer when the substrate are not totaly cover by the polyelectrolyte. So the number of layers deposited is function of the substrate and of the polyelectrolyte couples [10,23]. For the PAH/PSS on silica adsorbed at pH 9, as demonstrated by Dejeu et al. [10, 38], 5 layers are necessary to cover the substrate.

4.2.1. Film Terminating by PSS

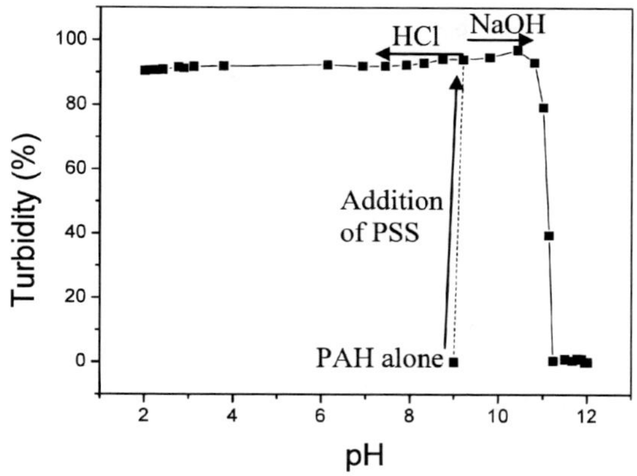

Figure 4. Turbidimetric followup of the influence of pH on the stability of PAH/PSS aggregates formed at pH 9 in solutions containing NaCl 10^{-2} M. Change of pH with NaOH 1 M or HCl 1 M.

Coacervats in liquid medium: To illustrate more clearly, the agregate stability domain, a mixture of the two polymers at pH 9 are prepared in the same conditions as the build-up in the reflectometric cell. As expected, the system was highly turbid, due to the formation of PAH/PSS aggregates. We did not characterize the precipitated units, we merely added a concentrated NaOH solution stepwise, under strong magnetic stirring, to raise the pH and measured the turbidity of the system continuously. Figure 4 shows that the suspension did indeed become perfectly clear at pH 12 (i.e., aggregates broke, and the two polymers dissolved completely).

SiO_2(PAH/PSS)$_{x_{1or2}}$ Layers: PSS terminating films with different numbers of bilayers, SiO_2(PAH/PSS)$_x$, with x varying between 1 and 2 were produced at pH 9 and subsequently washed with polymer-free solutions at different pH values between 4 and 12. The results were presented Figure 5 and Table 4.

An unstability of the multilayer film terminating by PSS, so when aggregates between the two polyelectrolyte were formed, has observed in basic solution and a stability of ones in acidic solution. These results were explained in the section 5.2.1..

4.2.2. Film Terminating by PAH

SiO_2(PAH/PSS)$_{x_{1or2}}$PAH Layers: Similar experiments were performed with films ending with PAH; they can be written as SiO_2(PAH/PSS)$_x$PAH, with x varying between 1 and 2. These films were also unstable (Figure 6 and Table 5) upon rinsing at a pH above and below the build-up pH. The remaining amount of polymer after rinsing at pH 12 showed no specific trend with the film thickness.

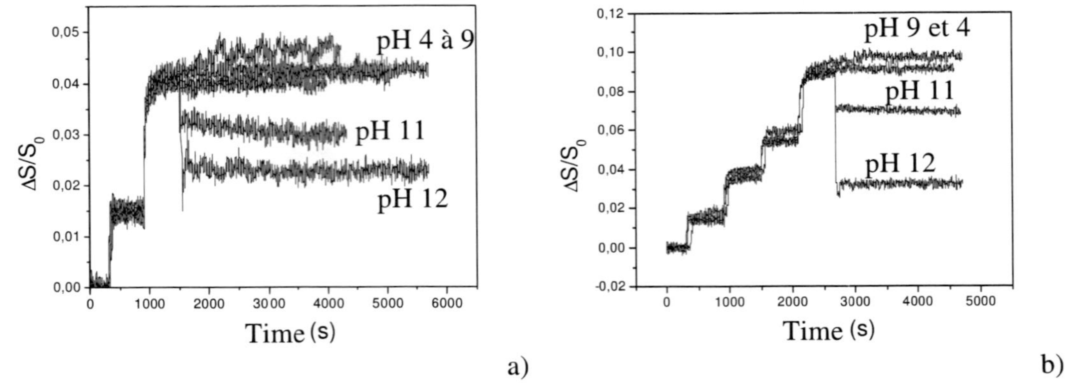

Figure 5. Reflectometric signal variation induced by rinsing with solution at different pH values. After construction, the films were rinsed at the different pH: a) $SiO_2(PAH/PSS)_1$, b) $SiO_2(PAH/PSS)_2$

Table 4. Polymer Release (wt %) and Remaining Weight (mg.m^{-2}) after Rinsing at Different pH Values of Films Built at pH 9 (all experiments in NaCl 10^{-2} M solution.)

	Rinsing pH				
Number of layers	4	5	9	11	12
2	0 %	0 %	0 %	29 %	45 % (0.69)
4	0 %		0 %	26 %	63 % (0.97)

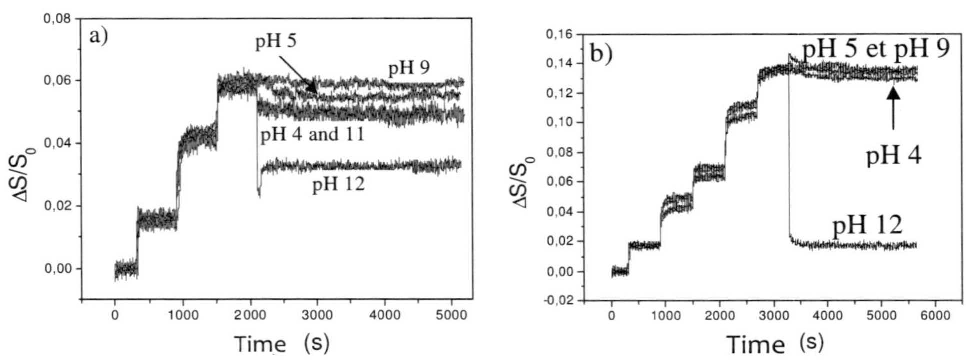

Figure 6. Reflectometric follow up of the build-up of films at pH 9 and consecutive injection of the solutions at different pH values (NaCl 10^{-2} M): (a) films containing three layers and (b) films containing three layers. The different traces correspond to the repetition of the experiments.

Although no bonds were expected between PAH and silica at pH 5 and below (shown previously by the total desorption of the single PAH layer), the film was only partially desorb of the surface. When PAH was the last layer deposited, a fraction of the molecules

Table 5. Polymer Release (wt %) after Rinsing at Different pH Values of Films Built at pH 9 (all experiments in NaCl 10^{-2} M solution)

	Rinsing pH				
Number of layer	4	5	9	11	12
1	100 %	34 %	0 %	0 %	0 %
3	12 %	2 %	0 %	18 %	34 %
5	2 %	0 %	0 %		87 %

had adsorbed as discrete units on the silica surface and majoritary on the aggregates top, as illustrated by step 3 (or step 5, not shown) in Figure 1. Thus, when the solution became more acidic, a fraction of the isolated PAH desorbed upon electrostatic repulsion since the charge density of silica decreased and that of PAH increased. The same phenomenon took place in the desorption from a single layer of PAH in Table 5. We are shown previsouly the unstability of the PAH layer on the silica wafer. So the reflectometric signal decreasing are as a consequence of the PAH desorption in interaction with the substrate. Indeed, as demonstrated in the previous section (section 4.2.1.) the hydrophobic PAH/PSS aggregates were stable and not released from the interface due to the cohesive strength of the whole polymer film. Conversely, the film broke at a pH higher than the build-up pH, due to the decrease in the charge density of PAH and the corresponding decrease in the bonds between PAH and PSS (more explaination in section 5.2.1.).

SiO_2PEI(PSS/PAH)$_{x_{1or2}}$ Layers: In this part, the influence of the pre-anchroring PEI layer on the film stability is study. After this first PEI layer, the multilayer film is composed of $(PSS/PAH)_1$ and $PSS/PAH)_2$. The variation of the reflectometric signal is presented Figure 7.

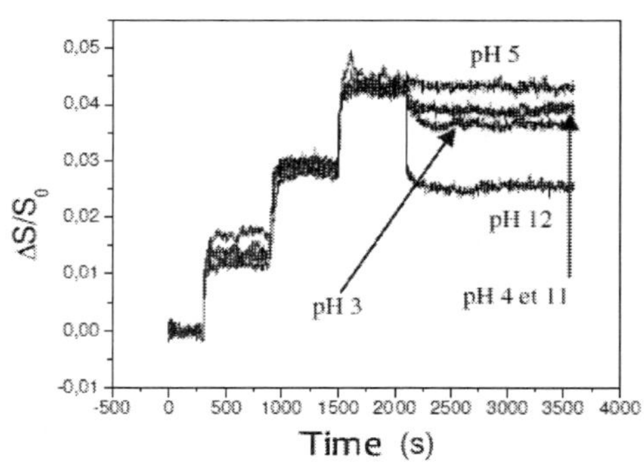

a)

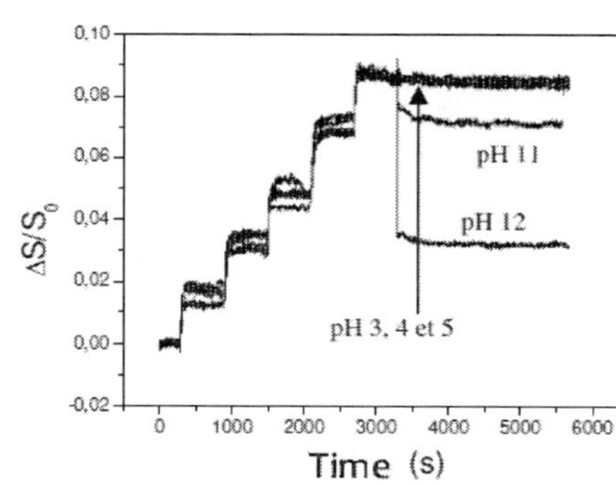

b)

Figure 7. Variation of the reflectometric signal for the build-up of self-assembled films at different pH values. After construction, the films were rinsed at the same pH as the build-up: a) SiO_2PEI$(PAH/PSS)_1$, b) SiO_2PEI$(PAH/PSS)_2$

In pH 5, the multilayer is stable but below this pH it is degraded. In pH more basic than

the adsorption, the unstability of the self-assembled layer is always present. However for all this pH the stability of the film is better than this consisted of three layers of (PAH/PSS) polyelectrolytes without PEI (Table 6).
The deviation of the reflectometric signals (figure 7.a) has explained by the unstability of the PAH layer adsorbed directly on the silica. Indeed, the PEI molecules, during the first layer were not able to cover perfectly the silica surface because of the electrostatic repulsion between the PEI molecules but also by the steric blocking resulting from molecules already adsorbed with high molecular weight (750 kDa). During the second injection of cationic polymer, here the PAH, it fill the space between the aggregate left free by the PEI/PSS adsorbed previously.
During rinsing in pH acid (figure 7.a), the decrease of the silica charge and the increase of that of the PAH involve a desorption of this last one. In pH basic item, aggregates (PEI/PSS/PAH) are destabilized by the charge loss of cationic polymers. However, the PEI charge is weaker than the PAH when the adsorption at pH 9 is performed. The (PEI/PSS/PAH) aggregates electroneutrality is assured by more important number of sodium counter-ions that (PAH/PSS/PAH) coaservats. During the change of pH, the charge variation inside the (PEI/PSS/PAH) aggregates is less important than that in (PAH/PSS/PAH) and they are thus less destabilized.

Table 6. Comparaison of the reflectometric signal unstability for 3 and 5 layers with or without an anchroring PEI layer (all experiments in NaCl 10^{-2} M solution)

Rinsing pH	desorption for 3 layers		desorption for 5 layers	
	PAH	PEI	PAH	PEI
3	100 %	100 %	89 %	100 %
4	100 %	100 %	79 %	22 %
5	98 %	34 %	70.5 %	22 %
11	9.5 %	0 %	3.4 %	18 %
12	0 %	0 %	0 %	100 %

A perfect stability is observed for rinsing from pH 3 to 9. The stability of the multilayer film is so no function of the substrate charge variation. Only two cationic polyelectrolyte layers are necessary to cover totally the silica surface when the first layer is realized with PEI. Also the Table 6 shows an improvement of the film stability in the basic pH for the same reasons as to explain to the above.
A first PEI layer allows to improve the stability of the multilayer film. Indeed, only three polycationic adsorptions are necessary for the obtaining of a good stability towards a acid pH change. This observation are in accordance with the results of Kolasinska et al. [26].

5. Polymer-Polymer Interaction

5.1. Rinsing with the Same Build-up Condition (pH and Ionic Strenght)

Films containing five bilayers were deposited on silica at different pH values in the reflectometric cell by the successive and alternate adsorption of PAH and PSS from solutions of each polymer containing also NaCl at a concentration of 10^{-2} mol.L^{-1}. These films were further rinsed by circulating a polymer-free solution at the same pH and salt concentration as the build-up system. All experiments were repeated 3 times and yielded similar results within a range of 5%. Figure 8 presents the stepwise output of the reflectometer after each addition of a polymer solution.

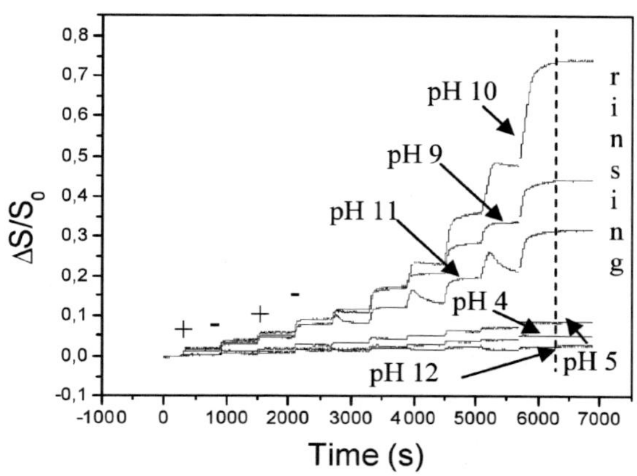

Figure 8. Variation of the reflectometric signal for the build-up of self-assembled films at different pH values. After construction, the films were rinsed at the same pH as the build-up.

Since silica is negatively charged, the first polymer added was PAH. Considering that the step height is approximately proportional to the deposited amount [44], we can note that the deposited weight of the polymer varied considerably with the pH, exhibiting a maximum at about pH 10. Again, this result is explained by the pH dependence of the charge densities of PAH and silica. PSS is a strong acid that is fully ionized at all pH values explored in this investigation. In addition, PSS only binds with the positive PAH molecules; it does not adsorb directly on silica. Therefore, since the adsorption of PSS is limited to approximately the level of charge neutralization of the previous deposited film, it is actually determined by the adsorption of PAH, which means that the dependence of the film growth on pH follows from the dependence of the adsorption of PAH on silica in the first step.

Another feature in Figure 8 is the progressive increase of the deposited weight in the successive adsorption steps. This is clearly seen in higher steps for PSS at pH 9, 10, and 11 and it is illustrated and discussed in a previous work at pH 9 [38]. This phenomenon has been reported in several papers where distinction was drawn between a linear and an exponential growth of the film with the step number [45–47]. In the first case, each polymer adsorbed the same amount in successive steps; in the second case, at least one of the two polymers progressively adsorbed an increasing amount. These different growth rates have been explained differently [4]. Several investigations support that the progressive increase in adsorption of one polymer is caused by the diffusion of this polymer throughout the deposited film [48], rather than a straight deposition on the existing film, which would produce a constant deposited weight. In many cases, the stepwise growth of the film changes from exponential to linear [49]. Our former investigation addressing the charge balance of PAH/PSS films on colloidal silica showed that PSS macromolecules bind with amine groups far inside the formerly deposited PAH layers, thus demonstrating the penetration of the last deposited PSS in the film [45]. The same phenomenon did not appear to take place with PAH.

In the case of PAH at pH 11, the reflectometric output sometimes presented a peculiar shape with a sharp increase and a slow decrease that lasted for most of the recording, about 10 mn (Figure 8). These overshoots have been described in the literature, mentioning two principal causes: either a fast adsorption process followed by the dissolution of a polyelectrolyte complex [50] or a slow rearrangement of the macromolecules in the film consecutive to the initial adsorption step [51]. Both explanations make sense; however, parallel measurements using reflectometry and a quartz microbalance did not show any decrease in film weight during the overshoots [51, 52]. Moreover, recent calculations have revealed that the drift of the reflectometric output may result from an optical artifact related to the change of the refractive index and/or the thickness of the polymer film [40, 53]. In our opinion, no desorption took place during the overshoot.

When rinsed with polymer-free solutions, at the same pH and salt concentration as the build-up, no alteration in reflectometric output was detected. All films were stable, which is reminiscent of the known irreversibility of the adsorption of single polymer layers [39]. Similar observations have been reported in other investigations, which also showed that although the output signal was not affected by rinsing, it did induce a much higher adsorption in the subsequent steps, and the overshoots vanished [23].

5.2. Rinsing with Different Conditions to Ones of Build-Up

5.2.1. Stability of $SiO_2(PAH/PSS)_x$ and $SiO_2(PAH/PSS)_{x\geq3}PAH$ Layers

Reflectometry analyse: PSS terminating films with different numbers of bilayers, $SiO_2(PAH/PSS)_x$, with x varying between 3 and 5, and PAH terminating film with more than 5 layers, $SiO_2(PAH/PSS)_xPAH$, with x superior to 5 were produced at pH 9 and subsequently washed with polymer-free solutions at different pH values between 4 and 12. The washing solutions contained the same concentration of NaCl (10^{-2} M) as the build-up solution. To facilitate comparison, the outputs for a given value x were set graphically at the same level at the end of the build-up (i.e., before beginning the rinsing experiments). In fact, the build-up experiments were highly reproducible: differences between the output in the last adsorption step did not exceed 5% for any film sequence.

The results in Figure 5.2.1. may be summarized as follows. Films were not stable when

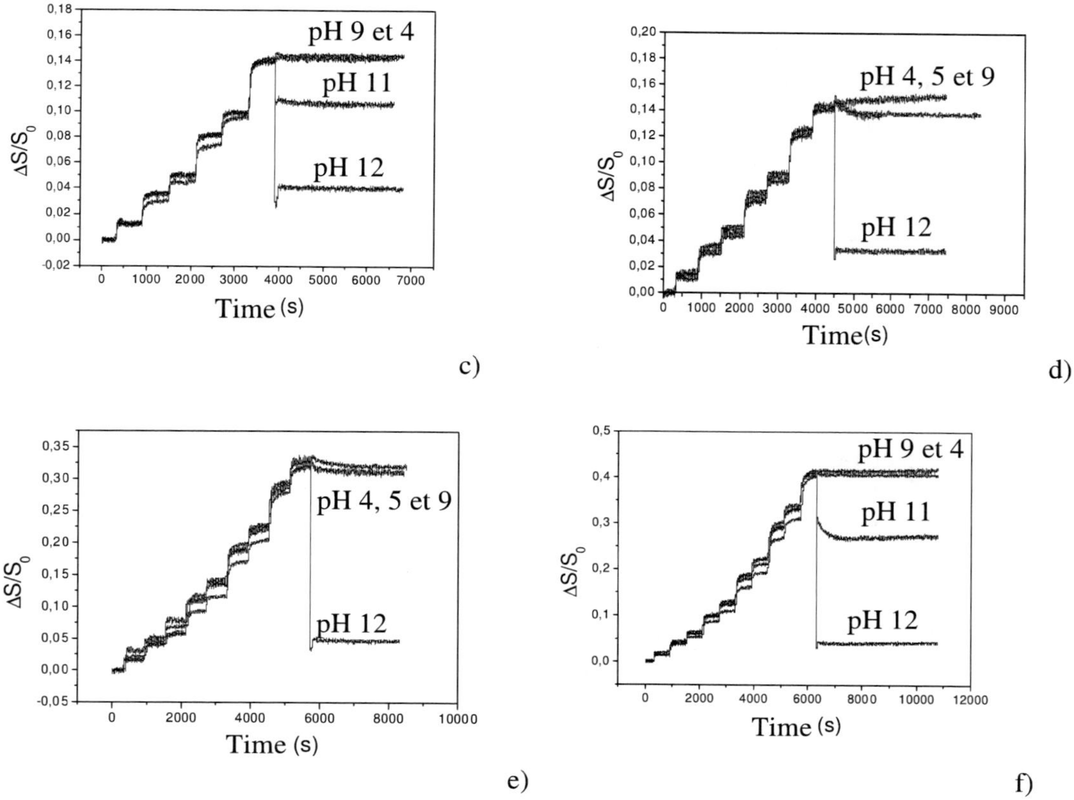

Figure 9. Variation of the reflectometric signal for the build-up of self-assembled films at different pH values. After construction, the films were rinsed at the same pH as the build-up: c) $SiO_2(PAH/PSS)_3$, d) $SiO_2(PAH/PSS)_3PAH$, e) $SiO_2(PAH/PSS)_4PAH$, f) $SiO_2(PAH/PSS)_5$

the pH of the rinsing solution was more basic than the build-up pH; they released most of the macromolecules. On the contrary, no alteration in the films was detected upon rinsing with a solution that was more acidic than the build-up pH. Assuming that the remaining polymer film was a single layer, after rinsing at pH 12, the remaining polymer density was calculated from the final output and equation **??**. Values collected in Table 7 and 4 show a progressive increase with the number of bilayers in the film (0.69, 0.97, 1.24, and 1.72 $mg.m^{-2}$).

AFM analyses: AFM pictures were consistent with the reflectometric observations. Figure 10 shows the typical morphology of a polymer film containing five bilayers deposited from polymer solutions at pH 9 and NaCl 10^{-2} M. Films were characterized by a continuous polymer background with a mean thickness of about 13 ± 3 nm and small thicker islands [38]. In a former study, we showed that the islands contained aggregates of PAH/PSS complexes [38]. The origin of the particular features of the film will be discussed in the last section. After rinsing at pH 4, the picture showed no significant difference, as was to be expected from the stable reflectometric measurements, whereas rinsing at pH 12 caused a

Table 7. Polymer Release (wt %) and Remaining Weight (mg.m^{-2}) after Rinsing at Different pH Values of Films Built at pH 9 (all experiments in NaCl 10^{-2} M solution.)

	Rinsing pH				
Number of layer	4	5	9	11	12
6	0 %		0 %	25 %	71 % (1.24)
7	0 %	0 %	0 %		77 % (0.92)
8	0 %		0 %	27 %	75 % (1.72)
9	0 %	0 %	0 %		86 % (1.43)
10	0 %		0 %	33 %	90 % (1.24)

considerable alteration in which most of the islands disappeared.

Films of higher ranks (four adsorption steps and above) built at pH 9 were stable when the pH was lowered, but they released part of the macromolecules in more basic media, which was contrary to the behavior of the PAH monolayer. The stability in acidic media is explained by the cohesive strength of the macromolecular aggregates. When the pH was switched to lower values, the charge density of PAH increased and so, consequently, did the number of electrostatic bonds between PSS and PAH (for isolated macromolecules, the amine groups are totally protonated at about pH 7 and below). Although no bonds were expected between PAH and silica at pH 4 and below (shown previously by the total desorption of the single PAH layer), the hydrophobic PAH/PSS aggregates were not released from the interface due to the cohesive strength of the whole polymer film. Conversely, the film broke at a pH higher than the build-up pH, due to the decrease in the charge density of PAH and the corresponding decrease in the bonds between PAH and PSS. In addition to the lower binding strength, the negative charge of PSS turned in large excess of the positive charge of PAH, which induced electrostatic repulsion between PSS molecules and with the negative silica surface. Therefore, one is inclined to think that the signal's decrease reveals the desorption of both kinds of molecules. The remaining amount of polymer showed no particular trend; it varied erratically between 0.5 and 1.4 mg.m^{-2}, which was the same or significantly higher than the single layer of PAH adsorbed at pH 9 (0.53 mg.m^{-2}). We did show that the single layer of PAH formed at pH 9 was stable upon rinsing at pH 12, which we attribute to the electrostatic bonds remaining at this pH, owing to the charge induced on PAH by its interaction with silica and to the long time required to rearrange and release the excess of entangled macromolecules. In comparison, selfassembled polyelectrolyte films were not stable upon rinsing at pH 12, but the remaining amount was at least that found for a single PAH layer. We presume that the PAH/PSS aggregates forming the film (Figure 1) broke up when the electrostatic bonds became too low, due to the decrease in charge of PAH at higher pH values. As shown elsewhere [4, 21, 23, 24, 27, 28], there is a critical polymer charge density below which polyelectrolyte films or aggregates do not form; therefore, in the case of weak polyelectrolytes, which are pH/charge dependent, there should be a critical pH for the stability domain.

An interesting feature was observed in the earlier steps of growth (see Table 7). Films ending with a PSS layer were completely stable in acidic media, whereas films ending

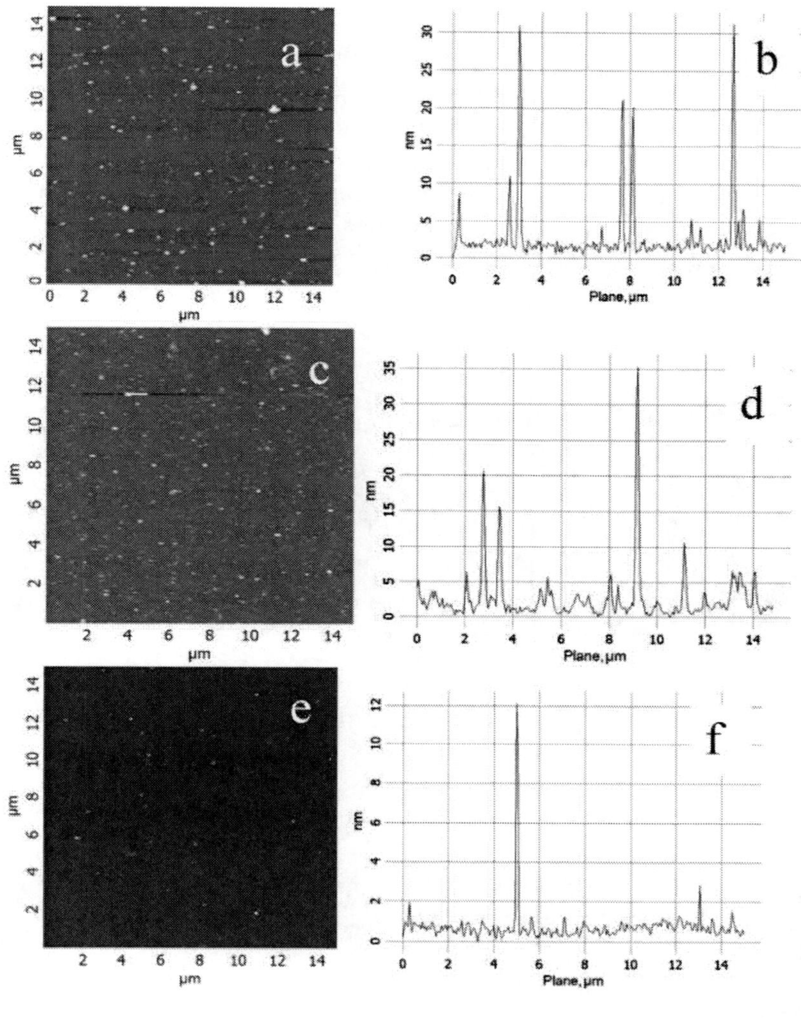

Figure 10. AFM contact mode topographic images of $SiO_2(PAH/PSS)_5$ films. Effect of rinsing with NaCl 10^{-2} M solution at different pH values and the corresponding z-profiles along an arbitrary horizontal line. (a and b) Film after build-up at pH 9; (c and d) film rinsed at pH 4; and (e and f) film rinsed at pH 12.

with PAH released part of their content: 12 wt % after step 3 and 2 wt % after step 5. No release was detected upon acidification at step 7 and above. This is explained by the growth mechanism sketched in Figure 1. In steps 2 and 4, there was no release of polymer from the film since all PAH molecules formed insoluble aggregates with PSS. As stated previously, polyelectrolyte aggregates are stable upon a decrease in pH since the binding strength of the polyelectrolyte complexes increases with the related increase of the charge density of PAH.

Confocal analyses [31]: Dejugnat et al. [31] always show the unstability of the hollow PAH/PSS capsules in basic pH. These capsules were prepared in two stages: first the assembly of PAH/PSS multilayers on the surface of $MnCO_3$ or PS particles was performed by the LbL technique. The pH of the polyelectrolyte solution was pH 6. Washing out nonadsorbed polymer molecules followed each adsorption step. After the desired number of layers was reached, the sacrificial core was removed to obtain hollow microcapsules. The decomposition and removal of the MnCO3 cores were achieved by a 0.2M citric acid solution in 0.1 M HCl for 1 h. In the case of covered PS particles, the core was not destroyed but only dissolved overnight using THF. The stability of the core was followed directly by confocal microscopy (Figure 11 and 12).

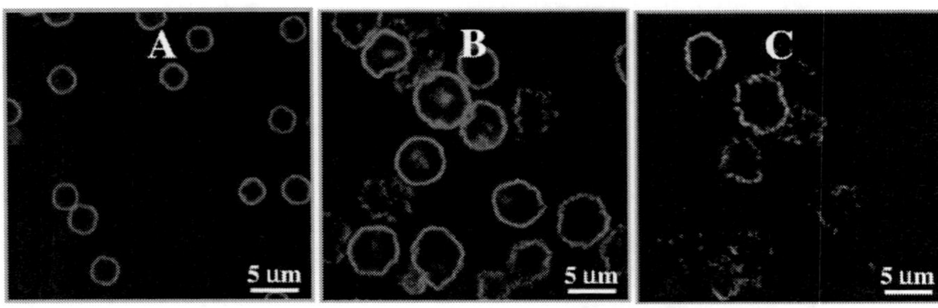

Figure 11. Dramatic swelling and dissolution of capsules templated on MnCO3 in the presence of 0.1 M NaOH: (A) in water, (B) 1 s after NaOH addition, (C) 3 s after addition, as recorded by confocal laser scanning microscopy (CLSM).

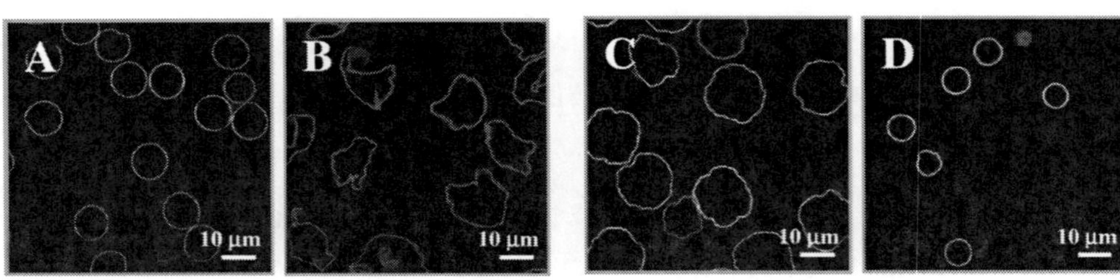

Figure 12. Influence of pH on $(PAH/PSS)_6$ (after dissolution on PS core): (A) in water, (B) 1 s after NaOH addition, (C) 1 min after NaOH addition, (D) after 0.1 M HCl addition, as recorded by CLSM.

When 0.1 M NaOH (pH = 12.5) was added to capsules templated on $MnCO_3$ cores, they dissolved after a remarkable swelling (at least 3 times increased size) regardless of the number of adsorbed polyelectrolyte layers (Figure 11). The majority of the ammonium groups of PAH become deprotonated (e.g.,RNH_2) upon raising pH, and only a certain limited number of the NH_3^+ and SO_3^- charge couples are maintained. This kind of pH-induced disintegration of capsules has already been described for systems based on poly(methacrylic acid), in which the initial cohesion between the multilayers ensured by hydrogen bonds was reduced by deprotonation of the carboxylic groups [54]. Similar pH sensitivities have been reported for peptide microcapsules and microspheres containing carboxylate surfaces, and they have been monitored using Raman spectroscopy [55–57]. The remaining charges are mainly the negative ones due to PSS. Because of electrostatic repulsion, the polyelectrolytes increase their mutual distance, leading to the swelling of the structure. In addition, the few remaining stabilizing electrostatic interactions are not strong and numerous enough to keep the integrity of the structure.

In the case of PS-templated capsules, swelling in the presence of 0.1 M NaOH was also observed (Figure 12), but it was less pronounced, as their size only increased by 1.5-2 times (depending on the number of layers). These swollen structures did not dissolve within seconds, enabling one to record images. However, in approximately 30 min, the capsules disappear due to complete dissolution because the stabilizing electrostatic interactions were removed.

Finally, we see that the stability of the PAH/PSS films is easily explained by the stepwise growth sketched in Figure 1, which is also consistent with AFM pictures in Figure 10 and other film characteristics described in former investigations [10, 38]. As expected, we also see that the stability of self-assembled polyelectrolyte films is intimately related to that of complexes formed in the solution (coacervates), as has been stated in the literature [4] but seems largely ignored in many investigations.

5.3. Other Polyelectrolyte Couples

5.3.1. Multilayer Formed with Strong Polyelectrolyte

The strong polyelectrolyte charge do not change when the pH solution is modified. So, the interaction between the two polymers into multilayer film is not modified during the rinsing step (Figure 13 when ϕ is near to zero). Indeed, with the couple PDADMA/PSS (Polydiallyldimethylammonium), Sui and Schlenoff [58] observed a non variation of the film thickness build at pH 2 and rinsing at pH 11 contrary to the result obtained with PAH/PSS.

5.4. Multilayer Formed with a Strong and a Weak Polyelectrolyte

PDADMA-co-PAA/PSS (poly(acrylic acide)) [58]: The multilayer was build by alternate polyelectrolyte adsorption and rinsing at pH 2 by spin coating deposition. To assess the stability of various combinations of polyelectrolyte multilayers, they were subjected to "pH shock" by immersion in pH 11 solution (adjusted by NaOH) for 60 min. Ellipsometric thicknesses were recorded before and after this treatment, and the loss in thickness is plotted in Figure 13 as a function of composition.

At the extreme of pure copolymer, the entire multilayer was lost because the PDADMA charge was compensated by the PAA charge in the copolymer and can not interact with the PSS charge to stabilise the multilayer. For the blends, intermediate behavior was observed, which showed an interesting dependence on the surface charge of the as-made polyelectrolyte multilayers: films terminated with the polycation mixture appeared to be more resilient to pH jumps than those terminated with PSS. Films with negative surface charges exhibited erosion of polymer with any copolymer present, while positive-capped

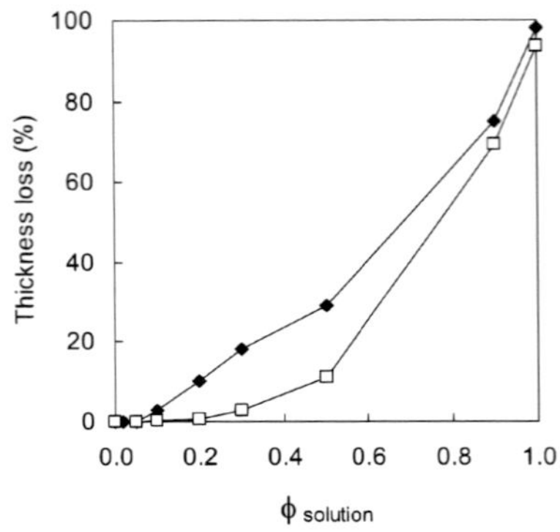

Figure 13. Thickness loss (%) vs mole fraction of the copolymer in the polycation solution ($\phi_{solution}$). (PDADMA-co-PAA$_{\phi}$PDADMA$_{1-\phi}$/PSS)$_{10}$ (black diamonds) and (PDADMA-co-PAA$_{\phi}$PDADMA$_{1-\phi}$/PSS)$_{10}$PDADMA-co-PAA$_{\phi}$PDADMA$_{1-\phi}$ (squares) were deposited on Si wafers (1.0 M NaCl, pH 2) and then exposed to pH 11 solution for 60 min. [58]

polyelectrolyte multilayers with up to about a nominal copolymer mole fraction of 0.2 were stable [58].

5.4.1. Multilayer Formed with Weak Polyelectrolyte

PAMA/PSS (poly((dimethylamino)ethyl methacrylate))) [23]: The PAMA/PSS multilayer film was analysed by laser reflectometry by Hoogeven and co-worker [23]. The film was build at pH 8 and rinsing at pH 4 (Figure 14).

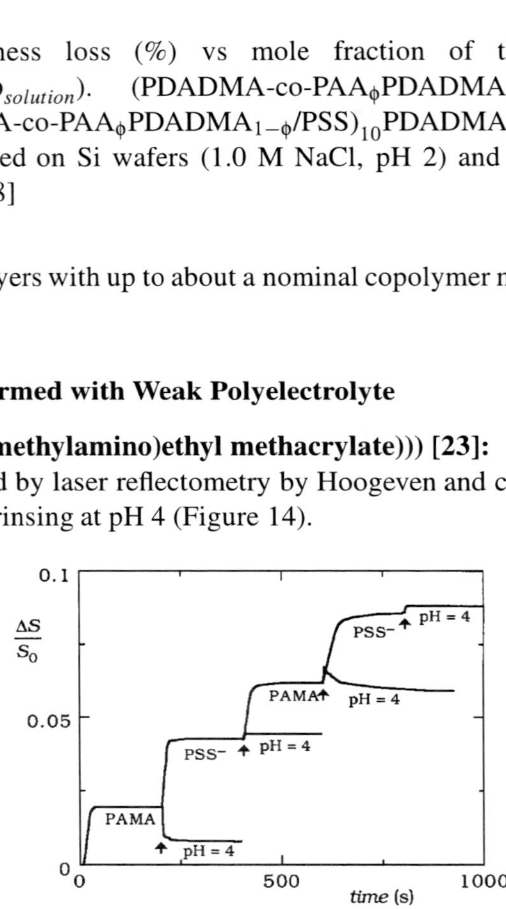

Figure 14. Stability of PAMA/PSS multilayers against changes in the pH. In this experiment multilayers of one, two, three, or four layers were formed on TiO_2 at pH 8. After that the pH was decreased from 8 to 4. The switch either to a new polymer solution or to the buffer solution at pH 4 is indicated by the arrows. The ionic strength in all solutions was 0.005 mol.kg^{-1} (barbital-acetate buffer), and the polymer concentrations were 10 mg.kg^{-1}. [23]

When the pH in the solution is switched from 8 to 4, the response of the system depends on the number of underlying layers. For multilayers with two, four, or eight layers

(the latter is not shown in Figure 14) a slight increase in the signal is observed, whereas for a multilayer of three layers the signal first increases and then decreases again. One is inclined to think that the increase in the signal indicates extra adsorption, but this is highly unlikely, since no polymer is supplied. Another reason for an increase in the signal could be an increase in dn/dc of the polymer (see equation **??**). It was reported before [59] that the dn/dc of PAMA depends on its degree of protonation. Its value ranges from 0.14 $mg.kg^{-1}$ (uncharged, high pH) to 0.21 $mg.kg^{-1}$ (fully charged, low pH). Therefore, an increase in the signal upon changing the pH from 8 to 4 is more likely to be due to a higher dn/dc of PAMA. The absence of a decrease in the signal (no desorption) when PSS is the outer layer (for an even number of layers) suggests that the multilayer is stable enough to overcome increases in the polymer charge of PAMA, so that PAMA is effectively trapped inside the multilayer [23]. When PAMA forms the outermost layer, some desorption is indeed observed as a decrease of the signal after an initial increase. The desorbed molecules are probably loosely attached PAMA molecules in the outer layer [23].
These observations were similar to the previous results on PSS/PAH.

PAH/PMAA (acide poly(methylmacrylate)) [24]: Mauser et al. [24] study the hollow PAH/PMA capsules stability versus the pH storage and the influence of the polyelectrolyte molecular weight. The hollow capsule were obtained after alternate polyelectrolyte adsorption on silica particles and dissolution of its with fluoridric acid. The pH-dependent behaviour of $(PAH/PMAA_{75kDa})_4$ and $(PAH/PMAA_{750kDa})_4$ was presented Figure 15.

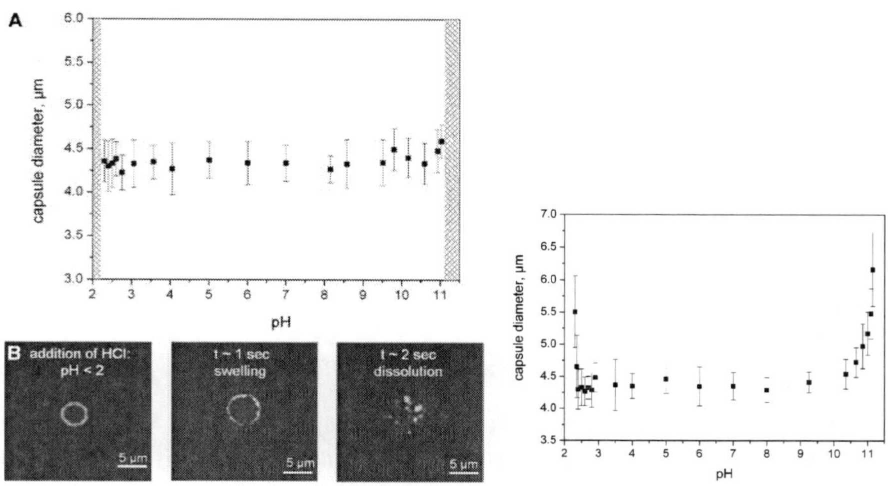

Figure 15. Diameter of $(PAH/PMAA)_4$ capsules as a function of pH. The average diameter was calculated from 20 individual capsules. The hatched areas indicate regions where the capsules are dissolved : **left** (A) Diameter of $(PAH/PMAA_{75kDa})_4$, (B) CLSM images of the dissolution process at low pH; **right** Diameter of $(PAH/PMAA_{750kDa})_4$. [24]

The $(PAH/PMAA_{75kDa})_4$ capsule are stable over a broad pH regime but dissolve at extreme high or low pH. The stability of the capsules as a function of pH as well as some CLSM images of the dissolution process is highlighted in Figure 15 (left). Shells made of 8 layers of PAH and low molecular weight PMAA are stable within the pH range from 2.3 to

11. In this pH range, the size of the capsules is nearly constant with a mean diameter of 4.4 $\pm$ 0.3 μm, corresponding to the size of the silicon oxide templates. If the pH is increased or decreased to extreme values, the shells swell and dissolve within seconds. The dissolution at low pH is shown in Figure 15.b [24].

The stability in the intermediate pH range can be explained by a global compensation of charges between the ammonium groups of PAH and the carboxylic groups of PMAA. At low pH, the carboxylic groups get protonated and thereby loose their charges. Because the ionization of the ammonium groups is not reduced, this results ii a lowered electrostatic interaction of the polymers as well as in an increased electrostatic repulsion between uncompensated positive charges of PAH. In addition small counter-ions can penetrate the layer structure to compensate the positive charges. The higher ionic concentration compared to the surrounding solution increases the osmotic pressure and results in the permeation of water into the membrane; thus the membrane starts to swell. At some point, the electrostatic interaction is not strong enough to keep the multilayers stable, and the capsules swell and dissolve. At high pH, the situation is complementary [24].

$(PAH/PMAA)_4$ capsules prepared from high molecular weight PMAA show a different pH dependent stability at the edges of the stability range than capsules with low molecular weight PMAA. The capsule size as a function of pH is shown in Figure 15. The capsules are stable within the pH range from 2.3 to 11.1 and dissolve in extreme conditions. From pH 2.4 to 10.4, the mean diameter is nearly constant with 4.4 $\pm$ 0.3 μm. Below pH 2.4, the capsules swell by 20% to a mean diameter of 5.5 $\pm$ 0.6 μm. In the pH range from 10.4 to 11.1, they swell by 35% to a diameter of 6.2 $\pm$ 0.6 μm. Both swollen structures are stable for several days. At low pH, the swelling occurs over only 0.1 pH unit, whereas at high pH, the increase in size is a more continuous transition over 0.7 pH units. One parameter influencing the stable swollen state at low pH can be attributed to intermolecular hydrophobic stabilization of uncharged PMAA that is accompanied by a conformational change of the polymer backbone from coil to globule [60–62]. In addition, uncharged polycarboxylic acids are stabilized by hydrogen bonds that can stabilize polyelectrolyte multilayers in acidic media [63].

At high pH, the entanglement of polymers seems to be the only stabilization, as no other efficient counteracting forces to the electrostatic repulsion are known for uncharged polyallylamine [24].

PMA/P4VP (poly-4-vynilpyridine) [27, 28]: The stability of PMAA/P4VP multilayer and the influence of the quaternization degree of the P4VP on the film stability were determined by Kharlampieva et al. [27] and Mauser et al. [28]. The last one observed a capsules destruction in extrem acidic pH (below 2) and a capsule skrinkage above pH 8 [28]. The P4VP ionisation studied by Karlempieva et al. [27] influences the multilayer stability in the acidic pH (figure 16).

They observed dissolution of the multilayers below pH 2, due to protonation of carboxylic groups and disruption of polymer/polymer ionic contacts. The stability of the multilayer in acidic conditions was dependent on the alkylation degree of P4VP. It is seen that the stability of the multilayers systematically improves as the alkylation degree of P4VP was increased from 12 to 37%. The exception was a PMAA/P4VP(Q90) multilayer whose stability was lower than that of the PMAA/P4VP(Q37) multilayers [27]. This

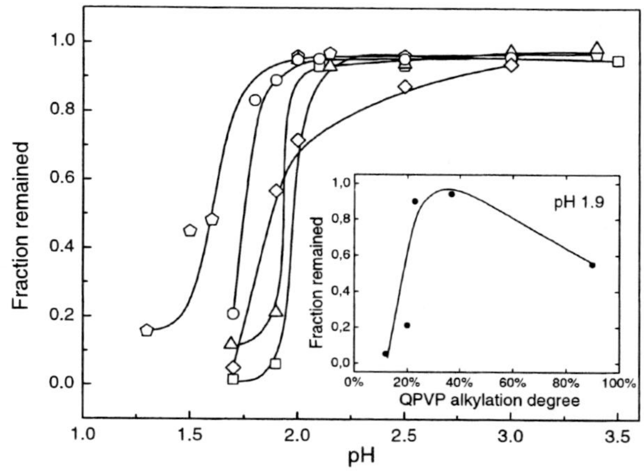

Figure 16. Thickness loss (%) vs mole fraction of the copolymer in the polycation solution ($\phi_{solution}$). (PDADMA-co-PAA$_\phi$PDADMA$_{1-\phi}$/PSS)$_{10}$ (diamonds) and (PDADMA-co-PAA$_\phi$PDADMA$_{1-\phi}$/PSS)$_{10}$PDADMA-co-PAA$_\phi$PDADMA$_{1-\phi}$ (squares) were deposited on Si wafers (1.0 M NaCl, pH 2) and then exposed to pH 11 solution for 60 min [27].

unusually low stability probably points toward the possible contribution of hydrophobic interactions on the multilayer stability. It is known that the hydrophobicity of the QP4VP chain decreases with alkylation degree [64]; this could have a negative effect on the multilayer stability.
They attributed the destruction of the multilayers to the protonation of a small ¡5% residual fraction of carboxylic groups, that was included into polymer/polymer ionic pairs and suggested that a cooperative effect in the formation of polymer-polymer ionic pairs might play an additional role in film stability [27].

These results are in according to Caruso and co-workers [3] where the films P4VP/PAA are stable up to pH 7, after which the desorbed amounts increase in the range from pH 7.5 to 9. Sukhishvili et al. [65] observed stability at neutral pH of P4VP/PMAA capsules is much higher, than usually observed for purely hydrogen-bond stabilized multilayers. For such systems the dissolution occurs at pH 6.9 (poly(vinylpyrrolidone)/PMAA), pH 4.6 (PEO/PMAA), and pH 3.6 (PEO/PAA).

6. Reconstruction of Films after Rinsing at pH 12

Films with the composition SiO_2(PAH/PSS)$_x$PAH and SiO_2(PAH/PSS)$_{x+1}$ were prepared and then rinsed successively with a solution NaCl 10^{-2} M at pH 12 and 9. After stabilization of the output, a new solution of polyelectrolyte at pH 9, either PAH or PSS, was introduced in the cell to attempt the reconstruction of the film broken at pH 12. Figure 17a-d shows the successive steps of the experiment for different films compositions (x = 2 or 3), the film construction, rinsing at pH 12, rinsing at pH 9, and the introduction of PAH or PSS. In the case of the film construction, the two stepwise curves corresponded to the repetition of the experiment, which aimed to test each of the polymer solutions.

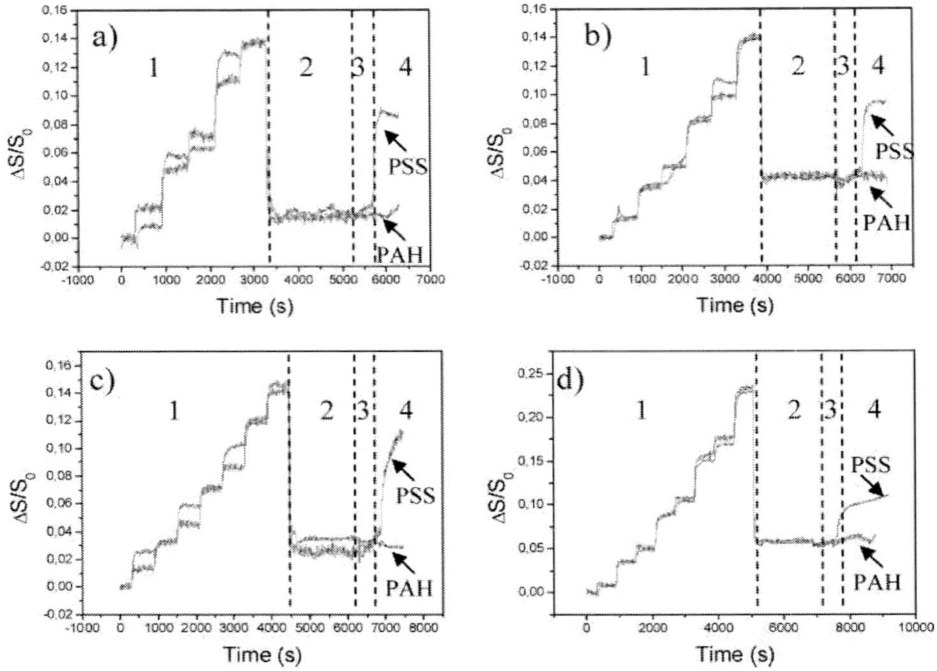

Figure 17. Reflectometic followup of the build-up of films at pH 9[1] consecutive rinsing at pH 12[2] and pH 9[3] and injection of PAH and PSS solutions at pH 9[4]. All experiments in NaCl 10^{-2} M solutions. (a) $SiO_2(PAH/PSS)_2PAH$, (b) $SiO_2(PAH/PSS)_3$, (c) $SiO_2(PAH/PSS)_3PAH$, and (d) $SiO_2(PAH/PSS)_4$. The different traces correspond to the repetition of the experiments.

After destruction of the film at pH 12 and rinsing at pH 9, the attempts of adsorption, showing that PSS only can adsorb on the remaining film, show that this film was positively charged (i.e., contained an excess of PAH) and very likely was a PAH layer directly bound with silica. In a more quantitative way, we see also that the output after PSS adsorption reached a roughly similar value in all experiments ($\Delta S/S0 = 0.10$, be aware of the different scales): approximately 3 $mg.m^{-2}$.

7. Conclusion

The stability of PAH/PSS films, and more generally the stability of polyelectrolyte films containing at least one weak polybase or polyacid, are strongly dependent on pH, due to the determining influence of electrostatic bonds between the two polymers. There is a critical pH, above or below which the film breaks, releasing a large part of the polymers. There are two reasons for the destruction of the film: (1) when one of the polymer charge densities becomes too low, the binding strength of the molecular aggregates, which act as the elementary bricks of the film, is no longer sufficient and the film breaks and (2) when the charge of one polymer in the film decreases, the uncompensated charge of the other polymer creates a repulsive interaction between the molecules of that polymer, which induces a fast release of molecules. After the film broke, we found that the fraction remaining in the interface was the polymer bound electrostatically to the substrate and that the amount was close to, and just above, the monolayer formed by direct uptake from the solution. Our results show that the stability of self-assembled polymer films depends on two major factors: the strength of the binding between the surface and the polymer of opposite charges, which dominates at earlier steps of the build-up, and the strength of the binding between the two polymers in aggregates, which allows the cohesion of the film in later steps. The multilayer film stability can be improve by increasing the layers number or the molecular weight.

Acknowledgment

This work was supported by the EU under Contract NMP4-CT-2003-001428: Nanocapsules for Targeted Controlled Delivery of Chemicals.

References

[1] G. Decher, J.D. Hong, and J. Schmitt. *Thin Solid Films*, **210-211**:831–835, 1992.

[2] A. Sukhisvili and S. Granick. *Macromolecules*, **35**:301–310, 2002.

[3] J. Cho and F. Caruso. *Macromolecules*, **36**:2845–2851, 2003.

[4] G. Decher and J.B. Schlenoff. *In Multilayer Thin Films, Sequential Assembly of Nanocomposite Materials*. 2003.

[5] A. Wu, D. Yoo, J.K. Lee, and M.F. Rubner. *J. Am. Chem. Soc.*, **121**:4883–4891, 1999.

[6] Y. Lvov. *Handbook for Surfaces and Interfaces, Nanostructured Materials*, volume 3. Academic Press, New York, 2001.

[7] J.H. Cheung, A.F. Fou, and M.F. Rubner. *Thin Solid Films*, **244**:985, 1994.

[8] E. Kharlampieva, Y.A. Izumrudov, and S.A. Sukhishvili. *Macromolecules*, **40**:3663–3668, 2007.

[9] P.J.G. Goulet, D.S. dos Santos, R.A. Alvarez-Puebla, O.N. Oliveira, and R.F. Aroca. *Langmuir*, **21**:5576–5591, 2005.

[10] Jérôme Dejeu, Lionel Buisson, Marie Céline. Guth, Cyrille Roidor, Francois Membrey, Daniel Charraut, and Alain Foissy. Early steps of the film growth mechanism in self-assembled multilayers of pah and pss on silica. polymer uptake, charge balance and afm analysis. *Colloids and Surfaces A: Physicochemical and Engineering Aspects*, **288**:26, 2006.

[11] N. Gupta, A.A. Patel, R. Nassar, Y.M. Lvov, M.J. McShane, and J.D. Palmer. *Colloids Surf., A*, **245**:137–142, 2004.

[12] M.K. Park, S. Deng, and R.C. Advincula. *Langmuir*, **21**:5272–5277, 2005.

[13] S.T. Dubas and J.B. Schlenoff. *Macromolecules*, **34**:3736–3740, 2001.

[14] G. Ibarz, L. Dähne, E. Donath, and H. Möhwald. *Chem. Mater.*, **14**:4059–4062, 2002.

[15] J. Chung and M.F. Rubner. *Langmuir*, **18**:1176–1183, 2002.

[16] T.J. Halthur, P.M. Claesson, and U. Elofson. *J. Am. Chem. Soc*, **126**:17009–17015, 2004.

[17] T. Serizawa, D. Matsukuma, K. Nanameki, M. Uemura, F. Kurusu, and M. Akashi. *Macromolecules*, **37**:6531–6536, 2004.

[18] F. Boulmedais, M. Bozonnet, P. Schwinte, J.C. Vogel, and P. Schaaf. *Langmuir*, **19**:440–445, 2003.

[19] S.E. Burke and C.J. Barrett. *Biomacromolecules*, **4**:1773–1783, 2003.

[20] J.A. Hiller, J.D. Mendelsohn, and M.F. Rubner. *Nat. Mater.*, **1**:59–63, 2002.

[21] P.M. Biesheuvel, T. Mauser, G.B. Sukhorukov, and H. Möhwald. *Macromolecules*, **39**:8480–8486, 2006.

[22] M. Kolasinska and P. Warszynski. *Appl. Surf. Sci.*, **66**:65–70, 2005.

[23] N.G. Hoogeeven, M.A. Cohen-Stuart, G. Fleer, and M.R. Böhmer. *Langmuir*, **12**:3675–3681, 1996.

[24] T. Mauser, C. Dejugnat, H. Mohwald, and G.B. Sukhorukov. *Langmuir*, **22**:5888–5893, 2006.

[25] S. Leporatti, C. Gao, A. Voigt, E. Donath, and H. Möhwald. *Eur. Phys. J. E*, **5**:13–20, 2001.

[26] M. Kolasinska, R. Krastev, and P. Warszynski. *J. Colloid Interface Sci.*, **305**:46–56, 2007.

[27] E. Kharlampieva and S.A. Sukhisvili. *Langmuir*, **19**:1235–1243, 2003.

[28] T. Mauser, C. Dejugnat, and G.B. Sukhorukov. *J. Phys. Chem. B*, **110**:20246–20253, 2006.

[29] G.B. Sukhorukov, A. Fery, M. Brunnen, and H. Möhwald. *Phys. Chem. Chem. Phys.*, **6**:4078–4089, 2004.

[30] A.A. Antipov, G.B. Sukhorukov, S. Leporatti, I.L. Radtchenko, E. Donath, and H. Mohwald. *Colloids Surf. A*, **198**:535, 2002.

[31] C. Dejugnat and G.B. Sukhorukov. *Langmuir*, **20**:7265–7269, 2004.

[32] A.W.M. de Laat, G.L.T. Van den Heuvel, and M.R. Bohmer. *Colloids Surf. A*, **98**:61, 1995.

[33] J.G. Göbel, M. Beeseling, M.A. Cohen Stuart, and C Poncet. *J. Colloid Interface Sci.*, **209**:129–135, 1999.

[34] T. Roques-Carmes, F. Membrey, M. Kaisheva, C. Filiatre, and A. Foissy. *J. Colloid Interface Sci.*, **299**:804–812, 2006.

[35] T. Roques-Carmes, F. Membrey, C. Filiatre, and A. Foissy. *J. Colloid Interface Sci.*, **245**:257, 2002.

[36] J. Dejeu, A. Foissy, F. Membrey, M. Balastre, and B. Bordes. Laser reflectometry: Examination of the influence of some assumptions and approximations to calculate the adsorption from the reflectometric signal. *Colloids and Surface A: Physicochem. Eng. Aspects*, **310**:78–85, 2007.

[37] A. Foissy and J. Persello. *The Surface Properties of Silicas*. John Wiley & Sons, 1998.

[38] S. Diziain, J. Dejeu, L. Buisson, D. Charraut, F. Membrey, and A. Foissy. Investigations in the build-up mechanism of self assembled polyelectrolyte films by laser reflectometry and atomic force microscopy. *Thin Solid Film*, **516**:1–7, 2007.

[39] G.J. Fleer, M.A. Cohen Stuart, J.M.H.M. Scheutjens, T. Cosgrove, and B. Vincent. *Polymers and Interfaces*. Chapman and Hall, London, 1993.

[40] C.C. Buron, C. Filiatre, F. Membrey, C. Bainier, D. Charraut, and A. Foissy. *J. Colloid Interface Sci.*, **314**:358–366, 2007.

[41] C. Picart, Ph. Lavalle, P. Hubert, F.J.C. Cuisinier, G. Decher, P. Schaaf, and J.-C. Voegel. *Langmuir*, **17**:7414–7424, 2001.

[42] T. Roques-Carmes, F. Membrey, A. Deratani, M. Boehmer, and A. Foissy. *J. Colloid Interf. Sci.*, **256**:273, 2002.

[43] M.A. Cohen-Stuart, C.W. Hoogendam, and A. de Keizer. Kinetics of polyelectrolyte adsorption. *Journal of Physics: Condensed Matter.*, **9**:7767–7783, 1997.

[44] J.C. Dijt, M.A. Cohen Stuart, and G.J. Fleer. *Advances in Colloid and Interface Science*, **50**:79–101, 1994.

[45] G. Decher. *Sciences*, **277**:1232, 1997.

[46] J. Ruths, F. Essler, G. Decher, and H. Riegler. *Langmuir*, **16**:8871, 2000.

[47] P. Schaaf, J.C. Voegel, and P. Lavalle. In *Proc. Natl. Acad. Sci.*, volume 99, pages 12531–12535, USA, 2002.

[48] D.L. Elbert, C.B. Herbert, and J.A. Hubbell. *Langmuir*, **15**:5355–5362, 1999.

[49] C. Porcel, P. Lavalle, G. Decher, B. Senger, J.C. Voegel, and P. Schaaf. *Langmuir*, **23**:1898–1904, 2007.

[50] D. Kovacevic, S. van der Burgh, A. de Keizer, and M.A. Cohen Stuart. *Langmuir*, **18**:5607–5612, 2002.

[51] L. Wagberg, G. Petterson, and S.J. Notley. *J. Colloid Interface Sci.*, **274**:480–488, 2004.

[52] C.C. Buron, C. Filiatre, F. Membrey, H. Perrot, and A. Foissy. *J. Colloid Interf. Sci.*, **296**:409–418, 2006.

[53] Jérôme Dejeu. *Films de deux polymères auto-assemblés* : chlorydrate de polyallylamine et polystyrène sulfonate de sodium; mécanisme de croissance et stabilité,. PhD thesis, University of Franche-Comte, 2007.

[54] V. Kozlovskaya, S. Ok, A. Sousa, M. Libera, and S. Sukhishvili. *Macromolecules*, **36**:8590–8592, 2003.

[55] R.J. Bergeron, G.W. Yao, G.W. Erdos, S. Milstein, F. Gao, J. Rocca, and W.R. Weimar H.L Price O. IV. Phanstiel. *Bioorg. Med. Chem.*, **5**:2049–2061, 1997.

[56] H. Matsui, B. Gologan, H. Schaffer, F. Adar, D. Seconi, and O. IV. Phanstiel. *Langmuir*, **16**:3148–3153, 2000.

[57] O. IV. Phanstiel, G.MW. Erdos, R.J. Lachicotte, D. Torres, M. Richardson, H. Matsui, H. Schaffer, F. Adar, and D. Seconi. *Chem. Mater.*, **13**:264–, 2001.

[58] Z. Sui and J.B. Schlenoff. *Langmuir*, **20**:6026–6031, 2004.

[59] N. G. Hoogeveen, M. A. Cohen Stuart, and G. J. Fleer. Polyelectrolyte adsorption on oxides: I. kinetics and adsorbed amounts. *Journal of Colloid and Interface Science*, **182**:133–145, 1996.

[60] M. Mandel, J.C. Leyte, and M.G. Stadhouder. *J. Phys. Chem.*, **71**:640–649, 1967.

[61] R. Arnold. *J. Colloid Sci.*, **12**:549–556, 1957.

[62] E.V. Anufrieva, T.M. Birshtein, T.N. Nekrasova, O.B. Ptitsyn, and T.V.T. Sheveleva. *J. Polym. Sci., Polym. Symp.*, **16**:3519–3531, 1968.

[63] V. Izumrudov and S.A. Sukhishvili. *Langmuir*, **19**:5188–5191, 2003.

[64] S.A. Sukhishvili and S.J. Granick. *J. Chem. Phys.*, **109**:6868, 1998.

[65] S.A. Sukhishvili and S. Granick. *J. Am. Chem. Soc.*, **122**:9550–9551, 2000.

In: Polymer Films: Properties, Performance and Applications ISBN 978-1-61209-201-0
Editors: Stefan A. Romano and George P. Somners

Chapter 6

INITIAL BUILD-UP OF SELF-ASSEMBLED POLYELECTROLYTES: MECHANISM, POLYMER UPTAKE, CHARGE BALANCE AND AFM ANALYSIS

Jérôme Dejeu*
University of Franche-Comte, FEMTO-ST Institute

Abstract

This work aims to investigate the initial build-up stages of polyelectrolyte multilayers formed by the successive deposition of the weakly charged Poly(Allyl amine, HCl), PAH and the strong acid Poly(Styrene Sulfonate), PSS, on silica at pH=9. Complementary tools were used: laser reflectometry to determine the step-by-step deposited weight of each polymer, Atomic Force Microscopy (AFM) to characterize the topography and measure the thickness of the film and adsorption on silica colloids to measure successive adsorption isotherms, the effect of maintaining a constant pH, the charge balance and the counterions co-adsorption in the successive steps. The experiments show that the deposited weight of PAH is the same at each step (around 0.5 $mg.m^{-2}$), whereas that of PSS increases more or less linearly along with the step number. Starting at the second polymer deposit (PSS), the structure of the film is very heterogeneous, with a thin polymer layer and bumps that are attributed to the presence of PSS/PAH hydrophobic complexes. The amount of bumps and their height were determined at each of the ten first polymer deposits. The film thickness between the bumps was evaluated by scratching small areas with an AFM tip. Our results support a former description of film growth, based on the electrostatic interactions between the two polymers and the substrate. They also reveal different the roles of the two polymers in the growth of the films: PAH is responsible for the progressive bi-dimensional coverage of the surface area and PSS for the increase in the thickness of the film.

PACS 68.55.-a, 68.55.A-, 68.37.Ps, 68.37.-d.

Keywords: Build-up mechanism, self-assembled, polyelectrolytes, laser reflectometry, AFM, colloids, charge compensation.

---

*E-mail address: jerome.dejeu@univ-fcomte.fr

1. Introduction

In the early nineties, self assembled polyelectrolyte film appeared and drawn great interest from the scientific community. It consists in alternately depositing oppositely charged polymers onto a substrate (the core) up to the desired thickness or shell permeability [1–3]. In addition to controlled release self-assembled polyelectrolyte multilayers have other potential applications in electronics [4] optical devices [5], membranes [6], electrocatalysis [7] and many other areas [8]. This method is advantageous in that the process is very easy to apply to a variety of substrates including solid particles [9, 10] and emulsions [11]. Moreover, recent reviews on the subject show that many theoretical and practical aspects of the interaction parameters during the growth of the polymer films remain ambiguous and actively debated [12, 13]. In order to progress in the control of the assembly of macromolecules, it is necessary to gain more and better knowledge on the dependence of film construction on the main experimental factors: polymer charge densities, pH, ionic strength and type of electrolytes. Fundamentally, the growth of the polyelectrolyte multilayer resides in the interaction forces between the two polyelectrolytes, which are electrostatic in nature. In general, it is thought that high ionization of polyions results in a lower growth rate and lower ionization gives thicker layers. Therefore, for weak polyelectrolytes, the pH is a determining parameter of the buildup mechanism, as are the type and the concentration of the salt [14].

In this chapter, the investigation in the reverse case of the strong anionic polymer poly(styrene sulfonate) (PSS) and a weak polybase, poly(allyl amine, HCl) (PAH) was improved. Investigations into the growth of PSS-PAH layers has already been reported in the literature, with contradictory results advocating either linear growth (constant weight deposit at increasing step number) [15, 16] or exponential growth (the weight uptake of one polymer increases constantly in successive steps) [17, 18]. The layer by layer growth of the film is related to the charge densities and the charge-charge stoechiometry in neighboring polycation and polyanion layers. In general, it is thought that high ionization of polyions results in a lower growth rate and lower ionization gives thicker layers. Therefore, for weak polyelectrolytes (as PAH in this investigation) the pH is a determining parameter of the growth. The other important factor in the neutralization stoechiometry is the incorporation of small electrolyte ions in the film, since they may neutralize part of the polymer charge densities. Moreover, we have formerly reported that the strong binding of small ions with one polyelectrolyte (at higher ionic strength) may hamper the binding of the other polymers [19].

2. Polyelectrolyte Adsorption

Polyelectrolyte adsorption on a charged substrate reaches more or less charge compensation, i.e. the charge density ($C.m^{-2}$) of the adsorbed polymer equals that of the surface. However, in the case of monolayer adsorption, it is generally found that the electrokinetic potential after adsorption has the same sign as the adsorbed polelectrolyte, which means that there is an excess of polymer charge density in comparison with the structural surface charge. Using the Gouy-Chapman relationship for diffuse layer charge density [20], one may calculate that the charge excess corresponding to a potential about 50 mV is very small in comparison with that of the polymer (less than 1 C.m^{-2}). Theoretical studies carried out by Cohen-Stuart et al. [21] and experimental illustrations [22] have shown that once the surface potential reaches a given value, an electrostatic repulsion is created in the interface that prevents additional molecules from reaching the substrate. In other words, polyelectrolyte adsorption is kinetically limited, with a threshold for surface saturation when the zeta potential reaches about $\pm$ 40 mV [23]. This all applies well to the present measurements.

2.1. Polymer Uptake by Laser Fixed Angle Reflectometry

2.1.1. Reflectometry Output

The polymer films were grown in the reflectometric cell described in detail previously [24]. Reflectometry delivers a signal that depends on the thickness and the refractive index of the adsorbed layer. In short, the relevant output of the reflectometer S, given by the equation $S = f\frac{R_p}{R_s}$, is an intensity ratio between the parallel (Rp) and the perpendicular (Rs) components of the polarized light beam that is reflected by the interface. The factor f in the previous equation is an apparatus parameter. Practically, all experimental data are calculated as

$$\Delta S/S_0 = (S - S_0)/S_0 \quad (1)$$

where S and S_0 are the signal ratios with and without the adsorbed layer respectively (S_0 is the reference signal). In the case of polyelectrolyte multilayers, the successive weight uptakes (Γ in mg.m^{-2}) from the change of the reflectometric output were calculated with a previously described procedure [25]. Practically, the polymer solution (100 mg.L^{-1}, 10^{-3}M NaCl) was introduced through a capillary set perpendicularly to the substrate (impinging jet cell). The reflectometric analyses were done at the stagnation point, at which the transport of molecules was purely diffusive. The liquid flow rate was 1.25 cm^3.mn^{-1}. No water rinsing was done between polymer deposits. The deposition time was set at 5 mn, although the reflectometric output was constant after 2 mn.

The reflectometer output for the successive deposition strokes of polyelectrolytes is presented in figure 2.1.1.. The first polymer layer was the cationic PAH, since the silica substrate was negatively charged. The steps in figure 2.1.1. show a steep increase upon polymer addition, followed by a plateau, which indicates that the adsorption reaction has been completed up to practical saturation.

2.1.2. Thickness and Refractive Index of the Polymer Layer and Polyelectrolyte Adsorption

From the reflectometry output, the refractive index and the thickness of the polymer film can be deducted by using the thickness method developped by Buron et al. [25], figure 2(a). The thickness increase rapidly during the four first deposited, before to stabelize near a value of 60 nm. The adsorbed polymer layer refractive index increase regulary, so the adsorption densify the polymer layer. An exclusion of the water molecules initially into the film are performed by the polyelectrolytes who replace the water molecules. With the thickness and the refractive index, the total adsorbed amount can be determined by the De Feijter relationship [26] and so the uptake for each polymer after their adsorption, figure 2(b).

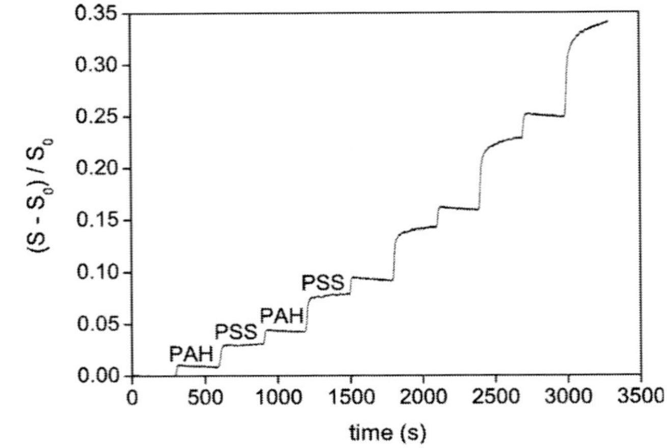

Figure 1. Reflectometric signal for alternate deposition of PAH / PSS on a silica substrate at pH=9 and with a ionic strength equal to 10^{-3} M.

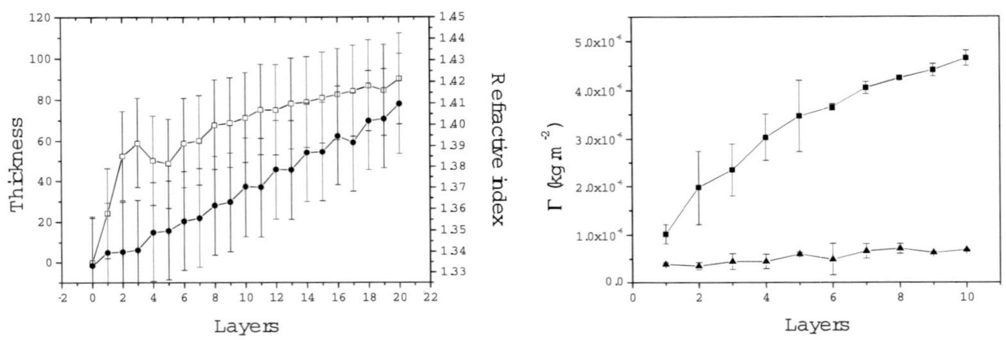

(a) (b)

Figure 2. Reflractive index, thickness (nm) (a) and amount of deposited (b) for alternate deposition of PAH / PSS on a silica substrate at pH=9 and with a ionic strength equal to 10^{-3} M.

The graph 2(b) also reveals different behaviours for PAH and PSS: PAH height remains roughly constant, whereas PSS height increases in successive steps. The output was converted into deposited weight [25], which allowed the quantitative comparison presented in figure 2(b). The deposited weight of PAH was about 0.5 mg.m^{-2} and that of PSS increased quite linearly between 1.25 mg.m^{-2} and 5.00 mg.m^{-2} during the 20 first layers.

2.2. Isotherm Adsorption by Layer by Layer Deposit on Silica Particles

2.2.1. Isotherm

The layer by layer (LbL) deposit of the two polymers was performed at pH 9 on the silica particles starting with PAH and further alternating solutions of PSS, PAH, etc... Up to 6

layers (3 bilayers) were deposited. No water washing was done between steps. In order to avoid the separation of the particles by filtration or centrifugation between successive steps to eliminate the residual polymer in the solution, we limited the addition of polymer so that little or excess remained in the solution. An indication for the choice of the appropriate amount of polymer was obtained from preliminary adsorption isotherms, as will be shown in the next section. Preliminary experiments certified that the small excess actually present (less than 10 ppm) did not alter the next polyelectrolyte deposit.
Adsorption experiments were performed by mixing 100 g of a silica dispersion (weight fraction of silica 5%) with 10 g of polymer solution at the same pH (pH 5 or 9) and different concentrations. The mixture was stirred for at least 1 h before analysis. The pH usually changed after mixing the polymer solution and the silica dispersion. Unless stated otherwise, the pH was restored to the initial value using 1M NaOH or HCl aliquots. Uptakes of polymer were calculated from the amount remaining in the solution after centrifuging the particles (Sigma 1-15 K, 15 mn, 14 000 rpm). The polymer concentration was determined using total carbon titration for PSS and PAH (Shimadzu, TOC 5050) and UV spectroscopy (255 nm, Beckman DU 520) for PSS.

Figure 3 compares the adsorption of PAH on silica between pH 5 and pH 9 as a function of the equilibrium concentration (unadsorbed) in the solution. These two pH were chosen in order to screen the deposit properties at different ionization ratios of PAH at pH 5 when the charge ratio (percentage of protonated groups) is close to 1 and pH 9 when it is close to 0.4. The pH in these experiments was not adjusted after mixing the silica and the polymer. The increase in adsorption with pH shown in figure 3 was to be expected for two reasons. First, the surface charge density of silica increases with pH and that of PAH decreases. Based on the charge compensation principle, which should roughly apply here, more polymer is required to neutralize the surface of silica at pH 9 than at pH 5. Second, it has been well demonstrated that polyelectrolyte adsorption is determined kinetically by an electrostatic barrier between formerly deposited and incoming macromolecules [27]. In the present case, adsorption at pH 9 is less prone to induce an electrostatic barrier since PAH is only slightly charged. The charge balance will be examined more thoroughly in the last section. Quantitatively the adsorption of PAH on silica resembles that of the weakly basic poly(vinylimidazole), which showed a maximum about pH 8 [28].

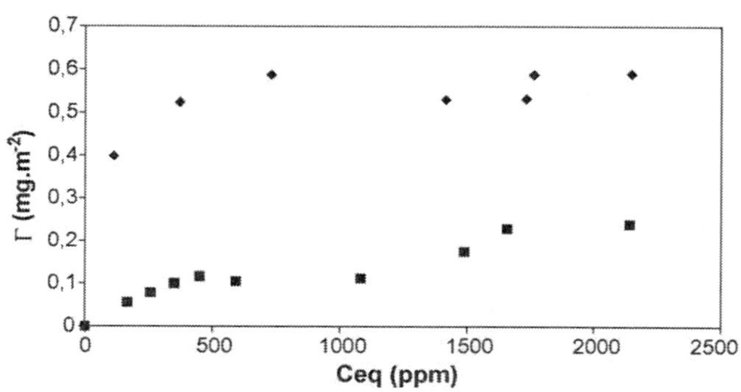

Figure 3. Adsorption isotherms for PAH at pH 5 (■) and 9 (♦) (pH not adjusted).

Figure 4 shows the successive adsorption isotherms of PAH and PSS on silica at pH 9. In this case the pH was corrected after mixing the silica and the polymer solutions. Considering the adsorption of PAH on SiO2, we see that PAH adsorbs more on SiO_2 when the pH is corrected after mixing (0. 75 mg.m^{-2} in figure 4) than when it is not (0.60 mg.m^{-2} in figure 3). This is understandable since the pH after mixing was 7.4, i.e. the surface charge of silica was lower and that of PAH was higher than when the pH was restored at pH 9. Data for the adsorption of PSS were obtained by using a PAH pre-covered silica substrate. The PAH pre-coverage was 0.7 mg.m^{-2}, which is somewhat below the saturation of the silica surface, in order to minimize the amount of unadsorbed polymer in the solution. In a separate experiment, the supernatant of the PAH pre-covered silica dispersion was analyzed and showed that the amount of dissolved PAH was below 10 mg.L^{-1} and that no precipitation was seen after addition of the PSS solution. Addition of PSS (preset at pH 9) in the dispersion of precovered silica (preset at pH 9) caused an increase in pH that was corrected before the adsorption measurements. This will be explained further on.

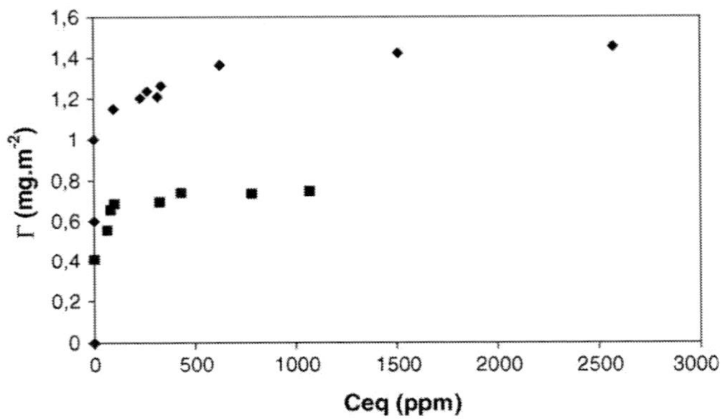

Figure 4. Adsorbption isotherms for PAH at pH 5 (■) and 9 (♦) (pH adjusted).

Similarly to the above example, further successive adsorptions of PAH and PSS were performed at controlled pH 9. Figure 5 summarizes the final data from the 1st (PAH) to the 6th (PSS) polymer adsorption. It is important to note (remember?) that each isotherm in figure 5, save the "first" one for PAH on silica, was measured on a former layer of the polymer with an opposite charge at a coverage corresponding to the top of the steep rise for that polymer. The successive polymer coverages prior to each isotherm in figure 5 are reported in Table 1. For instance, the isotherm for layer 6 in figure 5(b) (3rd PSS layer, 3 m^{-2} at the plateau) was obtained on a film made with the successive deposits of PAH (0.7 mg.m^{-2}), PSS (1.15 mg.m^{-2}), PAH (0.8 mg.m^{-2}), PSS (1.24 mg.m^{-2}) and PAH (0.88 mg.m^{-2}).

Adsorption isotherms presented in above have the usual profile with a steep rise and a saturation plateau. The steep rise is explained by the strong electrostatic attraction between the polymer and the substrate with an opposite charge. The plateau is determined by the kinetic barrier, as explained above. In the case of monolayer adsorption, the intermediate section is attributed to the selectivity of adsorption with respect to the molecular size in polydispersed samples [29,30]. In the case of PSS, the gradual broadening with the adsorption step number suggests a more complicated phenomenon that we attribute to an

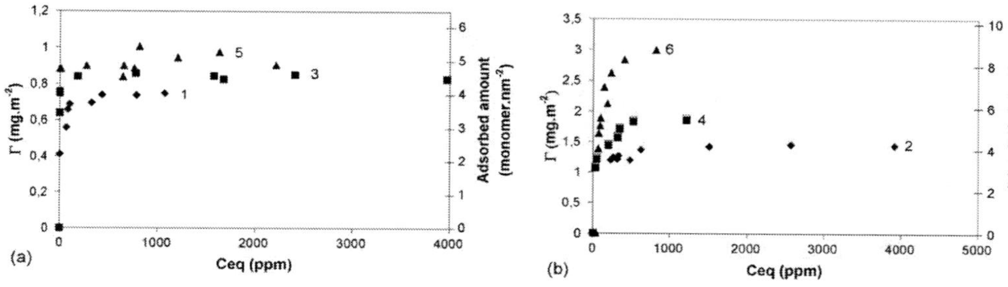

(a) PAH adsorption (b) PSS adsorption

Figure 5. Adsorption isotherms for PAH (a) and PSS (b). Odd and even numbers refer to PAH and PSS respectively. In 5(a), layer 1 of PAH was deposited on silica (♦), layer 3 was the 2nd layer of PAH deposited on SiO_2/PAH/PSS (■), layer 5 was the 3rd layer of PAH deposited on SiO_2/PAH/PSS/PAH/PSS (▲). In 5(b), layer 2 was the 1st layer of PSS deposited on SiO_2/PAH(♦), layer 4 was the 2nd layer of PSS deposited on SiO_2/PAH/PSS/PAH (■) and layer 6 was the 3rd layer of PSS depsoited on SiO_2/PAH/PSS/PAH/PSS/PAH (▲). Intermediate adsorbed polymer densities are given in Table 1.

adsorption induced rearrangement of the molecules in the film. We note that PAH isotherms did not produce the same figure, but kept a rectangular profile. Again, the two polymers did not react similarly with respect to adsorption.

Table 1 also indicates the plateau values for each adsorption isotherm (Γ_p). We see a larger difference for PSS than for PAH between the top of the steep rise and the saturation plateau. This difference increased with the step number, which will be attributed later to a significant rearrangement of the polymer film upon adsorption of PSS. In addition, plateau values revealed a progressive increase in adsorption for each polymer, but clearly a larger increase for PSS than for PAH. In the former case, the adsorption doubled between the 1st and the 3rd PSS layer, whereas the increase was only 25% for PAH. Similar figures were obtained with other polymer pairs, but with no evidence that the cationic or the anionic polymer growth was systematically higher. In the case of PSS-PAH, the former generally adsorbs more than it did here, but this may also depend on the pH in the experiment. In fact, PSSwas found to adsorb less than PAH at pH 5.5 and pH 7, which might be linked to the higher charge density of PAH at lower pH [31]. In a recent study involving the polymer pair MADQUAT (strong cationic polymer)-PAA(weak anionic polymer), growth was much higher for the fully charged polycation than for the weak polyacid PAA [23].

The adsorbed amount measured on colloids and on wafer of silica are similar: a constant PAH adsorption and increasing PSS uptake.

2.2.2. Increase of Adsorption with the Step Number

Figure 5 and table 1 show a stepwise increase of adsorption for both polymers, much more so for PSS than for PAH. A priori, the principal parameters, that may influence the pro-

Table 1. Summary of adsorption data for the 3 first PAH-PSS bilayers. Incorporated amount in the construction of the film (Γ), plateau values from figure 5 (Γ_p), volume of base or acid 1M to maintain pH 9

	Layer 1	Layer 2	Layer 3	Layer 4	Layer 5	Layer 6
Polymer added	PAH	PSS	PAH	PSS	PAH	PSS
Γ (mg.m^{-2})	0.7	1.15	0.8	1.24	0.88	1.89
Γ (monomer.nm^{-2})	4.5	3.4	5.2	3.6	5.7	5.4
Γ_p (mg.m^{-2})	0.75	1.45	0.85	1.85	0.95	3
V_{NaOH} 1M (mL)	2.05		1.20		1.00	
V_{HCl} 1M (mL)		1.70		1.60		2.60

gression of the uptake, are an increase in surface charge density, an increase in adsorption site density and a progressive change in the conformation of the adsorbing polymer. As seen above, a larger increase in PSS adsorption correlates with a larger uptake of protons by the underlying amine groups. Conversely, there is no opportunity for PAH to cause a symmetrical reaction since PSS is fully ionized. The different rates of growth for PSS and PAH presumably comes from the difference in their ionization ratio. A similar difference was observed in a previous study with MADQUAT-PAA, where the increase in adsorption was much higher for the fully charged cationic MADQUAT than for the weak polyacid PAA [23]. With regards to our results, we believe that the highly charged polymer determines film growth in the case of a strong-weak polyelectrolyte pair. However, as noted earlier, it is difficult to make any definitive statement from the reports in the literature since there is little information about pH control in deposit experiments.

3. Variation of pH upon Polymer Mixing

3.1. pH drift

The change in pH upon adsorption of a polymer is an interesting phenomenon that has generally been ignored in former studies. The direction of the pH drift provides information concerning the adsorption mechanism. Figure 6 summarizes the variation of pH upon the successive additions of each polymer. One set of data relates to the uncontrolled procedure and the other to the case in which the mixture was returned to pH 9 after the addition of each polymer. Starting at pH 9, the addition of PAH caused a decrease to pH 7.2. In the case where no adjustment was made, the next addition of PSS increased the pH to about 8.3. With further additions of PAH, PSS and PAH, respectively, the pH decreased to about 8.1, increased to 10 and eventually decreased to 8.9. When the mixture was returned to pH 9 after each addition of polyelectroyte (i.e. back to the horizontal line in Figure 6), the drift induced by the next addition varied more symmetrically with respect to pH 9. We see also that the drift diminished progressively, more so after the addition of PAH than of PSS. The amount of base or acid required to restore pH 9 is reported in Table 1.

The change in pH reveals the occurrence of several reactions in the adsorption process:

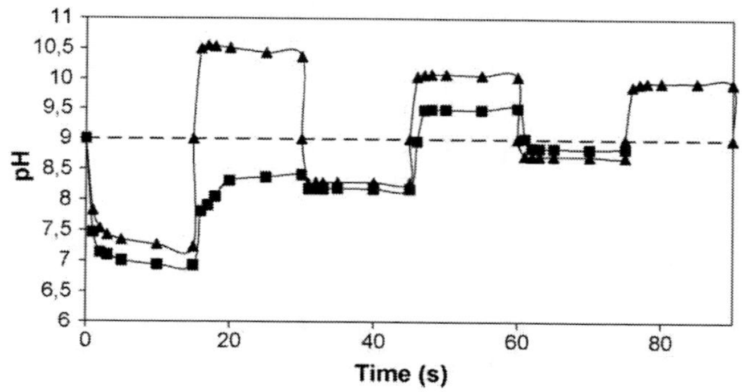

Figure 6. pH variation versus time and addition step number, pH 9 with adjustment after mixing (▲) and without adjustment (■).

- The first addition of PAH in the silica dispersion caused a decrease in pH (pH 9 to pH 7.2) due to the release of protons from the silanol groups. The reason was that the polycation neutralized the surface charge of silica, which induced increased ionization in the silanol groups. The phenomenon is described below by two binding equilibria for PAH:

$$\begin{cases} SiO^- + NH^+_{3_{PAH}} \rightarrow (SiO^-)\,NH^+_{3_{PAH}} \\ SiOH + NH^+_{3_{PAH}} \rightarrow (SiO^-)\,NH^+_{3_{PAH}} + H^+ \end{cases} \tag{2}$$

SiOH and SiO^- are the silanol group and its ionized form respectively, $NH^+_{3_{PAH}}$ is a protonated aminogroups on PAH. Although unprotonated NH_2 groups forming Hydrogen bond with SiOH groups are not to be excluded, they are not necessary for the purpose of this study.

- Adsorption of PSS on PAH-covered Silica caused an increase in pH (pH 9 to pH 10.4), which is attributed to an increase in the protonation of PAH:

$$\begin{cases} y - NH^+_{3_{PAH}} + PSS^- \rightarrow (NH^+_{3_{PAH}} PSS^-) \\ y - NH_{2_{PAH}} + PSS^- + H_2O \rightarrow \left((NH^+_{3_{PAH}})_y PSS^-\right) + yOH^- \end{cases} \tag{3}$$

- Adsorption of PAH on the formerly deposited PSS layer induced a decrease in pH, revealing an additional interaction of PAH with uncovered parts of silica. In this case as well, additional protonation of the amino groups may take place as above, but the dominant effect was that from the silica surface:

$$\begin{cases} y - \left((NH^+_{3_{PAH}})_y PSS^-\right) + zNH^+_{3_{PAH}} \rightarrow \left((NH^+_{3_{PAH}})_y PSS^-\right)(NH^+_{3_{PAH}})_z \\ SiO^- + NH^+_{3_{PAH}} \rightarrow (SiO^-)\,NH^+_{3_{PAH}} \\ SiOH + NH^+_{3_{PAH}} \rightarrow (SiO^-)\,NH^+_{3_{PAH}} + H^+ \end{cases} \tag{4}$$

These equations demonstrate that the pH drift, its magnitude and its direction, is an indication of the polyelectrolyte binding mechanism. For example, the data in figure 6 reveal that PAH still reached the silica surface at the 5th deposit. This will be discussed further on, in the last section.

3.2. pH Adjustement

As stated above, the pH drifts in Figure 6 revealed part of the binding mechanism for the polyelectrolytes, but more quantitatively, the amount of acid or base required to return to pH 9 measured the change of ionization of the polyelectrolyte and (or) the substrate upon adsorption at constant pH. Note that when the pH decreased between 9 and 7.23 upon adsorption of the first PAH layer, the net uptake of protons was ($9.9\text{x}10^{-6}$ (H^+ mole.L^{-1}) / $5.02\text{x}10^{-2}$ (mole.L^{-1}) =) $2\text{x}10^{-4}$ mole per mole of PAH unit, which is negligible. However, when returning to pH 9, the increase in adsorption was (0.7-0.53) = 0.17 mg.m^{-2} and the concomitant addition of hydroxyl ions was 2.05 10^{-3} mole. Since the silica surface area was 1000 m^2 and the molar weight of PAH was 93 g, it was easy to calculate a ratio of about 1.1 mole of hydroxyl ion per additional mole of adsorbed PAH unit when restoring pH 9. Part of these hydroxyl ions increased the surface charge of silica and the other part deprotonated the polymer. In other words, there was a considerable exchange of protons and probably a significant rearrangement of the polymer in the second step (return to pH 9) for the adsorption of the first PAH layer. These observation points to a possible cause of experimental uncertainties and discrepancies in the literature, since many investigations have not explicitly mentioned the pH control procedure and many have ignored the pH drift . As stated above, the shape of the intermediate part of the isotherms (between the steep rise and the plateau) may result from the rearrangement of the polymer film upon pH adjustment. This was more so for PSS than for PAH, as attested by the larger amount of acid (1.7, 1.6, 2.6 mL after PSS adsorption) than base (2, 1.2, 1 mL after PAH adsorption) required to correct the pH. In the case of PAH, the first value (2 mL) was exceptionally high due to the large increase in the silica surface charge density. After adsorption of the first PSS layer, the pH was 10.4, the addition of acid ($1.7\text{x}10^{-3}$ mole) to return to pH 9 increased the charge density of PAH and caused an additional uptake of PSS. The latter increase of adsorption was 0.16 mg.m^{-2} (0.8 10^{-3} mole). Therefore, the increase of ionization of the amine groups (equation 2b above) was about twice the additional negative charge of PSS ($1.7\text{x}10^{-3}/0.8\text{x}10^{-3}$). This is certainly due to the addition of acid, which caused a reorganisation of the whole polymer layer. In the first step of adsorption (when the pH increased from 9 to 10.4), PSS formed macromolecular complexes with PAH and most of the sulfonate groups were neutralised with Na^+ counterions. When returning to pH 9, protons were available as needed to reach equilibrium at this pH. In the course of pH adjustment, Na^+ ions were released from PSS molecules and new bonds were formed between PSS (sulfonate groups) and PAH (protonated amine groups). Overall, the increase of the binding ratio between the two polymers was 30% (i.e. the increase in protonation of PAH/total adsorbed PSS units = $1.7\text{x}10^{-3}$ mole x $100/5.8\text{x}10^{-3}$ mole). Naturally, a concomitant change in the conformation of the macromolecules and in the film structure took place so as to adjust to the additional binding. This phenomenon was seen to amplify for the next PSS layers. In our opinion, this is one reason for the presence and the broadening of the intermediate part in the isotherm. The above interpretation corroborates other indications that successive polyelectrolyte layers interpenetrate [32], although there is no evidence here that true layers form.

4. Charge Balance

4.1. Zeta potential

The electrophoretic mobility of the particles was measured using dynamic light scattering (Zetasizer Nano ZS, Malvern Instrument, England) on particles redispersed in air-equilibrated NaCl 10^{-3} M solution at pH 9. The scattering angle was 17 and the temperature 25C. Electrophoretic mobility was measured after each deposit step. The zeta potential was calculated from the mobility using the Smoluchowski relationship [33] and reported in Figure 7. It shows the usual alternation between positive and negative charge excess [23]. Values about $\pm$ 40 mV are known to correspond to surface saturation for polyelectrolyte adsorption [34].

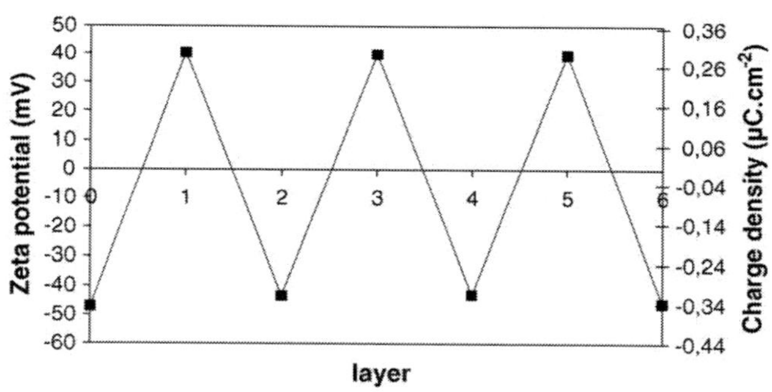

Figure 7. Zeta Potential of PAH/PSS covered silica particles in NaCl 10^{-3} M pH 9. PAH outer layer (1, 3, 5) PSS outer layer (2, 4, 6).

4.2. Layer Compensation on the Film

The driving force for the successive adsorption of the polymer is essentially electrostatic. Charges on polymer repeat units can be balanced by those on oppositely charged chains or by small ("salt") ions occluded within the film. The two scenari for charge balance are illustrated figure 8 below. In one case, which we term intrinsic compensation, a polymer positive charge is balanced by a negative charge, also on a polymer. In the alternative mechanism, extrinsic compensation, much of the polymer charge is balanced by salt counterions derived from the bathing solution used to construct the multilayers [35].

In order to distinguish more clearly the different ionic contributions in early growth, we computed the charge balance in the deposited film stepwise. For the first layer at pH 9, the values in table 2 show charge compensation between the silica surface and the adsorbed polymer (PAH). The initial charge of silica was -8.8 $\pm$ 0.5 μC.cm^{-2} [36], the charge density of the adsorbed polymer was 29 $\pm$ 1.5 μC.cm^{-2}, the increase in silica charge density

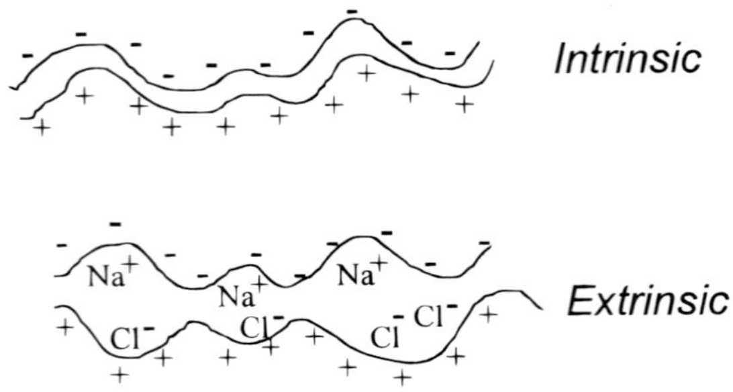

Figure 8. Intrinsic vs Extrinsic Charge Compensation [35].

(calculated from the known addition of NaOH) was - 20 ± 1.5 μC.cm^{-2} and the remaining charge corresponding to the zeta potential (+ 40 mV) was - 0.3 μC.cm^{-2}, which may be negligible in comparison with other values. Therefore, the sum of all charge densities was close to zero:

$$\sigma_{Cl^-}+\sigma_{SiO_2}+\sigma_{PAH}+\sigma_{OH^-}+\sigma_{Zeta}=0 \tag{5}$$

Table 2. Charge balance for the first layer, all charge densities are in μC.cm^{-2}

σ_{SiO_2}	σ_{PAH}	σ_{OH^-}	σ_{Cl^-}	balance
- 8.8	29	- 19.8	- 0.4	0

The net release of H^+ (decrease in pH) was explained by the increase in ionization of the SiOH groups, due to electrostatic screening from PAH. For the same reason, the ionization of PAH certainly increased when binding with the silanol groups, causing a consumption of protons. Practically, however, the release of protons dominated the reaction. Since the charge balance was close to zero, we conclude that a negigible amount of Cl^- counterions co-adsorbed within the first PAH layer.

Conversely, in the case of the second polymer layer (adsorption of PSS on PAH-covered silica at controlled pH 9), the balance shows a large excess of negative charge. The sum of values reported in table 3 (charge densities of the silica substrate, the adsorbed PAH, the adsorbed PSS and the additional charge of PAH calculated from the quantity of HCl to return from pH 10.5 to pH 9) amounts to - 37 μC.cm^{-2}. The charge complement to produce an electrically neutral film is attributed to the coadsorption of the PSS counterions, Na^+.

$$\sigma_{SiO_2}+\sigma_{PAH}+\sigma_{OH^-}+\sigma_{PSS}+\sigma_{H^+}+\sigma_{Zeta}=-\sigma_{Na^+} \tag{6}$$

Table 3. Charge balance for the second layer, all charge densities are in μC.cm^{-2}

σ_{SiO_2}	σ_{PAH}	σ_{OH^-}	σ_{PSS}	σ_{H^+}	σ_{Na^+}	balance
- 8.8	29	- 19.8	- 53.9	16.4	37	0

The charge balances for the 3rd and the 4th polymer layers are summarized in table 4. As stated earlier, the deposit of the 3rd layer (2nd PAH layer) caused a decrease in pH, due to an increase in the surface charge of silica (- 11.6 μC.cm^{-2}, calculated from the required addition of HCl to return to pH 9), which means that PAH still bound significantly with the silica surface. Consecutively to the adsorption of PAH, the calculation also reveals a decrease in the overall negative charge excess of the film (from - 37 to - 15 μC.cm^{-2}). The explanation is that the protonated groups of the new PAH layer displaced Na+ ions from unbound sulfonate groups of PSS. Overall, the unbound fraction of PSS groups (neutralized by Na^+ ions) decreased from 68 to 20% after PAH adsorption. Although the sum of the polymer + surface charge was negative after adsorption of PAH, the electrokinetic potential of the film was positive. This is understandable since the negative charge of PSS within the film was neutralized by Na^+ ions; whereas the electrokinetic potential resulted from a small charge excess of the outer PAH layer which was compensated by the charge in the adjacent diffuse layer (< 1 μC.cm^{-2}).

Table 4. Charge balance for the three first bilayer, all charge densities are in μC.cm^{-2}

Layer	0	1 (PAH)	2 (PSS)	3 (PAH)	4 (PSS)	5 (PAH)	6 (PSS)
σ_{SiO_2}	-8.8						
$\sigma_{polymer}$	0	29.0	53.9	33.2	58.1	36.5	88.2
σ_{pH}	0	- 19.8	16.4	- 11.6	15.4	- 9.65	25
$\Sigma\sigma_{ions}$= $F(\Gamma_{Na^+}-\Gamma_{Cl^-})$	- 8.8	0.47	- 37	- 15.25	- 42.5	- 15.8	- 63.3
Counter ions	100 % Na^+	0 % Cl^-	69 % Na^+	0 % Cl^-	73 % Na^+	0 % Cl^-	72 % Na^+
Counter ions left adsorbed		0% Na^+	0% Cl^-	29 % Na^+ in bilayer 1	29 % Na^+ in bilayer 1	29 % Na^+ in bilayer 1 26 % in Na^+ in bilayer 2	29 % Na^+ in bilayer 1 26 % in Na^+ in bilayer 2

Table 4 gives the lowest value for the coadsorption of sodium ions, since the calculation ignores the possible co-adsorption of Cl^- ions with PAH. In fact, even if part of the PAH charge was neutralized by chloride ions, the coadsorption of Na^+ ions would still be higher. Measurements for the second deposit of PSS (4th polymer layer) were similar to those in the first deposit step, but the deficit of positive charge was still higher. The second adsorption of PSS brought about - 58 μC.cm^{-2} (table 4), and the corrected pH, a positive charge (increase in the protonation of the PAH groups) + 16 μC.cm^{-2}. Overall the excess in charge was - 42 μC.cm^{-2} (i.e. about 75% of the charge of the last adsorbed PSS layer), that were compensated for by coadsorbed Na^+ ions. The next deposits of PAH and PSS (3rd bilayer) resembled the first two, although even more protons were required to maintain the pH (2.6 ml of HCl 1 M) upon adsorption of PSS, thus indicating a strong interaction with the underlying PAH layer. Note again that the adsorption of PSS increased significantly in the 3rd bilayer, contrary that of PAH. Thus, PAH and PSS behaved differently in the adsorption process. PAH adsorbed with an apparently negligible amount of Cl^- counterions, as opposed to the charge of PSS, which was largely compensated for with Na+ counterions. In addition, the uptake of PAH increased only slightly in comparison with that of PSS. As suggested previously, the difference in behavior is attributed to the difference in charge densities: the protonation ratio of PAH was about 40% at pH 9 [37] whereas PSS was fully ionized. Other results in the literature suggest similar differences in the behavior of the two polymers, although the charge balance has rarely been computed explicitly. With respect to washing with water or salt solutions, the coadsorption ratio of counterions in the polymer layers (so-called extrinsic charge compensation ratio) is an important factor in both film growth behavior and film stability. An absence of small ions in the polymer film (intrinsic charge compensation) reveals a very good matching of charge between the two polymers. This was the case for PSS with PAH at pH 5 [38] and with PDADMAC [35, 39]; it was apparently associated with a linear growth of the film with the step number. However, this question remains open since a reversible exchange of salt ions was found with PAH-PSS films in contact with solutions of different NaCl concentrations [40] and small ions were found to partially neutralize the polyions in PSS-PAH films [31, 41, 42].

5. AFM Analyses

AFM experiments were performed using a stand-alone SMENA scanning probe microscope (NT-MDT). AFM images were recorded on polymer films formed at the estimated stagnation point, in contact mode-more precisely in constant-force mode. During the scan, the deflection of the cantilever was maintained to a preset value by the feedback loop and the topographic map of the surface was obtained by the vertical displacement of the scanner. Conical silicon tips on the extremity of AFM cantilevers with a force constant from 0.1 to 0.3 $N.m^{-1}$ were used to probe the samples (NT-MDT).
Two series of samples, containing 1 to 10 successive polymer deposits (i.e. 5 bilayers), were submitted to AFM examination. One series was first dried (Nitrogen, room temperature) and the other one was kept wet for examination. The general morphology of several dried samples is presented in Figure 9.

It shows a rather regular distribution of islands, starting at the second layer (first PSS deposit), whereas the bare silica and the PAH covered silica substrate show no specific feature. Note the different scales in Figure 9, between the z-axis of the profile and the horizontal line. Due to the ≈ 60 magnification of the z-axis, the bump-like islands resemble peaks in the figure. The average height of the bumps and their amount were determined as a function of the deposition number (Figure 10) with a grain analysis software (Nova, NT-MDT). This analysis consisted in defining a threshold plane at a fixed distance $z=z_0$ to the base plane (z=0). Each area above the threshold plane, corresponding to a closed domain, was interpreted as a separate bump. The height of the bumps was calculated from the z_0 position while the mean height of the bumps shown in Figure 10(a) was the statistical average calculated from all bump heights on the sample. Bump number density (Figure 10(b)) was the grain number for which the height was above the threshold plane. Obviously, the measurements depend on the choice of the z_0 plane, which was selected to avoid any artefact due to electronic noise from the AFM. Overall, the confidence interval of the measurements was estimated at about 15%. Figure 10 shows that both the average height and the bump density increase, but not regularly, with the number of polymer deposits. Minima in height

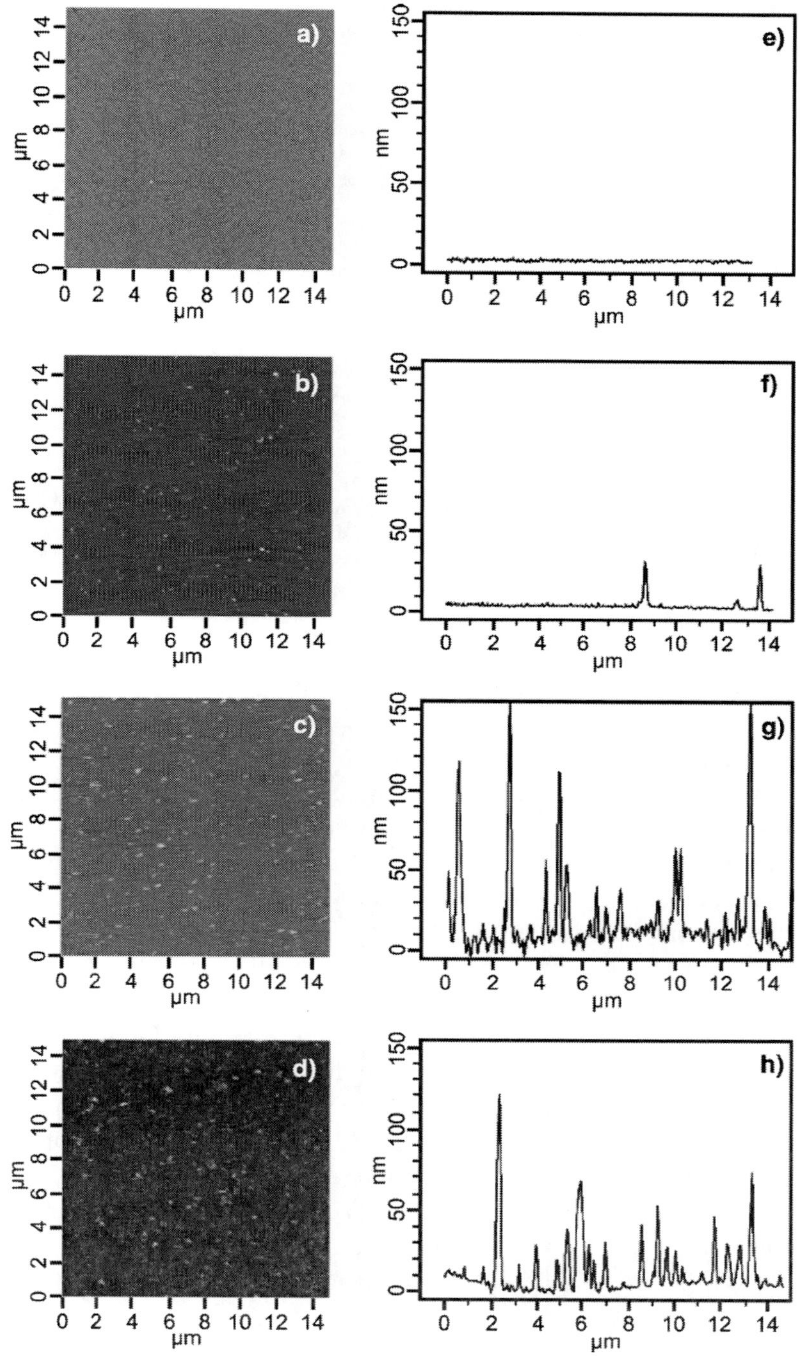

Figure 9. AFM contact mode topographic images of PAH/PSS layers adsorbed on a silica substrate (dried samples) and the corresponding z-profiles along an arbitrary horizontal line. a), e): one layer; b), f): two layers; c), g): seven layers; d), h): eight layers.

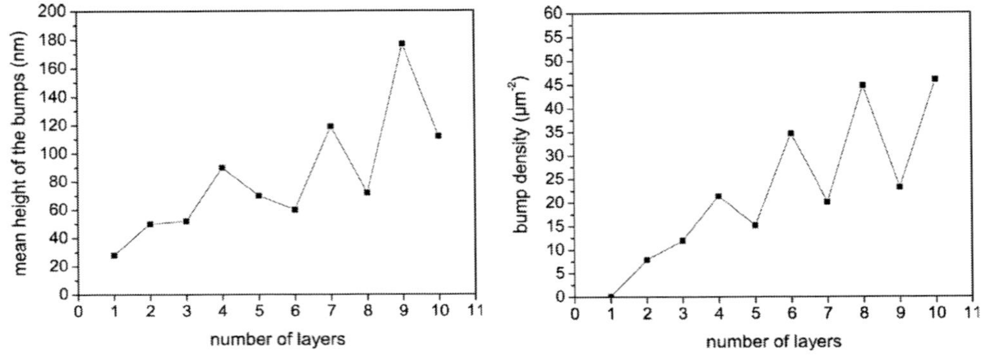

(a) Mean height (b) Bumps density

Figure 10. Mean height(a) et density (b) of the bumps on the surface of PAH/PSS PEM as a function of the number of deposited layers.

at numbers 6, 8 and 10 correspond to PSS as the last deposited polymer and maxima at numbers 7 and 9, to PAH. Height maxima correspond to minima for the bumps number density.

The scratching procedure described by Lobo et al. [43] was carried out to take off the polymer and to measure the film thickness between the bumps. The method consists in "digging" an area of the film with the AFM tip until the substrate is reached. Scratching experiments were done on bare silica substrates, proving that the substrate itself does not suffer any abrasion by the tip. Practically, the chosen area (about 500x500 nm^2) was scanned repeatedly while increasing both the set point and the scan rate to their maximum value. After removing the polymer film, a larger zone containing the scratching area was scanned in contact mode, which allowed to measure the thickness of the film. Figure 11 shows an example of a scratched area on a sample covered with 5 polyelectrolyte layers, and the topographical cross-section plotted along the dashed line. Quantitatively, the film thickness was calculated by a statistical procedure taking into account two areas for each sample; the relative mean heights were evaluated in the scratched area on the substrate (square 1 in Figure 11(a)) and on the polymer film (in square 2). The thickness of the layers results from the subtraction of these two values. The accuracy of this measurement is the sum of the root mean square values in both squares.

Figure 12 displays all measurements of the film thickness between the bumps, both for the wet and the dry samples. We see that a low amount of macromolecules adsorbed between the bumps and that the polymer film thickness increased more or less linearly at each successive deposit. In this figure, a slope rupture, in the film thickness variation, is observed near the fifth or the sixth layer. At the beginning, the homogeneous film thickness grows to 0.74 nm.layer^{-1}, then 2 nm.layer^{-1}. This phenomena ca be explained by a modification of the growth mechanism. So the fifth or sixth layer is the number of polyelectrolyte layer necessary to cover the silica surface. This result is in accordance with this found on colloids (Figure 6) et these reveal in the literature [23, 40, 44].

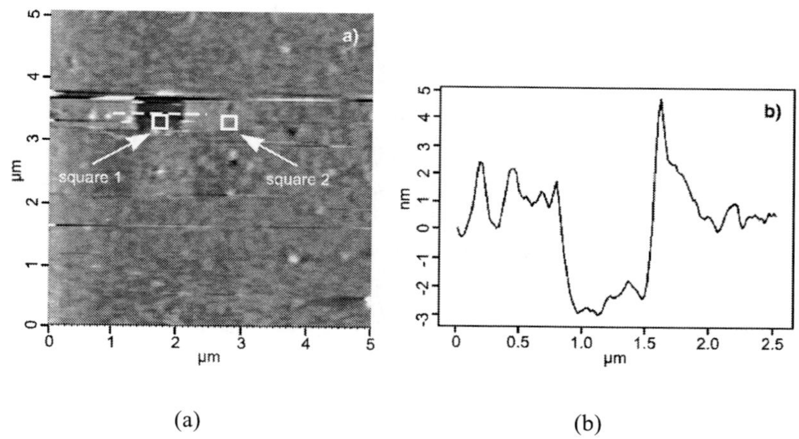

Figure 11. a): AFM contact mode topographic images of five PAH/PSS layers after scratching a (500x500) μm^2 zone. b): z-profile above the scratched area (along the dashed line).

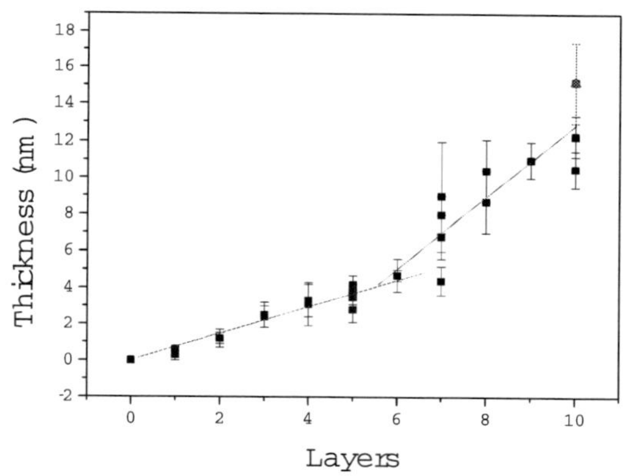

Figure 12. Mean thickness of PAH/PSS PEM between bumps as a function of the number of deposited layers.

6. Comparaison between Film Thickness Detrmined by AFM and Laser Reflectometry

The thicknesses determined par the scratting technic (figure 12) are inferior to these found by laser reflectometry (figure 2(a)). Indeed, for eight deposited layer, by scratching, the thickness is 10 nm whereas 70 nm by reflectometry. However by scratching, only the homogeneous layer between the PAH/PSS agregates was measured. The film was performed by a homogeneous layer covered by agregates. In order to determined the total thickness, the z-profile (figure 9, picture h) must be analysed. In this picture, agregates with a thickness of 70 nm was scanned that is in according with the reflectometry. During the polyelectrolytes adsorption, reflectometry measures the maximal film thickness which increase slowly. The reflectometric signal variation provides from the density agregates increasing, which is in according with the regular increasing of the refractive index (figure 2(a)). This refractive index is an average between the adsorbed polyelectrolytes, the water molecules and the ions include into the layer.

From the 20th, a thickness difference was observed between the reflectometry (Figure 2(a)) and the AFM (Figure 13), respectively 95 nm against 70. This difference is explained by the fact that the homogeneous thickness layer was not measured by the AFM. Indeed the zero in z-axis (in figure 13(b)) correspond to the beginning of the agregates and not the silica wafer. At the agregates height, the homogeneous layer thickness must be added. This layer, obtained by extrapolation of the Figure 12 slope to the twentyth layer, measured 30 nm. So the total film thickness is 30 + 70 = 100 nm, near the reflectometry data.

7. Mechanism for the Multilayer Growth

The adsorption measurements and the AFM pictures in this investigation show that the self-assembled multilayer film did not grow as a successive superposition of interacting polymer layers, such as puff pastry. However, the principal findings in this work can be helpful in attempting to rationalize the data. Figure 14 presents the successive deposit steps as described below.

The charge balance showed that in the first layer, PAH adsorbed up to the neutralization of the silica surface. At this point, a large part of the silica surface remained free of polymers. In the second step, the deposit of PSS took place exclusively through electrostatic interaction with the predeposited PAH molecules. In separate experiments, we found that PSS did not adsorb directly on silica. Interaction between PSS and PAH molecules on the surface formed polymer complexes that probably aggregated to form units about 200 nm wide, resembling coacervates in a solution [45]. The agregates formation was also demonstrated and simulated by Abu-Sharkh [46], Figure 15. Figure 15 shows the mechanism of deposition of the PE chains on a previously deposited layer. The deposition proceeds through the following steps. The first step is attraction of part of a PE chain in solution to tail and loop segments of the previously deposited layer. This step is followed by gradual flattening of part of the newly deposited chain to become part of the second layer. The other part of the newly deposited chain extends into solution forming the segments that overcompensate the charges. In addition, it is observed that chains of the new layer penetrate deep into the first layer and form a molecular multilayer that is one atom thick. Furthermore, adsorption of the new layer results in rearrangement of the previous layer. A dominant form of rearrangement is folding of the tail segments as shown in Figure 15. Tail segments from the first layer that protruded out of the surface before deposition flatten and become oriented parallel to the surface. The tail segments from the first layer fold over the oppositely charged chains to contribute to formation of a new layer [46].

The adsorption of PSS caused a reversal from a positive to a negative charge. In fact, as shown in Figure 7, the surface always takes the charge sign of the last deposited polymer. In the next step, when PAH was introduced for the second time, the molecules could adsorb both on the bare silica areas and on the PAH-PSS aggregates. In our experiments, two results support this assumption: (i) the decrease in pH reveals the binding of PAH with silica (equilibrium 1) and (ii) from the charge balance, we deduce that about half of the

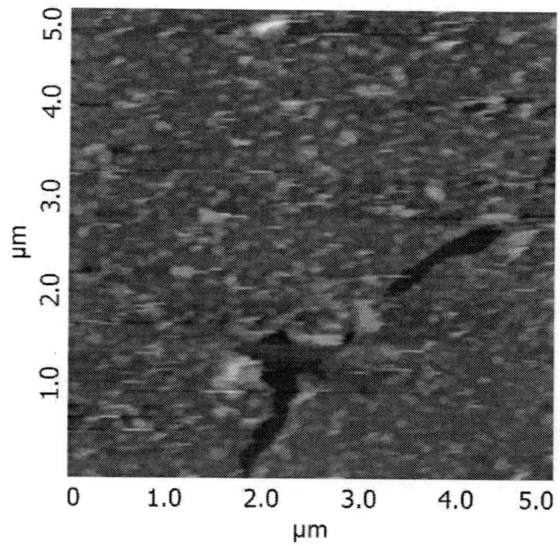

(a)

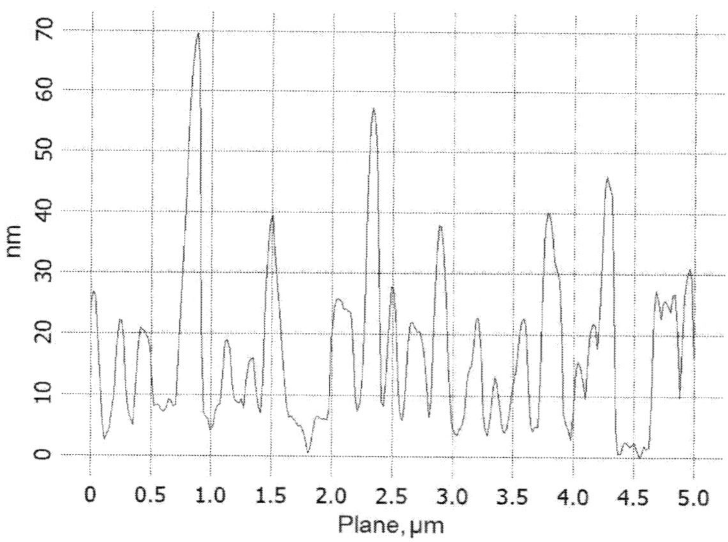

(b)

Figure 13. AFM contact mode topographic images of $(PAH/PSS)_{10}$ layers adsorbed on a silica substrate and the corresponding z-profiles along an arbitrary horizontal line.

Na^+ counterions of the sulfonate groups are displaced by the positive PAH. The presence of a 1-2 nm thick polymer layer (presumably PAH) between the bumps is also confirmed by the scratching experiments. The overall charge of the film after adsorption of PAH (deposit

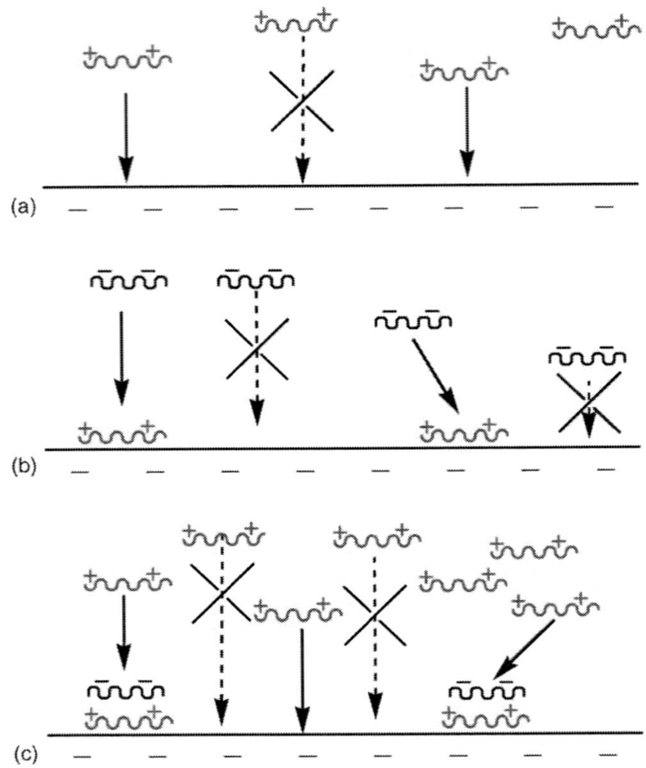

Figure 14. Scheme of multilayer deposition. First bilayer (a), second bilayer (b), third bilayer (c).

sequence SiO_2/PAH/PSS/PAH) was positive, although the charge balance showed an excess of about 30% of the negative charge of the deposited PSS. We previously explained that the negative excess was neutralized by codeposited Na^+ ions (Table 4). At the second stroke of PSS, the molecules could bind on the preformed aggregates (the bumps) and form new aggregates of PAH-PSS complexes between the bumps.

The accuracy of the AFM analysis does not yet allow a reliable evaluation of a possible change in width. The same mechanism was involved in the build up of the third bilayer comprised a significant binding of PAH with bare silica sites (as shown by the decrease of pH) and with the PSS covered aggregates, as revealed by the correlated displacement of Na^+ ions as well as, the adsorption of PSS molecules which formed new aggregates with PAH molecules. A detailed investigation of the growth process by AFM strengthen this description.

The findings in the present study strengthen this description. In the course of the build-up we saw that the average bump height decreased significantly upon PSS adsorption and reversely with PAH, which we attribute to the combination of two phenomena: the difference between PSS and PAH in the conformation of the outer polymer layer and the diffusion of PSS within the aggregates. Since PAH was only charged at 40 % of full ionization, some segments of the macromolecules bound with the outermost sulfonate groups of PSS, and the remaining ones formed loops and tails that protruded into the solution, which caused the

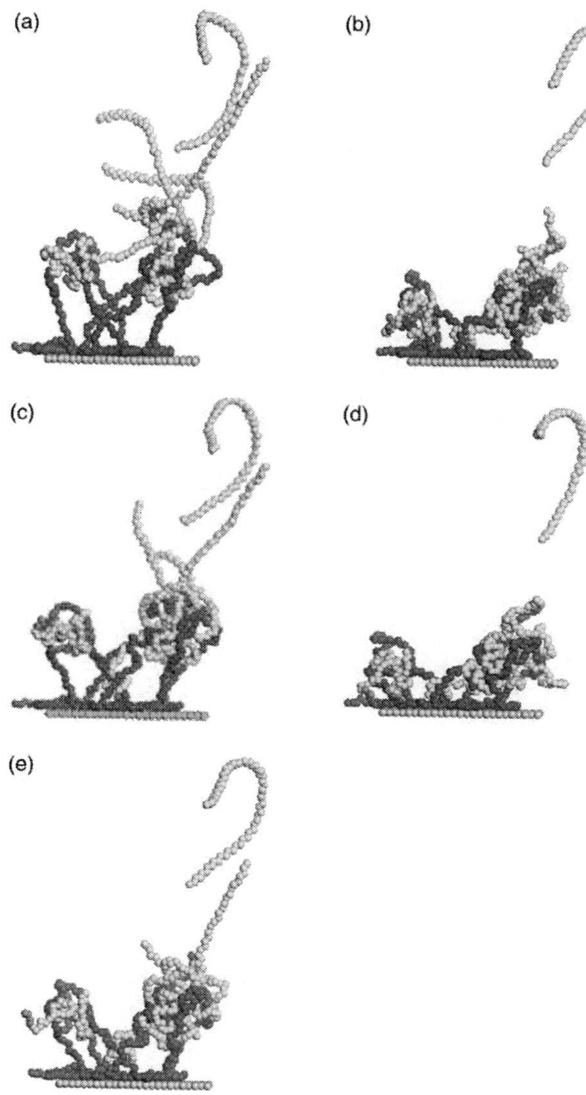

Figure 15. Mechanism of the second adsorption of PE molecules on a previously deposited layer [46].

significant increase in film thickness. Conversely, the fully ionized PSS molecules laid flat on the surface with a stretched rod-like conformation. Upon contact with PSS, the charge density of PAH increased, which reduced the size of the loops and therefore the thickness of the outer layer. In addition, as stated earlier, PSS molecules diffused into the aggregates, which caused the bumps to shrink upon release of water and counterions. Similar shrinking or swelling phenomena, depending on the outermost polymer, were observed upon heating PSS-PAH films [18]. Kolasinska et al. also showed an alternating change in conformation and hydrophobic-hydrophilic properties of the external film layer for PAH-PSS with water contact angle measurement [16].

Although the number of aggregates increased, on average, with the step number, Figure 10(b) shows a slight decrease between the deposition of PSS and PAH. With PAH, all of the substrate was available for adsorption and the adsorption area remained roughly the same during the whole build-up process. The deposited weight was limited only by the electrostatic barrier raised by the excess in charge, which prevented molecules of the same charge from reaching the surface. This explains why the deposited weight of PAH was roughly constant (0.5 $mg.m^{-2}$). In the case of PSS, adsorption took place only on the formerly deposited PAH molecules, some of which were at the top of the bumps and the others directly bound with the surface. Since we know that isolated PAH molecules were not detected by the AFM tip (Figure 14a), it is understandable that new bumps appeared only for each alternate deposition of PSS, since they formed new polymer complexes with the last deposited and isolated PAH molecules. Presumably, at the following PAH adsorption, the small decrease in the number of bumps was the result of a coalescence phenomenon induced by the binding of PAH with neighbouring polymer complexes.

Interestingly, we found a very similar description of the growth mechanism in a report by Picart et al. [47]. The polymer pair was pol(l-lisyne)/hyaluronic acid (PLL^{+}/HA^{-}) and the deposit was done on quartz. Using optical waveguide lightmode spectroscopy (OWLS) and a quartz microbalance, they found that the positive PLL adsorbs over the entire surface of silica, whereas the next deposit of HA formed islands of PLL/HA complexes (Figure 16). They also observed the coalescence of islands and the increase of surface coverage during the successive and alternate deposits of the two polymers. In their work the islands in the first bilayer were larger (a few micrometers width and height) than those we observed (200 nm). This difference may be due to the fact that they used high molecular weight polymers (about 200.10^{5} Da) and made the deposits in high ionic strength solutions (0.15 M NaCl), which is known to increase the adsorption of polyelectrolytes significantly. These authors continued the deposit up to 13 bilayers and found a new growth regime after the deposit of eight PLL/HA layers. At this stage the surface was completely covered and exhibited a continuous and homogeneous film. Very similarly to our observations, they did not measure a significant increase in the average height of aggregates, which would suggest that the increase in coverage is more likely to come from the coalescence of aggregated PAH/PSS complexes.

In addition to the general scheme of growth described above, our measurements bring some insight to the interaction mechanism between PSS and PAH. We have already shown that the change in pH and the consumption of reactant (acid or base) to restore the initial pH are indicative of the type and the extent of a binding mechanism. With respect to the binding of PAH, and as stated previously, the decrease in pH provided an unequivocal indicator measuring the direct contact between the polymer and the bare surface of the silica. As for PSS, we gave the increase in pH upon addition of PSS as evidence for the increase in the charge density of PAH (i.e. higher protonation ratio) caused by binding with this polymer. Quantitatively, it is interesting to note that the amount of acid required to return to pH 9 is far in excess of the last deposited amount of PAH. This could be a strong indication that the last PSS molecules also bound with PAH molecules deposited in earlier layers, which are reminiscent of the so-called interdigitation of the polymer molecules in succession. We conclude that the growth of the polyelectrolyte multilayers proceeds via a complex mechanism in which the two macromolecules have a different role, depending essentially on their

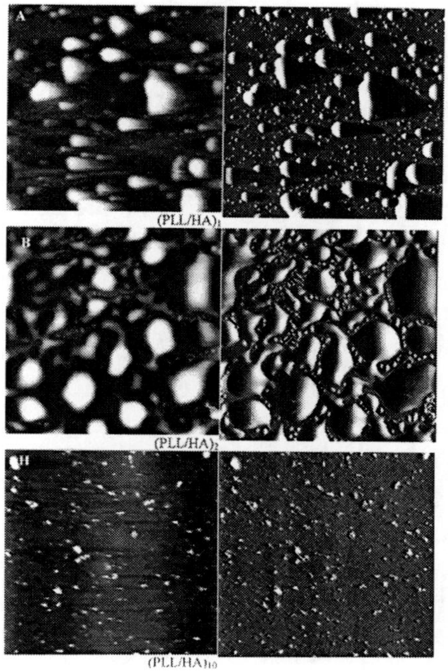

Figure 16. AFM images obtained at the different steps of the alternate deposition of PLL and HA layers. The polyelectrolytes PLL ($3.26x10^4$ Da) and HA ($1.5x10^5$ Da) were used at a concentration 1 mg.mL[1] in 0.15 M NaCl (same solutions as in Figure 1B). The surface is observed at different steps during the buildup process, corresponding to a given deposition cycle, until the 10th (PLL/HA) layer pair: (A) $(PLL/HA)_1$, (B) $(PLL/HA)_2$, (H) $(PLL/HA)_{10}$. In the left column of part G is plotted a cross section of the height image along a dashed line. Images dimensions are 50x50 μm^2 and Z scales of three images (left column) are 1000 nm for images A-B and 30 nm for image H [47].

charge density and that of the substrate. Schematically, in the present case, PAH, with a low charge density opposite that of the substrate, was responsible for the bi-dimensional increase in surface coverage, due to its adsorption both on the preformed polymer complexes and on the bare surface. Conversely, the highly charged PSS caused thickening and densification of the layer, thus increasing the number of bumps and the weight uptake, both of which are explained by the formation of hydrophobic complexes and the diffusion of PSS into the aggregates.

8. Conclusion

This chapter addressed the initial build-up stages of polyelectrolyte multilayers made on silica with a strong (highly charged) and a weak (low charged) polyelectrolyte, PSS and PAH respectively. Complementary analyses were done step by step measurements of the adsorbed polymer weight (by reflectometry), the charge balance and counter ions adsorption(by pH drift analyses) and the evolution of the film morphology (by AFM). Together, these analyses demonstrated a comprehensive mechanism of the early growth, with each polymer behaving differently. Separate adsorption isotherms demonstrated different behaviors between the strongly charged PSS and the moderately charged PAH. PSS uptake increased with the step number significantly more than PAH uptake did. PSS also showed a broadening of the intermediate part of the isotherm that was attributed to a progressive modification of the film structure. While PAH could adsorb on the entire surface of silica, PSS could only form hydrophobic aggregates with the pre-adsorbed PAH molecules. PAH formed an outermost polymer layer with loops and tails protruding in the solution, but the rod-like PSS laid flat on the surface and diffused into the preformed aggregates. Considerable attention was paid to the change in pH consecutive to the adsorption of each polymer, which allowed identification of the binding sites and mechanism. A significant difference in uptake was found, depending on whether the pH was returned or not to the initial value at pH 9. Although the comparison was not continued to the end of the construction, we observed that not controlling the pH has a strong influence on both the depositing process and the properties of the film. From the charge balance, we deduced that PSS adsorbed with a large number of Na^+ counterions, which was not the case for Cl^- with PAH. We also found that from the amount of acid required to restore the initial pH, PSS macromolecules bound with amine groups far inside the formerly deposited PAH layers. We pointed out that the charge excess corresponding to the zeta potential of the film was quite negligible in comparison with the total ionic charge of a polymer layer. Oppositely, PAH did not seem to diffuse inside the PSS layer to bind with sulfonate groups; this is probably a consequence of the lower charge density of the molecule. This is also an explanation to the low increase of the PAH uptake with the step number. With regards to the film growth, each polymer had a specific role: PAH was responsible for the increase in the surface coverage by adsorbing on the remaining bare silica areas whereas PSS determined the thickening and densification of the film by forming polymer complexes with PAH and by diffusing into the polymer aggregates. These studies provide tools to rationalize the choice of the physico-chemical parameters in order to control the growth and the polyelectrolyte multilayers properties.

Acknowledgment

This work was supported by the EU under Contract NMP4-CT-2003-001428: Nanocapsules for Targeted Controlled Delivery of Chemicals.

References

[1] G. Decher, J.D. Hong, and J. Schmitt. *Thin Solid Films*, **210-211**:831–835, 1992.

[2] G. Decher. *Sciences*, **277**:1232, 1997.

[3] G. Decher and J.B. Schlenoff. *In Multilayer Thin Films, Sequential Assembly of Nanocomposite Materials*. 2003.

[4] J.H. Cheung, A.F. Fou, and M.F. Rubner. *Thin Solid Films*, **244**:985, 1994.

[5] A. Wu, D. Yoo, J.K. Lee, and M.F. Rubner. *J. Am. Chem. Soc.*, **121**:4883–4891, 1999.

[6] P. Stroeve, V. Vasquez, M.A.N. Coelho, and J.F. Rabolt. *Thin Solid Films*, **285**:708, 1996.

[7] M. Onda, K. Ariga, and T. Kunitake. *Biosci. Bioeng.*, **87**:69, 1999.

[8] Y. Lvov. *Handbook for Surfaces and Interfaces, Nanostructured Materials*, volume 3. Academic Press, New York, 2001.

[9] G.B. Sukhorukov, E. Donath, S. Davis, H. Lichtenfeld, F. Caruso, V.I. Popov, and H. Mohwald. *Polym. Adv. Technol.*, **9**:759, 1998.

[10] A.A. Antipov, G.B. Sukhorukov, S. Leporatti, I.L. Radtchenko, E. Donath, and H. Mohwald. *Colloids Surf. A*, **198**:535, 2002.

[11] O. Gaserod, A. Sannes, and G. Skjak-Braek. *Biomaterials*, **20**:773–783, 1999.

[12] P. Bertrand, A. Jonas, and R. Legras. *Macromol. Rapid Commun.*, **21**:319, 2000.

[13] M. Schonhoff. *Curr. Opin. Colloid Interf. Sci.*, **8**:86, 2003.

[14] Jérôme Dejeu, Lionel Buisson, Marie Céline. Guth, Cyrille Roidor, Francois Membrey, Daniel Charraut, and Alain Foissy. Early steps of the film growth mechanism in self-assembled multilayers of pah and pss on silica. polymer uptake, charge balance and afm analysis. *Colloids and Surfaces A: Physicochemical and Engineering Aspects*, **288**:26, 2006.

[15] V. Bosio, F. Dubreuil, G. Bogdanovic, and A. Fery. *Colloids Surf. A*, **243**:147, 2004.

[16] M. Kolasinska and P. Warszynski. *Appl. Surf. Sci.*, **66**:65, 2005.

[17] Ph. Lavalle, C. Gergely, F.J.G. Cuisinier, G. Decher, P. Schaaf, J.C. Voegel, and C. Picart. *Macromolecules*, **35**:4458, 2002.

[18] J. Ruths, F. Essler, G. Decher, and H. Riegler. *Langmuir*, **16**:8871, 2000.

[19] C.C. Buron, C. Filiatre, F. Membrey, H. Perrot, and A. Foissy. *J. Colloid Interf. Sci.*, **296**:409–418, 2006.

[20] R.J. Hunter. *Introduction to Modern Colloid Science*. Oxford University Press, 1993.

[21] M.A. Cohen Stuart, C.W. Hoogendam, and A. de Keizer. *J. Phys. Cond. Mater.*, **9**:7767–7783, 1997.

[22] C. Geffroy, M.P. Labeau, K. Wong, B. Cabane, and M.A. Cohen Stuart. Kinetics of adsorption of polyvinylamine onto cellulose. *Colloids and Surfaces A: Physicochemical and Engineering*, **172**:47, 2000.

[23] C. Buron. PhD thesis, University of Franche-Comte, 2004.

[24] T. Roques-Carmes, F. Membrey, C. Filiatre, and A. Foissy. *J. Colloid Interface Sci.*, **245**:257, 2002.

[25] C.C. Buron, F. Membrey, C. Filiatre, and 289 (2006) 163. A. Foissy. *Colloids Surf., A Physicochem. Eng. Asp.*, **289**:163, 2006.

[26] J.A. De Feijter, J. Benjamins, and F.A.Veer. *Biopolymers*, **17**:1759, 1978.

[27] M.A. Cohen-Stuart, C.W. Hoogendam, and A. de Keizer. Kinetics of polyelectrolyte adsorption. *Journal of Physics: Condensed Matter.*, **9**:7767–7783, 1997.

[28] T. Roques-Carmes, F. Membrey, A. Deratani, M. Boehmer, and A. Foissy. *J. Colloid Interf. Sci.*, **256**:273, 2002.

[29] C. Geffroy, J. Persello, A. Foissy, P. Lixon, F. Tournilhac, and B. Cabane. *Colloids Surf. A*, **162**:107–112, 2000.

[30] A.W.M. de Laat, G.L.T. Van den Heuvel, and M.R. Bohmer. *Colloids Surf. A*, **98**:61, 1995.

[31] H. Riegler and F. Essler. *Langmuir*, **18**:6694, 2002.

[32] P. Lavalle, C. Picart, J. Mutterer, C. Gergely, H. Reiss, J.-C. Voegel, B. Senger, and P. Schaaf. *J. Phys. Chem. B*, **108**:635–648, 2004.

[33] M. von Smoluckowski. *Bulletin International de l'Académie des Sciences de Cracovie*, **8**:182–199, 1903.

[34] D. Barten. PhD thesis, Wageningen University, 2003.

[35] J.B. Schlenoff, H. Ly, and M. Li. *J. Am. Chem. Soc.*, **120**:7626, 1998.

[36] A. Foissy and J. Persello. *The Surface Properties of Silicas*. John Wiley & Sons, 1998.

[37] Alexander I. Petrov, Alexei A. Antipov, and Gleb B. Sukhorukov. Base-acid equilibria in polyelectrolyte systems: From weak polyelectrolytes to interpolyelectrolyte complexes and multilayered polyelectrolyte shells. *Macromolecules*, **36**:10079–10086, 2003.

[38] J.M.C. Lourenco, P.A. Ribeiro, A.M. Botelho do Rego, F.M.B. Fernandes, A.M.C. Moutinho, and M. Raposo. *Langmuir*, **20**:8103–8109, 2004.

[39] J.B. Schlenoff and S.T. Dubas. *Macromolecules*, **34**:592, 2001.

[40] G. Ladam, P. Schaad, J.C. Voegel, P. Schaaf, and F. Cuisinier G. Decher and. *Langmuir*, **16**:1249, 2000.

[41] T. Radeva, V. Milkova, and I. Petkanchin. *Colloids Surf. A*, **240**:27, 2004.

[42] F. Caruso, H. Lichtenfeld, E. Donath, and H. Mohwald. *Macromolecules*, **32**:2317–2328, 1999.

[43] R.F.M. Lobo, M.A. Pereira da Silva, M. Raposo, R.M. Faria, M.A. Pereira-da Silva O.N.Oliveira Jr., and R.M. Faria. *Nanotechnology*, **10**:389, 1999.

[44] T. Radeva, V. Milkova, and I. Petkanchin. *Journal of Colloid and Interface Science*, **266**:141, 2003.

[45] R.N. Smith, M. McCormick, C.J. Barrett, L. Reven, and 37 (2004) 4830. H.W. Spiess. *Macromolecules*, **37**:4830, 2004.

[46] B.F. Abu-Sharkh. Structure and mechanism of ofrmation polyelectrolyte multilayers. *Polymer*, **47**:3674–3680, 2006.

[47] C. Picart, Ph. Lavalle, P. Hubert, F.J.C. Cuisinier, G. Decher, P. Schaaf, and J.-C. Voegel. *Langmuir*, **17**:7414–7424, 2001.

In: Polymer Films: Properties, Performance and Applications ISBN 978-1-61209-201-0
Editors: Stefan A. Romano and George P. Somners

Chapter 7

FILM GROWTH PROPERTIES OF SELF-ASSEMBLED POLYELECTROLYTE FILMS OF HIGHER THICKNESS

Jérôme Dejeu*
University of Franche-Comté, FEMTO-ST Institute

Abstract

This chapter addresses the growth mechanism of the self assembled film of higher thickness. Different weak polymer pairs are investigated as PAH/PSS, PLL/HA, PAH/(PGA_x-PSS_{1-x}), using the fixed-angle laser reflectometry, and the Quartz Crystal Microbalance to follow in situ the deposition of self-assembled polyelectrolyte films in a large domain of thickness. It is observed that a mixture of both components leads to a hybrid growth law. At the beginning of the construction, the thickness of the film increases exponentially with the number of deposited bilayers. Once a given thickness is reached (from 150 to 300 nm depending on the polymer pair), one observes the crossover to a linear growth regime. The experiments lead to refinement of the buildup mechanism model, introduced previously for the exponentially growing films. The mechanism no longer needs all the "in" and "out" diffusing polyanions or polycations to be involved in the buildup process to explain the linear growth regime but merely relies on the interaction between the polyelectrolytes with an upper zone of the film. This zone is constituted of polyanion/polycation complexes which are "loosely bound" and rich in the polyelectrolyte deposited during the former deposition step.
The morphology and the topography of the higher thickness film were analyzed by Atomic Force Microscopy (AFM) and Focused Ion Beam (FIB) technique. The film thickness was determined from the FIB, the ellipsometric data and the reflectometric output interpreted in terms of refractive index and thickness way above the usual range which covers the earlier deposition steps of the polymers (i.e., 30-50 layers, depending on the film composition).

PACS 68.55.-a, 68.55.A-, 68.37.Ps, 68.37.-d.

Keywords: Multilayer, Polyelectrolytes, Thick film characterization, Focused ion Beam.

---

*E-mail address: jerome.dejeu@univ-fcomte.fr

1. Introduction

The self-assembled polyelectrolytes films are obtained by successive and alternate adsorption of two polyelectrolytes of opposite charge. This procedurewas previously described by Decher in the early 90s [1]. Recent reviews on the subject show that many theoretical and practical aspects of the interaction between the two polymers along with the film growth remain ambiguous and debated [2]. The multilayer filmformed layer by layer is an efficient method for controlling the shell permeability [3] or the release of the active matter in electronics [4], electrocatalysis [5] and many other areas [6]. Controlled release was performed with an appropriate choice of the thickness and porosity of the shell, the polyelectrolytes properties (hydrophobic, hydrophilic) and the incorporation of counterions [7]. Two types of multilayer growth process were reported: films with thickness that increases linearly with the number of deposition steps [1] and other systems whose thickness growth is non-linear and usually called exponential [8]. Linear growth is due to the fact that during the build-up process the polyelectrolyte from the solution interact only with a polyelectrolyte of opposite charge forming the outer layer of the film [1]. For the "exponential" growth, on the other hand, it was stated that it results from the diffusion "in" and "out" of the whole filmof at least one of the polyelectrolytes which constitute the multilayer [9]. However, this view appears oversimplified since the recent observation that the exponential growth regime only holds in fact over a limited number of deposition steps, before the crossover to a linear growth regime [10]. The thickness increment is usually much smaller in the linear regime than in the exponential one (typically a few nanometers versus hundreds of nanometers). The thickness of the multilayer film is usually measured by ellipsometry [11], AFM [12], confocal spectroscopy [13], X-ray [14], laser reflectometry [15, 16], and quartz crystal microbalance [17] optical waveguide lightmode spectroscopy (OWLS) [18]. All these techniques can be used with any autoassembled films but have specific advantages: the two first methods require a sample with a small roughness, the confocal a transparent sample and the last one a thickness film inferior to near 200 nm [19]. In this study, the deposition of multilayers is realized on silica substrate. The roughness obtained for a great number of layers forbids the traditional techniques.

In this chapter, we presented the different strategy to analysed the thick multilayer film and the model modification to evaluated the film thickness build-up with weak polyelectrolyte. In many cases, a transition was observed between first an exponential and then a linear growth regime [19, 20]. The transition was said to take place at the film thickness corresponding to the diffusion length of the polymer [21]. This transition was determined by focused ion beam (FIB), elippsometry, reflectometry, quartz crystal microbalance, ... The focused ion beam (FIB) is an important tool in failure analysis, defect characterization, design modification, and process control in a variety of industries. In this work, it is used to cut the film and measure its thickness. Scanning ion microscopy (SIM) offers a spatial resolution of 5 nm, with different contrast mechanisms than those of electron microscopy. The usual reflectometric analysis to track the polymer deposition in the low thickness range was replaced by theoretical route to do the same in the higher thickness range, backed with an experimental assessment.

2. Thick Multilayer Morphology [22]

Films containing 10, 15, 20, 40, 80, and 160 bilayers of PAH (poly(allylamine hydrochloride)) and PSS (poly(styrene sulfonate)) were built on two different silica wafers with thicknesses of 114 and 298 nm, respectively. Figure 1 shows representative AFM pictures of films containing between 10 and 160 bilayers. These pictures reveal a rough relief, with a strong increase in heterogeneities with the thickness of the film. Be aware of the different horizontal (m) and vertical (nm) scales; peaks in the pictures are large bumps in reality. Bumps start forming at the second deposition step from the aggregation of PAH and PSS (see below). The average height between peaks and valleys approximates 200 nm in the early steps (10-15 bilayers) and 1.2 μm for 160 bilayers. The roughness varies between 20 and 180 nm in the same range of thickness. Details of the growth mechanism, the parameters, and the film morphology were given earlier [16, 23].

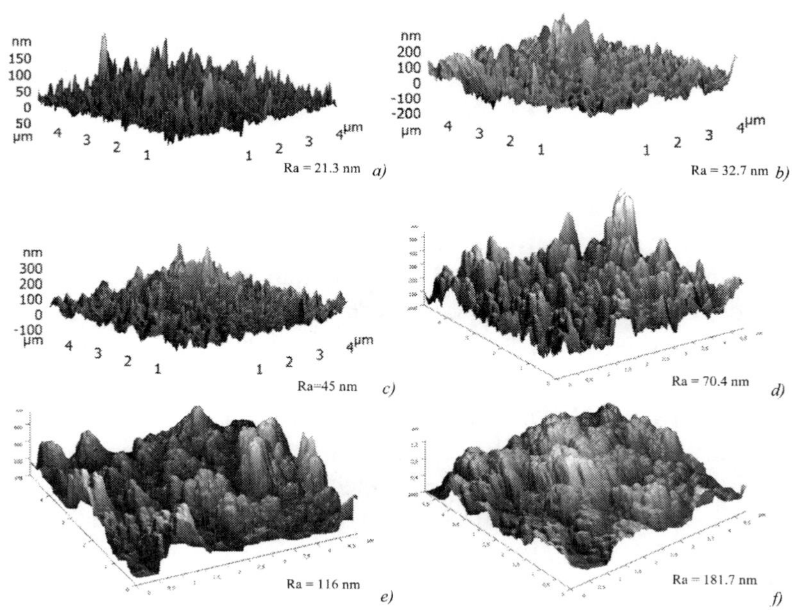

Figure 1. Topographic image of the polymer film of PAH/PSS on silica substrate: (a) 10, (b) 15, (c) 20, (d) 40, (e) 80, and (f) 160 bilayers.

All pictures reveal a rough relief. No leveling of the surface was observed, contrary to the study of Buron et al. with the pair MAQUAT(poly(trimethylammonium ethyl methacrylate chloride)/PAA (poly(acrylic acid)), Figure 2 [24] where"bumps" started to coalesce after about 30 layers.

We see that the peaks become much wider (10 bilayers) and the films smooth over from filling the holes, but a few cavities remain. Similar pictures by Picart et al. [8] showed a progressive transition from an island like structure to an uniform film after six bilayers with the pair poly(l-lysine)-hyaluronic acid on glass.

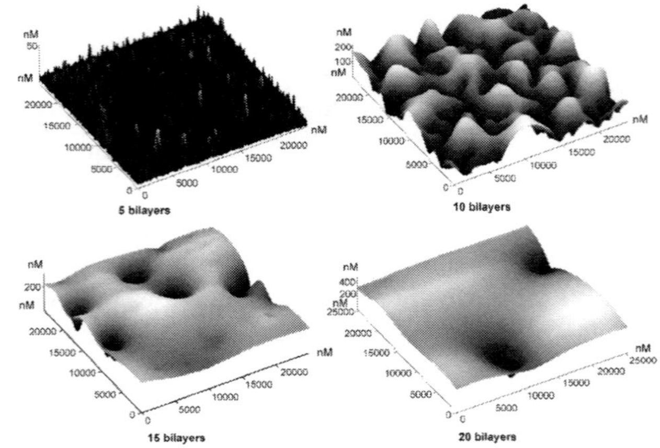

Figure 2. Evolution of the surface morphology of MAQUAT/PAA in function of the number of deposited bilayers [24].

3. Thicker PAH/PSS Film

3.1. Focused Ion Beam [20]

The polymer film deposit on the wafer was covered with a chromium layer by cathodic sputtering. The aim of this layer is to increase the picture contrast and to avoid charge accumulation on the deposit which would reduce the stability of the ion beam. The cathodic sputtering was done on a Plassys apparatus with a pressure of 7.10_3 mbar, a continuous current of 1 A, deposition rate of 120 nm.min^{-1} during 1 min and 15 s. The 30 keV focused Ga^+ ions (FIB Orsay Physics Canion 31) was used to etch polymer films of various thickness and their Si/SiO_2 substrate. After FIB irradiation, scanning ion microscopy (SIM) was used to observe the surface. The wafer was tilted and viewed at an incidence angle of 40 or 35: the same ion beam, with a current limited to 30 pA, was used as a scanning probe to image the crosssection of the film. Different layers of materials can be clearly identified (Figure 3 due to good material contrast in the ion-induced secondary electron image. The thickness of SiO_2 and polymer film can be measured directly from the cross-section view by compensating the tilt angle, the apparent width of the layers being divided by sinus of the tilt angle.

On the image of the wafer, the different layers, silica and chromium, are clearly differentiated. The thickness of the layers deduced after angular correction from the scale on Figure 3 is, respectively, of 150 and 330 nm for chromium and silica. The discrepancy with the measurements given by ellipsometry remains within the uncertainty. This reference experiment gives the thickness of the chromium layer deposited on the multilayer films.

The FIB pictures of the film's cross section also showed a strong heterogeneity of the deposits (Figure 4). Thinner films exhibited bumps, and thicker ones showed a granular, cauliflower-like morphology (Figure 4). The question whether holes and cracks in the film could result from drying was considered previously [14]. A former study showed that drying did not modify the general morphology of films beyond five bilayers [19]. The high roughness may be a consequence of the growth mechanism. When PAH adsorbs on the

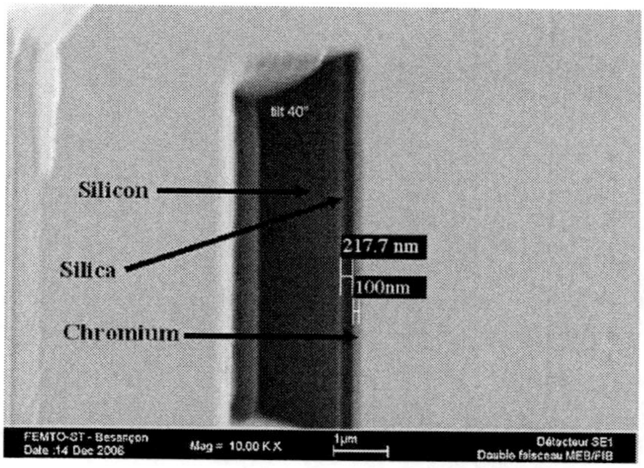

Figure 3. SIM image of the silica wafer covered of chromium with a tilt of 40 to study the cross-section. All the scales must be divided by sin (40) to give the real thickness.

top of the PAH/PSS aggregates, it induces a coalescence phenomenon that creates large cavities.

In Figure 4, it is impossible to differentiate the chromium layer from the polymers PAH/PSS film. This chromium layer is however present and must be subtracted from the total thickness. For 30 layers of polyelectrolytes, the thickness of total film (chromium + polymers) on the silica substrate is 270 $\pm$ 50 nm. The uncertainty in the thickness is evaluated from the discrepancies in repeated measurements. As the measured chromium layer deposited on polymer film was 150 nm, the polymer film is 270 - 150 = 120 $\pm$ 50 nm. The same calculation can be done for the other films and the average thickness of the polymer film was summarizes Table 1.

Table 1. Mean PAH/PSS polymer film thickness versus number of bilayer deposits

Bilayers	5	10	15	20	40	80	160
Thickness (nm)	66	83	107	145	400	1000	2350
uncertainties (nm)	15	20	50	100	100	150	400

The thickness found for the first bilayers are similar to these obtained by laser reflectometry (Figure 5) [22]. Indeed, Dejeu and co-wokers shown that the thickness exhibits a peculiar trend. It increases strongly during the first layers (near 60 nm after 4 layers) and increases slowly after. This quasi plateau is attributed to the polymer film concentration. After the formation of coacervates between PSS and PAH on the substrate in the first layer, there are densification of the aggregates by diffusion of the PSS into the aggregates [16,23]. So the focused ion beam can be used to determine the thickness of the polymer film when it is very difficult by the other methods

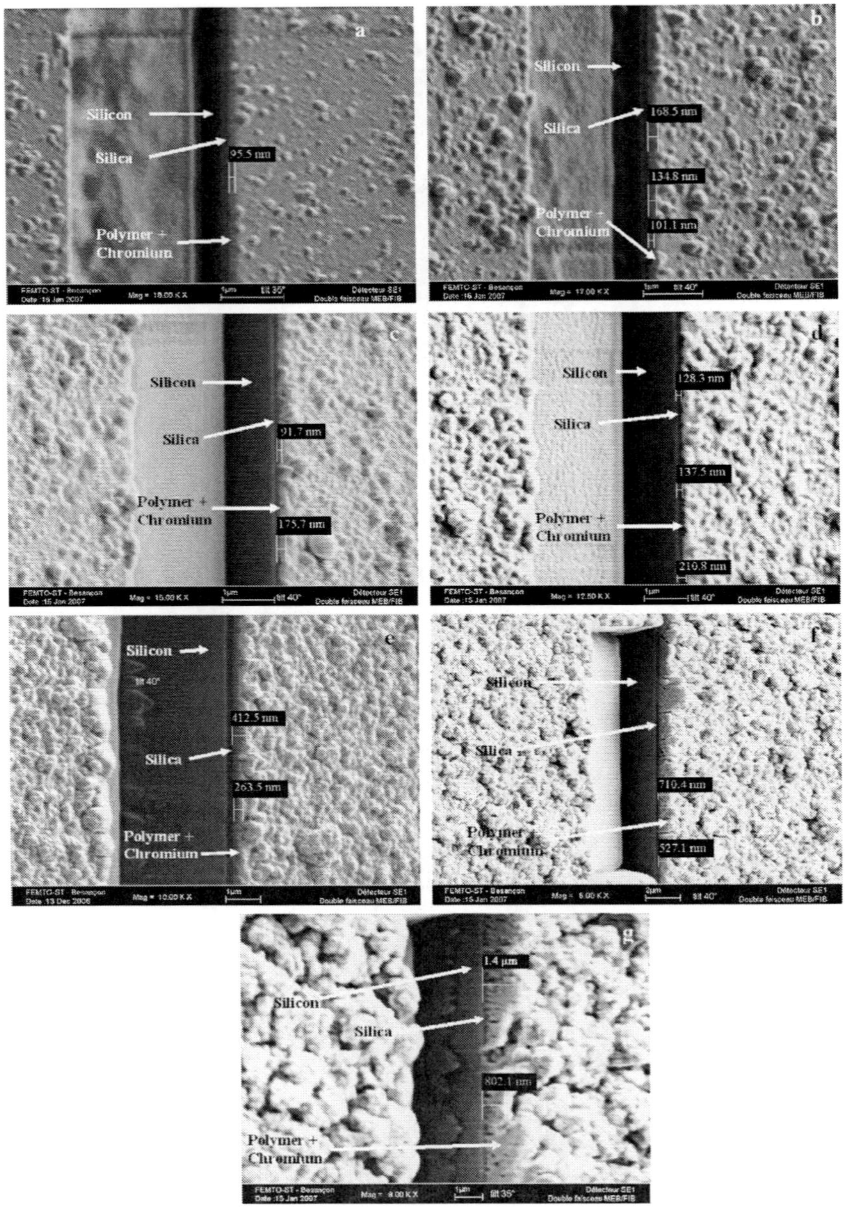

Figure 4. SIM images of film multilayer deposit on silica wafer and covered by chromium layer with a tilt of 40 (images b-f) or 35 (image a and g) to study the cross-section. (a) 5 bilayers, (b) 10 bilayers, (c) 15 bilayers, (d) 20 bilayers, (e) 40 bilayers, (f) 80 bilayers, (g) 160 bilayers. All the scales must be divided by sin 40 (image b-f) or sin 35 (image a and g) to give the real thickness.

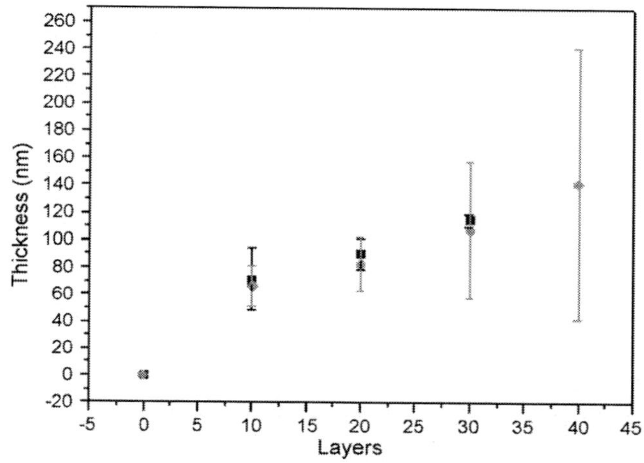

Figure 5. Thickness variation vs. the number of layers obtained by FIB (grey) and laser reflectometry (black).

3.2. Laser Reflectometry Analyses

3.2.1. Reflectometric Curve [22]

Representative examples of the reflectometric data collected during the growth of the deposit are presented in Figure Reflectoepais. As observed in a former study with the polymer pair MAQUAT/PAA [25], the signal shows a complex oscillation profile with an opposite alternation for the two substrates of different thickness (114 and 298 nm).

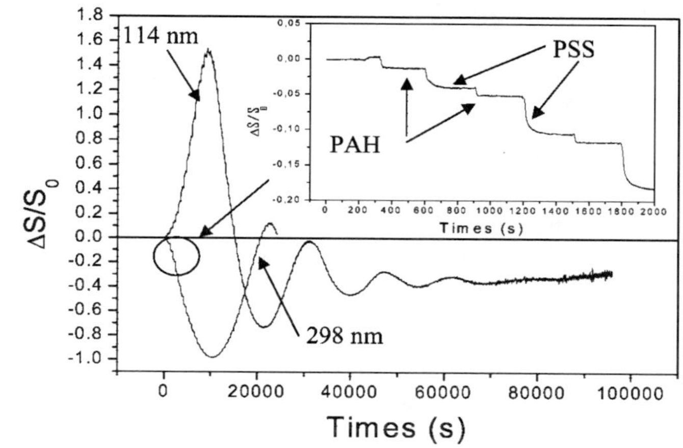

Figure 6. Variation of the reflectometric signal ($\Delta S/S_0$) collected during the formation of 40 and 160 bilayers on substrates with thicknesses of 298 and 114 nm, respectively.

The oscillations result from the variation of the cosinus of the phase term ϕ in the adsorbed layer, with ϕ reading [26]:

$$\phi = \frac{4\pi . d_{ads} . n_{ads} . \cos\theta_3}{\lambda} \tag{1}$$

where d_{ads} and n_{ads} are the thickness and the refractive index of the polymer layer, respectively, λ is the laser wavelength in vacuum, and θ_3 is the incidence angle of the laser in the polymer layer, Figure 7.

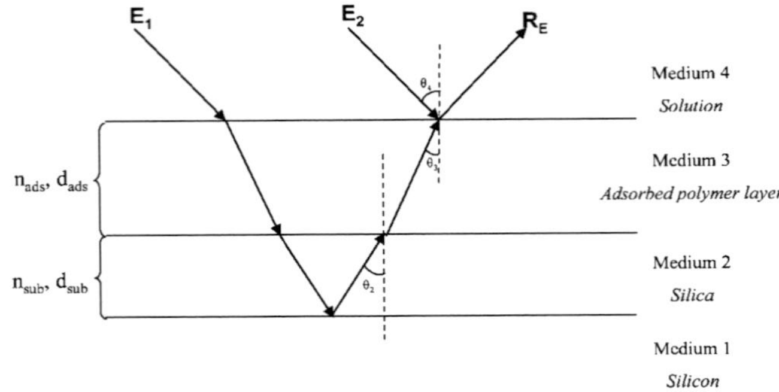

Figure 7. Sketch of the reflective interface used in the reflectometry laser beam.

The opposition of maxima and minima for the two substrate thicknesses results from the phase term of the substrate itself (the silica layer), which reads the same as equation 1, but with the appropriate parameters for the silica layer (d_{sub}, n_{sub}, Θ_2). The reflectometric response can be modeled and calculated rather easily for the lower series of deposits (i.e., up to about the first extremum) [26]. These calculations show that the output data develop in the right direction for each substrate, and initial values are positive and go through a maximum for the silica thickness 114 nm and reversely for the thickness 298 nm.

The signal damping, observed Figure 6, arise from the influence of heterogeneities and roughness in the reflection properties. This point will be considered in the next section.

3.2.2. Reflectometric Properties of Thick Polymer Deposits: A New Approach [22]

If the film is homogeneous with regard to the refractive index (i.e., the composition) and with a negligible rugosity, calculations show that the signal should oscillate regularly with the film thickness. This is shown in Figure 8 for the two silica substrates and three different values of the refractive index. In addition to the regular alternation between positive and negative values, the figure shows two specific influences of the refractive index: (i) a decrease of the refractive index induces an increase of the distance between extrema, and (ii) a decrease of the refractive index does not change $\Delta S/S_0$ at the minima ($\Delta S/S_0 = -1$), whereas it causes a decrease at the maxima.

The value -1 at the minima results from the definition of S which is the ratio between the reflected intensities of the parallel and the perpendicular components ($S = f.R_p/R_s = I_p/I_s$; see cite). Since S is necessarily positive and may take the value 0, the minimum for $\Delta S/S_0$ is -1 ($\Delta S/S_0 = S/S_0 - 1$).

In contradiction with the calculations, the experimental output in Figure 6 shows a progressive decrease of the distance between extrema, together with a damping of the signals. The theoretical curves in Figure 8 show that the decrease in the distance between maxima might result from a progressive increase of the refractive index, but this would contradict the correlated decrease of $\Delta S/S_0$ at the maxima. In addition, we expect that the compo-

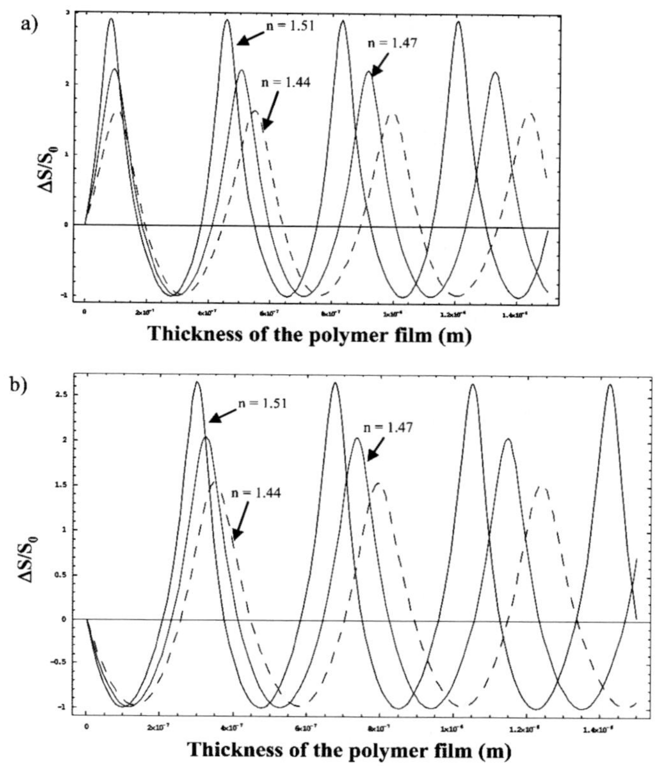

Figure 8. Calculated variation of the reflectometric signal versus the thickness of the polymer film for different values of a constant refractive index of the film. Silica thickness: 114 nm (a) and 298 nm (b).

sition and the refractive index of the film reach a steady value after a certain number of steps, which is the case for the polymer ratio after 15 bilayers for PAH/PSS citeDejeu2008. Indeed, the refractive index may not exceed

The principal reason for the damping of the signal is actually that a significant part of the light is diffused by the polymer film, the more so the thicker the film. The diffusion phenomenon results from heterogeneities in the film (interfaces between areas of different refractive index, including the rugosity) with a size close to that of the wavelength of the laser beam. Practically, the increase of the diffusion phenomenon during the film construction was also revealed visually in the reflectometric cell by the progressive increase of the diffused laser light intensity at the solution-film interface.

Although the output $\Delta S/S_0$ at the maxima is affected by the diffusion phenomena, it is reasonable to assume that their position is not. Thus, considering a homogeneous film with a given refractive index, the distance between two maxima should correspond to a phase difference π. Consequently, the increase of film thickness between two maxima can be calculated from [26]:

$$\Delta d = \frac{\lambda}{4 n_{ads} \mathrm{Cos}\theta_3} \tag{2}$$

where n_{ads} is the refractive index of the film, λ is the wavelength of light in vaccum, and Θ_3 is the incidence angle of the laser beam (Figure 7).

This calculation requires a choice of refractive index. As explained earlier, the refractive index calculated from the reflectometric output at the first maximum (15 bilayers) was n = 1.44, which revealed a significant amount of water (37 wt %). The refractive index of a pure and dense organic film, containing an equal amount of PAH and PSS, would be n = 1.53. Thus, calculations of the film thickness were attempted with three constant values of the refractive index (1.44, 1.47, and 1.51), as if the film composition was the same during the whole buildup. However, the progressive decrease of the distance between extrema (Figure 3) demonstrated a progressive increase of the refractive index. It was about 1.44 (water content 37%) after 15 bilayers and increased with the number of bilayers up to a value between 1.47 (water content 22%) and 1.51 (water content 0 wt %). In comparison, Jaber and Schlenoff found 23 and 7 % of water in films (PAH/PSS) containing 50 and 250 bilayers, respectively.35 We postulated 1.49 as a reasonable maximum value of the refractive index in the later stage of the growth (water content 11 wt %), which led to the thickness 1452 ± 75 nm for 115 bilayers. Table 2 gives an idea of the influence of the choice of nads in the calculated thickness.

Table 2. Location of maxima and calculated mean thickness (nm) of the polymer film using data in figure 6 (silica thickness 114 nm)

Bilayers	15 max	36 min	51 max	67 min	80 max	92 min	102 max	115 max
Thickness (nm) n = 1.44	122	317	547	762	993	1208	1438	1653
Thickness (nm) n = 1.47	110	300	525	719	917	1130	1324	1525
Thickness (nm) n = 1.51	79	278	474	649	828	1020	1195	1377

The approach and the calculations above extend interestingly the range of applicability of reflectometry to characterize the fabrication of interfacial polymer films. Practically, we see that the evolution of the refractive index, the film composition (polymer ratio, water content), and the film thickness may be fairly deduced from the detailed analysis of the experimental data. As a confirmation, we may now compare the FIB and the reflectometric measurements.

3.3. Glow Discharge Optical Emission Spectroscopy (GDOES)

GDOES has been widely used for depth profiling analysis of relatively thick films, in the tens micron range, such as galvanized or painted steels. More recently, however, it has been demonstrated that Rf-GDOES has enormous potential, not only for the analysis of thick film, but also for the analysis of ultra-thin layers, conducting or non-conducting, less than 10 nm thickness, where AES and SIMS are frequently employed.

The sample is analysed in GDOES instrument (Figure 9 at an applied power of 15 W and Argon pressure of 650 Pa. The area of analysis is 4 mm diameter and different spectra wavelengths (nm) are employed for detection of chloride, oxygen, nitrogen, silicon, sodium and sulphur. For all the results presented in this paper, a multiplicative coefficient at 100 will be applied for the oxygen, the sodium, the nitrogen, the sulphur and the chlorine. Only the values of the silicon will not be multiplied.

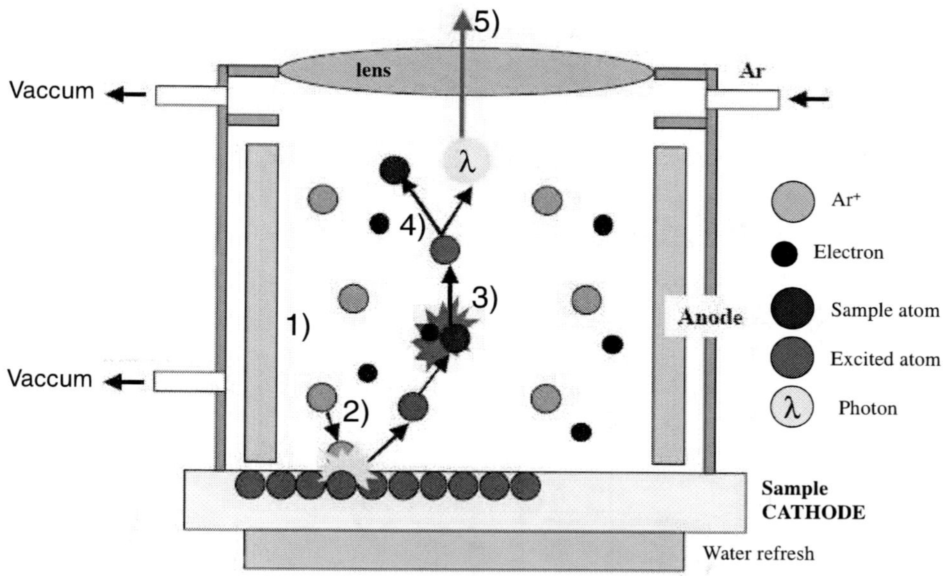

Figure 9. Glow Discharge Optical Emission Spectroscopy principle.

The GDOES results for the silica/ silicon wafer and the polymer film with different number of layer are presented Figure 10.

In figure 10.a, several important features are evident: it is easy to separate the silica layer to the silicon layer. Indeed in the silica layer, we observe the oxygen atoms whereas it is absent in the silicon layer and the most importantly, it is evident that the signal intensities for the four probing atoms, namely nitrogen, sulphur, sodium and chloride are at the background noise level except the sodium at the interface silica-silicon (6s). This element is not present in the silicon layer and come from the silica layer formation by thermo-oxydation. In the other graph of the Figure 10, The sodium peaks Na_2 correspond to it observes at the the silicon/silica interface in figure 10.a. We observe a decrease of the height of peaks but a extension halfway up with the increasing number of layers. The quantity of polymers, represented by the peaks integrated, increases with the number of layers. The maximum position was observed in the middle of the time exploration of the polymer film. This observation was due to the fact the sample was an important roughness [] and the height of the peaks are near the middle of the film thickness. When the Argon beam arrives on the sample, it shells the top of the aggregates (Argon 1) and the homogeneous film (Argon 2). After that the Argon 1 ions were each the aggregates, they arrives on the homogeneous film whereas the argon 2 ions finished to each the homogeneous film. It is at this moment when the maximum is reached. Then the argon 1 ions finished ejecting the atoms of the films while the Argon 2 analysed the silica layer. This explication was comforted by the fact that the Rf-GDOES results shown the silicon atom in the self-assembled film (Figure 10.d.) The maximum of each peaks are not observed is the same position. Indeed, the sequence nitrogen, sulphur, and silica are noticed. The maximum of the sulphur was more close of the silica because the PSS diffuse on the polymer film, whereas PAH adsorbs only on the top of the aggregate.

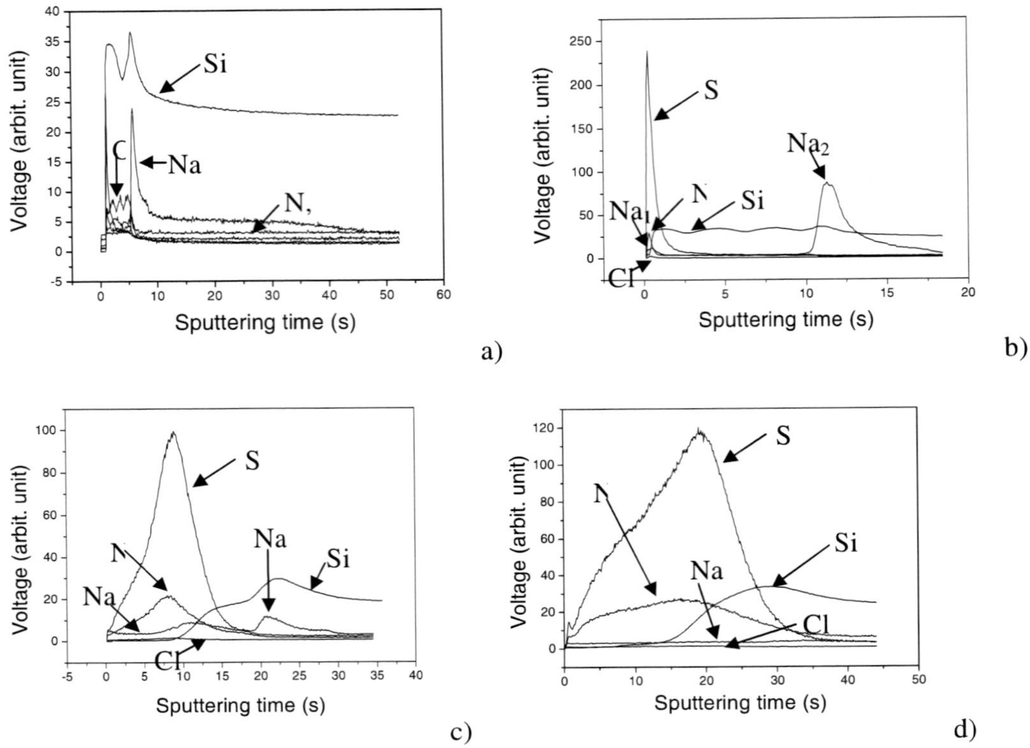

Figure 10. Rf-GDOES profile at the stagnation point of the wafer : Silica wafer (a) 10 bilayers (b) 40 bilayers (c) and 80 bilayers (d).

The time penetration of argon beam in the silica layer and the self assembled film can evaluated the polymer thickness thickness with the hypothesis that this time is independant of material nature. This time was summarized Table 3.

Table 3. Time penetration of the Argon beam in the silica layer and in the self assembled film versus the number of bilayers deposits)

Bilayers	0	0	5	10	15	20	40	80
Time penetration (s)	4.6	11.5	1.9	4.2	4.3	5.5	18.6	80
Uncertainties (s)	0.2	0.3	0.5	1.1	1.3	1.8	4	6
Thickness (nm)	113	300	50	110	115	145	490	902
Uncertainties (nm)	10	14	20	25	25	50	100	150

The time necessary to analyse the silica layer allows to estimated an average speed of penetration of the ions bundle. This value is 25.9 ± 0.4 nm.s^{-1}. With the value and the analysis time of polymer films, it is possible to determine the thickness of the polymer films. The thicknesses obtained was summarized Table TableGDOES and is valid only in the hypothesis where the speed of attack is the same, for the silica layer and the polymer film.

3.4. Comparison of the Polymer Film Thickness Found by FIB, GDOES and Reflectometry

As seen in Figure 11, the FIB and the reflectometric data for the film thickness compare well. In the case of reflectometry, the error bars take into account the measurements made with two silica substrates with different thicknesses (114 nm for a 160 bilayers film and 298 nm for a 80 bilayers film). For clarity, the error bars for the FIB data (of the order of 15 %) are not shown.

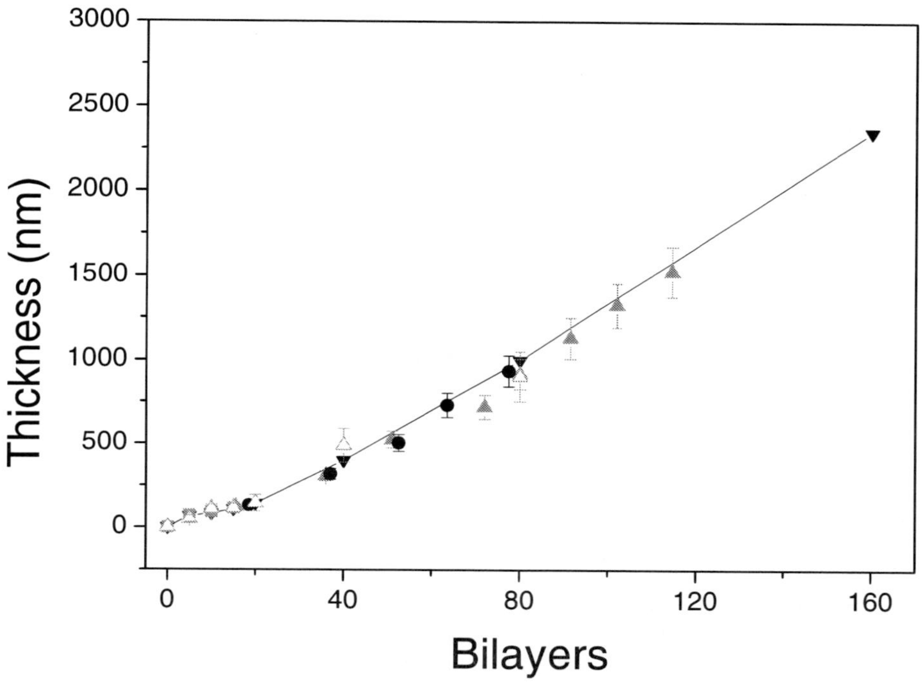

Figure 11. Thickness of the polymer film. Measured by FIB (▼), GDOES (△)calculated from reflectometric data using the regular procedure for thin films (■), the present method for the 80 bilayers buildup (•), and the 160 bilayers buildup (▲).

A change of growth rate is observed at about 40 bilayers, which corresponds to an approximate thickness of 200 nm. The change of growth regime, sometimes designated as a transition between an exponential and a linear increase of thickness with respect to the number of adsorption steps, has been linked to the penetration depth of the polymers, at least one of the two [17].

In our case, the linear regime starts about 40 bilayers and the growth rate is 15 $\pm$ 2 nm/bilayer, which gives a thickness of 2.3 $\pm$ 0.4 μm for 160 bilayers, identical to the experimental FIB evaluation: 2.3 μm (Table 1). In the literature, we found also the same thickness variation for PAH/PSS [27] and HA/PLL [21].

4. $(PGA_x\text{-}PSS_{1-x}/PAH)_n$ Film [17]

The construction of multilayer films by exposing a charged surface consecutively to a solution of PAH and to a solution containing both PGA (L-glutamic acid) and PSS was investigate versus the mixture PGA-PSS composition. Hubsch et al. [17] follows the film build-up by quartz crystal microbalance, Figure 4.. Byassuming that the density of the film is close to 1 g.cm^{-3}, they deduce the evolution of the film thickness, d, during the buildup (Figure 4.).

[Resonance frequency shifts]

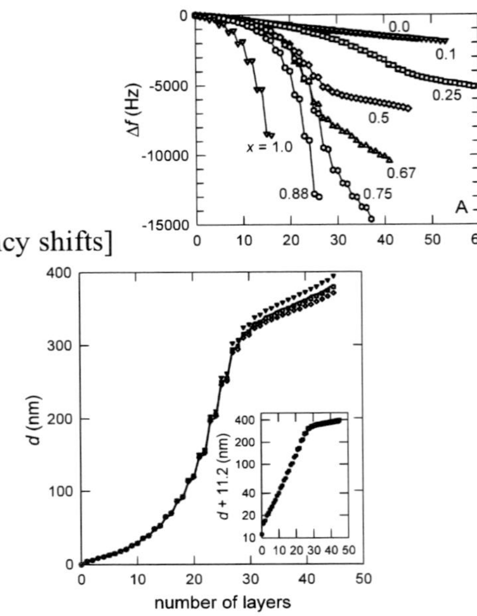

[thickness]

Figure 12. A. Resonance frequency shifts as obtained by QCM-D at 15 MHz. These experimental data correspond to the buildup of the multilayer film with $PGA_x\text{-}PSS_{1-x}$ polyanion mixtures, where x ranges from 0 to 1 as indicated by the labels on the curves. The results corresponding to x = 0 and x = 0.1 are practically indistinguishable; B. Thickness of the multilayer film built up with the $PGA_{0.5}\text{-}PSS_{0.5}$ polyanion mixture. The three sets of data point (triangles, squares, and diamonds) are derived from the resonance frequency shifts measured at 15, 25, and 35 MHz, respectively. The solid line represent the thickness derived by means of the monolayer model described in refs 28-29. The three sets of data points derived using Sauerbrey's relationship have been fitted with -11.2 + 12.68 exp(0.119 i) up to the 26th layer (i is the number of layers). The insert shows d + 11.2 for the three frequencies (as in the mainframe) on a logarithmic scale to reveal the exponential growth up to about the 26th layer [17].

In Figure 4., the thickness increases exponentially during the first deposition cycles. After 26 layers the thickness evolution is modified and the film growth linearly. The thickness evolution transition is function to the film composition (see x = 0.25, 0.50, 0.67, 0.75, Figure 4.). Interestingly, for each mixture composition where the two regimes are visible, the slope observed in the linear growth regime is smaller than that observed at the end of the exponential phase. Moreover, the slope increases as the polyanion solution becomes richer in PGA and the difference between the slope at the end of the exponential phase and that in the linear regime decreases. One can expect that for the (PGA/PAH) film built from a purePGA solution (x = 1), the exponential regime ultimately also goes over into a linear one but without change in the slope. Unfortunately, we could not observe this behavior because the change of the deposition regime is expected to occur only at a film thickness exceeding the detection limit of both OWLS and QCM-D [17].

5. PLL/HA Film [19, 21]

The growth of the PLL (Poly(L-lysine)) and HA (hyaluronic acid) film was studied by Porcel and co-workers [19, 21]. They analysed the dipping time [19] and the molecular weight [21] influences on the exponential-linear transition, Figure 13.

To verify that the transition from the exponential to the linear growth regime for the HA/PLL system is not a peculiarity of the spraying method, we performed several experiments with the conventional dipping method. Figure 5. represents two cases where the films were built up to 20 pairs of layers, the contact time of the PE solutions with the substrate being equal to 3 and 10 min, respectively.

To investigate the influence of the already deposited multilayer on the thickness growth in the linear domain, we performed experiments where the (HA/PLL)n film construction began with a first couple of polyelectrolyte molecular weights and then, during the buildup, one of the polyelectrolytes was replaced by the same polyelectrolyte but of different molecular weight, Figure 5..

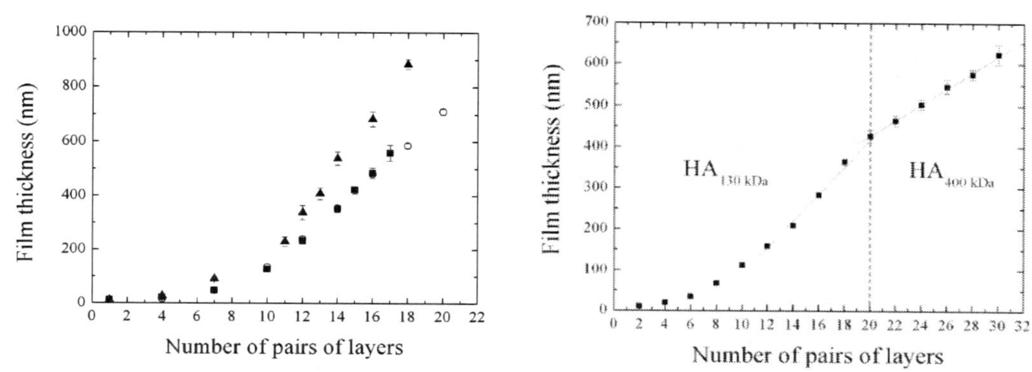

Figure 13. a. Spraying versus dipping: comparison of the evolution of the PEM film thickness as a function of the number of added pairs of layers between films made by dipping or spraying. (■) PEI_{3s}-$(HA_{15s}/PLL_{9s})_{17}$ film obtained by spraying, (O) PEI-$(HA/PLL)_{20}$ obtained by dipping for 3 min per PE, and (▲) PEI-$(HA/PLL)_{18}$ obtained by dipping for 10 min per PE. All of the PE concentrations were equal to 3.10^{-3} M. In all experiments, the film thicknesses were measured by ellipsometry after drying []; b. Buildup of an HA/PLL film by spraying with a change of the HA molecular weight during the construction, $(HA_{130}/PLL_{360})_{20}$-$(HA_{400}/PLL_{360})_{10}$. The slopes in the linear regime are respectively equal to $\Delta d_1 = 32.7$ nm/pair of layers and $\Delta d_2 = 19.1$ nm/pair of layers [19, 21].

In Figure 5., the transition from the exponential to the linear growth regime sets in after the deposition of 10-12 bilayers for dipping and spraying methods. A very similar evolution of the film thickness for a PEM film constructed by dipping the surface for 3 min in the PE solutions and by spraying the PLL solution for 9 s and the HA solutions for 15 s is observed, with the concentrations of the PE solutions being equal to 3.10^{-3} M in monomer units. The buildup processes performed with the dipping and the spraying methods do not seem to be fundamentally different from each other even though the characteristic time scales involved in the two approaches are not the same. Moreover, as will be seen below, the exponential-to-linear transition always takes place at about n) 12 for films constructed by the spraying method over a very large range of buildup conditions (i.e., different spraying times and spraying rates) [19].

Porcel et al. [] observed that the thickness increment for an $(HA/PLL)_n$ film is not influenced by the presence of the underlying $(HA/PLL)_n$ film built with a different HA molecular weight: the thickness increment is similar to the value found when the film is directly deposited on a bare silica substrate. Moreover, one can observe that the change of the thickness increment takes place as soon as the first HA is replaced by the second HA (Figure 5.): there is no progressive smooth transition of the increment from Δd_1 to Δd_2 over several deposition steps. This is observed if one starts with a multilayer first constructed with the Mw = 130000 HA and then pursued with the Mw = 400000 HA; a similar observation is made if one uses the reverse buildup procedure [21].

6. Exponentail-Linear Transition Explication

This paragraph is aimed at explaining (i) the origin of the transition from the exponential to the linear regime, which can potentially occur in any exponentially growing polyelectrolyte system, and (ii) why this transition is accompanied by a change of slope of the thickness as illustrated in Figure 12 and 13.

Aslong as the thickness of the film is smaller than a certain value, all polyanion chains diffuse out of the film during the time of exposure to the PAH solution. The number of polyanion chains available for the formation of complexes withPAHchains at the film/solution interface is therefore proportional to the total thickness of the film (or size of the reservoir). This relationship between thickness and number of complexes formed (or additional thickness) leads immediately to the observed exponential growth law for the thickness. However, because of the finite value of the diffusion coefficient, the film eventually reaches a thickness, which no longer allows all "free" polyanions to diffuse out of the multilayer during the contact time with the polycation solution (10 min for the experiments described here). At this point, the in and out diffusion affects predominantly a zone beneath the film/solution interface whose thickness is about (Dpt)1/2, where Dp is the diffusion coefficient of both polyelectrolytes and t is the contact time of the multilayer with the polycation solution. Because the number of polyanion chains that participate in the formation of new complexes is now constant, the film thickness grows linearly from this point on and the thickness increment per layer pair deposition is the same at the end of the exponential regime and in the linear regime (for constant contact times) [17]. So the non-linear film growth cann be resume as follow [10, 17, 28]: the films may be subdivided in three main domains (Figure 14). Domain I corresponds to the initial buildup phase and is in contact in PGA and the difference between the slope at the end of the exponential phase and that in the linear regime decreases. One can expect that for the (PGA/PAH) film built from a purePGA solution (x = 1), the exponential regime ultimately also goes over into a linear one but without change in the slope. Unfortunately, we could not observe this behavior because the change of the deposition regime is expected to occur only at a film thickness exceeding the detection limit of both OWLS and QCM-D [17].

5. PLL/HA Film [19, 21]

The growth of the PLL (Poly(L-lysine)) and HA (hyaluronic acid) film was studied by Porcel and co-workers [19, 21]. They analysed the dipping time [19] and the molecular weight [21] influences on the exponential-linear transition, Figure 13.
To verify that the transition from the exponential to the linear growth regime for the HA/PLL system is not a peculiarity of the spraying method, we performed several experiments with the conventional dipping method. Figure 5. represents two cases where the films were built up to 20 pairs of layers, the contact time of the PE solutions with the substrate being equal to 3 and 10 min, respectively.
To investigate the influence of the already deposited multilayer on the thickness growth in the linear domain, we performed experiments where the (HA/PLL)n film construction began with a first couple of polyelectrolyte molecular weights and then, during the buildup, one of the polyelectrolytes was replaced by the same polyelectrolyte but of different molecular weight, Figure 5..

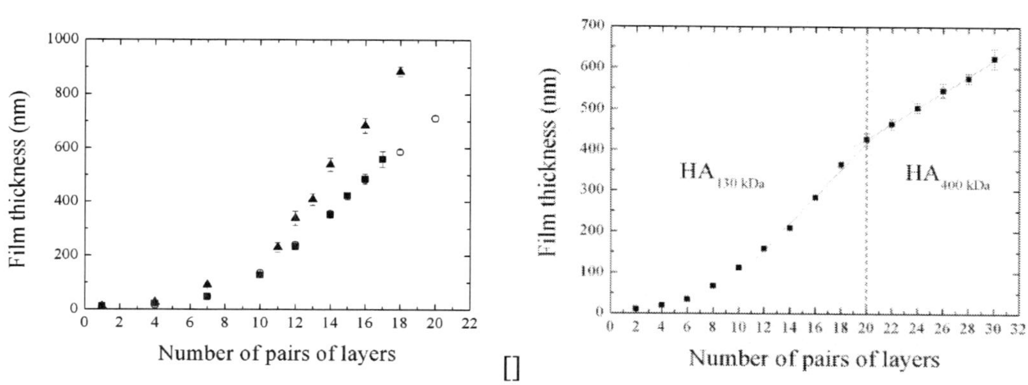

[] []

Figure 13. a. Spraying versus dipping: comparison of the evolution of the PEM film thickness as a function of the number of added pairs of layers between films made by dipping or spraying. (■) PEI_{3s}-$(HA_{15s}/PLL_{9s})_{17}$ film obtained by spraying, (O) PEI-$(HA/PLL)_{20}$ obtained by dipping for 3 min per PE, and (▲) PEI-$(HA/PLL)_{18}$ obtained by dipping for 10 min per PE. All of the PE concentrations were equal to 3.10^{-3} M. In all experiments, the film thicknesses were measured by ellipsometry after drying []; b. Buildup of an HA/PLL film by spraying with a change of the HA molecular weight during the construction, $(HA_{130}/PLL_{360})_{20}$-$(HA_{400}/PLL_{360})_{10}$. The slopes in the linear regime are respectively equal to Δd_1 = 32.7 nm/pair of layers and Δd_2 = 19.1 nm/pair of layers [19,21].

can be evaluated by focused ion beam when the other methods are affected by roughness, mainly for thick polymer film.
The thicker film was initally a non-linear growth and after it reaches a certain value, it continues to grow linearly linearly, the transition between both regimes being quite abrupt. The characteristic feature of this transition is that the growth rate in the linear regime is smaller than that at the end of the non-linear regime. This transition was reported around 12 bilayers with the polymer pair poly(L-lysine) (PLL)/hyaluronic acid (HA) by Porcel et al. [19] where only the polyelectrolyte HA diffused into the predeposited film. More recently, the same authors showed that the transition took place in the thickness range 150-250 nm [21], in agreement with the present result (200 nm). The changeover from an exponential to a linear regime is explained by the a model proposed by Hubsch et al. [17] and later by Salomaki et al. [10] in which it is assumed that the exponential-to-linear transition is due to a film restructuring that progressively forbids the diffusion of one of the polyelectrolytes constituting the film over part of the film. This forbidden zone then grows with the number of deposition steps so that the outer zone of the film that is still concerned with diffusion keeps a constant thickness and moves upward as the total film thickness increases.

Acknowledgment

This work was supported by the EU under Contract NMP4-CT-2003-001428: Nanocapsules for Targeted Controlled Delivery of Chemicals.

References

[1] G. Decher. *Sciences*, **277**:1232, 1997.

[2] M. Schonhoff. *Curr. Opin. Colloid Interf. Sci.*, **8**:86, 2003.

[3] G. Decher and J.B. Schlenoff. *In Multilayer Thin Films, Sequential Assembly of Nanocomposite Materials*. 2003.

[4] J.H. Cheung, A.F. Fou, and M.F. Rubner. *Thin Solid Films*, **244**:985, 1994.

[5] A. Wu, D. Yoo, J.K. Lee, and M.F. Rubner. *J. Am. Chem. Soc.*, **121**:4883–4891, 1999.

[6] Y. Lvov. *Handbook for Surfaces and Interfaces, Nanostructured Materials* , Academic Press, New York, 2001.

[7] N. Gupta, A.A. Patel, R. Nassar, Y.M. Lvov, M.J. McShane, and J.D. Palmer. *Colloids Surf., A*, **245**:137–142, 2004.

[8] C. Picart, Ph. Lavalle, P. Hubert, F.J.C. Cuisinier, G. Decher, P. Schaaf, and J.-C. Voegel. *Langmuir*, **17**:7414–7424, 2001.

[9] C. Picart, J. Mutterer, L. Richertand Y. Luo, G.D. Prestwich, P. Schaaf, J.C. Voegel, and P. Lavalle. In *Proc. Natl. Acad. Sci.*, volume 99, pages 12531–12535, USA, 2002.

[10] M. Salomaki, I.A. Vinokurov, and J. Kankare. *Langmuir*, **21**:11232–11240, 2005.

[11] O. Mermut and C.J. Barrett. *J. Phys. Chem. B.*, **107**:2525–2530, 2003.

[12] R.F.M. Lobo, M.A. Pereira da Silva, M. Raposo, R.M. Faria, M.A. Pereira-da Silva O.N.Oliveira Jr., and R.M. Faria. *Nanotechnology*, **10**:389, 1999.

[13] T. Mauser, C. Dejugnat, H. Mohwald, and G.B. Sukhorukov. *Langmuir*, **22**:5888–5893, 2006.

[14] D. Korneev, Y. Lvov, G. Decher, J. Schmitt, and S. Yaradaikin. *Phys. B*, **213-214**:954–956, 1995.

[15] C.C. Buron, F. Membrey, C. Filiatre, and 289 (2006) 163. A. Foissy. *Colloids Surf., A Physicochem. Eng. Asp.*, **289**:163, 2006.

[16] S. Diziain, J. Dejeu, L. Buisson, D. Charraut, F. Membrey, and A. Foissy. Investigations in the build-up mechanism of self assembled polyelectrolyte films by laser reflectometry and atomic force microscopy. *Thin Solid Film*, **516**:1–7, 2007.

[17] E. Hubsch, V. Ball, B. Senger, G. Decher, J.C. Voegel, and P. Schaaf. *Langmuir*, **20**:1980–1985, 2004.

[18] F. Boulmedais, M. Bozonnet, P. Schwinte, J.C. Vogel, and P. Schaaf. *Langmuir*, **19**:440–445, 2003.

[19] C. Porcel, P. Lavalle, V. Ball, G. Decher, B. Senger, J.C. Voegel, and P. Schaaf. *Langmuir*, **22**:4376–4383, 2006.

[20] J. Dejeu, R. Salut, M. Spajer, F. Membrey, A. Foissy, and D. Charraut. Multilayer thickness determination by focused ion beam. *Applied Surface Sciences*, **254**:5506–5510, 2008.

[21] C. Porcel, P. Lavalle, G. Decher, B. Senger, J.C. Voegel, and P. Schaaf. *Langmuir*, **23**:1898–1904, 2007.

[22] J. Dejeu, F. Membrey, S. Diziain, C. Bainier, M. Spajer, D. Charraut, and A. Foissy. Study of film growth properties of self assembled polyelectrolyte films of higher thickness: reflectometric and focussed ion beam analysis. *Journal of physical chemistry C*, **112**:10531–10537, 2008.

[23] Jérôme Dejeu, Lionel Buisson, Marie Céline. Guth, Cyrille Roidor, Francois Membrey, Daniel Charraut, and Alain Foissy. Early steps of the film growth mechanism in self-assembled multilayers of pah and pss on silica. polymer uptake, charge balance and afm analysis. *Colloids and Surfaces A: Physicochemical and Engineering Aspects*, **288**:26, 2006.

[24] C.C. Buron, C. Filiâtre, F. Membrey, C. Bainier, D. Charraut, and A. Foissy. Effect of substrate on the adsorption of polyelectrolyte multilayers: Study by optical fixed-angle reflectometry and afm. *Colloids and Surface A: Physicochemical and Engineering Aspects*, **305**:105–111, 2007.

[25] C. Buron. PhD thesis, University of Franche-Comte, 2004.

[26] J.C. Dijt, M.A. Cohen Stuart, and G.J. Fleer. *Advances in Colloid and Interface Science*, **50**:79–101, 1994.

[27] T. Radeva, V. Milkova, and I. Petkanchin. *Colloids Surf. A*, **240**:27, 2004.

[28] G. Ladam, P. Schaad, J.C. Voegel, P. Schaaf, and F. Cuisinier G. Decher and. *Langmuir*, **16**:1249, 2000.

In: Polymer Films: Properties, Performance and Applications
Editors: Stefan A. Romano and George P. Somners
ISBN 978-1-61209-201-0

Chapter 8

CHEMICAL SENSORS AND BIOSENSORS BASED ON CONDUCTING POLYMER THIN FILMS

Nada F. Atta, Ahmed Galal and Maher F. El-Kady
Department of Chemistry, Faculty of Science, Cairo University, Giza, Egypt

ABSTRACT

Conducting polymers (CPs) are organic polymers that exhibit electronic, magnetic and optical properties of metals and semiconductors while retaining the attractive mechanical properties and processing advantages of polymers. These features, along with chemical sensitivity, room temperature operation, and tunable charge transport properties, have launched conducting polymers as a major class of chemical transducers, creating powerful thin/thick film sensors for over two decades. Additionally, their synthetic routes are compatible with several functionalization schemes, including covalent bonding to the monomer, entrapment during synthesis, and surface chemistry, which can impart a high degree of selectivity. CPs can be used as the selective layer in sensors (sensor receptor) or as the transducer itself. Physical transducers varying from optical, mass sensitive and electrochemical are all discussed in this chapter. In comparison with most of the commercially available sensors, based usually on metal oxides and operated at high temperatures, the sensors made of CPs have many improved characteristics. They have high sensitivities and short response time and more interestingly they work at room temperature. As a result, more and more attentions have been paid to the sensors fabricated from CPs.

This chapter covers main applications of conducting polymers in chemical sensors and biosensors. The properties, synthesis and processing of CPs into films are reviewed and discussed in the first part. Following this, the concept of chemical sensors and biosensors along with a commercial perspective are presented. There is always a desire to further optimize CPs when targeting a specific application by increasing their affinity to particular analytes. In this respect, the modification strategies of CPs with synthetic or natural receptors to meet specific application are introduced. Follows, the applications of CPs to sensors classified by transducing mode (optical, mass, and electrochemical) and application mode (pH, humidity, gas/odor and ions selective). Additionally, CPs have attracted much interest in the development of biosensors. The developing role of CPs as health care biosensors, immunosensors, DNA biosensors, biosensors for environmental monitoring of hazardous chemicals and biosensors for agriculture and food industry are extensively reviewed. We will basically consider the emerging trends in CPs based sensors and biosensors during the last 5 years. Finally, a section is presented on challenges and future directions.

1. Conducting Polymers: Promising Materials for Sensor Applications

Conducting polymers (CPs) are polyconjugated polymers with electronic properties resembling those of metals, while retaining properties of conventional organic polymers. The most widely investigated conducting polymers include polyaniline, polypyrrole, polythiophene and their derivatives. The award of the Nobel Prize in Chemistry in 2000 to H. Shirakawa, A. MacDiarmid and A. Heeger for their pioneering work on conducting polymers widely recognized the importance of these materials and has prompted vigorous research in the field.

CPs have attracted much interest as potential materials for chemical sensors and biosensors [1]. A cursory summary of the main properties of CPs is necessary to understand the reason of why they are good for sensing applications:

- CP films can be easily synthesized by chemical or electrochemical methods, and they are known to have considerable flexibility in their chemical structure that the molecular chain structure can be modified conveniently by copolymerization or structural derivations. This enhances the selectivity of CPs towards specific analytes.
- CPs are remarkable materials that possess the electrical properties of metals yet retain the flexibility of polymers. This makes possible the fabrication of flexible, complex shape sensors based on CPs.
- CPs can be reversibly doped and undoped using chemical/ electrochemical techniques accompanied by significant changes in conductivity. This makes them attractive transducers materials in various sensing devices. Suitable transducing systems can be adapted in a sensor assembly depending on the nature of interaction with the species of interest. The physical transducers vary from electrochemical, spectroscopic, thermal, piezoelectric and surface wave technology.
- The electrical conductivity of conducting polymers changes over several orders of magnitude by the interaction with various analytes. This results in changes in parameters such as resistance, current or electrochemical potential/work function. This interaction is the basis of detecting the analyte.
- CPs are ideally suited for sensor applications because they not only exhibit high conductivity and electroactivity but they could also be used as a general matrix and can be further modified with other compounds in order to change selectivity.
- CPs have attracted much interest as suitable matrices of biomolecules (nucleic acids, antibodies, enzymes) and can be used to enhance stability, speed, sensitivity and hence are finding increasing use in medical diagnostics.

Those are just some of the potential properties of CPs as candidates for sensor applications. We have addressed in this chapter many other properties that make CPs promising for various specific applications. In general, CP sensors are based on the variation of their capacity, optical properties, redox properties, etc. The sensitivity of CP-based sensors is primarily determined by the difference in a specific property before and after exposure to a test target molecule. The variation of the conductivity caused by doping/de-doping processes is widely used in the conducting polymer-based sensors [2].

2. Chemical Sensors And Biosensors: Commercial Perspective

Global Industry Analysts, Inc. announced in February 2010 the release of a comprehensive global report on chemical sensors markets. According to the report,the world market for chemical sensors is projected to reach $17.28 billion by the year 2015. New product innovations and expanding applications will offer considerable growth opportunities in spite of relatively slower gains in mature markets such as pH testing, industrial safety, and flue gas emissions monitoring.

Chemical sensors are used to improve point-of-care diagnostics in medical applications, control and monitor environmental pollution, provide compact, fast, low power, and sensitive tools for quality and process control in industrial applications, as well as improve or implement warfare threat detection and security. Advances in the medical/diagnostic test market and escalating use of chemical sensors in large-scale industrial and environmental applications will drive the market for chemical sensors, including biosensors and gas sensors. Poor water quality and the urgent need to conserve water resource will stir up demand for dissolved-gases sensors. Demand for chemical sensors will be fuelled by newer applications for established sensor products; product innovation through the use of microfabrication and other high tech manufacturing techniques; and falling cost of high-performance sensor types. Biosensors and optical sensors are a hot spot for generating revenue, which is especially true for mass volume applications in the industrial, consumer and motor vehicle sectors. The decline in end-use markets such as automotive and biometrics will to a large extent be counterbalanced by the indispensable use of chemical sensors in medical diagnostics. Biosensors market stands immensely benefited by the rapidly aging populating, increased spending in the healthcare sector, increasing demand for point-of-care testing & monitoring solutions, shift towards alternate sites of healthcare, and increased occurrences of a number of chronic diseases. The demand for dissolved-gases sensors is expected to witness steady growth due to rising environmental concerns such as water quality as well as growth witnessed in aquaculture industry. Global market for gas sensors is driven by demand from OEM and replacement market needs. There is an increasing demand for carbon dioxide (CO_2) sensors in bulk food storage sector, due to the proven ability of CO_2 sensors in detecting incipient spoilage as well as assessing CO_2 levels in storage structures and controlled atmosphere packages. The United States and Europe collectively account for a major share of the world chemical sensor market. The medical market is expected to be the largest, as well as the fastest growing end-use segment for chemical sensors.

Despite the many technological advances in biosensor research and development and the introduction of many different products, glucose biosensors still account for approximately 85% of the current world market for biosensors [3]. The reasons why the glucose market was particularly receptive to the introduction of biosensors are numerous, but the single greatest factor was the prevalence of diabetes in developed nations. The disease is a major world health problem. It was estimated in 2004 that there are over 170 million diabetics worldwide [4]. Through 2012, optical sensors and biosensors are expected to provide the best opportunities. Optical sensors -- including infrared sensors and products based on fiber optic, photoionization, fluorescence, chemiluminescence, light emitting diode, laser and ultraviolet technologies -- will see the fastest gains of all sensor types. Optical sensors will continue to benefit from their high sensitivity, stability, immunity to interference and product improvements such as smaller size and enhanced ruggedness.

Key players dominating the chemical sensors market include Alphasense Limited, City Technology Ltd, Delphi Corporation, Emerson Process Management, Figaro Engineering Inc, Freescale Semiconductor Inc, Mine Safety Appliances Company, NGK Spark Plug Company Limited, Nova Biomedical Corporation, Ocean Optics Inc, RAE Systems Inc, Smiths Detection, Teledyne Monitor Labs Inc, and Thermo Fisher Scientific Inc.,

When asked by an editor from a prestigious analytical chemistry journal on how to distinguish the best biosensors papers, Peter Kissinger of Purdue University set the following seven criteria for a really good biosensor manuscript [5]:

1. Do enough people want or need to have a sensor for the analyte of interest?
2. Are there existing sensors for the same analyte? Does the new sensor present clearly defined advantages over the older technology, or is it simply a small variation on a well-established theme? Unless this new sensor is so innovative that it is interesting for its conception alone, the following five issues should be addressed.
3. Has the chemical stability of each component used to prepare the sensor been considered?
4. Has the stability of the sensor itself been thoroughly tested both in use and in storage? At what temperatures? Six months stability of a prepared sensor is perhaps the absolute minimum for any practical commercial application.
5. If the sensor is intended for biological samples, has it actually been tested with biological samples and not just aqueous buffers?
6. If a sensor is intended to be used in tissue, have biocompatible components been chosen? Has tissue reaction to the foreign implant been considered both for its effect on the sensor and on the organism? Has a means of sterilization and sterile packaging been considered?
7. Has the dynamic range of the sensor been tested appropriate to the anticipated analyte concentration in "real" samples?

Nanotechnology generally is defined as the creation and use of materials, devices and systems through the manipulation of matter at scales of less than 100 nanometers or 0.0000001 of a meter (i.e., about the size of three atoms). At this scale, the properties that characterizelarger systems do not necessarily apply. Nanostructuredmetals are generally harder than macroscale metals,nanoceramics are softer, etc. Nanoscale materials lightabsorption properties also differ from macroscalematerials, e.g., some materials become transparent if theparticles are small enough.Since many of the properties that sensors are supposed tomeasure are formed on the molecular or atomic level, nanotechnology has obvious sensing applications. Sensorsconstructed at the molecular scale would be extremelysensitive, selective, and responsive and thus, the impact ofnanotechnology on sensors potentially is huge.NanoMarkets believes that the market opportunities for nanosensors will stem, in large part, from unique features that surpass competing technologies. Nanosensors for military and homeland defense applications, where they may be used for example to detect the presence of biotoxins, such as anthrax and smallpox, were estimated to reach $827 million in 2008 and grow to $3.9 billion in 2012. The overall Nanosensor market was projected to reach $2.7 billion in 2008 and climb to $17.2 billion in 2012.

3. Synthesis of Conducting Polymers

Conducting polymers have been synthesized in literature by various methods, including chemical [6-12], electrochemical, template, enzymatic, photo, plasma, and a number of other special methods. Chemical polymerization is usually subdivided into heterophase [6], solution [7], interfacial [8], seeding [9], metathesis [10], self-assembling [11], and sonochemical [12] polymerizations. Of these methods, chemical and electrochemical methods are the most commonly used.

Chemical polymerization is usually performed by mixing the monomer with oxidant in solution, see. The most widely used oxidants are ferric chloride, ammonium persulfate, hydrogen peroxide, etc. (see Fig. 1 for a typical example of chemical polymerization). The electrochemical synthesis of conducting polymers, first demonstrated with polypyrrole, has proven important in the development of the field. Today, electrochemical polymerization is performed using a three-electrode configuration (working, counter, and reference electrodes) in a solution of the monomer, appropriate solvent, and electrolyte (dopant) (Fig.2). Electrochemical polymerization is generally performed with galvanostatic[1], potentiostatic[2] or potentiodynamic[3] methods.

Chemical synthesis not only provides many different possible routes to synthesize a variety of CPs, but also permits the scale-up of these materials, which is currently not possible with electrochemical synthesis. Electrochemical polymerization, on the other hand, is becoming of technological importance. First, because in electrochemical polymerization, doping and processing take place simultaneously with polymerization. However, in conventional methods, polymer synthesis and doping are first carried out, followed by processing. Second, the electrochemical oxidation route also has greater control over the electrochemical potential of the system (thereby limiting over-oxidation) and in turn the rate of polymerization as well as the film thickness. Third, The electrochemical reactions are often much cleaner, and the CP obtained is expected to be in a relatively purer form (as no additional chemicals such as surfactant, oxidant, and so on are used here) compared to that obtained from chemical polymerization. Furthermore, the use of limited chemicals reduces the problem of pollution.

The most significant difference between electrochemical and chemical methods of CP synthesis is that very thin CP films on the order of 20 nm can be produced using the electrochemical technique, whereas powders or very thick films are typically produced with chemical polymerization [13]. All CPs can be synthesized chemically, but electrochemical synthesis is limited to those systems in which the monomer can be oxidized in the presence of a potential to form reactive radical ion intermediates for polymerization. The standard CPs (i.e., PPy, PT, PANI, and PEDOT) can be polymerized both chemically and electrochemically; however, several novel CPs with modified monomers are only amenable to chemical polymerization.

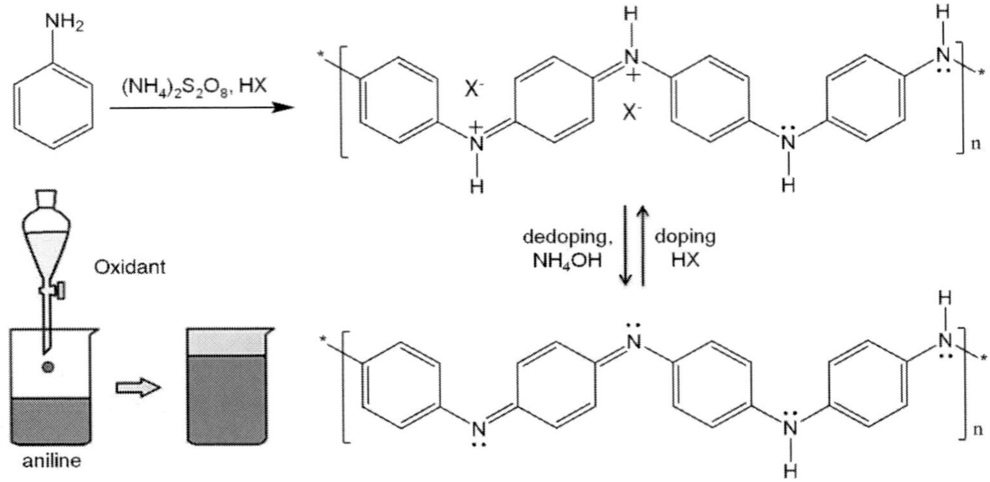

Figure 1. The conventional chemical synthesis of polyaniline: The oxidative polymerization reaction of aniline is typically carried out in an aqueous solution in which aniline is dissolved in a strong acidic solution (e.g., 1 M HCl) at 0°C and the polymerization is initiated by slowly adding an oxidant (e.g., ammonium peroxydisulfate) into the solution. The as-prepared polyaniline is in its doped emeraldine form, which can be dedoped by a base to its emeraldine base form. The bottom left scheme illustrates a typical reaction for making polyaniline. Reproduced with permission from ref [14].

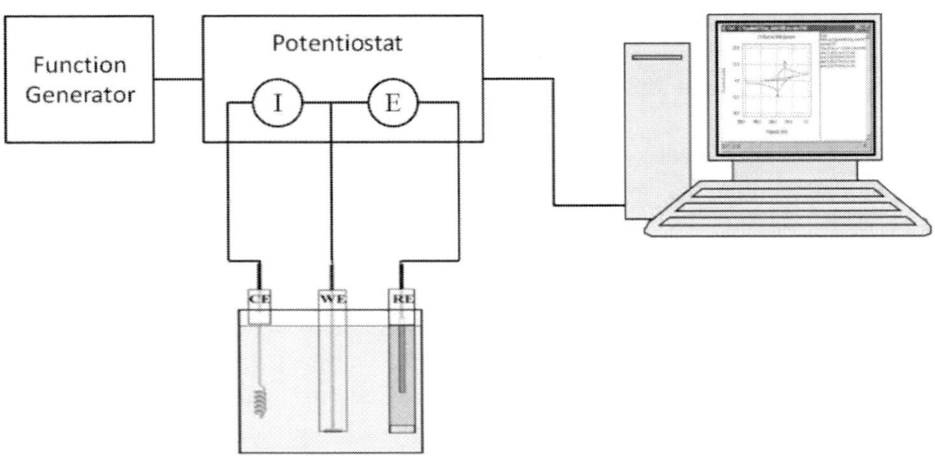

Figure 2. Schematic diagram for the electrochemical arrangement used in controlled potential electropolymerization: a measurement system with a potentiostat that uses three electrodes (the most commonly used setup in electrochemical studies[1]), WE— working electrodes; RE—reference electrode; CE—counter electrode. The electrodes are inserted through holes in the cell cover. The systems work in such a way that a desired potential is imposed to the working electrode vs. the reference electrode in accord with the program defined by a function generator. A variety of shapes and designs for both the electrodes and the cell are usually used [15]. The polymer film forms at the positively charged working electrode or anode.

4. Techniques of Formation of Conducting Polymer Films

4.1. Electrochemical Deposition

Electrochemical deposition (ECD) is probably the most convenient method to deposit CP films and is the most reported technique for the design of sensors based on CP films [16]. Again, ECD presents several advantages over other chemical methods as cleanness because no extraction from the monomer–solvent–oxidant mixture is necessary, doping and thickness control via electrode potential and simultaneous synthesis and deposition of the polymer thin layer. The thickness of the film can be controlled by the total charge passed through the electrochemical cell during film growing process. Moreover, the film can be deposited on patterned microelectrodes. Of course, the deposition must be carried out on a conducting substrate. However, if the insulating gap between the neighboring electrodes is close enough (few micrometers), the growing film can cover the insulated gap and connect electrodes. This is especially important in fabricating chemiresistors.

4.2. Spin Coating

Spin coating is a very widely used technique for making uniform thin polymer films [17]. It is simple, cheap and fast method to produce homogeneous thin layers of conducting polymers on flat substrates. In this process, the conducting polymer is dissolved in the appropriate solvent and spread on a rotating substrate. After evaporation of solvent, a thin film is formed (see figure 3 for the detailed process). Repeating above process is feasible, which can control the thickness of the film. This method can coat conducting polymers on both conducting and insulating substrates. Final film thickness and other properties depend on the nature of the conducting polymer fluid (viscosity, drying rate, surface tension, concentration, etc.) and the parameters chosen for the spin process. The combination of spin speed and time contribute to how the properties of the coated films are defined. In general, higher spin speeds and longer spin times create thinner films. Although spin-coating method is an easy processing technique for the production of nanothin film, it is limited to the use of soluble conducting polymers.

Polymeric conductive films manufactured by spin coating exhibit improved electrical conductivity and relatively good optical properties [18]. In addition, the all-spin-coating processing provides a possibility of high-throughput all-solution-processed fabrication of polymer solar cells [19]. The active layers in most experimental semiconducting polymer-based devices, such as light-emitting diodes, photovoltaics, and polymer-filed effect transistors, are made by spin coating [20]. Spin coating has also been used for the fabrication of glucose sensor based on organic thin film transistors (OTFT) with a channel consisting of glucose oxidase immobilized on conducting polymer film. Other potential applications of spin coated conducting polymer sensors include gas sensors [21], optical sensors [22] and more [vide infra].

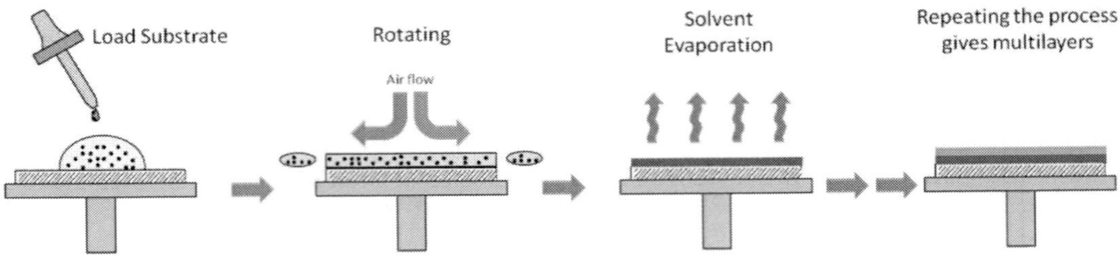

Figure 3. Typical spin process consists of (a) Loading the substrate with the coating fluid, (b) Acceleration of the substrate up to the desired final rotation speed (c) Spinning of the substrate at a constant rate; fluid viscous forces dominate the fluid thinning behavior (d) Solvent evaporation dominates the coating thinning behavior. The process can be repeated to get multilayer deposition.

4.3. Drop-Coating

This is a simple technique in which the solution of the material is put on the substrate surface by use of a precision pipette and the solvent is allowed to evaporate in air, in a controlled environment or even in a rough vacuum. The latter is often preferable because the rate of solvent evaporation can be increased at a given sample temperature [23]. The amount of material deposited on the substrate and hence the average film thickness is easily controlled by the amount of solution deposited on the substrate and by the concentration of the solution. Researchers reported drop coating of conducting polymers either from a conducting polymer solution [23,24] or from monomer/oxidant solution followed by polymerization on the substrate [25]. This technology is rather simple. However, the resulting film is usually not uniform. Conducting polymer films prepared by this method proved good for humidity sensors [25], gas sensors [24] and biosensors [23].

4.4. Dip-Coating

Dip-coating is a simple coating technique where the object to be coated is lowered into the solution and withdrawn at a suitable speed [26]. It does not require any specialized apparatus, and objects of various shapes can be dip-coated (see figure 4). Both internal and external surfaces of an object wetted by the solution are coated. An object can be dip-coated with a conducting polymer film by dipping the object in the chemical polymerization solution. Polymerization takes place in the solution bulk as well as on the object surface producing a film with a thickness that can be controlled by dipping time and temperature [26]. CP films deposited by this method have found applications in humidity and gas sensors [27].

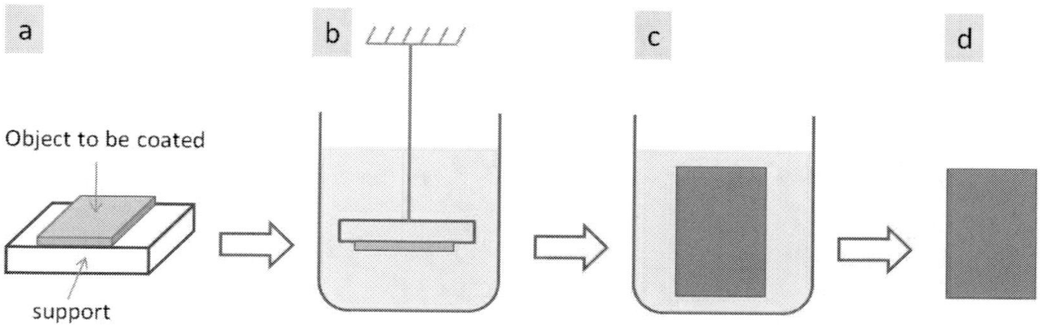

Figure 4. Schematic of the process of the deposition of a CP film with the dip coating technique: (a) the object to be coated is fixed onto a support. (b) The object is immersed into the polymerization solution which contains the monomer(s) and an oxidant; a polymer film spontaneously forms on the surface of the object. (c) Washing the coated object in the appropriate solvent, and (d) evaporating off the solvent. (Reproduced with permission from ref [28])

4.5. Layer-by-Layer Assembly

Layer-by-layer (LBL) assembly, first reported by Decher [29], is a method for fabricating thin films of polymers by the sequential adsorption of two or more polymers with complementary functional groups. Typically, thin films of oppositely charged layers are deposited via the sequential adsorption of positive and negative species by alternatively dipping into the solutions. The excess or remaining solution after each adsorption step is rinsed with solvent and thus we get a thin layer of charged species on the surface ready for next adsorption step (see Figure 5). There are many advantages of layer-by-layer assembly technique and these include simplicity, universality and control of the film thickness in the nanoscale. Further the LBL assembly process does not require highly pure components and sophisticated hardware. For almost all-aqueous dispersion of even high molecular weight species, it is easy to find an LBL pair that can be useful for building thin layer. Actually in each adsorption step, we get either a monolayer or a sub monolayer of the species and therefore we can find the number of adsorption steps needed for a particular nanoscale layer.

The most important reaction parameters or deposition conditions for LBL assembly include solvent, concentration of adsorbing species, adsorption time, temperature, nature and concentration of added salt, rinsing time, humidity of the surrounding air, drying, agitation during adsorption or rinsing, dipping speed and so forth. While the LbL technique generally works very well due to the fact that the processing window is rather large, it is highly recommended to keep the deposition conditions as constant as possible in order to get highly reproducible results.

Conducting polymers are known to have interesting chemical, electrochemical, mechanical, optical, electrical properties [30]. These properties make CPs potential candidates for numerous electronic applications such as organic light emitting diodes, organic solar cells, rechargeable batteries, microelectronics devices, biosensors, and chemical sensors [31]. For many device applications, it is desirable to have CPs in a thin film structure, preferably with known thickness and molecular packing. LBL is an excellent technique for making films with such properties. In fact, Molecular-level processing of conducting polymers by LBL technique has been widely reported in literature; polypyrrole, polythiophene, poly(3,4-ethylenedioxythiophene) (PEDOT), polyaniline, poly(phenylene vinylene), poly(*o*-anisidine), polyacetylene. LBL polymer films were used in gas, humidity and optical pH sensors.

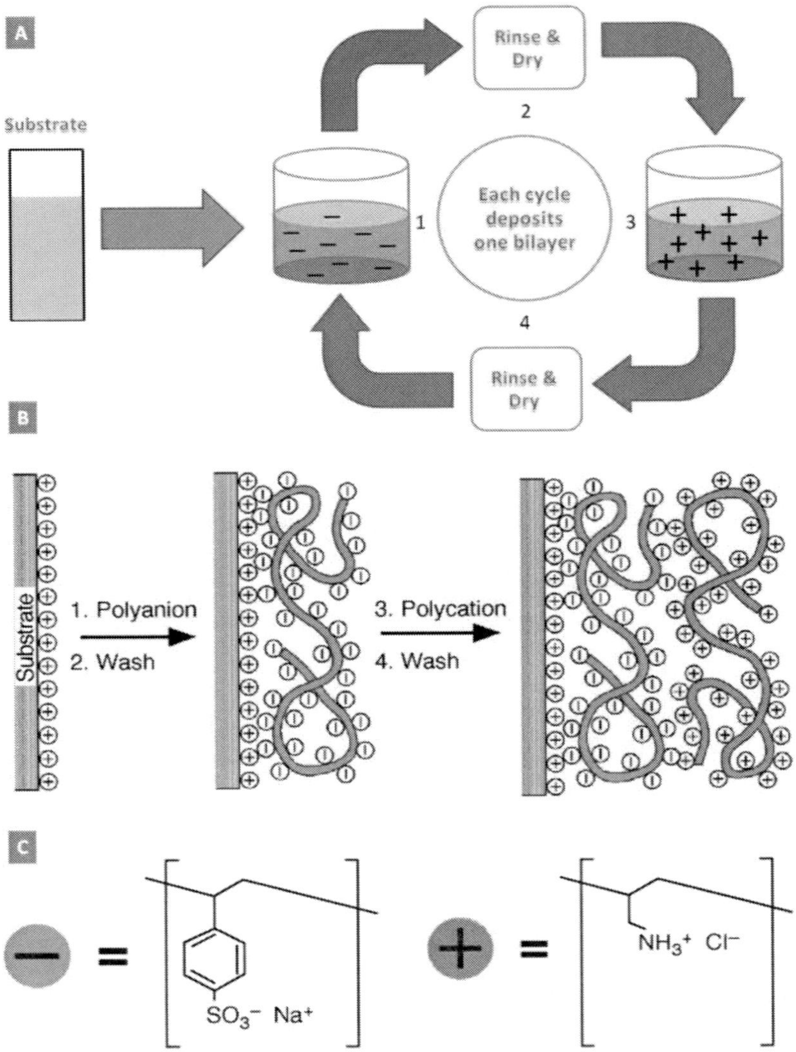

Figure 5. (A) Schematic of the film deposition process using slides and beakers. Steps 1 and 3 represent the adsorption of a polyanion and polycation, respectively, and steps 2 and 4 are washing steps. The four steps constitute one bilayer and are the basic buildup sequence for the simplest film architecture, (A/B)n, n is the number of bilayers (also called cycles). The construction of more complex film architectures requires only additional beakers and a different deposition sequence. (B) Simplified molecular picture of the first two adsorption steps, depicting film deposition starting with a positively charged substrate. Counterions are omitted for clarity. The polyion conformation and layer interpenetration are an idealization of the surface charge reversal with each adsorption step. (C) Chemical structures of two typical examples of polyions, the sodium salt of poly(styrene sulfonate) and poly(allylamine hydrochloride). Conducting polymer multilayers have been widely reported by LBL technique (see text for details).

4.6. Langmuir Blodgett Films

In recent years, a great deal of interest has been devoted to the Langmuir–Blodgett (LB) films. This is due to the fact that the LB technique is a useful way of forming sequential layers of ultrathin organic films [32] and can precisely control the thickness and order of a film at the molecular scale. Thus, LB films have inherent suitability to functional molecular devices such as metal-insulator-semiconductor (MIS) devices, nonlinear optical, electro-optic switches, modulators, memories. In LB process, monolayers (usually composed of amphiphilic molecules) are formed on a water surface and subsequently transferred onto a solid support (see Figure 6). LB films have been extensively used over the years in connection with biomolecules and biosensors. This technique has been extended to conducting polymers in the past two decades and successfully used as sensors and biosensors. Two different approaches have been reported for the conducting polymer LB films. The first involves the deposition of LB film from preformed amphiphilic polymers which are spread as polymers at the air/water interface and are subsequently deposited on a substrate by the LB technique [33]. The second approach, on the other hand, involves the deposition of monomer followed by polymerization on the substrate [34]. To have a control over the conducting polymer LB deposition, researchers have reported various processing techniques. For example, use of a processing aid such as stearic acid, mixed solvent, processing solvent , and formation of nanostructures [35] to successfully transfer the polymer to the substrate.

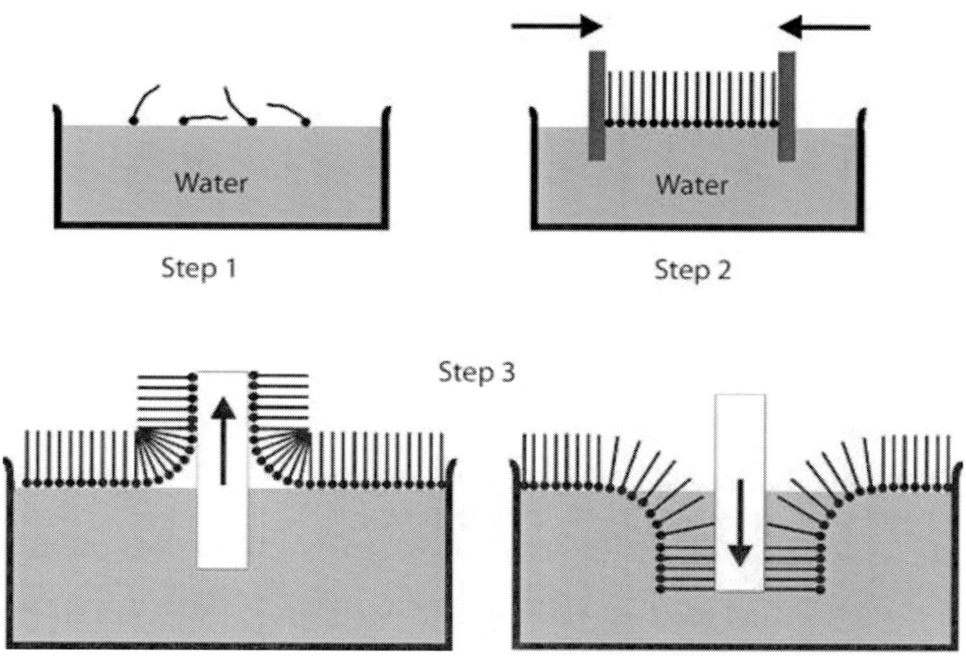

Figure 6.The Langmuir-Blodgett technique: step 1, deposition of the amphiphiles on water. Step 2, the film is pressed by moving the slide in the Langmuir trough until a pressure meter indicates that the state of a condensed film is attained. Step 3, the substrate is dipped slowly into the solution, from either the top or bottom, depending on whether the substrate is hydrophilic or hydrophobic. A multilayered structure is built up by repeated up and down strokes of the substrate. Reproduced with permission from ref [36].

4.7. Vapor Phase Polymerization

An alternative to the wet synthesis of conducting polymers is the oxidative chemical vapor deposition (*oCVD*) polymerization [37]. In *oCVD*, the monomer vapors flows into the reactor where it meets the oxidant pre-deposited on the substrate, beginning a step polymerization resulting in conducting polymer thin films (Figure 7) [38]. The synthesis of conducting polymers with *oCVD* has been termed vapor phase polymerization (VPP) [39]. By eliminating the need to dissolve macromolecules, *oCVD* enables insoluble polymers to be coated and prevents solvent damage to the substrate. Since de-wetting and surface tension effects are absent, *oCVD* coatings conform to the geometry of the underlying substrate. Hence, *oCVD* polymers can be readily applied to virtually any substrate: organic, inorganic, rigid, flexible, planar, three-dimensional, dense, or porous [37]. This route was first used for PAn's in 1998 for the deposition of conducting PAN films on cotton fibers, by impregnating the thread with ammonium peroxydisulfate oxidant and then exposing the surface to aniline vapor. Other conducting polymers, poly(3,4-ethylenedioxythiophene) (PEDOT) , polypyrrole (PPy) and poly(3-thiopheneacetic acid) (PTAA) have been deposited by *oCVD* [40]. VPP conducting polymer films have been successfully applied to gas, alcohol sensors [41].

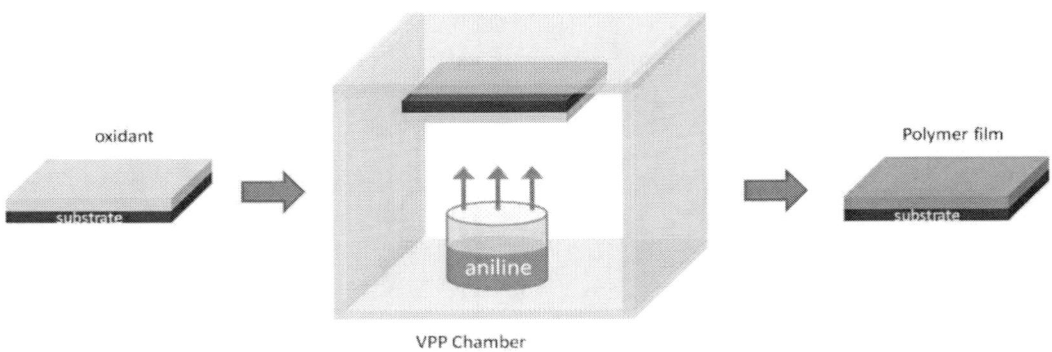

Figure 7. Schematic of a typical *oCVD* reactor. The oxidant is first deposited on the substrate by vacuum evaporation in the reactor chamber forming a thin layer. Monomer vapors are then allowed to flow into the chamber. Monomer diffuses into the oxidant layer resulting in a spontaneous polymerization and a polymer film forms.

4.8. Physical Vapor Deposition

Physical vapor deposition (PVD) is a variety of vacuum deposition and is a general term used to describe any of a variety of methods to deposit thin solid films by the condensation of a vaporized form of the solid material onto various surfaces [42]. PVD involves physical ejection of material as atoms or molecules and condensation and nucleation of these atoms/molecules onto a substrate. PVD processes include many techniques [43], of which the following was used for the deposition of CP thin films:

4.8.1. Thermal Evaporation

In this technique, polymer thin films are obtained by heating conducting polymer under vacuum, polymer vapors are then allowed to deposit on the target substrate. The polymer film thickness can be controlled by the deposition time. The vacuum-deposited polymer films are electrochemically active and show the same chemical and electrochemical responses as those of the original polymers [44]. They are also usable in making devices such as electrochromic devices, diodes, and electroluminescent devices [44]. Recently, Adhikari and Banerji employed thermal evaporation technique to deposit pristine polyaniline thin films and the subsequent thermally evaporated doping with iodine. The electrical conductivity increased by seven orders of magnitude upon iodine doping. Thermally deposited CP films find applications in gas sensors.

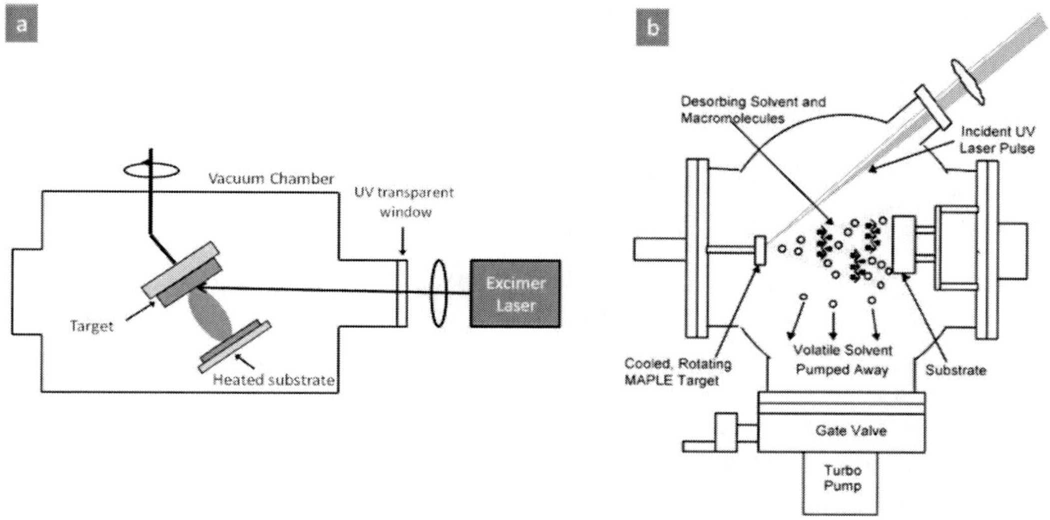

Figure 8. (a) Schematic showing the basic elements of a PLD deposition system: An intense laser pulse passes through an optical window in a vacuum chamber, and it is focused onto the target. The laser/target interaction produces a plume of material that is then deposited onto a substrate placed at a small distance forming a thin film. (b) Schematic diagram showing the basic elements of a MAPLE deposition system: The laser beam is focused on the polymer/solvent target which heats up as a result. This leads to the evaporation of the solvent and polymer molecules. The evaporated solvent molecules are pumped away whereas the polymer molecules are deposited on the substrate forming thin film. Reproduced with permission from ref [45].

4.8.2. Matrix Assisted Pulsed Laser Evaporation

Matrix assisted pulsed laser evaporation (MAPLE) was developed at the Naval Research Laboratory in the late 90's for the deposition of functional organic materials for chemical sensor applications, specifically directed at the detection of nerve and mustard gases [45]. It's the development of the pulsed laser deposition (PLD), largely applied for inorganic thin film deposition (Figure 8a). In MAPLE, an excimer laser is used to evaporate a frozen target consisting of polymer or other organic compound dissolved in a volatile solvent matrix. The laser pulses vaporize the solvent and heat the polymer or organic molecules, causing them to enter the gas phase. A coating from the evaporated polymer or organic molecules is formed on the substrate surface, while the volatile solvent molecules are evacuated from the deposition chamber by the vacuum pump (Figure 8b). The advantage of using MAPLE compared to other laser techniques is that deposition can be accomplished with little or no damage to the transported material [45]. This technique has been recently used for the formation of CP thin films of polyaniline, polypyrrole, and polythiophene. The polymer films thus formed are porous which is advantageous for sensor applications. Moreover, MAPLE allows the deposition of well oriented films with accurate thickness control and good surface morphology.

5. Modification Strategies of Conducting Polymers

CPs offer many advantages over other materials, with electrical activity being the most attractive property. However, there is always the desire to further optimize a material when targeting a specific application. Affinity of CP to analytes can be modified by incorporation of synthetic or natural receptors [46].

5.1. Incorporation of Metals, Metal Oxides and Other Metal Containing Components

The resonance-stabilized structure of CPs allows, for example, incorporation of a secondary component such as ions, nanoparticles, or nanowires of metals, metal oxides, carbon, or molecular species such as metal phthalocyanines or biologically active components such as enzymes, antibodies, and antigens. In some cases, the CP will simply act as a template for the incorporation of the secondary component. In that case, the secondary component will impart the chemical properties required for chemical sensing. In other cases the materials are linked through electrostatic interactions which influence the electronic and physical properties of the materials used to prepare the composite. The combined physical/chemical properties of the components are exploited for chemical sensing when the materials are joined. The chemical sensing applications of CPs composites that utilize CPs and at least on secondary component which can be inorganic or organic materials are reviewed in details by researchers from Georgia Institute of Technology [48]. We present a summary of these composites, their properties and sensing applications in Table 1.

5.2. Incorporation of Synthetic Receptors

Introduction of some synthetic receptors such as cyclodextrines, ferrocenes, crown ethers, calixarenes and phenylboronic acid can improve the sensitivity of CPs to ions, organic molecules and gases. Examples of immobilization of some synthetic receptors and their sensing applications are presented in Table 2.

Table 1. Characteristics of CP composite sensors that utilize CPs and at least one secondary component that can be inorganic or organic materials; Summarized from [48]

Composite type	Principles	Advantages	Sensing Applications
CPs with carbon black (CB) and carbon nanotubes (CNT)	These composites are obtained by: (i) direct mixing of CB, dispersion of CNT with CP, (ii) Polymerization of CP on the dried CNT networks. CNT act as an active substrate for monomer polymerization.	-Addition of CNT tend to drastically influence the electrical and thermal conductivity of CPs. -CNT produces composites with enhanced properties relative to CP/CB composites primarily due to the higher surface area.	-Most of the sensor applications of CP/CNT composite have focused primarily on PANI, PPy, PT, PEDOT -CP/CNT composites have been used for various sensors including gas, electrochemical and biosensors.
CPs with metals (M), metals oxides (MO) and organometallic (OM) species	-Metallic species can be introduced into CP by controlled chemical or electrochemical reduction of anions like $AuCl_4^-$, $PtCl_6^{2-}$, $PdCl_4^{2-}$ -In some cases, the preformed clusters or nanoparticles of M, MO or OM are dispersed in solution containing the monomer. Chemical oxidation of the monomer results in encapsulation of these species to form the CP composite. The resulting materials can then be cast onto substrates for subsequent applications. -direct chemical synthesis of CP/M composites using polymer precursors and oxidizing anions such as $AuCl_4^-$, $PtCl_6^{2-}$	-Incorporation of M, MO, and OM (e.g. metallocenes) species in CPs can enhance electron transfer through a direct or mediated mechanism with improved conductivity and enhanced stability. -CP/MO composites have high thermal stability; minimize the need to operate at high temperature (300-500□ C), typical of bulk MO.	-CP/Au composites are of interest for sensing gases and vapors and as binding sites for enzymes, proteins, and nucleotides. It's also used in amperometric, optical and surface plasmon resonance sensors -CP/Pt, CP/Pd composites are of interestin for hydrogen, methanol and glucose biosensors. -CP/MO composite sensors have enhanced stability and respond to both oxidizing (NO_2) and reducing gases (CO_2, CO, H_2). CP/SnO_2 or TiO_2 have shown to be interesting in gas sensing applications. -CP/phthalocyanies used in gas sensors and ion-selective electrodes.
CPs with nonconducting polymers (NCPs)	-Various CPs have been prepared in the presence Of (i) Water-soluble NCPs including poly(vinylalcohol) (PVA), poly(methylmethacrylate) (PMMA), etc. (ii) Water-insoluble NCPs such as polystyrene (PS) and others. -Ternary polymer blends show reversible adsorption/desorption kinetics which is favorable for sensors.	-It overcomes the limited processability, mechanical thermal properties of CPs. The resulting composites combine both the electrical, redox and optical properties of CPs with the mechanical and processability of the NCPs.	-CPs/NCPs have been utilized as humidity sensors. -The gas sensitivity of CP/NCPs is considerably larger (1 or 2 order of magnitude) than that of the CPs for the hydrophobic vapor analytes (xylene, benzene, acetone).

Table 2. Selected examples of sensors with synthetic receptors immobilized into CP; reproduced with permission from [58]

Analyte	Polymer or monomer/way of polymerization	Receptor	Type of Transducing	Ref.
Immobilization by polymerization of monomer covalently modified by receptor				
Anions/$H_2PO_4^-$	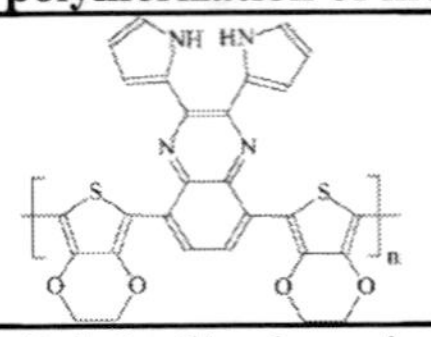	Quinoxaline	Potentiometric	46
Glycoproteins (GOx, HRP)	Poly(aniline boronic acid) electrochemical polymerization	Boronic acid	Photometric	47
Toluene, acetone vapor	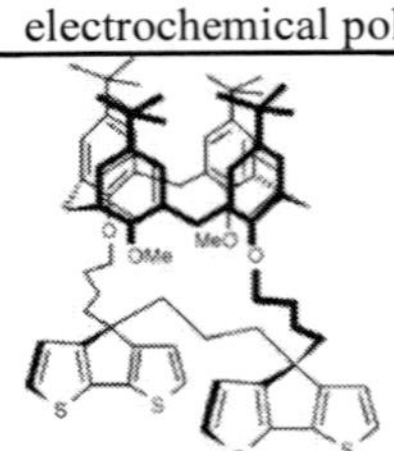	5,11,17,23-Tetra-*tert*-butyl-25,26,27,28-tetrahydroxycalix[4]arene	Gravimetric, QCM electrode	49
Li^+	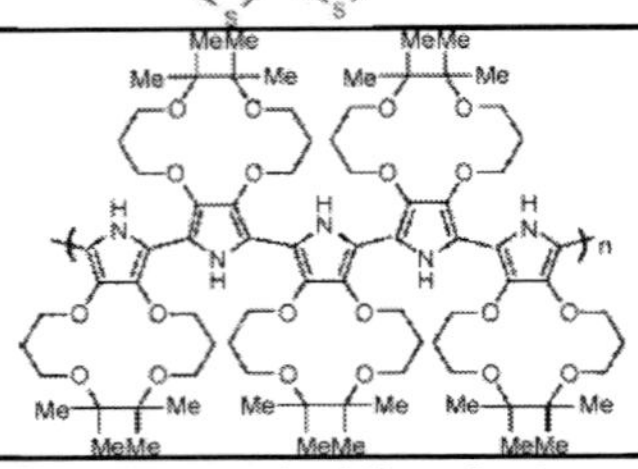	Substituted 14-crown-4 macrocycle	Amperometric	50
Immobilization by doping or embedding during polymer deposition				
Cu^{2+}, Eu^{3+}	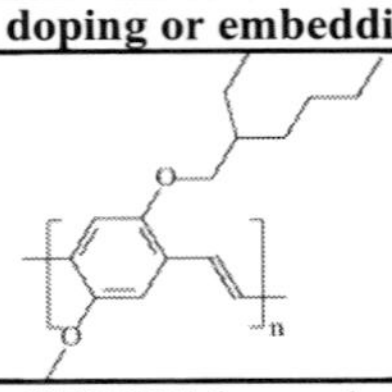	5,17-Bis(4-nitrophenylazo)-26,28-dihydroxy-25,27-di(ethoxycarbonylmethoxy)-calix[4]arene	Impedimetric	51
Ag^+	PEDOT/electrochemical polymerization	*p*-Tetrasulfonated calix[4]arene	Potentiometric	52
Na^+, K^+, Mg^{2+}, Ca^{2+} Mn^{2+}, Fe^{3+}, Co^{2+}, Ni^{2+}, Cu^{2+}, Zn^{2+}	PPy/electrochemical polymerization	α-Cyclodextrin, sulfated-β-cyclodextrin	Potentiometric	53
Dopamine neurotransmitters	Poly(3-methylthiophene)/ electrochemical polymerization	γ-Cyclodextrin	Amperometric	54
Catechol and pyrogallol neurotransmitters	PPy/electrochemical polymerization	β-Cyclodextrin	Amperometric	55
Na^+, K^+, NH_4^+	Overoxidized PPy/electrochemical polymerization	Nonactin, valinomycin, dibenzo-18-crown-6, bis[(12-crown-4)methyl]-2-dodecyl-2-methyl malonate, cisdicyclohexano-18-crown-6 (	Impedimetric	56
Zn^{2+}	PPY/electrochemical polymerization	Tetraphenylborate	Potentiometric	57

5.3. Immobilization of Biocomponents

The two common properties desired for all biomedical applications, aside from conductivity, are biocompatibility and redox stability, but beyond these needs, CP modification tends to be specific for the application. In particular, most research has focused on biological and physical modification of CPs. For biosensors, it is important to tune the hydrophobicity, conductivity, and reactive functionalities for modification of CPs to successfully incorporate/immobilize biomolecules (e.g. nucleic acids, antibodies, enzymes) and to improve detection of binding events. The immobilization procedures are based on non-covalent interactions (physical adsorption, electrostatic assembly, hydrophobic interactions) or covalent binding of a receptor to a matrix. The principles of these immobilization methods, advantages and limitations are summarized in Table 3. Detailed analysis of these immobilization techniques are reported by many researchers [59,60].

Table 3. Immobilization techniques of biomolecules on conducting polymers for biosensing devices; reproduced with permission from [61]

Immobilization technique	Principles of Immobilization	Advantages	Limitations
Non-covalent techniques			
Adsorption	Electrostatic forces, hydrogen bonding, Van der Waal's forces, etc.	-simple	- Biomolecule loss (desorption) over time - Limited control over immobilization - Random orientation on surface
Entrapment	Molecule incorporation during electropolymerization	- Simple - Good proximity between elements	- Potential loss of biomolecule activity - Steric and diffusion constraints - Requires high biomolecule concentration
Affinity binding	High affinity interactions such as avidin-biotin	- Control over molecule orientation - High accessibility of analytes - Minimal loss of biomolecule activity	- Requires pre-immobilization of one of the affinity molecules (e.g., biotin)
Covalent techniques			
Chemical conjugation	Surface chemical reaction between functional Groups	- Tighter control over immobilization - High accessibility of analytes - Minimal biomolecule loss over time - Control over biomolecule orientation	- Complex - Conditions are not always appropriate for biomolecules - Potential loss of biomolecule activity

5.4. Molecularly Imprinted Conducting Polymers

A promising way of introducing artificial receptors into CPs is by molecular imprinting. Molecular imprinting is a technique to create template-shaped cavities in polymer matrices with memory of the template molecules to be used in molecular recognition [62] (see a typical example in Figure 9). The molecularly imprinted polymers (MIPs) have high selectivity and affinity when compared to naturally occurring systems for sensor applications. For reasons of their recognition properties, stability, reproducibility, low cost, robustness, and manufacturing potential, MIPs are regarded as promising candidates for the replacement of biomolecules as the recognition element in a range of chemical sensors [63]. This research has seen MIPs used with a number of transducer technologies, including piezoelectric elements, optical methods, and in electrochemical sensing.

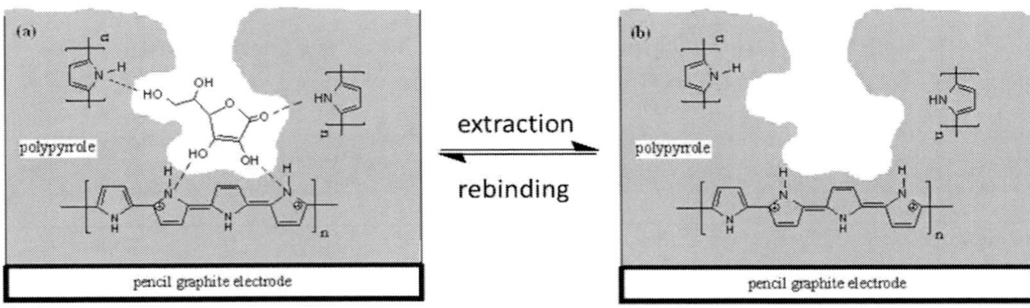

Figure 9. Schematic representation of (a) molecular imprinting of PPy with ascorbic acid (AA), (b) removal of AA from the molecularly imprinted PPy modified pencil graphite electrode. The rebinding process is induced by the hydrogen bonding between AA and NH of PPy. Reproduced with permission from [64].

6. Main Concepts of Chemical Sensors and Biosensors

6.1. Terminology and Classification

The technological progress in contemporary world in all branches of human activity is driven by pursuing a goal of better living conditions as well as learning and understanding of surrounding world. In both spheres of this activity, a very significant role is played by analytical chemistry, as a discipline of science and technology indispensable in tracing and control – when necessary – of functioning of living organisms, technological processes, changes in natural environment, and quality of various materials. The progress in analytical instrumentation is related to discoveries of new phenomena and materials, with increasing needs for analytical instrumentation and their functional abilities. This progress is observed in many directions, but certainly one of them is development of sensors for chemical analysis. Table 4 shows 20 inventions which can be considered as milestones in the field of chemical sensors. In August of 2010, authors of this chapter found almost 498 thousands papers with keyword "*sensor*," and almost 51 thousands papers with keyword "*biosensor*" Since the year

2000 using Scifinder search [65]. This seems to illustrate well the scientific and technological research potential involved in this area.

Table 4. Milestones in the development of chemical sensors and biosensors; Reproduced with permission from [66]

Year	Invented sensor	Reference
1909	Glass pH electrode	Haber and Klemensiewicz
1954	Amperometric oxygen electrode	Clark Jr.
1962	Amperometric enzyme biosensor for glucose	Clark Jr. and Lyons
1964	Piezoelectric sensors	King Jr.
1966	Solid-state fluoride selective electrode	Frant and Ross Jr.
1967	Liquid-state ion-selective electrode with molecular carriers	Stefanac and Simon
1969	Potentiometric enzyme biosensor for Urea	Guilbault and Montalvo
1970	Ion-sensitive field effect transistor	Bergveld
1974	Enzyme thermistor	Mosbach and Danielsson
1975	Microbe-based biosensor	Davies
	Immunosensor	Janata
	pCO_2/pO_2 optode	Lubbers and Optiz
	Gas-sensitive semiconductor sensor	Lundstrom, Shivaramn, Svenson, and Lundkvist
1980	Fiber optic pH optode	Peterson, Goldstein, Fitzgerald, and Buckhold
1983	Surface plasmon resonance biosensor	Liedberg, Nylander, and Lundstrom
1986	Tissue-based biosensor	Belli and Rechnitz
1987	Receptor-based biosensor	Taylor, Marenchic, and Cook
	Screen-printed biosensor	Matthews, Bown, Watson, Holaman, Steemson, Hughs, and Scott
1993	DNA biosensor	Millan and Mikkelson
1996	Aptamer biosensor	Drolet, Moon-Dermott, and Roming

According to definitions of Analytical Division of IUPAC “a chemical sensor is a device that transforms chemical information, ranging from the concentration of a specific sample component to total composition analysis, into an analytically useful signal. The chemical information mentioned above may originate from a chemical reaction of the analyte or from a physical property of the system investigated.”[67].

Analytical Division of IPUAC goes on with the description of the two basic functional units of a chemical sensor as follows:

“Chemical sensors contain two basic functional units: a receptor part and a transducer part. Some sensors may include a separator which is, for example, a membrane.

In the ***receptor*** part of a sensor the chemical information is transformed into a form of energy which may be measured by the transducer.

The ***transducer*** part is a device capable of transforming the energy carrying the chemical information about the sample into a useful analytical signal. The transducer as such does not show selectivity.

Table 5. Classification of chemical sensors based on the transducing mechanism as recommended by the Analytical Chemistry Division of IUPAC (IUPAC 1991) [67]

Classes of sensors	Subclasses	Operating principles
Optical	Absorbance	Caused by the absorptivity of the analyte itself or by a reaction with some suitable indicator; Measured in a transparent medium
	Reflectance	Measured in non-transparent media, usually using an immobilized indicator
	Luminescence	Based on the measurement of the intensity of light emitted by a chemical reaction in the receptor system
	Fluorescence	Measured as the positive emission effect caused by irradiation. Also, selective quenching of fluorescence may be the basis of such devices
	Refractive Index	Measured as the result of a change in solution composition. This may include also a surface plasmon resonance effect.
	Optothermal	Based on the measurement of the thermal effect caused by light absorption
	Light scattering	Based on effects caused by particles of definite size present in the sample
Electrochemical	Voltammetric, including amperometric	In which current is measured in the d.c. or a.c. mode. This subgroup may include sensors based on chemically inert electrodes, chemically active electrodes and modified electrodes.
	Potentiometric	In which the potential of the indicator electrode (ion-selective electrode, redox electrode, meta/metal oxide electrode) is measured against a reference electrode
	Chemically sensitized field effect transistors (CHEMFET)	In which the effect of the interaction between the analyte and the active coating is transformed into a change of the source-drain current
	Potentiometry with solid electrolytes for gas sensing	They work in high temperature solid electrolytes and are usually applied for gas sensing measurements
Electrical	Metal oxide semiconductor	Based on reversible redox processes of analyte gas components
	Organic semiconductor	Based on the formation of charge transfer complexes, which modify the charge carrier density
	Electrolyte conductivity	
	Electric permittivity	
Mass sensitive	Piezoelectric	Based on the measurement the frequency change of the quartz oscillator plate caused by adsorption of a mass of the analyte at the oscillator
	Surface acoustic wave	Based on the modification of the propagation velocity of a generated acoustical wave affected by the deposition of a definite mass of the analyte
Magnetic		Based on the change of paramagnetic properties of a gas being analyzed.
Thermometric		Based on the measurement of the heat effects of a specific chemical reaction or adsorption which involve the analyte
Others		α-, β-, γ-radiation may form the basis for a chemical sensor in case they are used for determination of chemical composition

The receptor part of chemical sensors may be based upon various principles:

- Physical, where no chemical reaction takes place. Typical examples are those based upon measurement of absorbance, refractive index, conductivity, temperature or mass change.
- Chemical, in which a chemical reaction with participation of the analyte gives rise to the analytical signal.
- Biochemical, in which a biochemical process is the source of the analytical signal. Typical examples are microbial potentiometric sensors or immunosensors. They may be regarded as a subgroup of the chemical ones. Such sensors are called ***biosensors.***"

In the same report, a classification of chemical sensors is given according to the operating principle of the transducer (Table 5). Chemical sensors can also be classified depending on the intended application of the sensor as chemical/industrial sensors and biosensors. Chemical sensors can be further classified into gas, pH, ion-selective, alcohol, humidity sensors [67,68]. Biosensors also find applications in various areas like health care biosensors, immunosensors, DNA sensors, biosensors for environmental monitoring and food analysis. Both types of classification, according to transducing mechanism and applications, will be considered in this chapter for the discussion of the different aspects of chemical- and bio-sensors based on conducting polymers.

Following another IUPAC recommendations, "biosensor is an analytical device incorporating a biological material or a bio-mimic e.g. tissue, microorganisms organelles, cell receptors, enzymes, antibodies, nucleic acids etc. intimately associated with or integrated with a physicochemical transducers or transducing micro-system using optical, electrochemical, thermometric, piezoelectric or magnetic properties etc." [69] Biosensors are often considered to be an independent group in sensor science. In this chapter, however, we will regard them as a special type of chemical sensors. Consequently, they will be integrated into the appropriate sections of this chapter. This concept corresponds to that of the responsible IUPAC commission which has expressed in an official document (IUPAC 1999): Biosensors are chemical sensors in which the recognition system utilizes a biochemical mechanism [70].

6.2. Elements of Chemical Sensors

Figure 10 shows a schematic of a typical chemical sensor. In the majority of chemical sensors, the receptor interacts with analyte molecules. As a result, its physical properties are changed in such a way that the appending transducer can gain an electrical signal. In some cases, one and the same physical object acts as receptor and as transducer. This is the case e.g. in metallic oxide semiconductor gas sensors which change their electrical conductivity in contact with some gases. Conductivity change itself is a measurable electrical signal. In mass sensitive sensors, however, receptor and transducer are represented by different physical objects. A piezoelectric quartz crystal acts as transducer. The receptor is formed by a sensitive layer at the crystal surface. The latter is capable of absorbing gas molecules. The resulting mass change can be measured as a frequency change in an electrical oscillator circuit.

6.2.1. Receptor

The receptor function is fulfilled in many cases by a thin layer which is able to interact with analyte molecules, catalyze a reaction selectively, or participate in a chemical equilibrium together with the analyte. Receptor layers can respond selectively to particular substances or to a group of substances. The term molecular recognition is used to describe this behavior.

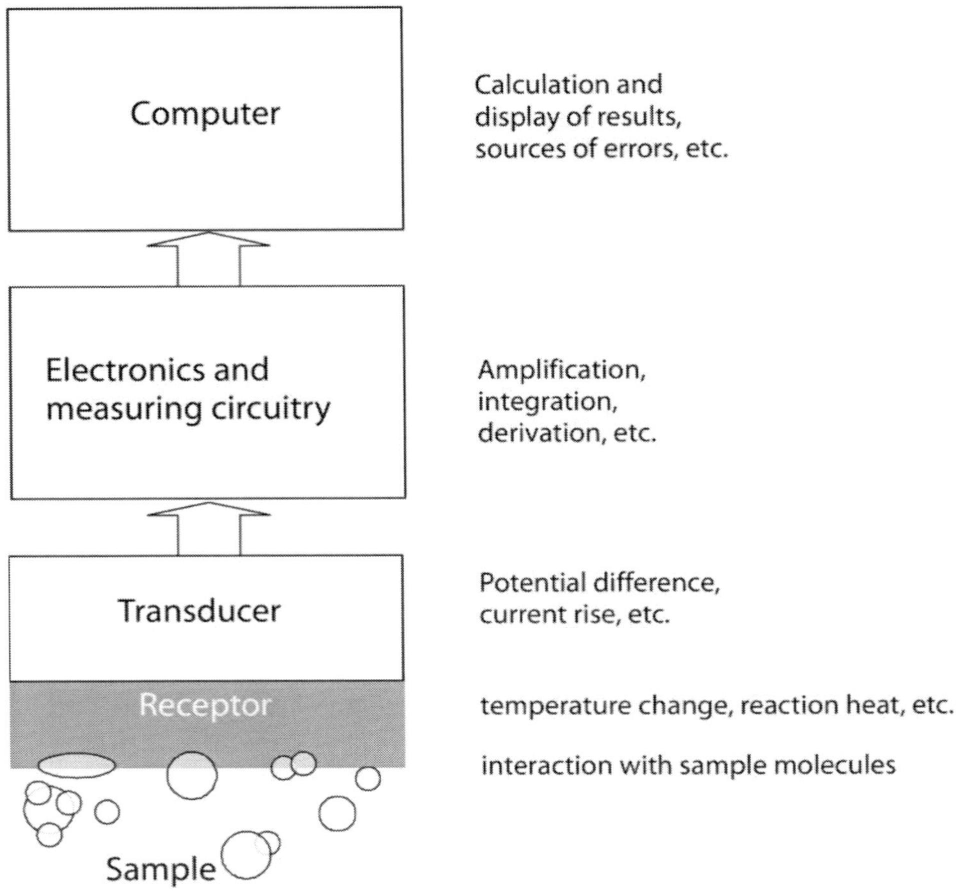

Figure 10. schematic of a typical chemical sensor. Reprinted with permission from [71].

Typical for biosensors is that molecules are recognized by their size or their dimension, i.e. by steric recognition. Among the processes of interaction, most important for chemical sensors are adsorption, ion exchange and liquid–liquid extraction (partition equilibrium). Primarily these phenomena act at the interface between analyte and receptor surface, where both are in an equilibrium state. Instead of equilibrium, a chemical reaction may also become the source of information. We find this, for example, in receptors where a catalyst accelerates the rate of an analyte reaction so much that the released heat from the reaction creates a temperature change that can be transduced into an electrical signal. Processes at the receptor-analyte interface can be classified into interaction equilibria and chemical reaction equilibria. The differences are not significant for work with sensors. A true chemical equilibrium is formed, for example, in electrochemical sensors where receptor and analyte are partners of the same redox couple.

6.2.2. Transducer

The transducer part of the sensor serves to transfer the signal from the output domain of the recognition system to, mostly, the electrical domain. Because of the general significance of the word, a transducer provides bidirectional signal transfer (non-electrical to electrical and vice versa); the transducer part of a sensor is also called a detector, sensor or electrode, but the term transducer is preferred to avoid confusion. Examples of transducers, which are often used for the listed types of measurement, are shown in table 5.

7. CONDUCTING POLYMER SENSORS

Chemical sensors based on CPs could be discussed based on the operating principle of the transducer (Table 5) or the intended application of the sensor in the chemical/industrial processes.

7.1. Sensors Based on Transducing Mode

7.1.1. Potentiometric Sensors

Potentiometry is based on the measurement of a potential under no current flow. The measured potential may then be used to determine the analytical concentration of some components of the analyte gas or solution. The signal is measured as the potential difference between the indicator electrode (ion-selective electrode ISE, redox electrode, metal-metal oxide electrode) and the reference electrode, Figure 11. The interest in the fabrication of all-solid-state ISEs that do not require internal filling solutions lead to the introduction of the so-called coated-wire electrodes, Figure 11. These devices, however, suffered from the ill-defined transduction of the ionically conducting ion-selective membrane and the electronic conductor, which made them rather instable. Another approach involves the use of a hydrogel based electrolyte instead of the liquid internal electrolyte of the classical ISE. Hydrogel-contact ISEs suffer, however, from shortcomings originating from the water uptake/release and the resulting volume changes of the hydrogel layer which is a function of the salt concentration of the hydrogel.

Conducting polymers have been used in the construction of ISEs according to general approaches, which have led to the development of solid contact ion selective electrodes (SCISEs), and single-piece ion selective electrodes (SPISEs) and conducting polymer based ion selective electrodes (CPISEs), Figure 11. In all these, the conducting polymer is deposited directly on the solid surface of electronic conductor (metal or graphite). This configuration resembles ion-selective electrodes with polymer membrane without inner solution, i.e., coated wire electrode. The typical problem of the coated wire electrodes is the instability of the potential drop at the metal/polymer interface: if there is no electrochemical reaction (and therefore no electrochemical equilibrium) at the interface, this potential drop cannot be fixed. A deposition of CP having intrinsic electrochemical activity, which maintains the electrochemical equilibrium of electrons at this interface, allows one to overcome this problem. This is the reason to use CP as an intermediate layer between metal and ion selective membrane [72]. A potential drop on the second border of CP is defined through ion exchange. But some CP films are selectively permeable for some ions or can bind some ions; therefore they can be used not only as an intermediate layer between ion selective membrane and electrode but also as an ion selective membrane itself [73]. The ion selectivity can be modified by introduction of receptors. However, electrochemical activity of CP results in interference of ionic and electronic equilibrium on the interface with electrolyte, which may lead to a strong interference in the presence of redox active compounds.

Potentiometric ion sensors (ISEs) are very attractive, especially due to their small-size and portability, and low cost. Neutral or charged carrier based ISEs, which are available for about 60 different analytes, have found widespread applications in routine uses like clinical analysis. Their great advantage is that they work in colored or turbid solutions where photometric methods do not work. However, errors caused are also proportionately large during measurement. The response depends upon the temperature of the solution. To get good results periodic calibration is a must. Impurities in solutions cause serious errors. On the other hand, to fully use the potentials of this technique, more durable and maintenance-free ISEs that are easily miniaturized are needed.

7.1.2. Voltammetric And Amperometric Sensors

Voltammetric and amperometric sensors use three electrode electrochemical cell (see Figure 2). Voltammetric sensors refer to the measurement of current resulting from the application of potential. While, in amperometric sensors a constant potential is applied and the change in current is monitored as a function of time. The current that flows between the indicator electrode and the auxiliary electrode in solution is measured and related to the concentration of the analyte. Voltammetric and amperometric transducers based on CPs are used for detection of current caused by either electrocatalysis or by ion flux into/from CP during its oxidation/reduction [74].

Recently, the use of CPs modified electrodes in the voltammetric/ amperometric quantitative analyses of organic compounds and compounds of biological and biochemical interest have received considerable attention. When compared to conventional electrodes based on carbon and noble metals, CPs modified electrodes have interesting properties. This can be ascribed to their efficient electrocatalysis, sensitivity, selectivity, low detection limit and other interesting properties.

From now on, dopamine (DA) is used as an example to illustrate the enhanced performance of CPs-based sensors and to present the development of selectivity, sensitivity and detection limits of CPs modified electrodes. Dopamine (DA) is one of the most significant catecholamines and belongs to the family of excitatory chemical neurotransmitters. It plays a very important role in the functioning of the central nervous, cardiovascular, renal and hormonal systems, as well as a key role in drug addiction and Parkinson's disease. Tremendous consideration has been given in biomedically oriented research, to design selective, sensitive and reliable methods for the direct measurement of DA without cross-interference. Three primary challenges to measure DA under physiological conditions utilizing electrochemical methods still remain: (i) the very low concentration levels of DA (0.01–1μM), (ii) the intensive interference arising from typical biological matrices which contains (AA) that is present at relatively high concentrations (about 0.1mM) and uric acid (UA) which results in poor selectivity and sensitivity for DA detection, and (iii) AA and DA oxidize at almost the same potential as DAwhich results into an overlapping voltammetric response for the oxidation of DA, AA and UA (Figure 12b).

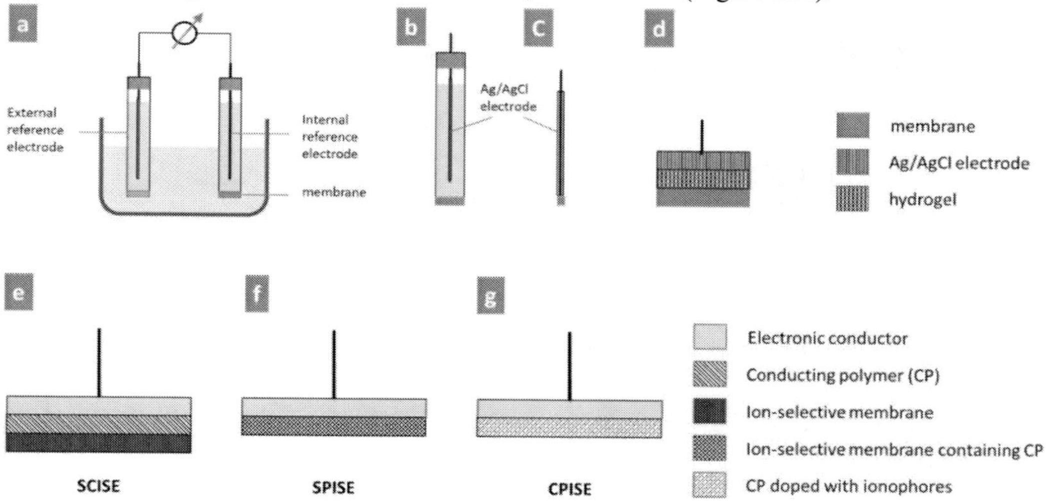

Figure11.Schematic of the different construction principles of the ion-selective electrodes with emphasis on solid state conducting polymer potentiometric sensors: (a) classical potentiometric ion-selective (ISE) electrode sensor. (b) Conventional ISE with an internal reference electrode and internal filling solution, (c) coated wire or graphite coated ISE, (d) ISE with a hydrogel contact. (e) Solid contact ion selective electrodes (SCISEs): the conducting polymer is electropolymerized on the surface of an electronic conductor (metal or graphite) and then, coated with the ion-selective membrane. (f) Single-piece ion selective electrodes (SPISEs): the selective membrane contains conducting polymer which is emanated by its solubility in an organic solvent. (g) Conducting polymer based ion selective electrodes (CPISEs): the conducting polymer layer is doped with a suitable ionophore which enhances the ion selectivity of the CP. This way, both the ion-exchange and redox characteristics of the conducting polymers are successfully combined with the selectivity of the ionophore-based ion-selective membranes.

High performance liquid chromatography (HPLC) with electrochemical detection is most often used for the analysis of DA and other catecholamines [75]. Electrochemical methods have attracted great interest because of their simplicity, rapidness and high sensitivity, the ability of sensing neurotransmitters in living organisms and in vivo real time analysis. However, electrochemical analysis on conventional unmodified electrodes such as glassy carbon (GC), Pt and Au electrodes has limitations because of irreversibility and surface fouling of the electrode.

The use of conducting polymer modified electrodes based on poly(3-methylthiophene) (PMT), catalyzes the oxidation of DA and eliminate the usual fouling problems[76]. Even interference from ascorbic acid and electro-inactive large proteins such as hemoglobin is substantially removed. Work with PMT electrode suggested the possible presence of ''active sites,'' which are believed to be the polymer's center of electrocatalytic activity. The interaction of 1,5-anthroquinone-disolfonic acid (1,5-AQDS) at the PMT electrode showed a nonreversible behavior resulting in the blocking of ''the active sites,'' suggesting the specific electcatalytic activity of this polymer is limited to catechol and similar compounds [77].

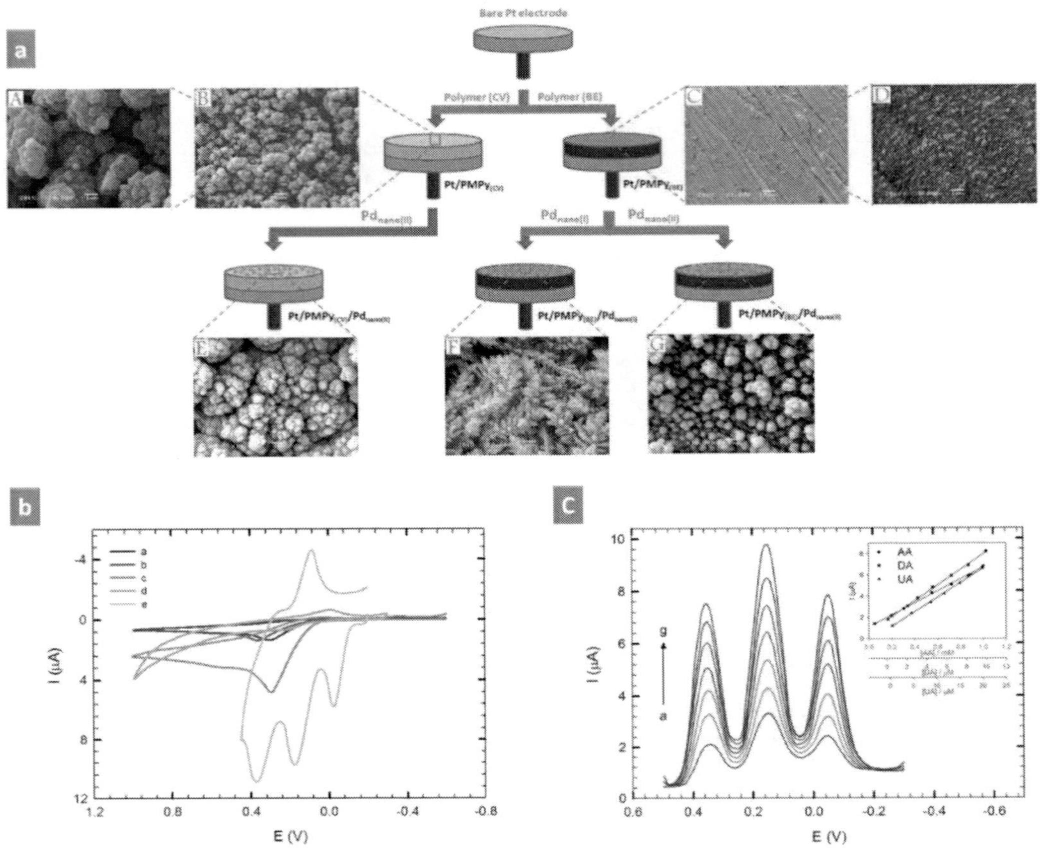

Figure 12. conducting polymer/metal nanoparticles composites, highly selective and sensitive sensors; (a) Schematic showing two-step process for the modification of Pt electrode. In the first step, a polymer film is electrodeposited on Pt electrode using two electrochemical methods (BE and CV). In the second step, palladium nanoparticles (Pd_{nano}) are deposited on the polymer film using two approaches. Obviously, the different structures and morphologies are obtained from each method (b) Separation of electrochemical responses of AA, DA, and UA at polymer/Pd_{nano} modified electrodes. The conventional electrodes produce two broad and overlapped oxidation peaks so that the peak potentials for AA, DA, and UA are indistinguishable. However, polymer/Pd_{nano} modified electrode resolve the overlapped voltammetric peaks into three well-defined peaks corresponding to AA, DA, and UA. (c) Simultaneous determination of AA, DA, and UA at polymer/Pd_{nano} electrode using differential pulse voltammetry. Inset shows calibration curves. Reproduced with permission from [86].

The performance of most of the voltammetric and amperometric selective electrodes depends primarily on the selectivity that is introduced at the electrode–electrolyte interface. This requires either the attachment of the selectivity element onto the electrode surface or its incorporation into a thin inert matrix. The incorporation of the organometallic complex, ferrocene, into PMT film resulted in enhanced charge transfer towards the oxidation of DA and some organic molecules of biological interest as reported by Galal [78]. Bouchta et al. synthesized functionalized film of PMT combined with γ-cyclodextrin which is eight glucose units with conical structure with hydrophobic internal cavity and hydrophilic exterior due to the presence of hydroxyl groups [79]. This composite offers fast and linear response towards DA, L-DOPA and chlorpromazine and is able to discriminate between AA and DA. The hydrophobic interior cavity of the γ-CD may play a key role in the selectivity process. DA (pK_a, 8.93) exists predominantly in the cationic form in solutions of pH 7 or less. Therefore, the use of Nafion, a cation-exchanger polymer could be suitable to improve the selectivity of DA detection. Huong et al. successfully prepared Nafion/PMT film [80]. The film showed high sensitivity and selectivity for DA detection and allowed its determination in the presence of high concentration of AA and resulted in two well separated (240 mV) oxidation peaks for DA and AA. Wang el al., on the other hand, used PMT coated with Nafion/single-walled carbon nanotubes film [82]. This hybrid film modified electrode combining the advantages of PMT, carbon nanotubes with Nafion exhibits dramatic electrocatalytic effect on the oxidation of DA and results in a marked enhancement of the current response. Moreover, the interferences of AA and UA are effectively diminished.

Conducting polymer incorporated metallic nanoparticles provide an exciting system and hold potential application in electronics, sensors and catalysis. It is well known that noble metal-based catalysts are widely used in industry. However, the catalytic activity of these metals is known to depend highly on their dispersion and surface properties. The porous structure of conducting polymer allows dispersing the metal particles into the polymer matrix and generates additional electrocatalytic sites (see an example of this trend in Figure 12). This represents one of the most recent approaches for the sensitive and selective determination of DA and results in the lowest voltammetric/amperometric limit of detection reported to date. Atta and El-Kady reported a promising modified electrode obtained by dispersing Pt or Pd nanoparticles into PMT. This hybrid material was used for the simultaneous determination of AA and DA in physiological pH 7.4 PBS. It was observed that in the presence of AA at millimolar level (0.1 mM), the Pd nanoparticle-modified PMT electrode can sense the increase of DA at micromolar concentration (0.05–1 μM) which is typical of the physiological conditions.

Poly(3,4-ethylenedioxythiophene) has been recently used in sensors because of its excellent stability and conductivity. Kumar et al. reported the possibility of detecting DA in the presence of excess of AA using PEDOT coated GC electrode by achieving good separation (170 mV) between the oxidation peak potentials of DA and AA [83]. The peak potential separation was explained using the argument that the cationic PEDOT film interacts with the negatively charged ascorbate anion through favorable electrostatic interaction, which results in pre-concentration at a less anodic value. The positively charged DA tends to interact with the hydrophobic regions of PEDOT film through hydrophobic–hydrophobic interaction thus resulting in favorable adsorption on the polymer matrix. The peak separation has been increased to 260 mV by introducing ferricyanide ion into PEDOT film [84]. This was possible since PEDOT coatings can be prepared by electrochemical polymerization in aqueous solution, which allows the direct incorporation of water soluble anions, like ferricyanide. In another approach, Mathiyarasu et al. incorporated gold nanoparticles into PEDOT film and used this composite film for the analysis of AA, DA and UA [85]. PEDOT matrix is responsible of the peak separation (selectivity) and also favoring catalytic oxidation of the above compounds while gold nanoparticles allow the nanomolar sensing of DA and UA (sensitivity). Thus, it is possible to detect nanomolar levels of DA and UA in presence of excess of AA. DA, AA and UA exist in different ionic form in solution at physiological pH 7.4. Basically, DA (pK_a 8.93) predominantly exists in the positively charged form whereas AA and UA (pK_a 4.1, 5.4 respectively) are negatively charged. A negatively charged surfactant, sodium dodecyl sulfate (SDS), can be introduced to PEDOT film assisted by the hydrophobic-hydrophobic interaction between PEDOT film and the hydrophobic tail of SDS. Atta et al. found that SDS forms a monolayer on PEDOT surface with a high density of negatively charged end directed outside the electrode [86]. It has been shown that this modified electrode can be used to selectively detect DA in the presence of large concentrations of AA and UA. This was explained by the enhanced accumulation of protonated DA at the electrode surface via favored electrostatic interactions with SDS whereas deprotonated negatively charged AA and UA get repelled from the surface. Thus, the DA current signal increases while the signals of AA and UA are quenched.

Table 6. characteristics of some dopamine (DA) voltammetric and amperometric sensors. Abbreviations: LDR, linear dynamic range; LOD, limit of detection; PMT: poly(3-methylthiophene); NR, not reported; FC, ferrocene; γ-CD, γ-cyclodextrin; SWCNT, Single-walled carbon nanotube; PNP, palladium nanoparticles; GNP, gold nanoparticles; SDS, sodium dodecyl sulfate; PMPy, poly(N-methylpyrrole); OPPy, overoxidized polypyrrole; MWCNT, multi-walled carbon nanotube; LC-ECD, liquid chromatography with electrochemical detection; SWCNTs, single-walled carbon nanotubes; PABA, poly(anilineboronic acid)

Electrode	Technique	Interference studied	pH	LDR (μM)	Sensitivity (μA/ μM)	LOD (nM)	Ref
PMT	Voltammetry, amperometry	AA, Hemoglobin, surfactants	6.9	NR	NR	10 ppb	76
PMT/FC	Voltammetry	-	0.1 M H_2SO_4	500-10,000	0.007	312	78
PMT/γ-CD	Voltammetry	AA	0.1 M H_2SO_4	1-8	NR	200	79
PMT/Nafion	voltammetry	AA	0.1 M H_2SO_4	0.5-200	26.7	100	80
PMT/SWCNT/Nafion	voltammetry	AA, UA	7.0	0.020–0.10, 0.10–1.0 and 1.0–6.0	7.765, 3.257 and 1.995	5	81
PMT/PNP	Voltammetry	AA, UA, glucose	7.4	0.05–1	1.37	9	82
PEDOT	Voltammetry	AA	7.4	1-30	0.2651	1000	83
PEDOT/ferricyanide	Voltammetry	AA	7.0	NR	NR	NR	84
PEDOT/GNP	Voltammetry	AA, UA	7.4	0.002–0.02	200	2	85
PEDOT/SDS	Voltammetry	AA, UA	7.4	0.5 – 25 and 30-100	0.673 and 0.288	61 and 86	86
PMPy/PNP	Voltammetry	AA, UA	7.4	0.1–10	0.71	12	87
OPPy/MWCNT	LC-ECD	AA	7.0	0.0025–5	-	0.75	88
SWCNTs/PPy/tyrosinase	Amperometry	AA	6.5	5-50	0.767	5000	89
AgCl@PANI core-shell	voltammetry	AA	7.0		0.49	NR	90
PANI/GNP	Amperometry, voltammetery	AA	6.8	0.3-200	3.5×10^{-6}	100	91
PANI/Ag composite nanotubes	Voltammetry	-	0.1 M H_2SO_4	NR	NR	NR	92
PABA/SWCNTs	Voltammetry	AA	7.4	NR	NR	0.6 and 1.5	93

Polypyrrole (PPy) is one of the most extensively used conducting polymers in design of bioanalytical sensors[3] S.B. Adeloju and G.G. Wallance, Analyst (2000), p. 699.. Poly(N-methylpyrrole)/Pd nanoclusters modified electrode sensor has been constructed in two electrochemical steps by Atta et al. [87] (Figure 12a). The sensor exhibits strong electrocatalytic activity toward the oxidation of a mixture of DA, AA and UA with obvious reduction of overpotentials. The simultaneous analysis of this mixture at conventional electrodes usually struggles. However, three well-resolved oxidation peaks for AA, DA, and UA with large peak separations allow this modified electrode to individually or simultaneously analyze AA, DA, and UA by using differential pulse voltammetry (DPV) with good stability, sensitivity, and selectivity (Figure 12c). The practical application of the modified electrode was demonstrated by measuring the concentrations of AA, DA, and UA in injection sample, human serum, and human urine samples, respectively, with satisfactory results. In another approach, Wen et al. have developed an overoxidized polypyrrole/multi-walled carbon nanotubes (OPPy/MWNTs) modified electrode for the sensitive detection of DA [88]. When combined with liquid chromatography, it not only obtains a low detection limit of 0.75 nM for DA, but also improves the selectivity of DA detection. The chemical polymerization of PPy onto single walled carbon nanotubes (SWCNTs) results in 3-dimensional composite as reported by Min et al. [89].The composite had many surface pores and crevices which could enhance mass transfer. As a biosensor application, tyrosinase was immobilized on SWNTs/PPy composite and used for the amperometric detection of DA. Other conducting polymers have also been used for analysis of DA as summarized in table 6.

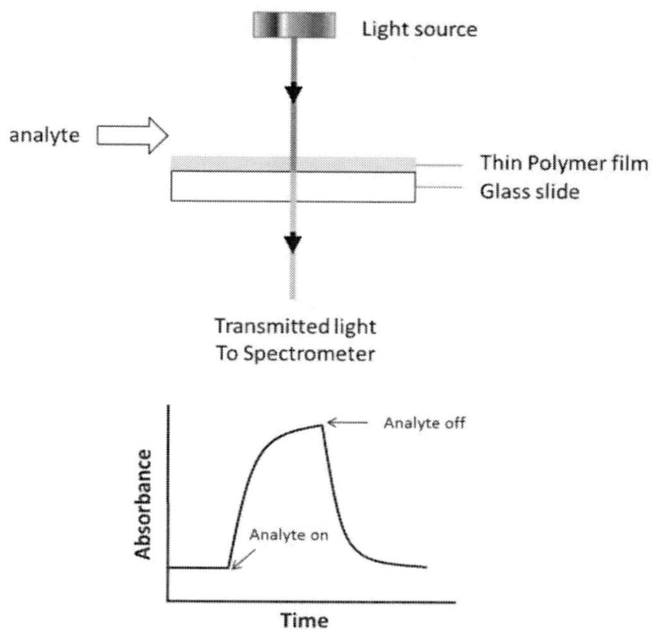

Figure 13. (a) Schematic of the simplest optical sensor in which a thin conducting polymer film deposited on glass is placed in the path of a light source. Doping/dedoping due to analyte interaction with the conducting polymer changes its electronic absorption. The UV-Vis and NIR spectra of the transmitted light reflect these changes indicating the presence of the analyte as shown in the bottom graph. Various optical sensors with different set-up have been reported [94-95].

7.1.3. Optical Sensors

7.1.3.1. Absorbance/Transmittance Based Sensors

Changes in the redox or protonation state of a CP lead to a strong modification of its electronic band structure, therefore an interaction of analytes which oxidize/reduce or protonate/ deprotonate CP can be measured by UV–vis spectroscopy. This was used in chemical sensors for HCl, NH_3 (see gas sensors) and pH (see pH sensors). An ultra-thin film is suitable for fabricating an optical sensor, because the spectrum of a thin film is easy to be recorded by using commercial spectrometers. In fact, the simplest senor is just a glass covered with an ultra-thin conducting polymer film. The response of the sensor is usually the transmittance or absorbance changes of the sensing films. However, the sensitivity of this kind of sensors is low, and the need of special spectrometers is expensive and not convenient.

7.1.3.2. Surface Plasmon Resonance Sensors

More sensitive detection of optical changes in CP can be realized by application of surface plasmon resonance (SPR). The measurable parameter is in this case either refractive index or optical absorbance at the used wavelength. The excitation of surface plasmons by light is denoted as a surface plasmon resonance (SPR) for planar surfaces or localized surface plasmon resonance (LSPR) for nanometer-sized metallic structures. It is the fundamentals behind many color based biosensor applications and different lab-on-chip sensors.A SPR sensor has a typical configuration illustrated in Figure 14 (Kretschmann-type configuration). On exposing to analytes, the minimum in the reflectance curve will shift, which indicates the existence of the analytes. The combination of electrochemistry with SPR (EC-SPR) has been used to characterize the growth of PANI on a gold electrode surface during polymerization as reported by Kang et al. [96].

Mirsky and colleagues at the University of Rogensburg observed that the optical properties of electrochemically polymerized N-methylaniline are changed in the presence of HCl gas [97]. This effect was used for the preparation of SPR sensor for HCl gas. The interaction of hydrogen chloride with a two nm layer of poly-(N-methylaniline) on gold leads to the shift of the surface plasmon resonance. The analysis of the resonance spectra demonstrates that the effect is caused by an increase of the imaginary component of the refractive index. This approach was used to detect changes in ultrathin films of CPs upon interaction with different analytes such as HCl [97], H_2S, NO_2, H_2O_2 and Hg^{2+}.

Moreover, any changes in dielectric properties of the interfacial region influence the excitation of surface plasmons, thus forming the basis for SPR sensing. Thus, SPR can be used to detect the binding of biomolecules on polymer modified gold surfaces. This approach has been recently used for detecting oligosaccharide-protein interactions and binding of ochratoxin A. EC-SPR was used for in situ monitoring polymer formation, probe immobilization, antigen-antibody intercation and protein immunosensing process.

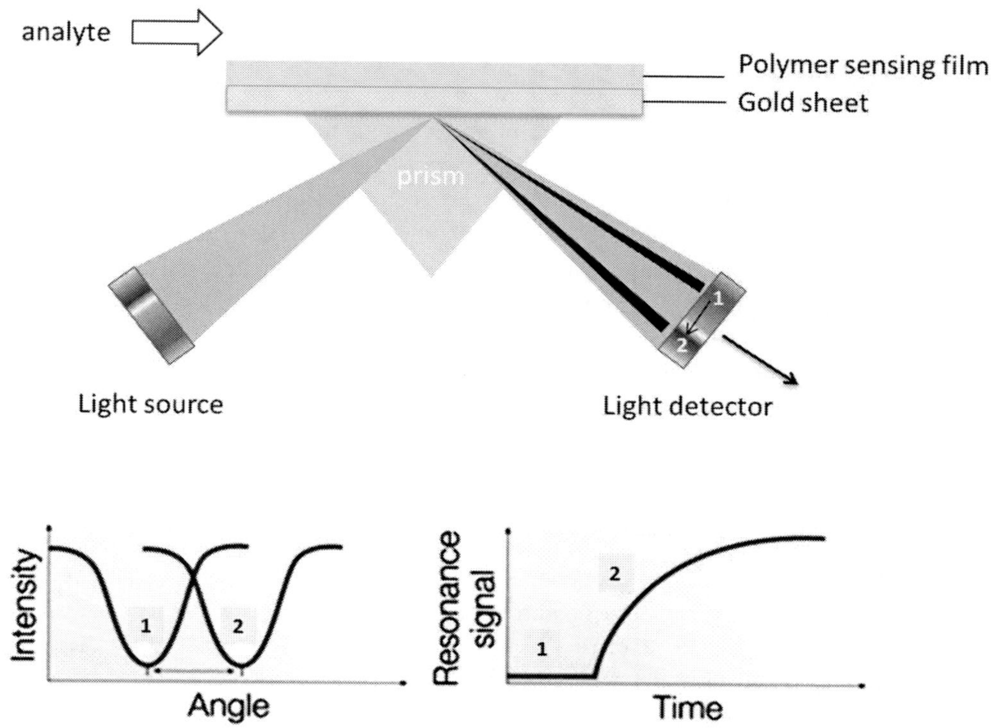

Figure 14. Schematic of surface plasmon resonance setup. The sensor surface is gold coated with an ultrathin CP film. The interaction of the analyte molecules with the polymer film shifts the refractive index and the SPR-dip moves to larger angles i.e. SPR angle shifts from 1 to 2 in the lower left-hand diagram. The movement of the SPR-dip is the actual monitored signal, and the movement over time forms the sensorgram, lower right.

7.1.3.3. Amplifying Fluorescent Polymer Sensors

Conjugated polymers are polyunsaturated compounds with alternating single and double bonds along the polymer chain in which all backbone atoms are sp- or sp^2-hybridized. This electronic conjugation between each repeat unit creates a semiconductive ''molecular wire''. The resulting interaction between orbitals creates a semiconductor band structure having a valence band (filled with electrons) and a conduction band (devoid of electrons). The semiconductive nature of these organic polymers gives them very useful optical and optoelectronic properties.

The group of Swager demonstrated that ''wiring molecular recognition sites in series'' leads to ultra-high sensitivity [98]. This sensitivity arises from the collective optical and conducting properties of the CP. These polymers are extremely sensitive to minor external structural perturbations or to electron density changes within the polymer, due to their ability to self-amplify their fluorescence quenching responseupon perturbation of the electronic network upon binding of analytes. Hence, they are often referred as amplifying fluorescent polymers (AFP).

Depending on the system, a CP can exhibit a strong luminescence; the luminescence efficiency is related to the delocalization and polarization of the electronic structure. These polymers are good candidates as materials for fluorescent sensing. Figure 15a shows schematically how conjugated polymers amplify the molecular recognition signal via migration of electrons along the polymer chain. It shows a basic band diagram illustrating the mechanism known as photoinduced electron transfer fluorescence quenching. Irradiation of the polymer with a photon causes promotion of an electron to the conduction band (which is now of a much higher energy), which then migrates along the polymer backbone. Analyte binding produces a trapping site whereby the excitation is effectively deactivated by electron transfer quenching. The low energy LUMO can in an exergonic process accept the electron from the excited state of the polymer. This destroys the polymer based excited state, and the polymer can no longer fluoresce. Examples include intrinsic fluorescence polymers caused by backbone units (e.g. PP, PPV, PPE, polyfluorene and PTH derivatives) or a fluorescence of fluorophores introduced into polymer.

Figure 15(b,c) compares schematically the CP and the classical chemosensor approach. Whereas complete fluorescence quenching would be observed in the case of the conjugated polymer upon interaction with an analyte, in the case of non-conjugated sites exposed to the same analyte concentration only a small percentage of quenching would be observed. Each analyte is confined to its particular molecule and can only sample one binding site. Therefore, the emission is observed from those molecules which did not bind an analyte. In the CP approach one single interaction can quench a large number of fluorophores and the signal obtained in the presence of the analyte is amplified. For example, nitroaromatic explosives act as electron acceptors for photoexcited electrons of the CP and quench fluorescence. In contrast, hydrazine enhances fluorescence of by reduction of oxidative traps in the polymer chains. Also the binding of Hg^{2+} by a mercury specific oligonucleotide in solution increases fluorescence of poly(3-(3_-N,N,N-triethylamino-1_propyloxy)-4-methyl-2,5-thiophene). In comparison with single fluorescent groups, fluorescent polymers provide a higher sensitivity [98]. This was demonstrated in fluorescence quenching by transition metals in bipyridyl-containing CP.

AFP can be used for quantitative analysis using the dependence of the fluorescence emission intensity, *F* on quencher concentration *[Q]* given by the Stern-Volmer equation [99]:

$$F_0/F = 1 + K_{sv}\,[Q]$$

Where F_0 is the fluorescence intensity when no quencher is present, *F* is the fluorescence intensity in the presence of quencher. K_{sv} is the Stern-Volmer constant (Figure 1d).

Wang and co-workers reported that phosphonate-substituted poly(fluorene)s show strongly amplified and reversible fluorescence quenching by Fe^{3+} in dichloromethane (Figure 1d) [100]. The authors also reported strong selectivity for Fe^{3+} over a wide variety of other metal ions. Zhao et al. demonstrated the direct detection of pyrophosphate (PPi) in aqueous solution using a cationic poly(phenylene-ethynylene) with polyamine side chains [101]. Pyrophosphate-induced polymer aggregation causes a significant spectroscopic change, which in turn allows quantification of dissolved pyrophosphate. Xu et al. designed a sensitive pH sensor on the basis of awater-soluble polyfluorene derivative (PFP-aa) [102]. Different fluorescent responses of the PFP-aa were demonstrated as the environmental pH changed from 3 to 12. It was also demonstrated that PFP-aa can be used as a pH sensor to detect D-fructose. Maiti et al. reported the use of a polythiophene derivative as sensor for acids and some metal ions [103]. The fluorescence quenching of polymer in acids is due to the protonation of imine nitrogen in quinolinyl group. Quenching as well as enhancement of fluorescence was found in the same polymer by addition of different metal ions. The fluorescence of polymer is quenched by the addition of metal ions because of formation of a nonluminescent complex between the polymer and the metal ions.

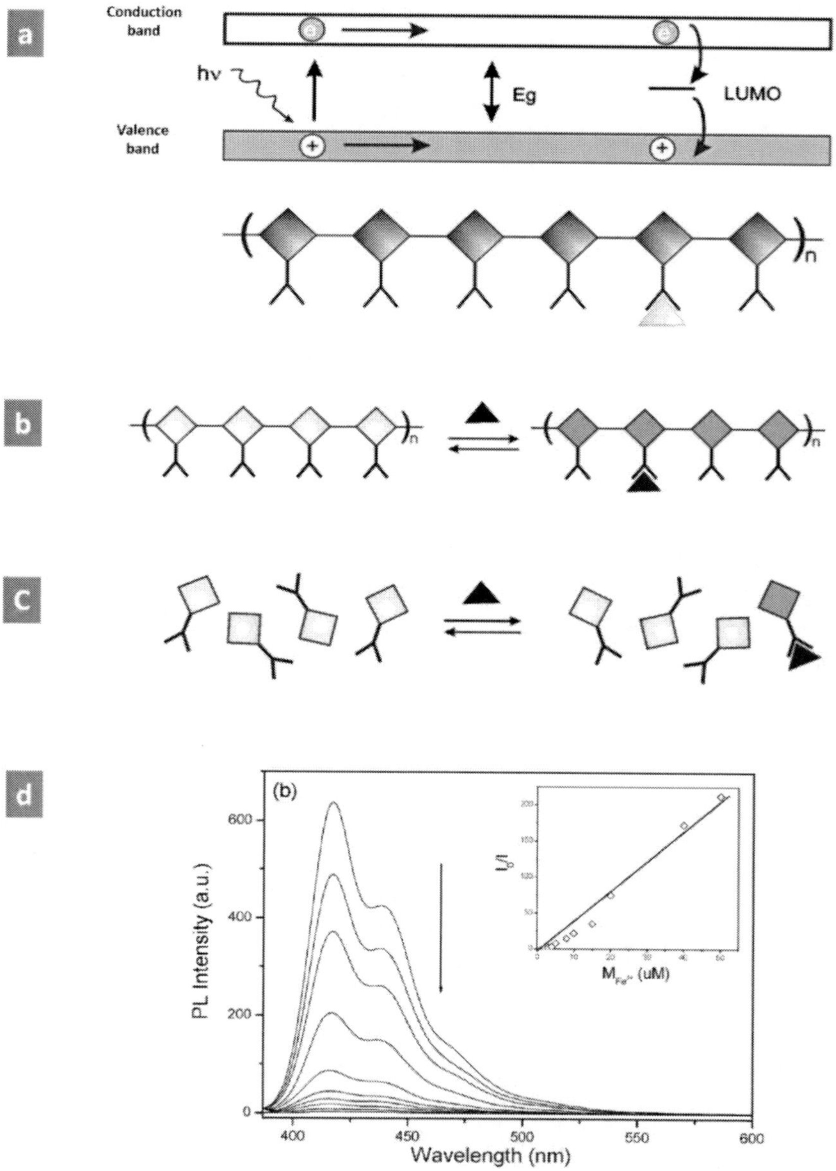

Figure 15. (a) Band diagram illustrating the mechanism for exciton transport and electron transfer fluorescence quenching of a conjugated polymer upon interaction with the analyte. (b) Conjugated polymer interacting with a small concentration of quencher giving complete quenching of the polymer chain. (c) Individual receptor-chromophore molecules exposed to a small concentration of quencher. This situation results only in a partial quenching. (a-c) reproduced with permission from [104] (d) Fluorescence spectra of phosphate substituted poly(fluorene) in CH_2Cl_2 solution upon addition of Fe^{3+}. The Stern-Volmer plots are shown as an inset. (d) Reproduced with permission from [100].

The fluorescent CP-based biosensors have attractive merits as a sensory system and have expanded their applications to various biomolecular detections. An excellent sensory system must provide high sensitivity and high selectivity together with the capability for convenient label-free detection. The signal amplification principle by means of the macromolecular response of a CP chain has been well established to achieve high sensitivity. This 1-dimensional amplification scheme has been further expanded to 2-dimensional and even 3-dimenstional amplification schemes through directed molecular assembly of CPs into nanofibers, well-defined films, and three dimensional objects. When it comes to high selectivity, devising rationally designed receptors that have specificity toward a target analyte is essential.

7.1.4. Conductometric Sensors

In conductometric sensors, an interaction of electron acceptors or donors with CP causes changes in the doping state of the polymer due to oxidation/reduction or protonation/deprotonation reactions, leading to changes in conductivity. In most cases the simplest two-point conductivity measurement technique is used, Figure 16. Such a device was named chemiresistor. The less used four-point conductivity measurement technique can provide a higher sensitivity.

A chemiresistor [105] is a resistor, whose electric resistance is sensitive to the chemical environment. Chemiresistors can be fabricated through a cheap and convenient process. It consists of one or several pairs of electrodes and a layer of conducting polymer in contacting with interdigitated electrodes, as illustrated in Figure 16. The electrical resistance change of the sensing material is measured as the output, so a simple ohmmeter or a multimeter is enough to collect the data. Usually, a constant current or potential is applied on the sensor, and the measuring signal is potential or current change, respectively. This type of sensor is simple but efficient. The related technologies, including fabrication and measurements, have been maturely developed. Thin films, fibers and bulk materials can be utilized as the sensing elements of chemiresistors and their output signals are resistances.

pH changes as well as modification of the redox state of CP can be induced by substrates, products or intermediates of enzymatic reactions, this is the main principle of CP application as conductive transducers in enzymatic biosensors (Section 8). Now there is an increased interest in conductometric immunosensors in combination with nanostructures, and especially nanowires, for biosensing. Conductometric transducers have been largely employed in gas sensing (section 7.2.3.).

Using conducting polymer as the sensing film brings several advantages in device fabrication and operation. Conducting polymers are easy to be processed into films by many techniques. The disadvantage of chemiresistor is that the resistance of a device is influenced by many ambient factors, and not only determined by the resistance of the conducting polymer sensing film, but also the contact resistance of the electrodes. Moreover, little information other than resistance can be obtained; this is unfavorable in distinguishing different analytes.

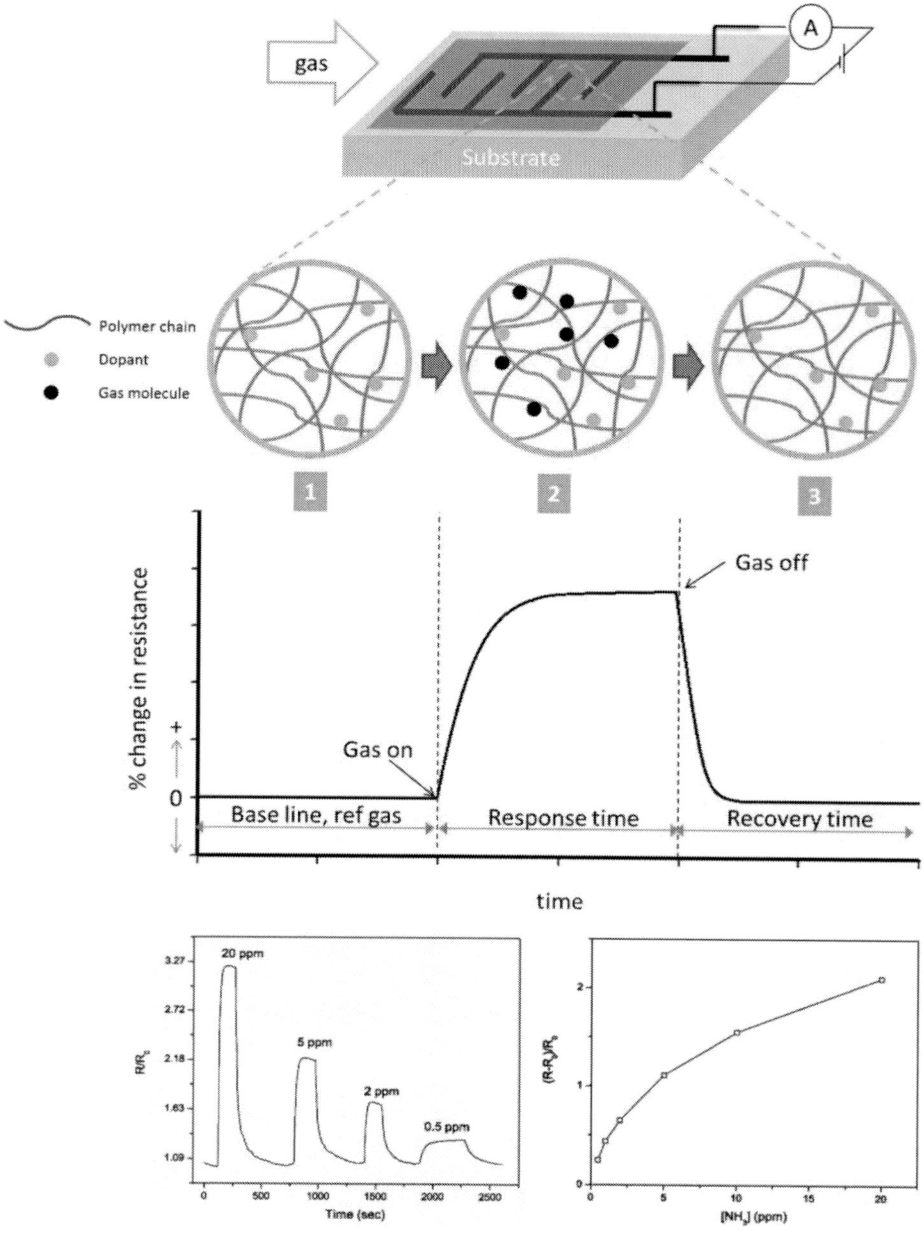

Figure 16. the mechanism of operation of a conductometric gas sensors: top, a schematic of a typical chemiresistor used in gas sensing. The device comprises a film of the sensing material (here CP film) deposited on an interdigitated electrode. In general, polar molecules were found to dope/dedope (oxidize/reduce) CPs resulting in a decrease/increase in resistance. Three distinct regions in figure are to be described: region (1) chemiresistor is in a steady state in clean air. Region (2), when the chemiresistor is exposed to vapor, gas molecules absorb into the polymer chains causing resistance change. The resistance change shown corresponds to a reducing gas. Region (3) when vapor is turned off, the resistance change decrease back to the baseline and the initial state is recovered especially if the interaction is reversible. Bottom left, the response curve of a polyaniline sensor on exposure to different concentrations of NH3 [106]. Bottom right, dependence of the relative response of the polyaniline sensor on NH3 concentration.Bottom curves, reproduced with permission from [106].

7.1.5. Field Effect Transistor Devices

Conducting polymer resistors possess good sensitivity to organic vapors, but they can suffer from temperature and humidity dependence. The well-known organic thin-film transistors (OTFTs) have been applied in sensing field just after it was first developed. Janata et al. [107], found that electrochemically deposited polypyrrole as a field-effect transistor gate material has demonstrated devices sensitive to many volatile organic compounds (VOCs). Since then, there has been much interest in combining conducting polymers with FET devices for sensor applications. Such a combination of technologies could lead to a new generation of low-cost, low-power gas sensors for handheld monitors.

TFT consists of a semiconductor active layer in contact with two electrodes "source" and "drain", and a third electrode "gate" which is separated with the active layer by an insulating film. When it works, a source-drain voltage is applied and a source-drain current is measured. The source-drain current is changed when sensing film interacts with analyte. Two types of conducting polymer transistor, classified by whether the current flow through the polymer, were used to detect gases, as illustrated in Figure 17. Figure 17A shows the configuration of a thin film transistor (TFT), its active layer is made of conducting polymer, and Figure 6B represents the structure of insulated gate field effect transistor (IGFET), whose gate electrode is made of conducting polymer and the current flows through the other semiconducting layer, e. g. silicon.

The performance of organic FET has been greatly improved over the last decade and has rapidly approached that of a-Si:H TFTs. It is known that the charge carrier mobility and on/off ratio are two key parameters that are of high significance in the performance of OFETs. Currently, most experimental studies are aiming at improving OFETs' performance including carrier mobility, on/off ratio and device stability. A wide variety of organic semiconductor materials, including pentacene, thiophene oligomers, and the regioregular poly(3-hexylthiophene) (P3HT) polymer, have been investigated as the channel materials of the OFETs. The P3HT-based OTFT sensors have been shown to provide good sensing responses to a range of gases, with sensitivities in the low-ppm levels, good repeatabilities, and stable operation in an air environment. Organic field-effect transistors (OFETs) have recently been developed as low-cost electronic products for use in flexible devices, light-emitting displays, pressure sensors and chemical sensors (see gas sensors).

OFETs generally have low detection limits and, if carefully designed, can also have low threshold voltages, short response and recovery times, high signal-to-noise ratios and good selectivity. Additionally, when compared with chemiresistors, OFETs provide more parameters for measurements, thus, may give more detailed information about the semiconductor natures of the conducting polymer sensing films. The beneficial of using conducting polymer as the sensing layer is the convenience in fabrication. Besides, easy modification in conducting polymer structures provides facile route to sensing materials with different work functions and selectivities to analytes, which insure high performance of transistor configured sensors. The disadvantages of these devices are, as other semiconductor devices, their preparation is slightly complicate, and the characterization of a transistor is more difficult than a chemiresistor.

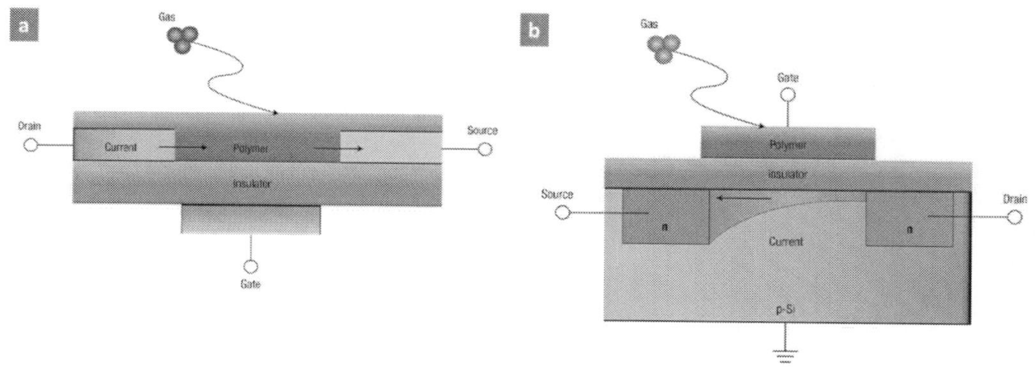

Figure 17. Conducting polymers in field-effect transistors (a) Thin film transistor; (b) insulated gate field-effect transistor. Reproduced with permission from [108].

7.1.6. Mass Sensitive Sensors

7.1.6.1. Quartz Crystal Microbalance Sensor

A typical schematic diagram of QCMB is shown in Figure 18a. It consists of a conducting polymer coated piezoelectric quartz crystal and a pair of electrodes. When an ac voltage is applied across the piezoelectric quartz crystal the material oscillates at its resonant frequency, normally between 10 and 30MHz. The three-dimensional wave produced, travels through the entire bulk of the crystal. The conducting polymer layer adsorbs gas when exposed to the vapor, which results in an increase in its mass. This increase in mass alters the resonant frequency of the quartz crystal and this change in resonant frequency is therefore used for the detection of the vapor. QCMB sensors coated with conducting polymers as sensing material has been used in measuring humidity with relative humidity range biosensors and gas sensors (see gas sensors).

7.1.6.2. Surface Acoustic Wave Sensors

Surface Acoustic Wave (SAW) gas sensors operate on the same principle as QCMB devices. The SAW device is composed of a piezoelectric substrate with an input (transmitting) and output (receiving) interdigital transducer (IDT) deposited on top of the substrate as shown in Figure 18b.The polymer film is coated on the gap between these two transducers and an ac signal is applied across the input transducer creating an acoustic two dimensional wave that propagates along the surface of the crystal at operating frequencies between 100 and 400 MHz. The mass of the gas sensitive membrane of the SAW device is changed on interaction with a compatible analyte and causes the frequency of the wave to be altered (see gas sensors).

The detection limits of above two types of devices are very low (<1 ppm). Because piezoelectric crystal is rather sensitive to mass uptake, adsorbing a very small amount of analyte can be detected through frequency change. For the same reason, the selectivity of piezoelectric crystal devices is poor. Conducting polymers here act only as active layers, which tune the surface property of crystal and improve both detection limit (by adsorbing more analyte molecules) and selectivity (by introducing special interaction with analytes) of the sensors [16].

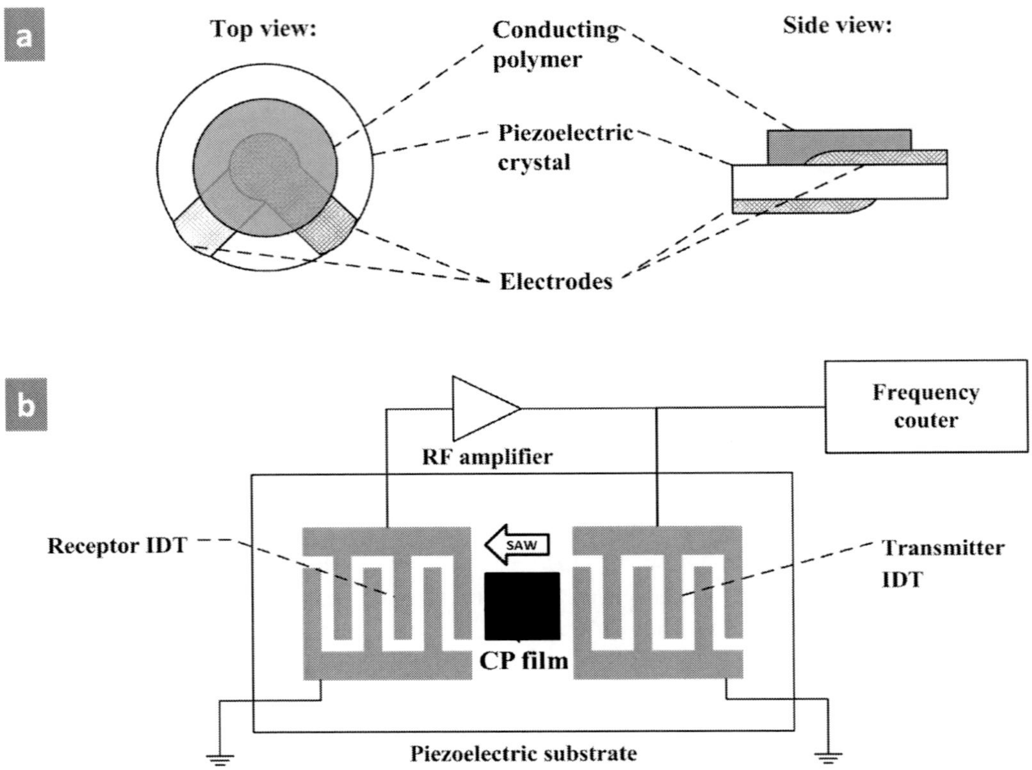

Figure 18. schematic of the typical configuration of quartz crystal microbalance (QCMB) sensor device, (b) surface acoustic wave (SAW) sensor device. Reproduced from [16] with permission.

7.2. Sensors Based on Chemical/Industrial Applications

7.2.1. pH Sensors

Monitoring pH is essential not only in finding the chemical characteristics of a substance but also as the first step toward managing chemical reactions. Currently, pH measurement is used in various fields, including nearly all industries that deal with water, not only the chemical industry, but public organizations, agriculture and fishery-related industries and biological industries, as well. The most common systems for pH sensing are based upon either amperometric or potentiometric devices. The most popular potentiometric approach utilizes the glass bulb pH electrode which is a well-established tool in the measurement of pH; however, there are some disadvantages with this type of electrode. Glass pH electrodes do not provide accurate data at very high and very low pH values because of alkaline error and dehydration, respectively. The electrode also needs special care and treatment before and after use. Ion selective membranes, ion selective field effect transistors, two terminal microsensors, fiber optic and fluorescent sensor, metal oxide and conductometric pH-sensing devices have also been developed [109]. However, these types of devices can often suffer from instability or drift and, therefore, require constant re-calibration. Considerable research interest is now focused on the development of chemical or biological sensors using functional polymers. The ion-exchange properties of conducting polymers are of special interest for potentiometric- sensor development. Conducting polymers are ideally suited for sensor applications because they not only exhibit high conductivity and electroactivity but they could also be used as a general matrix and can be further modified with other compounds in order to change selectivity.

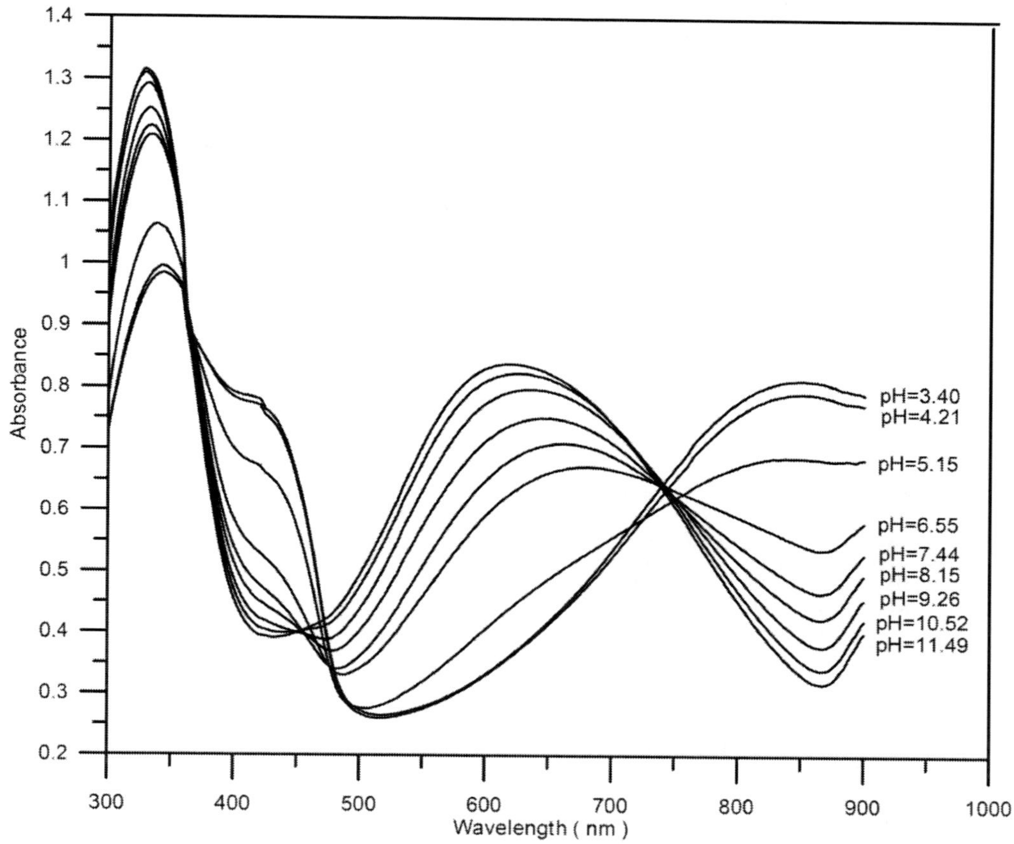

Figure 19. Set of absorption spectra recorded for PANI film deposited onto ITO glass in buffer solutions in the pH range from 2.15 to 12.54. These pH-dependent spectral variations of PANi film were explained by their transformation from the protonated to the unprotonated form. Reproduced with permission from [110].

Many CP possess acidic or/and basic groups which can be protonated or deprotonated. To this family of CP belong PANi, PPy, polyindole (PI), polycarbazole and polyazines. Deprotonation of these CP results in a decrease of charge carriers along the polymer chains and is accompanied by changes in the electrical and optical properties [Figure 19]. These changes lead to modification of redox properties of CP, this effect is very strong for PANi and PI but almost not recognizable in PPy .

Conductivity of CP can be caused by ions and/or electrons. Affinity of amino groups to protons may be the reason for high proton permeability of these polymers in aqueous solutions: films of PANi and PPy on glassy carbon and platinum electrodes show almost

Nernstian behavior relative to pH. The potentiometric response of PPy is linear between pH 2 and pH 11 and PANi between pH 2 and pH 10. This fact as well as an electronic component of the conductivity of these polymers makes them potential for applications in solid state pH sensors. The described pH dependent property changes in CP have therefore been used to develop potentiometric, conductometric and optical pH sensors [111].

Conductometric CP based pH sensors are usually prepared by depositing a thin layer of a CP film between two or four metallic or graphite contacts. Optical pH sensors are prepared by deposition of thin polymer films on the surface of ITO-electrodes, microtiterplates, cuvettes or other transparent support. Absorption of PANi is measured at 800–900nm or at 500–600nm (Figure 19), while the measurements for optical sensors based on PPy are performed at 650nm.

Problems with potentiometric CPs based pH sensors include the response which is not quite stable and the response time was relatively long (mostly more than 10 min) due to slow protonation and deprotonation of the CP. Miniaturization of pH sensors and fabrication of pH sensor arrays have enhanced the performance in response time and signal generation. a pH response of -50.54 ± 1.67 mV/pH (28°C) with a linear working range of pH 2–12 and a response time less than 100 s is recently reported for PPy based sensor [112]. Further research is still needed to provide the state of the art of CP pH sensors and to satisfy the market demands.

7.2.2. Humidity Sensors

Humidity sensors have been attracting much attention because of the growing demand for accurate measurement and control of environmental humidity in many fields. Fields like, quality control of food products in a wide range of industries, control of air conditioning, paper and textile industries, optimal functioning of modern solid state electronic equipment, civil engineering etc. are finding increasing interest in humidity sensors. Humidity sensors available in the market include dew point, infrared, catalytic and tin oxide sensors, which may be expensive, or require high temperature operation and consume significant amount of power and high cost of maintenance. Much research has been focused on the development of humidity sensitive material. Among these are the investigation of using conducting polymers such as polypyrrole, polythiophene and polyaniline for humidity and gas sensing. Advantages with polymers as sensing materials are light weight, flexible, low cost and simple fabrication process.

Miniaturized humidity sensors are generally categorized into several types in terms of transduction techniques, namely capacitive, resistive, hygrometric, gravimetric and optical techniques [113]. The principles of work of each of these techniques can be simply described here. In summary, capacitive sensors are based on dielectric changes of thin films upon water vapor uptake. While, resistive humidity sensors are devices that transduce air humidity into an impedance change. This can be measured by a current, a voltage or a resistance. Hygrometric humidity sensors transduce air humidity via a mechanical domain. Gravimetric humidity sensors transduce air humidity into a weight change. The most widely known gravimetric humidity sensor is the quartz crystal microbalance (QCMB). Optical humidity sensors make use of the absorption wavelength of water: when light is passed through a gas, the absorption of certain wavelength provides information about the composition.

Among the different types of humidity sensors, resistive-type humidity sensors are most interesting (Figure 20). Here, there are three groups of materials which are of particular importance; ceramics, polymers and polyelectrolytes. However, these resistive sensors cannot detect the low relative humidity due to the high impedance involved. Many methods have been taken explored to detect low humidity. Among various methods, conducting polymers have attracted considerable attention for their intrinsic high conductivity. When conducting polymers were added into the sensing humidity polymer matrix, the impedance of sensors was reduced at low humidity. Pure polymer, polymer blends and polymer–inorganic composites have also been studied for the purposes, resulting in different degree of advancements in this area [114-117]. Generally, conducting polymers behave quite similarly on exposure to humidity. A resistance decrease (or conductivity increase) is observed due to the formation of H-bonding between water molecules and the nitrogen center of the polymer backbone. This facilitates the proton exchanges to increase the doping level of the conducting polymer and so the number of charge carriers increases, resulting in an increase in conductivity.

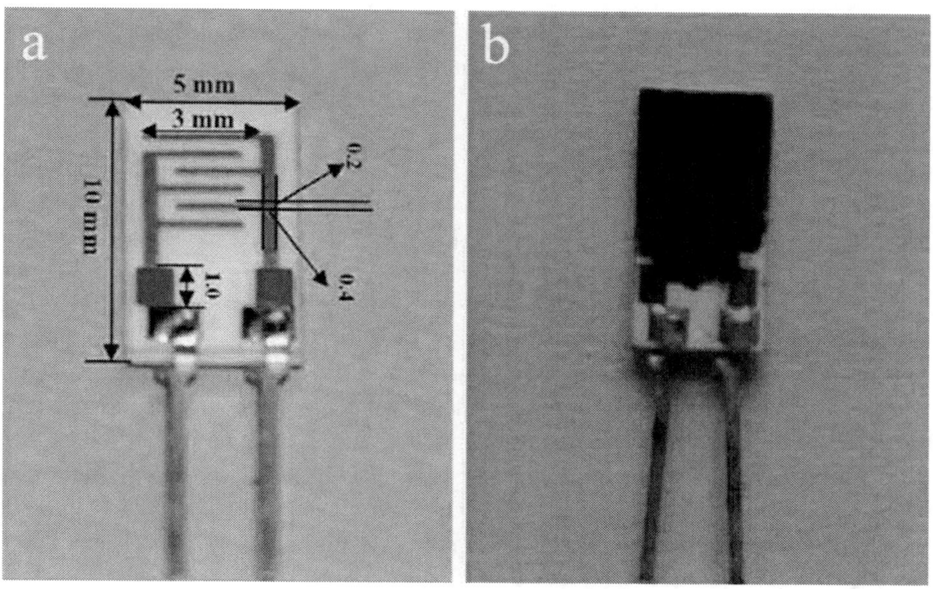

Figure 20. The structure of a typical resistive humidity sensor (a) the native and (b) polypyrrole as-coated sensors. Reproduced with permission from [115].

The concept of a miniature humidity sensor based on electrochemical growth of PEDOT conductive polymer onto two microfabricated conducting pads has been experimentally demonstrated by Li et al. [116]. The dimension of the whole device is around 100 × 200 μm and the resistance was found to vary from 37 to 62 ohm in different relative humidity environments at room temperature. Matsuguchi et al [117] used blended films of electron-conducting polyaniline and polystyrene to effectively depress the sensor drift caused by polarization and/or electrolysis upon dc operation, which is a serious problem for resistive-type humidity sensors using polyelectrolytes. Sun et al. [115], on the other hand, employed polypyrrole further quaterized with 1,4-dibromobutane as a humidity sensor at room temperature. The results showed that the humidity sensor has low impedance at low humidity, good sensitivity and quick response. Moreover, the logarithm of impedance of the polypyrrole composite film varied linearly with relative humidity in the range of 11–95%.

7.2.3. Gas and Odor Sensors

Gas sensors have become ubiquitous components in our lives, actively involved in the operation of the vehicles we drive, continuously monitoring air quality in our homes and place of work, and enabling industrial process control/emission compliance [118]. Indoor air quality has become an important community concern due to the increased amount of personal time spent in indoor environment. Nowadays, people generally spend more than 80% of their time in an indoor environment such as home, office, car and shopping center. Some studies showed that the level of pollutants in indoor environment is actually higher than that in outdoor environment. Indoor air pollutants mainly include nitrogen oxides (NOx), carbon oxides (CO and CO_2). Volatile organic compounds (VOCs) are well-known indoor pollutants. VOC refers to organic compounds which have significant vapor pressure and which can affect the environment and human health. VOCs are typically not acutely toxic but have chronic effects. Because the concentrations are usually low and the symptoms slow to develop, analysis of VOCs and their effects is a demanding area [119]. In this respect, CP sensors can be used independently for detection and measurement of individual vapors or jointly in the form of arrays for measurement of complex odors. This has resulted in the development of commercial instruments with applications in the chemical and food industries, environmental monitoring, and biomedical diagnosis.

In comparison with most of the commercially available sensors, based usually on metal oxides and operated at high temperatures, the sensors made of conducting polymers have many improved characteristics. They have high sensitivities and short response time; especially, these feathers are ensured at room temperature. Additionally, conducting polymers have good mechanical properties, which allow a facile fabrication of sensors. Besides, conducting polymers exhibit high porosity and rich collection of morphologies those are not only highly dependent on the monomer and synthesis route but were also shown to significantly affect their sensitivity and response times. As a result, more and more attentions have been paid to the sensors fabricated from conducting polymers [16].

The operation of conducting polymer gas sensors is based on the phenomenon that the electrical conductivity of the polymer film is changed when exposed to analyte vapors. The changes can be correlated quantitatively to the concentration of the vapors and can be readily reversed when the vapors are removed (see figure, conductometric sensors). The different concepts that are applied include electrochemical cells, gas-sensitive field-effect transistors, quartz crystal microbalances, surface acoustic wave devices, and chemiresistors. Among these, chemiresistors, or so-called resistive gas-sensing elements, are most commonly applied and many of the commercialized devices are based on it. All these techniques are discussed in detail in the following sections:

Chemiresistor gas sensors based on films of polyaniline nanofibers as the sensing element have been extensively reported by Kaner group [120-121]. The conjugated polymer polyaniline is a promising material for sensors as its conductivity is highly sensitive to chemical vapors. Polyaniline sensor research has focused on chemically modifying the polymer structure to facilitate interactions between vapor molecules and polymer. There were efforts to minimize the diffusional path length for vapor molecules; yet poor diffusion can readily outweigh any improvements made to the polymer chains because most of the material other than the limited number of surface sites is unavailable for interacting with the vapor, thus degrading sensitivity (Figure 21, left). This problem was solved by using nanofibers of polyaniline as they have much greater exposed surface areas, as well as much greater penetration depths for gas molecules (Figure 21, right), leading to superior performance in both sensitivity and response time. For example, improvements in both sensitivity and time response of many orders of magnitude are now observed by this simple morphological change to the polymer [120-121].

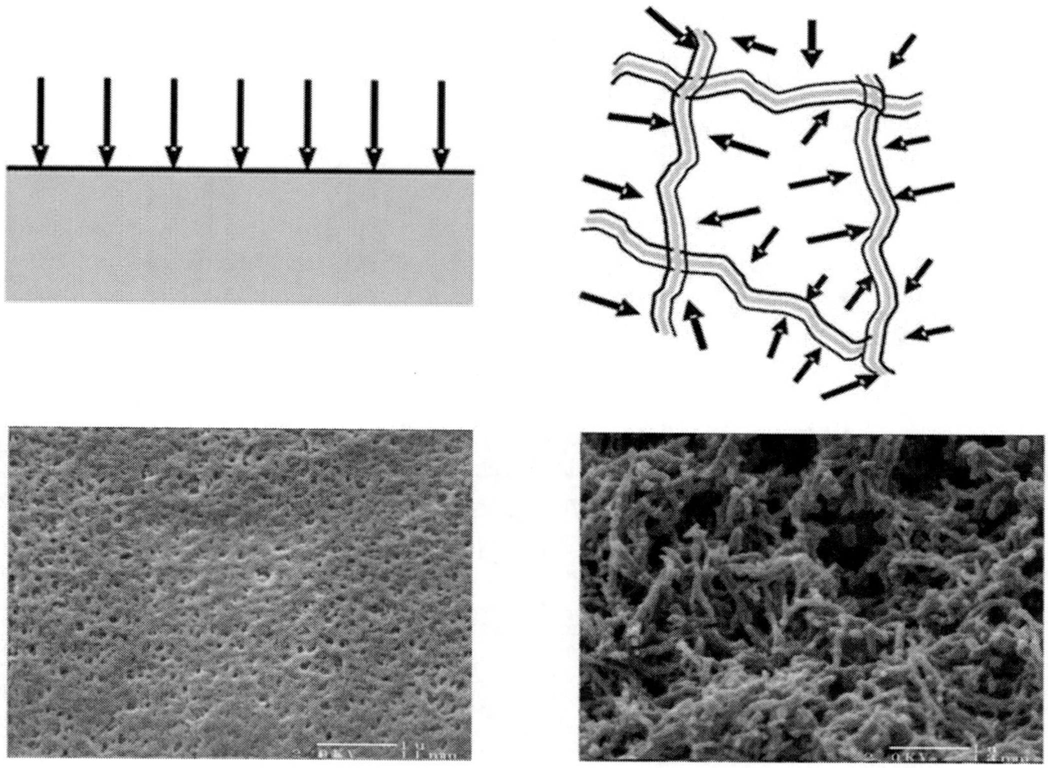

Figure 21. Schematic diagram of a conducting polymer thin film (left) and nanofibers (right) exposed to gas molecules (arrows). Compared to a thin film, nanofibers have a much higher fraction of exposed surface (black), and much shorter penetration depth (gray) for gas molecules. Bottom, SEM images of polyaniline film (lower right) and polyaniline nanofiber film (lower right). Reproduced with permission from [121], [120].

The ease of fabrication of chemiresistors is a major advantage. However, the resistance of the device is influenced by ambient factors and not only determined by the resistance of the CP film. Moreover, little information other than resistance can be obtained which is unfavorable in distinguishing different analytes.

Conducting polymer based field effect transistor devices hold promise for gas sensing (section 7.1.5.). In CP based Schottky-like structures, the current through the junction depends on thework function of the CP which can be modified by interaction with analytes. This principle was used for detection of NOx, ammonia, methane, alcohol and water vapors. Aluminium and/or gold were used as metals while PANI, P3OTH and PPY were used as CP in these diodes. Chemosensitive diodes formed by a junction of n-doped silicium and PPY or PANI were also reported and used as gas sensors.

A quartz crystal microbalance (QCMB) gas sensor based on polythiophene and its derivatives has been developed and had a linear frequency shift on interaction with various

VOCs (acetic acid, toluene, acetone, *p*-xylene, ethanol, 1-octanol, acetonitrile and water vapors) [122]. The QCM sensor array has a good and qualitative ability to separate the low-polar/nonpolar VOCs gases from polar VOCs gases. Ayad et al. [123] used polyaniline film as the sensing element in QCMB sensor for the on-line detection of primary aliphatic amine vapors. The differences in the diffusion rates and the adsorption kinetics of the various amine vapors made it possible to selectively detect the various amines. Lu et al [124] took QCMB further to produce six chips sensor module coated with synthetic polypeptides together with conducting polymers for the simultaneous detection and identification of VOCs such as acetic acid, butyric acid, ammonia, dimethyl amine, benzene, chlorobenzene and their mixtures. The sensor module demonstrated sensitivity and selectivity for detection, and discrimination via characterized odor profiles of all tested VOCs.

Surface acoustic wave (SAW) gas sensors have been extensively reviewed by Kalantar-zadeh group [125-126]. They employed a polyaniline/In_2O_3 nanofibre composite based layered SAW as a sensor for H_2, NO_2 and CO gases [125]. Fast response and recovery times with good repeatability were observed at room temperature. In another work, they featured polypyrrole nanofibers as the active component in a SAW sensor for hydrogen (H_2) and nitrogen dioxide (NO_2) detection at room temperature. A H_2 sensor based on polyethylaniline SAW sensor has also been made [126]. Groves et al. [127]used an array of SAW devices with polymer coatings 20-30nm thick to detect 16 different solvents using four different polymers. SAW devices can detect a broad spectrum of gases due to the wide range of gas sensitive coatings available.

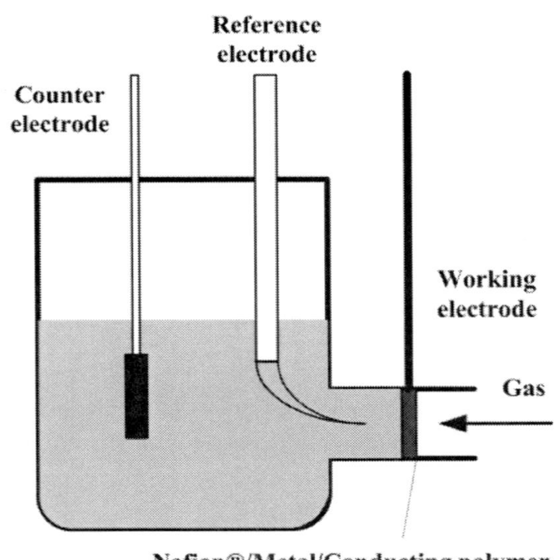

Figure 22. schematic of the typical configuration a type of electrolytic cell for amperometric gas sensing. Reproduced from [16] with permission.

Amperometric method has also been to detect hazardous gases. This sensor usually has a divided cell, with conducting polymer/metal/Nafion® as working electrode [16]. The supporting electrolyte usually is an inorganic acid to provide H^+ ions for ionic conductance in

Nafion® film. The configuration of the cell is presented in Figure 22. The sensing mechanism is the same as other amperometric sensors used in solutions. When an appropriate potential is applied on the electrode, the analyte molecules are anodic oxidized or cathodic reduced, resulting in a current change. Variation in the electrodes and the electrode potentials can be utilized to identify the gases present. Do et al. employed polyaniline/Au/Nafion® composite based Amperometric cell to detect NO_2 gas and good reproducibility and sensitivity was achieved [128].

Table 7. Abbreviations: PPy, polypyrrole; PANi, polyaniline; PEDOT, poly(3,4-ethylenedioxythiophene); PSS, poly(styrenesulfonate); ppm, parts per million; ppb, parts per billion; ppt, parts per trillion; P3HT, poly(3-hexylthiophene); OTFT, organic thin film transistor; SAW, surface acoustic wave; OFET, organic field effect transistor; QCMB, quartz crystal microbalance

Gas	Device/techniques/ principles	Polymer	Sensor characteristics	Ref.
NH_3	Electrical property measurement	PPy nanoribbons	Shows sensitivity to sub ppm NH_3 at room temperature	129
	Electrical property measurement	PANi-TiO_2 nanocomposite	Sensitive to 50 ppt NH_3, a thousand times more sensitive than the best PANi sensor.	130
	Electrical property measurement	Pd/PEDOT:PSS nanocomposite	the nanocomposite form stable colloid that makes it easy to process into devices	131
	Electrical property measurement	PANi nanojunction	Capable of in situ detection of ppb levels of NH_3 in human breath	132
	OTFT	P3HT	can effectively monitor for NH_3 gas in the atmosphere	133
HCl	Electrical property measurement	PANi nanofibers	Response time less than 2 seconds	134
H_2	SAW	PANi nanofibers	Dopant and nanofiber diameter have a significant effect on the sensitivity, response and recovery times.	135,136
	SAW	PPy nanofibers	The sensor performance was assessed during a five-day period and repeatable results were obtained	137
CO		PPy	Sensitive to100 ppm CO	138
		PANi	shows high sensitivity of order 800 and response time of 5 seconds	139
NO	Electrical property measurement	PEDOT-PSS	The limit of detection, response time, and recovery time were about 350 ppb, 527 s, and 1780 s, respectively	140
NO_2	OFET	P3HT	The gas response indicates that the device can outperform a chemiresistor in terms of detection performance	141
	SAW	PANi/ZnO composite	It has also been applied to SO_2 detection	142
CH_4	QCMB	PANi/PdO composite	has a better response at room temperature than at a high temperature	143

Table 7 gives a picture of the sensor characteristics of different polymers used in gas sensors based on different working principles. Conducting polymers showed promising applications for sensing gases having acid–base or oxidizing characteristics. Conducting polymers for gas sensing are commercialized by Osmetech and Fibron Technologies. Fibron has developed polyaniline nanofibers and other conducting polymer nanostructures that are water and organic solvent processable without additives, and are cost effective alternatives to the current water processable CPs.

7.2.4. Ion-Selective Electrodes

CPs are electroactive materials with mixed electronic and ionic conductivity; they can transduce an ionic signal into an electronic one in the solid state. These multifunctional properties of conducting polymers are excellent concerning ion-to-electron transduction in the solid state. Ion selectivity can be introduced in two different ways. One possibility is to use the CP only as ion-to-electron transducer (solid contact) in combination with classical ion-selective membranes (SCISE, Figure 11e). By using a conducting polymer that is soluble in the same solvent as the components of the ion-selective membrane it is possible to construct a so-called single-piece ISE (SPISE, Figure 11f, potentiometric sensor section). Another possibility to induce selectivity is to incorporate the ion-recognition sites directly into the conducting polymer matrix, e.g., by doping the conjugated polymer with counter ions containing ion-complexing groups or by covalent binding of ion-recognition sites to the conjugated polymer chain (CPISE, Figure 11g). Examples of ion recognition sites include salen and acetophenone groups which are known as selective ligands for divalent cations. In addition, crown ethers have been used for the selective determination of K^+ ions. Ionic selectivity can be also introduced in a CP by doping of this material with metal complexing anions. PPy films were doped with calcion, eriochrome black T, Kalces and ATP to obtain Mg^{2+} and Ca^{2+} selective membranes [144].

Solid-state ISEs (SCISE) were constructed for the determination of inorganic (Cl^-, K^+, pH, Ca^{2+}, CO_3^{2-}, Cu^{2+}, Ag^+, Pb^{2+}) and organic (N-methylpyridinium, bupivacaine, dimedrol, chlordiazepoxide) ions by using conducting polymers as the solid ion-to-electron transducer. The conducting polymer materials were based mainly on derivatives of polypyrrole, polythiophene or polyaniline. Several research reports exist on the use of SPISE for determination of Li^+, pH, K^+ and dodecylbenzene sulfonate. Here, the conducting polymer was dissolved in the PVC-based ion-selective membrane and cast on electronic conductors based on carbon, platinum, silver or gold. Solid-state ISEs using conducting polymer-based sensing membranes (CPISEs) were developed for determination of inorganic ions (K^+, Zn^{2+}, Ca^{2+}, Ag^+, Mg^{2+}, Cu^{2+}, pH, NO_3^-, CO_3^{2-}, Cl^- and $H_2PO_4^-$) and organic compounds (dodecyl sulfate, amino acids, ascorbic acid, aniline, saccharides), as summarized in Table 8.

Although conducting polymers have been known in the field of potentiometric ion sensors for almost two decades now, there is still a continuing development going on. In fact, the analytical performance of solid-state ion-selective electrodes where conducting polymers are used as ion-to-electron transducer (solid-contact ISEs) has been significantly improved over the last few years. Of particular interest is the large improvement of the detection limit of such solid contact ISEs down to the nanomolar level. Two important reviews detailing the progress in the use of CP in potentiometric sensors have been published recently which shows the interest in this research area [144]. Further optimization of the solid contact as well as the ion-selective membrane will most certainly bring this technology to many commercial applications. A deeper knowledge about the response mechanism of conducting polymer-based sensing membranes is needed in order to further optimize their analytical performance.

Table 8. Solid-state ISEs with conducting polymer-based sensing membranes, based on date reported in [144]

Ions	Conducting polymer	Linear Dynamic range (M)	Detection Limit (M)	Slope (mV/decade)	Ref
H^+	PANI ultrathin films	pH 3-9	1.0×10^{-9}	55-59	145
K^+	PPy	10^{-7}-10^{-1}	$10^{-7.4}$	65.9	146
K^+, Cu	Poly(3-octylthiophene) (POT)	10^{-6}-10^{-1} and 10^{-7}-10^{-1}	5.8×10^{-5} and 6.8×10^{-5}	58 and 54	147
Mg^{2+}, Ca^{2+}	PEDOT	10^{-5}-10^{-1}	1.0×10^{-5}	29.1 and 28.6	148
Ca^{2+}	PPy	0.1-10^{-10}	10^{-5}	27.3	149
Pb^{2+}, Ca^{2+}	PEDOT	10^{-5}-10^{-2} and 10^{-4}-10^{-1}	5.0×10^{-5} and 5.0×10^{-4}	29.9 and 27.3	150
Ag^{+1}	PEDOT and PPy doped with sulfonated calixarenes	10^{-5}-10^{-1}	$10^{-5.1}$	55.7	151
Zn^{2+}, K^+	PPy	$10^{-5.6}$-10^{-1}	10^{-6}	58	152
CO_3^{2-}	[poly(1-hexyl-3,4-dimethyl-2,5-pyrrolylene)	10^{-4}-10^{-1}	2.6×10^{-4}	-29.5	153
Cl^-	POT	10^{-4}-10^{-1}	5.0×10^{-4}	-55.1	154
F^-, $H_2PO_4^-$	PANI	10^{-3}-10^{-1}	NR	-45 and -39	155
Amino acids	PANI	NR	NR	59.8	156
Ascorbic acid	PANI	NR	5×10^{-5}	NR	157
aniline	PANI	NR	NR	NR	158
saccharides	poly(aniline boronic acid)	NR	NR	NR	159
dodecyl sulfate	PPy	1.58×10^{-5} - 3.88×10^{-3} and 6.31×10^{-6} - 1.0×10^{-3}	1.50×10^{-5}	36.4 54.49	160

8. Biosensors

8.1. Introduction to Biosensors

The unprecedented interest in the development and exploitation of analytical devices for detection, quantification and monitoring of specific chemical species has led to the emergence of biosensors. The estimation of metabolites such as glucose, urea, cholesterol and lactate in whole blood is of central importance in clinical diagnostics. Biosensors represent a new trend emerging in the diagnostic technology [161].

A biosensor is an analytical device, which converts a biological response into an electrical signal. It consists of two main components: a bioreceptor or biorecognition element, which recognizes the target analyte and a transducer, for converting the recognition event into a measurable electrical signal. A bioreceptor can be a tissue, microorganisms, organelles, cells, enzymes, antibodies, nucleic acids and biomimic and the transduction may be optical, electrochemical, thermometric, piezoelectric, magnetic and micromechanical or combinations of one or more of the above techniques (see section on main concepts of chemical sensors and biosensors) [162].

Figure 23 shows schematic diagram of a biosensor. The bioreceptor recognizes the target analyte and the corresponding biological responses are then converted into equivalent electrical signals by the transducer. The amplifier in the biosensor responds to the small input signal from the transducer and delivers a large output signal that contains the essential waveform features of an input signal. The amplified signal is then processed by the signal processor where it can later be stored, displayed and analyzed. Biosensor instruments are specific, rapid, simple to operate, can be easily fabricated with minimal sample pretreatment involved. The apparently alien marriage of two contrasting disciplines combines the specificity and sensitivity of biological systems with the computing power of microprocessor. Biosensors have been widely applied to a variety of analytical problems in medicine, the environment, food, process industries, security, and defense.

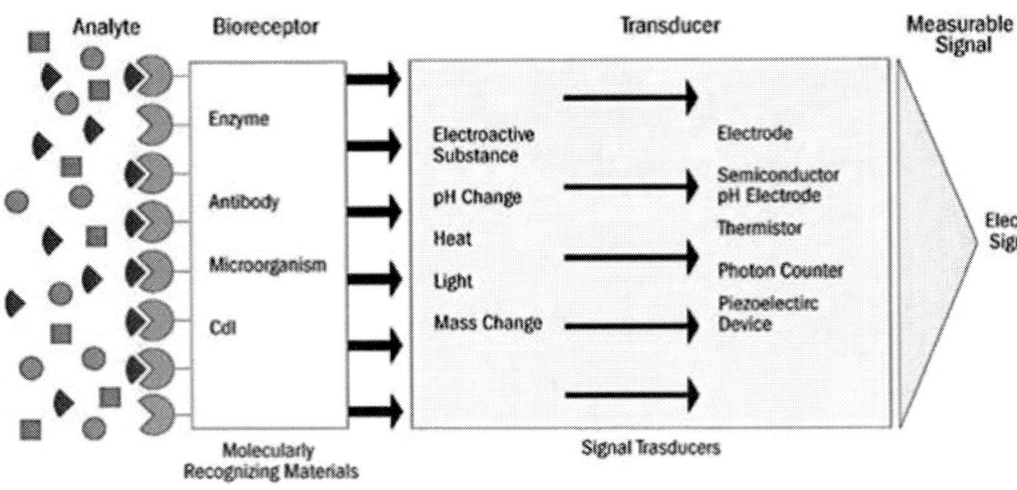

Figure 23. Schematic presentation of biosensor; Reprinted with permission from [162].

8.2. Generations of Biosensors

Depending on the level of integration, biosensors can be divided into three generations, i.e., the method of attachment of the biorecognition element or the bioreceptor molecule to the base of the transducer element [163]. In the first generation, the biorecognition element or the bioreceptor molecule is either bound to or entrapped in a membrane, which in turn is fixed on the surface of the transducer. The mediated or second-generation biosensors use specific mediators between the reaction and the transducer to improve sensitivity. It involves the adsorption or covalent fixation of the biologically active component to the transducer surface and permits the elimination of semi-permeable membrane. In the case of third-generation biosensors or direct biosensors, it is the direct binding of the bioreceptor molecule to the sensor element, and thus the bioreceptor molecule becomes an integral part of the biosensor.

So no normal product or mediator diffusion is directly involved in this. Conducting polymer-based biosensors come under this category.

8.3. Components of Biosensors

A biosensor consists of two primary components: a bioreceptor and a transducer [162].

8.3.1. Bioreceptors

Bioreceptors are classified into five different major categories. These categories include antibody/antigen, enzymes, nucleic acids/DNA/RNA, cellular structures/cells, and biomimetic. The enzymes, antibodies, and nucleic acids are the main classes of bioreceptors which are widely used in biosensor applications. Though the enzymes are one of the biorecognition elements, they are mostly used to function as labels than the actual bioreceptor.Within a biosensor the recognition biomolecule incorporated possesses an exquisite level of selectivity but is vulnerable to extreme conditions such as temperature, pH and ionic strength. Most of the biological molecules have very short lifetime in solution phase. Thus they have to be fixed in a suitable matrix. A number of techniques such as physical adsorption, cross linking, gel entrapment, covalent coupling etc. have been used to immobilize biological molecules in carrier materials. Various matrices have been used for the immobilization of enzymes such as membranes, gels, carbon, graphite, silica, polymeric films etc. There is thus a great need to design the electrodes that are compatible with the biological component that can lead to rapid electron transfer at the electrode surface. Conducting polymers are attractive as possible materials for such applications.

8.3.2. Transducers

The transducer plays an important role in the detection process of a biosensor. In case of conducting polymer based biosensor, the conductive polymer acts as a transducer that converts the biological signal to an electrical signal. Biosensors can also be classified based upon the transduction methods they employ. Wide varieties of transduction methods have been developed in the past decade for the detection of foodborne pathogens. Although there are new types of transducers constantly being developed for use in biosensors, the transduction methods such as optical, electrochemical, and mass based are given importance here since these are the most popular and common methods. Each of these three main classes contains many different subclasses and they can be further divided into label and label-free (nonlabeled) methods, where, the labeled methods depend on the detection of a specific label (e.g., fluorescence) and the label-free detection is based on the direct measurement of a product developing during the biochemical reactions on a transducer surface.

Optical biosensors are based on the measurement of light absorbed or emitted as consequence of a biochemical reaction. In such type of biosensors, light waves are guided by means of optical fibers to suitable detectors. These types of biosensors have been used for the detection of pH, O_2 and CO_2 etc. The most common electrochemical transducer being utilized is amperomertic and potentiometric. An amperometric biosensor measures the current produced during the oxidation or reduction of a product or reactant usually at constant applied potentials. Such sensors have fast response times and good sensitivity. Potentiometric biosensors relate electrical potentials to the concentration of analyte by using an ion sensitive electrode or a gas sensing electrode as the physical transducer. These are selective, have high dynamic range and are non-destructive.

In mass sensitive transducers, the adsorption of an analyte causes the mass of the crystal to increase resulting in altered frequency of oscillation. Such type of biosensors has been utilized for the measurement of ammonia, hydrogen, methane, carbon monoxide, nitrous oxide and other organophosphorous compounds.

8.4. Importance of Conducting Polymers to Biosensors

Conducting polymershave attracted much interest as suitable matrices of biomolecules and can be used to enhance stability, speed, sensitivity and hence are finding increasing use in medical diagnostics. The interest in conducting polymers in biosensor applications has grown over the past decade and there are many excellent reviews discussing the importance of conducting polymers to biosensors. We will just summarize some these attractive features:

- Availability of varied range of monomer types.
- Availability synthetic analogues of monomers.
- Composites can be prepared combing conducting polymers with non-conducting polymers or with non-polymer materials such carbon, carbon nanotubes, metals, etc.
- It can be prepared both electrochemically and chemically.
- It can be prepared in a range of soluble and insoluble forms.
- It has unique electrical, electronic, magnetic and optical properties.
- Compliance with micro and nanoscale fabrication.
- Compatibility with diverse range of fabrication techniques such as electrochemical, optical, mass-based, etc.
- Biomaterials such as enzymes, antibodies, whole cells, and nucleic acids can be incorporated into the polymer matrix.
- Strong biomolecular interactions.
- Low detection limits.
- Enhanced sensitivity (when used as a composite material with nanoparticles).
- Reversible responses at ambient temperatures.
- Cost effectiveness.

8.5. Conducting Polymers Based Biosensors

Conducting polymers have been used in the fabrication of biosensors in various fields such as:

- Health care: In medical diagnosis (glucose, fructose, lactate, ethanol, cholesterol, urea etc.) Immunosensors: Can be used in medical diagnostics and environmental sensors
- DNA sensors: In the detection of various genetic disorders.

- Environmental monitoring: For control of pollution and detection of hazardous chemicals in biosensors (polyphenols, sulfites, peroxides, formaldehyde etc.)
- Food analysis: For detection of glucose, fructose, ethanol, sucrose, lactate, malate, galactose, citrate, lactose, urea, starch etc. in food industries.

We have adopted the areas of applications of conducting polymers to biosensors from [164] and we further extend the ideas and research directions.

8.5.1. Biosensors for Health Care

There is an urgent need in the medical diagnostic laboratories for accurate, fast and inexpensive devices which can be routinely used. In this context biosensors are considered to provide viable solution to the problems posed by the contemporary industry. This is because these biosensing devices offer considerable advantages such as specificity, small size, and faster response and cost. A number of biosensors based on metabolites are available for monitoring chemically important parameters viz. blood glucose, urea lactate, cholesterol and uric acid, etc. These biosensors offer an advantage to extra laboratory analysis of relevant substances for clinical analysis.

8.5.1.1. Glucose Biosensors

Diabetes mellitus is one of the principal causes of death and disability in the World, and is highly responsible for heart disease, kidney failure, and blindness. This metabolic disorder results from insulin deficiency and hyperglycemia and is reflected by blood glucose concentrations higher or lower than the normal range of 80-120 mg/dL (4.4-6.6 mM). About 200 million people in the world are afflicted with diabetes mellitus. This figure is expected to rise up to more than three hundred million by 2030. Frequent testing of physiological blood glucose levels to avoid diabetic emergencies, is crucial for the confirmation of effective treatment. Actually, millions of diabetics test their blood glucose levels daily; making glucose the most commonly tested analyte. Indeed, glucose biosensors account for about 85% of the entire biosensor market. Such huge market size makes diabetes a model disease for developing new biosensing concepts.

Therefore, the development of high sensitive, low-cost, reliable glucose sensors having an excellent selectivity has been the subject of concern for decades, not only in medical science but also in the food industries. Glucose oxidase (GOx)-based glucose biosensors have prevalently had a hold on the glucose sensor research and development over the last four decades and the market place as well. This is due to the high demand of sensitive and reliable blood glucose monitoring in biological and clinical aspects. There are still some disadvantages of enzyme-based glucose determination. Examples include complicated enzyme immobilization, critical operating conditions such as optimum temperature and pH, chemical instability, and high cost.

8.5.1.1.1. Historical Perspectives of Glucose Biosensors

Although a variety of glucose sensors are available, the glucose biosensor has changed little in principle over the years as explained in the following [165,166] (Figure 24):

8.5.1.1.1.1. First Generation Glucose Biosensor

The concept of the biosensor for measuring glucose levels was first proposed in 1962 by Clark and Lyons from the Children's Hospital of Cincinnati [167]. This glucose biosensor was composed of an oxygen electrode, an inner oxygen semipermeable membrane, a thin layer of GOx, and an outer dialysis membrane. Enzymes could be immobilized at an electrochemical detector to form an enzyme electrode. A decrease in the measured oxygen concentration was proportional to the glucose concentration.

Updike and Hicks significantly simplified the electrochemical glucose assay by immobilizing and thereby stabilizing GOx. They immobilized GOx in a polyacrylamide gel on an oxygen electrode for the first time and measured glucose concentration in biological fluids [168]. However, the main problem with the first-generation of glucose biosensors was that the amperometric measurement of hydrogen peroxide required a high operation potential for high selectivity.

8.5.1.1.1.2. Second Generation Glucose Biosensors

The limitations of the first-generation glucose biosensors were overcome by using mediated glucose biosensors, *i.e.*, second-generation glucose sensors. The improvements were achieved by replacing oxygen with non-physiological electron acceptors, called redox mediators that were able to carry electrons from the enzyme to the surface of the working electrode. A reduced mediator is formed instead of hydrogen peroxide and then reoxidized at the electrode, providing an amperometric signal and regenerating the oxidized form of the mediator. A variety of electron mediators, such as ferrocene, ferricyanide, quinines, tetrathialfulvalene (TTF), tetracyanoquinodimethane (TCNQ), thionine, methylene blue, and methyl viologen were used to improve sensor performance.

The first research on the amperometric determination of blood glucose using a redox couple-mediated, GOx-catalyzed reaction was demonstrated in 1970. However, this study did not lead to the rapid application of amperometry in self-monitoring of blood glucose (SMBG) in the home setting.

8.5.1.1.1.3. Third Generation Glucose Biosensors

The third-generation glucose biosensors are reagentless and based on direct transfer between the enzyme and the electrode without mediators. Instead of mediators with high toxicity, the electrode can perform direct electron transfers using organic conducting materials based on charge-transfer complexes. Therefore, third-generation glucose biosensors have led to implantable, needle-type devices for continuous *in vivo* monitoring of blood glucose. Conducting organic salts, such as tetrathiafulvalene-tetracyanoquinodimethane (TTF-TCNQ), are known to mediate the electrochemistry of pyrrole-quinolinequinone enzymes (GDH-PQQ) as well as of flavoproteins (GOx). And the absence of mediators provides the biosensors with superior selectivity. However, only a few enzymes including peroxidases have been proved to exhibit direct electron transfer at normal electrode surfaces.

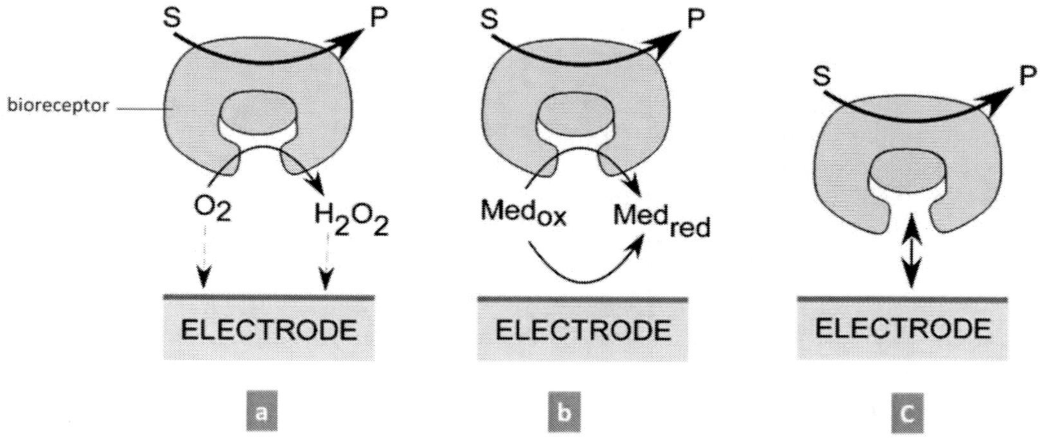

Figure 24. Three generations of glucose of biosensors based on (a) the use of natural cofactor, (b) artificial redox mediator, (c) direct electron transfer between GOx and the electrode. Abbreviations; S, substrate; P, product; Med_{ox}, mediator in the oxidized form; Med_{red}, mediator in the reduced form. Reproduced with permission from [169].

Enzymatic biosensors utilize the biospecificity of an enzymatic reaction, along with an electrode reaction that generates an electric current or potential difference for quantitative analysis. Enzymatic biosensors utilize the biochemical reactions, i.e. analyte and enzyme resulting in a product (hydrogen/hydrogen peroxide/hydroxyl/ammonium ion) that can be detected and quantified using a transducer (amperometric/potentiometric/optical thermal/piezoelectric). In general, many oxidoreductases including glucose oxidase catalyze the oxidation of substrates by electron transfer to oxygen to form hydrogen peroxide. These oxidoreductase enzymes can be immobilized on conducting polymer surfaces and the H_2O_2 formed as a result of enzyme and the corresponding analytes may be measured amperometrically. However, it has not been possible to discriminate between the direct electron transfer from the oxidation of hydrogen peroxide at polymer surface and that at the underlying electrode. Since conducting polymers are insoluble in aqueous solutions, electropolymerization has been frequently used to create a matrix for immobilization of enzymes at the electrode surface, and the sensor response was obtained by the oxidation of hydrogen peroxide. Belanger et al. suggested that the reaction of hydrogen peroxide and polypyrrole decreases the electrical conductivity [117]. Overpotential can be reduced by using mediators. Mediators are small molecules that shuttle between electrode and analyte to accelerate electron flows and the formal potential of the mediator should be close to or positive to that of the analyte.

In conventional glucose sensors without conducting polymer mediators, the reduced glucose oxidase drives the conversion of O_2 to H_2O_2, in the presence of O_2. H_2O_2 can then undergo oxidation to give H_2O. On the other hand, in the case of conducting polymer mediators, the oxidation of a reduced chemical species/polymer material can be directly measured by an electrode surface. Thus, conducting polymers can be considered as "wiring" materials for the electrochemical activation of redox enzymes. Redox enzymes usually have poor direct electrical conductivity contact and communication with metallic electrodes. Conducting polymer provides the matrix for electron transfer from the enzyme to the electrode, correlated with changes in the electrochemical (I-V) signal.

Attempts have been made to improve GOx entrapment by investigating the ion exchange capabilities, optimum temperatures, GOx concentration, pH, etc., and, particularly, increasing adsorption ability by increasing the effective surface area of the polymer film. Fabricating microporous films by ion exchange or by other means of chemical interference to increase the enzyme entrapment has been tried [171]. Another novel approach is to use a nanoporous electrode to fabricate a PPy film by introducing artificial nanoporosity into the film from the electrode [172]. This is a unique fabrication procedure for nanoporous PPy films and has led to a very simple thin-film fabrication method with all other requirements of a quality biosensor with a very high loading capacity of GOx. Thus, more enzymes are immobilized, and H_2O_2, which is produced as a result of the reaction catalyzed by the enzyme, can reach the electrode at a high rate for sensing.

Microencapsulation of enzymes into conducting polymer microtubules represent a good choice in view of generality and able to contain all kind of enzyme. It also reduces the degradation of enzyme activity in the immobilization process because enzyme can freely move in the pore of the capsule. Conducting polymer microcapsules were employed to take these advantages by Parthasarathy and Martin [173]. They showed that enzyme immobilization in the microcapsule is possible and the capsules are densely aggregated. They succeeded in measuring substrate like glucose with spectrophotometric method. Park et al. managed to design a microtubular array glucose biosensor based on PEDOT microtubules. The hollow tubules were loaded with glucose oxidase dispersed in aqueous buffer solution. To imprison the glucose oxidase into the cell, the opening of the tubule was sealed with PEDOT/PSS composite cap. This glucose sensor showed good selectivity and sensitivity by this enzyme imprisonment.

The surface plasmon resonance (SPR) optical technique is a powerful tool for characterizing the dielectric properties of surfaces, interfaces, and thin films. In biosensor applications, this technique allows for the investigation of biomolecular adsorption/desorption events at surfaces and provides in situ time-dependent surface coverage measurements without the need for labeling. The combination of the electrochemical and SPR techniques (EC-SPR) allows one to get the electrochemical and optical properties simultaneously on surfaces and ultrathin films at the nanometer scale. This involves the in situ monitoring of the film thickness and electrochromic properties during the anion doping/dedoping process of deposited conducting polymers. Baba et al. reported an EC-SPR glucose biosensor by immobilization of GO*x* (negatively charged) a water-soluble, conducting N-alkylaminated polypyrrole (positively charged) using the layer-by-layer (LbL) electrostatic self-assembly technique [174]. The electrochemical and optical signals were simultaneously obtained from the composite LbL enzyme electrode upon the addition of glucose as mediated by the electroactivity and electrochromic property of the polypyrrole layers.

The activity of enzymes is obviously affected by the temperature, pH, humidity, and toxic chemicals. Therefore, the determination of glucose with an enzyme free sensor has been a dream for many years.The interest in practical non-enzymatic glucose sensor has been centered on the efforts to find out the breakthrough in electrocatalysis. In this context, various trands have been investigated to improve the electrocatalytic activity and selectivity toward the oxidation of glucose. This can be done by using: (i) inert metals such as Pt, Au and Ni; (ii) metal alloys containing Pt, Au, Pb, Ir, Ru, Cu, and Pd; and (iii) metal-dispersed carbon nanotubes (CNTs) framework in which Pt, Pb, Pd, or Au are mixed with CNTs to form nanocomposites; (iv) nano-scale metal oxides, noble metal-doped metal oxides, metal oxides-CNTs nanocomposites. However, these materials are unsatisfactory with regards to sensitivity and selectivity, are high costing, and show quick loss of activity by adsorption and accumulation of intermediates or chloride ions.

In recent years, an increasing number of researchers have explored conducting polymer based non-enzymatic glucose biosensors [175]. The catalytic ability of conducting polymers for the nonenzymatic direct electrooxidation of glucose is one of the attractive properties in glucose detection allowing minimum fabrication cost and stable glucose sensors. Ozcan et al. investigated an enzymeless biosensor, based on electrodeposition of overoxidized polypyrrole nanofiber onto pencil graphite electrode and modified with cobalt(II) phthalocyanine tetrasulfonate (CoPcTS). CoPcTS showed electrocatalytic activity for the oxidation of glucose in alkaline solution. The resulting biosensor exhibited excellent performance for glucose determination with a wide linear range (0.25–20 mM), a highly reproducible response (R.S.D. of 2.7%), low percentage of the interferences and long-term stability. The results were satisfactory and agreed closely with those measured in a hospital. A novel enzymeless glucose sensor based on overoxidized polypyrrole(oPPy) modified on a self-made nickel microelectrode has been fabricated by Yu et al. [176].The electrocatalytic oxidation of glucose in alkaline medium directly at the modified microelectrode has been investigated and which shows excellent electrocatalytic activity towards glucose oxidation at low positive potential of 0.47 V(vs. AgAgCl). The calibration curve is linear in the concentration range of 5.0×10^{-6} -1.1×10^{-3} M with a detection limit of (SN=3) 2.4×10^{-6} M. The glucose sensor has short response time (within 3 s) and can be renewed easily in a reproducible manner by a simple polishing step.

8.5.1.2. Urea

Urea is one of the final products of protein metabolism. In clinical analysis and dairy industry urea is a very significant parameter as excess in blood serum from its permissible range causes dysfunction of the kidney. Hence, its analysis is significant and is carried out frequently in various laboratories [177]. The normal level of urea in serum is from 15 to 40 mg/dl (2.5–7.5 mM/l). In patients suffering from renal insufficiency, urea concentrations in serum vary from 180 to 480 mg/dl and, at elevated levels above 180 mg/dl, hemodialysis is required [177]. Apart from clinical applications, there is a growing demand for robust, reliable instrumentation toward the estimation of urea in other fields (e.g., food science and environmental monitoring).

Urea biosensors are collectively formed by biorecognition layer – biocatalyst i.e. urease, a biotransducer – to read change in physical environment and convert into electronic signal. For making biotransducer and choice of substrate, conducting polymers are most widely being researched. These materials have a key advantage in facilitating the conversion of physiochemical change into an electronic signal. The enzyme immobilized to the electrode surface catalyzes the hydrolysis of the urea, in an overall reaction leading to the formation of ammonium, bicarbonate and hydroxide ions as shown below:

The ionic products of the above reaction change the electronic properties of the biosensor electrode (modified with conducting polymers), which can be observed by various electrochemical techniques including conductometric, impedimetric, amperometric and potentiometric as will be shown.

Conducting polymers are most suitable for conductometric biosensor fabrication as their conductivity is sensitive to surrounding ions and their mobility. The change in film conductivity can be calculated by four probe or two probe DC techniques. These biosensors can be fabricated by putting enzyme layer over the polymeric transducer. Jha et. al. [178] used electropolymerized polyaniline-polystyrene (PANI-PSS) with brevibacterium ammoginase as a biocatalyst for urea biosensor. The immobilized cell works as a source of unpurified urease to develop conductometrc urea biosensor. The pH dependent change in resistivity of the polymer was also used as a basis for sensing. The response was linear over a range of 0-75 mM urea with a sensitivity of 0.125 mM^{-1}. Ortega et. al. [179] reported a urea biosensor based on NH_3 gas interaction by two probe method with DBSA-doped polyaniline-poly(styrene-co-potassium acrylate in a poly(n-butyl methacrylate) matrix. He also estimated urea concentration normal to pathological level.

Amperometry is most commonly used technique for biosensors based on conducting polymers. Urea biosensor is the typical example of biocatalytic amperometric biosensor where ammonium ion which is a product of biocatalytic reaction interacts with polymer to induce a change in conductivity of the polymer. Wallace and co-workers [180] developed an amperometric urea biosensor based on flow injection analysis (FIA) to get linear responses between 3-15 mg/l at the optimum condition of pH and potential of buffer and electrode respectively. The presence of interfering ions (KCl and KNO_3) at the concentration up to 0.01 M did not interfere with the estimation of the urea in blood and urine samples without any pretreatment. Rajesh et al [181] reported an amperometric urea biosensor with urease covalently immobilized with conducting copolymer poly(N-3-aminopropyl pyrrole-co-pyrrole) film. The electrode gave a linear response range of 0.16-5.02 mM.

Estimation of the urea using potentiometric techniques is also widely used because of a simpler mode of construction of the sensor electrode, detection and general availability of the instrumentation required for their utilization. Conducting polymers are widely studied for estimation of the urea by pH sensitive potentiometric technique. These polymers not only act as good transducers but also unique support to the immobilization of the enzyme due to having capability for direct electron transfer between enzyme and polymer. Ahuja et al. developed a potentiometric urea biosensor based on bovine serum albumin (BSA) embedded surface modified [182]. It was found that the electrode responds to low urea concentration with wider range of detection. The electrode showed a linear response range of 6.6×10^{-6} to 7.5×10^{-4} M urea. The response time is about 70–90 s reaching to a 95% steady-state potential value and 75% of the enzyme activity is retained for about 2 months

8.5.1.3. Lactate Biosensors

Lactate is the key metabolite of the anaerobic glycolytic pathway. Reliable measurements of l-lactate are of great interest in clinical chemistry, the dairy and wine industry, biotechnology, or sport medicine. In particular, blood lactate levels are indicative of various pathological states, including shock, respiratory insufficiencies, and heart and liver diseases. In general, determination of l-lactate is based on spectrophotometry and chromatography; but these methods involve complicated procedures and the cost is high because expensive enzyme must be used in each assay. Biosensors have been considered to be among the devices most suitable for biochemical analysis due to their good selectivity, fast response, miniature size, and reproducible results. Over the past few years, various approaches based on lactate oxidase (LOD) or lactate dehydrogenase (LDH) have been developed for the detection of lactate. They employ different methods to attach enzymes on the sensing layer. Among others, many studies dealing with lactate electrochemical biosensors based on conducting polymers have been reported. As such, the demand for simple, inexpensive and selective l-lactate sensors continues.

The biosensor response is based on LOD or LDH enzymes. LOD, for example, catalyzes the oxidation of lactate to pyruvate. In presence of dissolved O_2, the enzyme can be reoxidized, releasing H_2O_2. This last product can be oxidized at the electrode surface restoring the former concentration of O_2 and giving a current proportional to the amount of dissolved lactate. Korneyeva et al. reported a lactate biosensor based on LOD immobilized in polypyrrole layer on electrodes modified with Prussian blue [183]. The analytical characteristics of the resulting biosensor are as follows: a sensitivity of 190 ± 14 mA M^{-1} cm^{-2}, a linear dynamic range of 5×10^{-7} to 5×10^{-4} M, and high operational stability. The applicability of a lactate biosensor for food quality control (for example, quality control of kvass) was shown.

Carboxylic acid functionalized conducting polymer poly-5,2′-5′,2″-terthiophene-3′-carboxylic acid (PTTCA) has advantages to immobilize biomolecules through covalent bond formation and, consequently, offers stable and sensitive biomolecule detection. Rahman et al. developed an amperometric lactate biosensor based on PTTCA and multiwall carbon nanotube (MWNT) composite on a gold electrode [184]. LDH was subsequently immobilized onto the pTTCA/MWNT composite film. The results showed that the sensitivity, stability, and reproducibility of the sensor improved significantly using pTTCA/MWNT composite film. The applicability of the sensor for the analysis of l-lactate concentration in commercial milk and human serum samples was demonstrated successfully.

Haccoun et al. described a reagentless lactate biosensor based on an electropolymerized copolymer film poly(5-hydroxy-1,4-naphthoquinone-co-5-hydroxy-3-acetic acid-1,4-naphthoquinone) [185]. The quinone group, as part of the polymer backbone, is electroactive and very stable in neutral aqueous medium. It can therefore act as an immobilized mediator for the LOD enzyme recycling, at a working potential much lower than those commonly reported in the literature for other mediators. Some well-known interferents are shown to have no measurable effect on the amperometric curves.

8.5.1.4 Cholesterol Biosensors

The increasing incidences of cardiovascular diseases and cardiac arrest are major cause of death of humans all over the world. One of the most important reasons is hypercholesterolemia, i.e. increased concentration of cholesterol in blood. Hence estimation of cholesterol level in blood is important in clinical applications [186]. A high cholesterol level in human blood is related to arteriosclerosis, hypertension, myocardial infarction and many heart related disorders. The alarming rise in the rate of clinical disorders due to abnormal levels of cholesterol in blood have stimulated public concern about the determination of cholesterol level in the industrialized world. Various analytical methods such as colorimetric and spectrophotometric have been used for estimation of cholesterol. These methods involve complicated procedures and are expensive since high volume of enzyme is needed in each assay. There is considerable interest towards the application of conducting polymers to biosensors. This has been attributed to their many interesting properties such as biocompatibility, redox characteristics and the possibility of direct electron transfer between electrode and active sites of biomolecules.

Conducting polymer based cholesterol biosensors have recently attracted much attention. Singh et al. have reported a cholesterol biosensor based on PPy films and have shown that the electrode shows stability of 1 month [187]. Later, these authors have used the electrochemically prepared polyaniline (PANI) film for fabrication of cholesterol biosensor and have shown that electrode based on PANI is thermally stable up to 46 ◦C [188].

Malhotra and coworkers employed PANI monolayers formed by Langmuir-Blodgett technique as immobilization matrix for cholesterol oxidase [189]. The results revealed that this biosensor has linearity from 25 to 400 mg/dL of cholesterol concentration with sensitivity of 88.9 nA mg^{-1} dL. Increased interaction between cholesterol and cholesterol oxidase was observed and authors discussed the advantage of this biosensor as a potential candidate for efficient cholesterol sensing. Later on, the same group has developed a cholesterol biosensor by covalently coupling cholesterol oxidase onto polyaniline in the presence of a non-ionic surfactant [190]. The results of response measurements carried out on this biosensor using amperometric and photometric techniques, reveal detection limit as 5 mg/dl, linearity from 5 to 400 mg/dl of cholesterol and sensitivity as 131 μA/(mg/dl cm^{-2}). This biosensor can be used about 20 times and have a shelf-life of about 10 weeks when stored at 4 °C. Attempts have also been made to utilize this biosensor for estimation of free cholesterol concentration in serum samples.

Other conducting polymers have been used such as poly(3-thiopheneacetic acid), polypyrrole, poly(3,4-ethylenedioxythiophene and Poly (pyrrole-co-*N*-methyl pyrrole) and interesting results were obtained.

It is apparent that increasing incidences of cardiovascular diseases and cardiac arrest in contemporary society have necessitated the availability of cholesterol biosensors. However, only a few have been successfully launched in the market [186]. One of the reasons lies in the optimization of critical parameters, such as enzyme stabilization, quality control and instrumentation design. The cholesterol biosensor must be easy to use, self-testing, fast response and portable, etc. Increased understanding of the immobilized bioreagents, improved techniques for immobilization and technological advances in the microelectronics are likely to speed up commercialization of the much needed cholesterol biosensor.

8.5.2. Immunosensors

Immunosensors have generated great interest with expectations of providing fast and highly sensitive detection of proteins including antibodies that are generated during the immune response towards infections. For this reason, recently, immunosensors found widespread applications in clinical diagnostics, food safety, food quality, biological analysis, and environmental monitoring. In recent past years, researchin this field has evolved quickly with the aim of improving theperformance of the biosensors (specificity, stability, sensitivity,detection limit). Recent studies have focused on the use of conductingpolymers as tools for amplification of the response signal fromantibody–antigen interactions, which leads to the improvement ofthe performance characteristics of the biosensor.

Immunosensors act on the principle that the immune response of certain biological species (usually bacteria) to contaminants will produce antibodies, which in turn can be measured. Immunosensors are, thus, a combination of an immobilized immunological receptor used for detecting a target analyte by structural complementarity, and a transducer which converts a biological interaction into a measurable signal [191]. The term immunoassay is used for tests based on immunoreactions, while the term immunosensor is specifically employed to describe whole instruments, i.e. immunoreaction-based biosensors.

An effective combination of immunochemistry and electrochemistry in an analytical device could provide the basis for direct electrical detection of a wide range of analytes with great sensitivity and specificity. In this context, a number of immunosensors based on conducting polymers have been developed. Almost all materials that are more or less affine to analyte can be used in immunosensors. In immunosensors based on conducting polymers can be applied: (i) antibodies/antigens; (ii) animal tissues; (iii) whole cells; (iv) DNA; (v) molecular imprinting.

Antigen-antibody (Ag-Ab) complex formation in direct detection immunosensors can be detected electrochemically by potentiometric, amperometric, capacitive and amperometric. Because of simple miniaturization possibilities the electrochemical immunosensors are most suitable for the direct detection of analyte in the body of the patient investigated. A novel system device using multifunctional conjugated copolymer poly(5-hydroxy-1,4-naphthoquinone-co-hydroxy-2-thioacetic acid-1,4-naphthoquinone) acting both as immobilizing and transducing element for reagentless immunosensor has been constructed by Piro et al. Its functionality was evaluated in an antigen–antibody interaction model using ovalbumin–anti-ovalbumin. It was shown that the system specifically detects via electrochemical signal the antigen–antibody immune interaction in a reagentless context.

Out of the many biosensors that exist, the amperometric immunosensors with labeled enzymes is a highly sensitive and quick responsive biomolecule detecting method. Kim et al. fabricated an interesting amperometric chloramphenicol (CAP) immunosensor by covalently immobilizing anti-chloramphenicol acetyl transferase (anti-CAT) antibody on cadmium sulfide nanoparticles (CdS) modified-dendrimer that was bonded to the conducting polymer (poly 5, 2′: 5′, 2″-terthiophene-3′-carboxyl acid (poly-TTCA)) layer [192]. The AuNPs, dendrimers, and CdS nanoparticles were deposited onto the polymer layer in order to enhance the sensitivity of the sensor probes. The proposed immunosensors exhibited a linear range of CAP detection between 50 pg/mL and 950 pg/mL, and the detection limit was 45 pg/mL. The immunosensor was examined in real meat samples for the analysis of CAP. The glycoprotein, osteoproteogerin has been determined by an amperometric immunosensor formed by covalently immobilizing a monoclonal OPG antibody (anti-OPG) onto the gold nanoparticles (AuNPs) functionalized conducting polymer (5,2′:5′,2″-terthiophene-3′-carboxylic acid). The principle of immunosensor was based on a competitive immunoassay between free-OPG and labeled-OPG for the active sites of anti-OPG. HRP was used as a label that electrochemically catalyzes the H_2O_2 reduction. The catalytic reduction was monitored amperometrically at −0.4 V vs. Ag/AgCl. The immunosensor showed a linear range between 2.5 and 25 pg/ml and the detection limit was determined to be 2 pg/ml.

The use of fluorescence as an analytical technique in biochemistry, biology and biophysics has undergone a remarkable growth during the past 50 years. Fluorescence sensing of chemical and biochemical analytes is an active area of research. Ramanavicius et al. pioneered fluorescence-based immunosensors using polypyrrole matrix. These sensors were devoted to the detection of antibodies against bovine leukemia virus (BLV) protein gp51 (anti-gp51-Ab) [193].

Electrochemical impedance spectroscopy represents a powerful method for probing the interfacial reaction mechanisms of modified electrodes, providing a rapid approach for monitoring the dynamics of biomolecular interaction. The interaction between the proteins, mainly negatively charged at neutral pH, and the delocalized positive charges along the polymer chains induces changes capacitance of the material. Consequently, such interactions, evidenced from electrochemical measurement, are the basis of bioaffinity signal. Hafaid et al. explored the development of impedimetric immunosensors based on a conducting co-polypyrrole film to which a histidine tag antibody has been bound using a complex NTA chelator coordinated with copper as metal [194]. The results indicated that better immunosensor characteristics were obtained at an applied potential of −1400 mV where the reduction of copper occurs. The impedance obtained at this potential shows a linear variation of charge transfer resistance (Rct) with the specific antigen concentration in the range of 0.1–10 ng mL^{-1} with a detection limit of 0.1 ng mL^{-1}. Moreover, this study demonstrates that these immunosensors exhibit high sensitivity, stability and reproducibility.

Clearly, the use of conducting polymers in conjunction with bioaffinity reagents has provided a powerful route that expanded the range applications of electrochemical detection and its future development is expected to continue.

8.5.3. DNA Biosensors

The accomplishment of The Human Genome Project has stimulated our interest in identifying diseases based on the genetic information even before the abnormal symptoms can be noticed. Gene analysis plays an ever-increasing role in a number of areas related to human health such as diagnosis of infectious diseases, genetic mutations, drug discovery, forensics, and food technology. To fulfill these challenges, a new generation of gene sensors that are fast, reliable, and cost-effective needs to be developed. Conventional methods for the analysis of specific gene sequences are based on either direct sequencing or DNA hybridization. In the early 90s, gene array technologies appeared which rely on anchoring of specific probe DNA fragments or oligonucleotides (ODN) onto solid surfaces and detection of fluorescently or radioactively tagged analyte oligonucleotides that bind by Watson-Crick base pairing to the complementary probe sequence (hybridization) [195]. Such techniques have shortcomings arising from, for example, limited tagging efficiency, hazardous waste disposal, and complex mltistep analysis. A number of approaches that seek faster, sensitive, and label-free gene detection have been suggested using new detection techniques based on optical and electrochemical interactions.

Electrochemical DNA sensors are regarded as particularly suitable for direct and fast biosensing since they can convert the hybridization event into a direct electrical signal. Electrochemical DNA sensing approaches include the intrinsic electroactivity of DNA, electrochemistry of nanoparticles and conducting polymers.

The electronic structure of CPs is highly sensitive to changes in the polymeric chain environment and other perturbations in the chain conformation caused by, for example, a biological recognition event such as DNA hybridization. The changes in the delocalized electronic structure or in other CP properties are manifested in altered optical and electrical properties, and, when measured, can provide a signal for the presence of a target analyte molecule. A typical configuration for DNA sensors based on CPs is shown in Figure 25. Single-stranded DNA probes are immobilized on or within a conducting polymer layer. The target DNA is captured by base pairing to generate a recognition signal, which is recorded through an electrode (gold, platinum, glassy carbon, etc.). Conducting polymer is not only used as an immobilization matrix but also plays an active role in transduction. Because the recognition event takes place at the CP/electrolyte interface and the recognition signal generated reaches the transducer through the CP layer, the properties of the CP and the orientation of the immobilized DNA probes on the CP are crucial to the sensor performance. Amperometric, conductometric, impedimetric, and potentiometric based DNA sensors have been used.

In the amperometric approach the performance of biosensors is governed by the efficacy of the electron transfer between the biomolecules and the underlying electrode surface, involving the CP layer. DNA biosensor based on polyaniline (PANI)-polyvinyl sulphonate (PVS) has been fabricated using electrochemical entrapment technique for detection of organophosphorus pesticides (chlorpyrifos and malathion). These dsCT-DNA entrapped PANI-PVS/ITO bioelectrodes have been found to have response time of 30 s, stability of about 6 months and detection limit for chlorpyrifos and malathion as 0.5 ppb and 0.01 ppm, respectively.

Recent developments in nanomaterials and nanotechnologies create many opportunities to advance biomolecular sensing and gene detection. Particularly, use of nanomaterial as a medium of signal amplification attracts a great deal of attention. Change et al. [196] developed polyaniline nanotube array based ultrasensitive biosensor for DNA hybridization detection. This DNA electrochemical biosensor demonstrated an extremely high sensitivity that could detect the presence of a target nucleotide at a concentration as low as 1.0 fM, which corresponded to 300 zmol of target DNA molecules in 300 μL of sample and was much higher than other PANI biosensor with sensitivity of about 1 pM. This ultrasensitive property was interpreted in terms of the collective effect of PANI nanotubes, the enhanced PANI conductivity, and the faster hybridization kinetics of oriented nanostructure.

Conductometry measures the conductivity change of CPs arising between a pair of electrodes. Krishnamoorthy et al. electrochemically synthesized poly(3,4-ethylenedioxythiphene) microtubules in the presence of ssDNA using a polycarbonate membrane as a template, and developed a label-free conductometric DNA sensor based on these microtubules [197]. The authors investigated the effect of ssDNA length on the sensor response. The results showed that the sensor response, as well as the linear range, increased with an increase in the ssDNA length.

The immobilization of DNA probes and the hybridization event not only induce changes in the intrinsic properties of the CP film but also bring about changes in the various interfacial film properties, such as the capacitance and interfacial electron transfer resistance. EIS can detect these changes, which can be exploited in the development of truly label-less biosensing. Lien et al. described Multi-wall carbon nanotubes (MWCNTs)-doped polypyrrole DNA biosensor for label-free detection of genetically modified organisms by QCM and EIS [198]. An increase in DNA complementary target concentration results in a decrease in the faradic charge transfer resistance (Rct) and signifying "signal-on" behavior of MWCNTs-

PPy-DNA system. QCM and EIS data indicated that the electroanalytical MWCNTs-PPy films were highly sensitive

The unique electronic structures of CPs make them highly suitable materials for electrochemical label-free DNA detection. During the past ten years, numerous functionalized CPs have been developed and used in preparation of electrochemical DNA sensors with considerable sensitivity and selectivity. Conducting polymer-based electrochemical DNA sensors have shown applicability in a number of areas related to human health such as diagnosis of infectious diseases, genetic mutations, drug discovery, forensics and food technology due to their simplicity and high sensitivity [195].

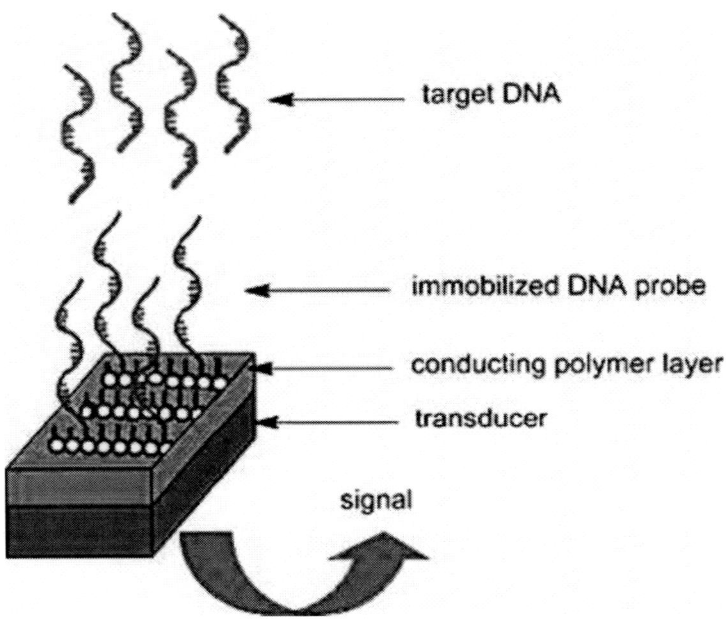

Figure 25. schematic, general DNA sensor design based on conducting polymers. Reproduced with permission from [195].

8.5.4. Biosensors for Environmental Monitoring

Environmental monitoring typically involves several steps such as sampling, sample handling, and sample transportation to a specialized laboratory and finally a laboratory to determine the chemical composition and to establish the toxic effect. These conventional approaches are expensive, time-consuming, and require highly trained personnel. Thus, the multiple steps involved frequently prevent rapid information about the composition and/or the toxicity of the sample to be obtained in efficient manner. Thus, remediation efforts could take months or years. Consequently, the need for fast, sensitive, selective, and cheap alarm systems is becoming more apparent. Presently, a great amount of scientific research has been devoted to designing such devices. Many reports emphasized the effectiveness of biological sensors for real-time monitoring of important environmental analytes [199]. The most studied classes of analytes include pesticides, polychlorinated biphenyls, and heavy metals.

The clear benefits derived from the use of a small, portable analyzer have prompted the application of electronic noses to environmental monitoring. The range of compounds posing significant hazards for humans as well as wildlife is very large. Researchers from the

University of Wollongong, Gordon Wallace and coworkers, have been interested in the development of a conducting polymer sensor array based on polypyrrole polymers to be used in an electronic nose designed to monitor a group of aromatic hydrocarbons called benzene, toluene, ethyl benzene, xylene (BTEX) [200]. A number of dopants have been investigated in terms of their ability to form suitable electrochemically deposited polymer films and to detect the analyte gases with sufficient sensitivity and reproducibility. A group of eight polymers has been found to meet most of the requirements for this application. Dojin Kim and coworkers from Chungnam National University explored polyaniline nanofibers for the detection of BTEX and chlorobenzene [201]. One-dimensional PANI nanofibers outperform the conventional PANI due to their high surface area, fast diffusion of gas molecules into and out of the nanofiber structures, and independence of film thickness due to porosity. The net results were the much faster response time and an order of magnitude higher response of the sensor. The sensor response to AOC was investigated in several hundred ppm ranges. The sensor showed an increase in conductivity, and the maximum response measured at 1000 ppm was several tens of percent.

Polychlorinated hydrocarbons (PCBs) are mixture of synthetic organic chemical commercially manufactured under the name Aroclors. PCBs are persistent organic pollutants and have entered the environment through both use and disposal. The environmental transport of PCBs is complex and nearly global in scale. The public, legal, and scientific concerns about PCBs arose from research indicating they were likely carcinogens having the potential to adversely impact the environment and therefore undesirable as commercial products. Despite active research spanning five decades, extensive regulatory actions, and an effective ban on their production since the 1970s, PCBs still persist in the environment and remain a focus of attention. Nowadays, there is an increased interest to detect these environmental chemicals from both scientific and regulatory communities. Consequently, many efforts are concentrating in order to develop a fast and reliable method for their determination.

The analysis of PCBs is currently carried out using traditional analytical methods such as gas or liquid chromatography. In spite of their performances, these methods are too expensive for screening purposes and generally require extensive sample preparation including extraction with organic solvents, clean-up, or separation. Fewer other expensive, but readily available techniques could be used for PCB detection. These include immunoassay (IA), enzyme (ELISA) and fluorescence. These techniques exploit the specific binding of a PCB to a unique antibody and are generally used for screening purposes since a large number of samples can be analyzed in a short time.

Immunosensors have been reported to exhibit considerable potential for PCB detection [202]. For instance, we have developed a PCB immunosensor, constructed by immobilizing anti-PCB antibody into a conducting polypyrrole (PPy) membrane. Pulsed-accelerated immunoassay for signal generation in stationary cell or FIA by applying a pulsed waveform between +0.60 and –0.60 V and a pulse frequency of 120 and 480 ms. With the optimized sensor, a linearity of 0.3–100 μg/l and a detection limit of 3.3, 1.56, 0.39, and 1.66 ng/ml, respectively, were obtained for Aroclors 1242, 1248, 1245, and 1016 [202].

Researchers at Nanjing University have recently prepared a novel PANI fiber by a simple electrodeposition method using fluorinated organic acid as dopant [203]. High extraction efficiency, high temperature resistance, reproducible synthesis, low cost and long life span are the main advantages of this fiber. The novel fiber has been demonstrated to be a suitable candidate for solid-phase microextraction (SPME) fiber for PCBs detection at ultra-trace level in water samples. Extraction efficiency of this fiber was comparable to commercial 30 μm PDMS but with thinner stationary, i.e. 10 μm. In another approach, Bachas and coworkers studied the reductive dechlorination of PCBs through the use of Pd nanoparticles immobilized in polypyrrole membranes [204]. This approach was endeavored in a hope to reduce the toxicity of PCBs. The toxicity results suggested that the dechlorinated PCB, unlike PCB, had no endothelial cell toxicity.

Heavy metals are metallic elements which have a high atomic weight and a density much greater (at least 5 times) than water. There are more than 20 heavy metals, but four are of particular concern to human health: lead (Pb), cadmium (Cd), mercury (Hg), and inorganic arsenic (As). According to the U.S. Agency for Toxic Substances and Disease Registry, these four heavy metals are four of the top six hazards present in toxic waste sites. They are highly toxic and can cause damaging effects even at very low concentrations. They tend to accumulate in the food chain and in the body and can be stored in soft (e.g., kidney) and hard tissues (e.g., bone). Being metals, they often exist in a positively-charged form and can bind on to negatively-charged organic molecules to form complexes. The analysis of heavy metals is performed with atomic absorption or emission spectrometry, mass spectrometry. Again, although these methods are sensitive and accurate, most of them require a tedious sample pretreatment, sophisticated performance and/or expensive equipment. Conducting polymers have emerged as promising materials in the development of compact and portable probes for the detection of biologically significant molecules [23].

Contractor and coworker from the Indian Institute of Technology developed a conductometric Hg^{2+} sensor using 2,3-dimercapto-1-propanol as receptor and polyaniline as transducer as well as immobilization matrix [204]. Using this approach, concentrations of Hg^{2+} ion as low as 10^{-12} M could be detected which is well below World Health Organization (WHO) limits in portable water for Hg^{2+} ion.

Michalska and coworkers at Warsaw University fabricated all-plastic ion-selective sensors for Ag^{+} and Pb^{2+} using poly(vinyl chloride)-based membranes and either PEDOT or PANI as transducer and electrical lead [205]. Both for Ag^{+} and Pb^{2+} selective electrodes, polyaniline results in super Nernstian type behavior observed for activities lower than 10^{-6} M and higher selectivity coefficients.

Researchers at Tokyo Institute of Technology and Bangladesh University of Engineering and Technology have developed nanogold-particle/polyaniline-modified glassy carbon electrode for the detection of As^{3+} in water [206]. The sensor showed a high sensitivity in detecting As^{3+} in water, and with stripping voltammetry, a limit of detection of 0.4 ppb arsenic was obtained, which is much lower than the arsenic guideline limit of WHO (10 ppb).

The Department of Energy's Pacific Northwest National Laboratory announced two years ago that they developed a new rapid, portable and inexpensive detection system that identifies personal exposures to toxic lead and other dangerous heavy metals [207]. The device can provide an accurate blood sample measurement from a simple finger prick, which is particularly important when sampling children.

8.5.5. Biosensors for Agriculture and Food Industry

In the agriculture and food industry, sensors are being used for process control, monitoring quality, and assessing safety. In this context, the application of biosensor technique may be an alternative to the present inefficient and expensive methods. Conducting polymers are promising materials for the fabrication of the sensitive elements for sensor devices required for the control of agriculture and food industries. In the following, we will present some example reported recently in literature.

In the process of preservation of animal protein food (milk, meat, fish products) under un-satisfactory conditions, a loss of freshness takes place. This process is accompanied by the evolution of gaseous substances such as ammonia, hydrogen sulfide, nitrogen oxides, and others.Tsizh et al. have proposed sensor devices based on conducting polymer films (polyaniline, poly-o-methoxyaniline, and poly-*o*-toluidine) for the visual express-control over animal product freshness [208]. The sensor action is based on a change in the spectral properties and the corresponding color transitions in conjugate polymer films on the transparent surface at a low partial pressure of ammonia which is evolved at the loss of protein product freshness.

John Lavigne and colleagues from the University of South Carolina described the need for better sensors that can detect food spoilage caused by E. coli, Salmonella, and other disease-causing bacteria [209]. Existing methods, such as "electronic noses" and "electronic tongues," require expensive equipment, are time consuming and involve complicated analyses. In the study, they describe development of a polymer material that raises a red flag, changing color in the presence biogenic amines, compounds produced by the bacterial decay of food proteins. In laboratory tests, the polymer identified and distinguished between 22 different kinds of key food-spoilage amines with 97 percent accuracy.

The world annual grain production is about 2 Gt (billion metric tonnes). Stored grain bulks are ecological systems where communities of insects, mites and microflora interact with abiotic variables to cause spoilage. While the importance of proper grain storage has been realised, post-harvest losses continue to range from 9% in North America to 50% in developing countries. Any loss in quality or quantity of the produced grain has far-reaching economic effects. During the spoilage of grain, CO_2, moisture and heat are produced by the metabolism of the grain, fungi, insects and mites, and distinct odours are released. Increased levels of CO_2 in a stored grain bulk indicate that insects, mould, or excessive respiration are present. Evolution of excessive CO_2 from a stored grain mass is a reliable indicator of deteriorating grain. Additionally, food packaging is often done under a modified CO_2 specific to a product and with a purpose to prevent microbial spoilage. Carbon dioxide is widely used in modified-atmosphere packaging, and a decrease in its concentration is a sign of leakage in a package. Freshness and safety of modified-atmosphere food packages can be determined by detection of CO_2 concentrations. Researchers at the University of Manitoba and Cereal Research Centre have developed a CO_2 sensor using polyaniline boronic acid (PABA) as the electrically conductive region of the sensor and was demonstrated for use in detecting incipient or ongoing spoilage in stored grain [210]. The developed sensor measured gaseous CO_2 levels in the range of 380–2400 ppm of CO_2 concentration levels. The sensor was able to respond to changes in CO_2 concentration at various humidity and temperature levels. The response of the PABA film to CO_2 concentration was not affected by the presence of alcohols and ketones at 1% of vapour pressure, proving that the developed sensor is not cross-sensitive to these compounds which may be present in spoiling grain. The methodology used for building the CO_2 sensor device is compatible with the MEMS process (printed circuit board process) and can be mass manufactured using screen printing technology.

As the demand for food quality, health benefits, and safety increases, more stringent scrutiny on the inspection of agri-food products have become mandatory. Also being increasingly demanded is “traceability”, which requires not only rigorous inspections, but also systematic detection, labeling and recording of quality and safety parameters while archiving the entire agri-food production chain, from farms to consumers' tables.

9. Challenges and Future Directions

Conducting polymers are promising candidates for fabricating state-of-the-art sensor devices and brought much attention over the last two decades. They possess the beneficial characteristics such as reversible signal transduction processes, low operating temperature, tunable sensitivity, and design flexibility. However, when bulk polymer is used to construct a sensor, the response time is relatively long due to slow penetration of the target molecules into CP. Often there is also an accompanying hysteretic effect. A special subclass of CPs, CP nanostructures, not only retain the unique optical, mechanical and chemical properties of CPs but also have the characteristics of nanomaterials (e.g. large surface area, size, and quantum effect), which further increases the merit of CPs in designing and making novel sensors. The response time of sensors constructed from CP nanostructures is expected to be significantly faster due to the porous structure of the nanotubes or nanofibers. For instance, a nanofiber thin film of PANI synthesized by an interfacial polymerization responds much more rapidly than does a conventional film for both acid doping and de-doping (section 7.2.3.). Especially, nanofiber films show no particular thickness dependence on their performance in the range of 0.2–2.5 μm. CP nanostructures with large specific surface area and porous structure are therefore predicted to be excellent sensing materials. The syntheses of various CP nanostructures have been demonstrated recently using chemical [211,212] and electrochemical methods [213]. CP nanospheres, nanorods, nanofibers, nanowires, nanoflowers, nanoplates, nanoribbons, etc. were all reported recently using facile chemistry [211-213]. This is a step further towards the fabrication of CP nanostructure-based sensors with high sensitivity and cycle number, as well as fast response time. CP nanostructures have brought much attention recently as sensing platforms for various applications [214]. Sensing technologies based on CP nanostructures are still in the early stages, requiring significant incubation for materials and device development. Additionally, from the viewpoint of large-scale synthesis, it is important to develop reliable routes to the manufacture of CP nanostructures with controlled morphology.

The next generation of CP materials for chemical sensors should be able to balance their processability with their sensing performance with regard to faster response times, sensitivity, selectivity, lower detection limits, and larger dynamic concentration ranges. In particular, the demand for miniaturization is related to the need for better portability, lower power dissipation, and better device integration. In this respect, CP nanostructures have considerable potential for fabrication of miniaturized multi-sensing arrays by using microcontact printing, surface-directed assembly, site-specific polymerization, inkjet printing, and so forth. Finally, CP nanostructures are highly compatible with a flexible substrate, which opens up the possibility of realizing all-polymer electronics [215-218]. Overall growth will also be supported by technological advances that allow for price reduction, sensor miniaturization and greater precision, all of which expand the use of chemical sensors into new markets or new applications within existing markets.

Many data on analyte/polymer interactions are based on empirical information. Physical and chemical explanations of interaction mechanisms are still relatively speculative. Detailed investigation of these processes will not only provide a better understanding of these processes, but can also indicate a way for further improvement of chemosensitive CPs. Also an implementation of techniques of high-throughput experimentation with subsequent data mining and analysis of empirical correlations between variation of parameters and chemosensitive properties will lead to essential acceleration of the sensor development. Notably, exactly in the case of CP the existing high-throughput technologies allow one to make not only an optimization of chemosensitive materials but also an optimization of complete sensing devices [215].

Increasing interest in and practical use of nanotechnology, especially, in CPs and CP composites have lead the researchers to the rapid development of sensors/biosensors with improved processability and functionality over previously developed sensors/biosensors [216-218].

Conclusion

In light of the sensor systems presented herein, it is clear that CPs represent one of the most useful chemical platforms for the design of chemical sensors due to their unique optical, electrical, mechanical and chemical properties. Principally, CPs can be used for creation of each component of a chemical sensor including a receptor layer, a transducer, a protective coating or even an electronic circuit for data proceeding. The chapter demonstrates the multi-functionality of CP for design and realization of receptors and transducers. In addition to these highly desirable properties, the ease of preparation and modification of CPs have made them a popular choice for many applications. Conducting polymers provide an excellent opportunity for highly selective, specific, stable, economic and handy biosensing devices. Many efforts are, however, required before these interesting conducting biomolecular electronic materials are widely used in biosensor industries. Despite the vast amount of research already conducted on CPs, for sensing applications, the field is still growing and many questions remain to be answered.

Acknowledgment

The authors acknowledge the financial support from Cairo University, the office of the vice president for Graduate Studies and Research.

References

[1] Malhotra, B. D.; Chaubey, A.; Singh, S.P. *Anal. Chim. Acta.* 2006, 578, 59–74.
[2] Xia,L.; Wei, Z.; Wan, M.*; J. Colloid Interface Sci.* 2010, 341, 1–11
[3] Newman, J.D.; Turner, A.P.F.; *Biosens. Bioelectron.* 2005, 20, 2435–2453.

[4] *Diabetes Action Now: an Initiative of the World Health Organization and the International Diabetes Federation*, WHO, Geneva, Switzerland, 2004.
[5] Kissinger, P. T. *Biosens. Bioelectron.* 2005, 20, 2512–2516.
[6] Kawaguchi H. In: Sugimoto T, ed., *Fine particles: synthesis, characterization and mechanism of growth.* New York: Marcel Dekker; 2000. p. 592–608.
[7] Kuramoto, N.; Tomita, A.; *Synth Met* 1997, 88,147–51.
[8] Dallas, P.; Stamopoulos, D.; Boukos, N.; Tzitzios, V.; Niarchos, D.; Petridis D. *Polymer* 2007, 48, 3162–9.
[9] Xing, S.; Zhao, C.; Jing, S.; Wang, Z. *Polymer* 2006, 47, 2305–13.
[10] Guo, Q.; Yi, C.; Zhu, L.; Yang, Q.; Xie, Y. *Polymer* 2005, 46, 3185–9.
[11] Yang, C. H.; Huang, L. R.; Chih, Y.K.; Lin, W.C.; Liu, F.J.; Wang, T.L. *Polymer* 2007,48,3237–47.
[12] Jing ,X.; Wang, Y.Y.; Wu, D.; Qiang, J. P.; *Ultrason Sonochem* 2007,14,75–80.
[13] Guimard, N. K.; Gomez, N.; Schmidt, C. E.*; Prog. Polym. Sci.* 2007, 32, 876–921
[14] Terji A. Skotheim, John R. Reynolds, editors. *Handbook of conducting polymers*, New York: CRC press; 2007, volume 1, chapter 7
[15] Zoski, C.G., ed.; *Handbook of Electrochemistry*; Elsevier B.V, The Netherlands, 2007; chapter 2, pp-33-55
[16] Bai,H.; Shi,G.; *Sensors* 2007, 7, 267-307.
[17] Norrman,k.; Ghanbari-Siahkali,A; Larsen,N.B. *Annu. Rep. Prog. Chem., Sect. C*, 2005, 101, 174–201
[18] Lai,S.L.; Chan,M.Y.; Fung, M.K.; Lee, C.S.; Lee, S.T. *Mater. Sci. Eng.* B104 2003, 104, 26.
[19] Dong,Q.; Zhou,Y.; Pei, J.; Liu,Z.; Li,Y.; Yao,S.; Zhang,J.; Tian,W. *Organic Electronics* 2010,11, 1327–1331.
[20] Heriot, S. Y.; Jones, R. A. L.; *nature materials*, 2005, 4, 782-786.
[21] Javadpour,S.; Gharavi, A.; Feizpour, A.; Khanehzar, A.; Panahi, F. *Sens. Actuators B* 2009,142, 152–158,
[22] Airoudj,A.; Debarnot, D.; Bêche,B.; Poncin-Epaillard, F. *Talanta* 2009, 77, 1590–1596.
[23] Lin, C-Y.; Chen,J-G.; Hu,C-W.; Tunney,J.J.; Ho,K-C. *Sens. Actuators B* 2009,140, 402–406.
[24] Chen,P-Y.; Vittal,R.; Nien,P-C.; Liou,G-S.; Ho,K-C. *Talanta* 2010, 80, 1145–1151.
[25] Zeng, F-W.; Liu, X-X.; Diamond, D.; Lau, K.T. *Sens. Actuators* B 2010, 143. 530–534.
[26] Rahaman, M.N. *Ceramic Processing*; Boca Raton: CRC Press. pp. 2007, 242-244.
[27] McGovern, S.T.; Spinks, G.M.; Wallace, G.G., *Sens. Actuators B* 2005, *107*, 657-665.
[28] Yang,C-H.; Huang, L-R.; Chih, Y-K.; Lin, W-C.; Liu, F-J.; Wang, T-L. *Polymer* 2007, 48, 3237-3247.
[29] Decher, G. *Science* 1997, 277, 1232–1237.
[30] Skotheim,T.A.; Reynolds, J.R. *Handbook of conducting polymers*, New York: CRC press; 2007.
[31] Sun,S.S.; Sariciftci, N.S. *Organic Photovoltaics: Mechanisms, Materials, and Devices*, CRC Press, Boca Raton, FL, 2005, 367–387.
[32] Fanucci,G.E.; Backov,R.; Fu,R.; Talham,D.R. *Langmuir* 2001, 17, 1660.
[33] Yang,Y.; Jiang,Y.; Xu,J.; Yu,J. *Colloids Surf. A: Physicochem. Eng. Aspects* 2007,302, 157–161.
[34] Yang,Y.; Jiang,Y.; Xu,J.; Yu,J. *Thin Solid Films* 2008, 516, 2120–2124.

[35] Manigandan,S.; Jain,A.; Majumder,S.; Ganguly,S.; Kargupta,K. *Sens. Actuators B* 2008, 133, 187–194.
[36] Gundler, Peter ;*Chemical Sensors: An Introduction for Scientists and Engineers*, Springer-Verlag Berlin Heidelberg 2007, chapter 2, 96.
[37] Asatekin,A.; Barr,M.C.; Baxamusa,S.H.; Lau,K.K.S; Tenhaeff,W.; Xu ,K.; Gleason,K.K. *Materials today* 2010, 13, 26-33
[38] Tenhaeff, W. E.; Gleason, K. K., *Adv. Funct. Mater.* 2008, 18, 979.
[39] Kim, J., *et al., Synthetic Metals* 2003, 139, 485.
[40] Choi,J.S.; Cho,K.Y; Yim,J.-H. Eur. Polym. J., 2010, 46, 389-396.
[41] Jang,J.; Bae,J. *Sens. Actuators B: Chemical* 2007, 122, 7-13.
[42] Martin, P.M.; ed.; Handbook of Deposition Technologies for Films and Coatings: *Science, Technology and Applications*, 3rd Ed. Elsevier Inc., Oxford, UK, 2010
[43] Mattox, D.M. *Handbook of Physical Vapor Deposition* (PVD) Processing, Second Edition, William Andrew; Oxford, UK, 2010.
[44] Yamamoto,T.; Wakayama,H.; Fukuda,T.; Kanbara,T. *J. Phys. Chem.* 1992, 96, 8677-8679.
[45] Chrisey,D.B.; Pique,A.; McGill,R.A.; Horwitz,J.S.; B. R. Ringeisen, *Chem. Rev.* 2003, *103,* 553-576.
[46] Anzenbacher,J.; Jursikova,K.; Aldakov,D.; Marquez,M.; Pohl,R. *Tetrahedron* 2004, 60, 11163.
[47] Liu,K.; Wei,W.Z.; Zeng,J.X.; Liu,X.Y.; Gao,Y.P. *Anal. Bioanal.Chem.* 2006, 385, 724.
[48] Hatchett,D.W.; Josowicz,M. *Chem. Rev.* 2008, 108, 746-769.
[49] Rizzo,S.; Sannicolo, F.; Benincori,T.; Schiavon,G.; Zecchin,S.; Zotti,G.; *J. Mater. Chem.* 2007, 14, 1804.
[50] Pernaut,J.M.; Zong,K.; Reynolds,J.R. *Synth.Met.*2002, 130, 1.
[51] Rouis,A.; Mlika,R.; Davenas,J.; Ouada,H.B.; Bonnamour,I.; Jaffrezic,N. *J. Electroanal. Chem.* 2007, 601, 29.
[52] V´azquez,M.; Bobacka,J., Luostarinen,M.; Rissanen,K.; Lewenstam,A.; Ivaska,A. *J. Solid State Electrochem.* 2005,9, 312.
[53] Reece,D.A.; Ralph,S.F.; Wallace,G.G. *J. Membr. Sci.* 2005 ,249, 9.
[54] Bouchta,D.; Izaoumen,N.; Zejli,H.; Kaoutit,M.E., Temsamani,K.R. *Biosens. Bioelectron.* 2005, 20, 2228.
[55] Izaoumen,N.; Bouchta,D.; Zejli,H.; Kaoutit,M.E.;Stalcup,A.M.; Temsamani,K.R. *Talanta* 2005, 66, 111.
[56] Cortina-Puig,M.; Munoz-Berbel,X.; Valle,M.D.; Munoz,F.J.; M.A. Alonso-Lomillo, *Anal. Chim. Acta* 2007, 597, 231.
[57] Pandey,P.C.; Singh,G.; Srivastava,P.K. *Electroanalysis* 2002, 14, 427.
[58] Lange,U.; Roznyatovskaya,N.V.; Mirsky,V.M. *Anal. Chim. Acta* 2008, 614, 1-26.
[59] Ahuja,T.; Mir,I.A.; Kumar,D.; Rajesh *Biomaterials* 2007, 28, 791.
[60] Hatchett,D.W.; Josowicz,M. *Chem. Rev.* 2008, 108, 746-769.
[61] Guimard,N.k.; Gomez,N.; Schmidt,C.E. *Prog. Polym. Sci.* 2007, 32, 876–921.
[62] Suryanarayanan,V.; Wu,C.-H.; Ho,K.-C. *Electroanalysis* 2010, 22, No. 16, 1795 – 1811.
[63] Hillberg, A. L.; Brain, K. R.; Allender, C. *J. Adv. Drug Delivery Rev.* 2005, 57, 1875–1889.
[64] Özcan,L.; Sahin,M.;Sahin,Y. *Sensors* 2008, 8, 5792-5805

[65] https://scifinder.cas.org/scifinder/view/scifinder/scifinderExplore.jsf

[66] Potyrailo,R.A.; Mirsky, V. M.; Editors; *Combinatorial Methods for Chemical and Biological Sensors,* Springer Science + Business Media, LLC, 2009, New York, USA

[67] Hulanicki, A.; Glab, S.; Ingman, F., Chemical sensors definitions and classifications, *Pure Appl. Chem.* 1991, 63, 1247–1250

[68] Gupta,N.; Sharma,s.; Mir, I. A.; Kumar,D. *J. of Scientific & Industrial Research*, 2006, 65, 549-557.

[69] Stephenson, G. R.; Ferris, I. G.; Holland, P. T.; Nordberg, M., Glossary of terms relating to pesticides, *Pure Appl. Chem.* 2006, 78, 2075–2154

[70] Thevenot, D. R.; Toth, K.; Durst, R. A.; Wilson, G. S., Electrochemical biosensors: Recommended definitions and classification, *Pure Appl. Chem.* 1999, 71, 2333–2348

[71] Gründler, P. *"Chemical sensors: An introduction for scientists and engineers",* Springer-Verlag Berlin Heidelberg 2007.

[72] Alaviuhkola,T.; Bobacka,J.; Nissinen,M.; Rissanen,K.; Ivaska,A.; Pursiainen,*J. Chem. A Eur. J.* 2005, 11, 2071.

[73] Bobacka,J. *Electroanalysis* 2006, 18, 7.

[74] Michalska,A.; Walkiewicz,s.; Maksymiuk,K. *Electroanalysis*, 2003, 15, 509.

[75] Venton,B.J.; Wightman,R.M. *Anal. Chem.* 2003, 75, 414A–421A.

[76] Mark,H.B.; Atta,N.; Ma,Y.l.; Petticrew,K.L.; Zimmer,H.; Shi,Y.; Lunsford,S.k.; Rubinson,J.F.; Galal,A. *Bioelectrochem. Bioenerg.*1995, 38, 229.

[77] Kelley,A.; Angolia,B.; Marawi,I. *J Solid State Electrochem* 2006,10, 397–404.

[78] Galal,A. *J Solid State Electrochem* 1998, 2, 7-15.

[79] Bouchta,D.; Izaoumen,N.; Zejli,H.; El Kaoutit,M.; Temsamani,K.R. *Biosens. Bioelectron.* 2005, 20 , 2228–2235.

[80] Huong,V.T.; Shimanouchi,T.; Quan,D.P.; Umakoshi,H.;Viet,P.H.; Kuboi,R. *J Appl Electrochem* 2009, 39, 2035–2042.

[81] Wang,H.S.; Li,T.-H.; Jia,W.-L.; Xu,H.-Y. *Biosens.nBioelectron.* 2006, 22, 664–669.

[82] Atta,N.F.; El-Kady,M.F. *Sens.Actuators B* 2010, 145, 299–310.

[83] Kumar,S.S.; Mathiyarasu,J.; Phani,K.L.N.;Yegnaraman,V. *J Solid State Electrochem* 2006, 10, 905–913.

[84] Vasantha,V.S.; Chen,S.-M. *Electrochim. Acta* 2005, 51, 347–355.

[85] Mathiyarasu,J.; Senthilkumar,S.; Phani,K.L.N.;Yegnaraman,V. *Mater. Lett.* 2008, 62, 571–573.

[86] Atta,N.F.; Galal,A.; Ahmed,R.A. *Bioelectrochem.* 2011, 80, 132-141.

[87] Atta,N.F.; El-Kady,M.F.; Galal,A. *Anal. Biochem.* 2010, 400, 78–88.

[88] Wen,J.; Zhou,L.; Jin,L.; Cao,X.; Ye,B.-C. *J. Chromatogr. B,* 2009, 877, 1793–1798.

[89] Min,K.; Yoo,Y.J. *Talanta* 2009, 80, 1007–1011.

[90] Yan,W.; Feng,X.; Chen,X.; Li,X.; Zhu,J.-*J. Bioelectrochem.* 2008, 72, 21–27.

[91] Zhang,Y.; Lin,L.; Feng,Z.; Zhou,J.; Lin,Z.*; Electrochim. Acta* 2009, 55, 265–270.

[92] Gao,Y.; Shan,D.; Cao,F.; Gong,J.; Li,X.; Ma,H.-Y.; Su,Z.-M.; Qu,L.-Y. *J. Phys. Chem. C* 2009, 113, 15175–15181.

[93] Ali,S.R.; Ma,Y.; Parajuli,R.R.; Balogun,Y.; Lai,W.Y.-C.; He,H. *Anal. Chem.* 2007, 79, 2583-2587.

[94] Castrellon-Uribe,J.; Nicho,M.E.; Reyes-Merino,G. *Sens. Actuators B* 2009,141,40–44.

[95] Elizalde-Torres,J.; Hu,H.; Garcıa-Valenzuela,A. *Sens. Actuators B* 2004,98, 218–226.

[96] Kang,X.; Jin,Y.; Cheng,G.; Dong,S. *Langmuir* 2002, 18, 1713-1718.

[97] Samoylov,A.V.; Mirsky,V.M.; Hao,Q.; Swart,C.; Shirshov,Y.M.; Wolfbeis,O.S. *Sens. Actuators* B 2005,106, 369.

[98] Zhou,Q.; Swager,T.M. *J. Am. Chem. Soc.*, 1995, 117, 12593–12602.

[99] E.L. Cabarcos, S.A. Carter, *Macromolecules* 38 (2005) 4409–4415.

[100] Zhou,G.; Qian,G.; Ma,L.; Cheng,Y.; Xie,Z.;Wang,L.; Jing,X.; Wang,F. *Macromolecules* 2005, 38, 5416-5424.

[101] Zhao,X.; Schanze,K.S. *Chem. Commun.*, 2010, 46, 6075–6077.

[102] [Xu,Q.; An,L.;Yu,M.;Wang,S. Macromol. *Rapid Commun.* 2008, 29, 390–395.

[103] Maiti,J.; Pokhrel,B.; Boruah,R.; Dolui,S.K. *Sens. Actuators B* 2009,141 447–451.

[104] Basabe-Desmonts,L.; Reinhoudt, D.N.; Crego-Calama,M. *Chem. Soc. Rev.,* 2007, 36, 993–1017.

[105] DH,R.; JE,M. *Anal. Chem.*, 2010, 82, 2150-2154.

[106] Sutar,D.S.; Padma,N.; Aswal,D.K.; Deshpande,S.K.; Gupta,S.k.; Yakhmi,J.V. *Sens. Actuators B* 2007,128, 286–292.

[107] Josowicz, M.; Janata, *J. Anal. Chem.*, 1986, 58, 514–557.

[108] JANATA,J.; JOSOWICZ,M. *nature materials* 2003,2, 19-24.

[109] Robinson, K. L.; Lawrence, N. S. *Anal. Chem.* 2006, 78, 2450-2455.

[110] Tsai,Y.-T.; Wen,T.-C.; Gopalan,A. *Sens. Actuators B* 2003,96 646–657.

[111] Ulrich Lange, Nataliya V. Roznyatovskaya1, Vladimir M. Mirsky, *Anal. Chim. Acta* 614 (2008) 1–26

[112] Prissanaroon-Ouajai,W.; Pigramb,P.J.; Jones,R.; Sirivat,A. *Sens. Actuators B* 2008,135, 366–374.

[113] Rittersma,Z.M. *Sens. Actuators A* 2002,96, 196-210.

[114] Zeng,F.W.; Liu,X.-X.; Diamond,D.; Lau,K.T. *Sens. Actuators B* 2010,143, 530–534.

[115] Sun,A.; Li,Z.; Wei,T.; Li,Y.; Cui,P. *Sens. Actuators B* 2009,142, 197–203.

[116] Li,L.; Vilela,F.; Forgie,J.; Skabara,P.J.; Uttamchandani,D. *Micro & Nano Letters*, 2009, 4(2), 84–87.

[117] Matsuguchi,M.; Yamanaka,T.; Yoshida,M.; Kojima,S.;Okumura,S. *J. Electrochem. Soc.* 2009, 156 (10) J299-J302 .

[118] Hangarter,C.M.; Bangar,M.; Mulchandani,A.; Myung,N.V. *J. Mater. Chem.,* 2010, 20, 3131–3140.

[119] Wang ,S.; Ang,H.M.; Tade,M.O. *Environmental International*,2007,33, 694-705.

[120] Huang,J.;Virji,S.; Weiller,B.H.; Kaner,R.B. *Chem. Eur. J.* 2004, 10, 1314-1319.

[121] Virji,S.; Huang,J.; Kaner,R.B.; Weiller,B.H.*; Lett.,N.* 2004, Vol. 4, No. 3, 491-496.

[122] Si,P.;Mortensen,J.; Komolov,A.; Denborg,J.; Møller,P.*J. Anal. Chim. Acta* 2007,597, 223–230.

[123] Ayad,M.M; Torad,N.L. *Sensors and Actuators B* 2010,147, 481–487.

[124] Lu,H.-H.; Rao,Y.K.; Wu,T.-Z.;Tzeng,Y.-M. *Sens. Actuators B* 2009,137, 741–746.

[125] Sadek,A.Z.; Wlodarski,W.; Shin,K.; Kaner,R.B.; Kalantar-zadeh,K. *Nanotechnology* 2006,17,4488–4492.

[126] Al-Mashat,L.; Kaner,R.B.;Tran,H.D.; Kalantar-zadeh,K.; Wlodarski,W. *Procedia Chemistry* 2009,1, 220–223.

[127] Groves, W.A.; Zellers, E.T.*The Annals of Occupational Hygiene* 2006 Vol. 45 No. 8, 609-23.

[128] Do,J.-S.; Chang,W.-B. *Sens. Actuators B* 2004,101, 97–106.

[129] Chartuprayoon,N.; Hangarter,C.M.; Rheem,Y.; Jung,H.; Myung,N.V. *J. Phys. Chem. C* 2010, *114,* 11103–11108.

[130] Gong,J.; Li,Y.; Hu,Z.; Zhou,Z.; Deng,Y. *J. Phys. Chem.* C 2010, 114, 9970–9974.

[131] Yin,K.; Zhu,Z. *Synth. Met.* 2010,160, 1115–1118.

[132] Aguilar, A.D.; Forzani, E.S.; Nagahara, L.A.; Amlani, I.; Tsui, R.; Tao, N.J.; *IEEE Sensors*, 2008,8,269 – 273.

[133] Jeong,J.W.; Lee,Y.D.; Kim,Y.M.; Park,Y.W.; Choi,J.H.; Park,T.H.; Soo,C.D.; Won,S.M.; Han,I.K.; Ju,B.K. *Sens. Actuators B* 2010,146, 40–45.

[134] LI,D.; HUANG,J.; KANER,R.B. *Accounts of Chemical Research*, 2009, 42, 135-145.

[135] Sadek,A.Z.;Wlodarski,W.; Kalantar-Zadeh,K.; Baker,C.; Kaner,R.B. *Sens. Actuators A*, 2007,139, 53-57.

[136] Sadek,A.Z.; Baker,C.O.; Powell,D.A.; Wlodarski,W.; Kaner,R.B.; Kalantar-zadeh,K. *IEEE Sensors Journal*, 2007,7, 213-218.

[137] Al-Mashat,L.; Tran,H.D.;Wlodarski,W.; Kaner.R.B.; Kalantar-zadeh,K. *Sens. Actuators B,* 2008,134, 826-831.

[138] Paul,S.; Chavan,N.N.; Radhakrishnan,S. *Synth. Met.* 2009,159, 415-418.

[139] Dixit,V.; Misra, S.C.K.; Sharma,B.S. *Sens. Actuators B*,2005,104, 90-93.

[140] Lin,C.-Y.; Chen,J.-G.; Hu,C.-W.; Tunney,J.J.; Ho,K.-C. *Sens. Actuators B,* 2009,140, 402-406.

[141] Liu,B.; Xie,G.; Du,X.; Li,X.; Sun,P. *Proc. SPIE* 2009, 13, 750813.

[142] Xie,G.; Sun,P.; Yan,X.; Du,X.; Jiang,Y. *Sens. Actuators B* 2010,145, 373–377.

[143] Moreira,F.; Sarry,F.; Nicolas-Debarnot,D.; Elmazria,O.; Poncin-Epaillard,F. *Eur. Phys. J. Appl. Phys. 2009, 47, 12702.*

[144] Faridbod,F.; Ganjali,M.R.; Dinarvand,R.; Norouzi,P. *Sensors* 2008, 8, 2331-2412.

[145] Ge, C. H.; Armstrong, N. R.; Saavedra, S. S. *Anal. Chem.* 2007, 79, 1401-1410.

[146] Pawlowski, P.; Michalska, A.; Maksymiuk, K. *Electroanalysis* 2006, 18, 1339-1346.

[147] Pandey, P. C.; Prakash, R.; Singh, G.; Tiwari, I.; Tripathi, V. S. *J. Appl. Polym. Sci.* 2000, 75, 1749-1759.

[148] Paczosa-Bator, B.; Peltonen, J.; Bobacka, J.; Lewenstam, A. *Anal.Chim. Acta* 2006, 555, 118-127.

[149] Michalska, A. *Electroanalysis* 2005, 17, 400-407.

[150] Kisiel, A.; Michalska, A.; Maksymiuk, K. *Bioelectrochem.* 2007, 71, 75-80.

[151] Vazquez, M.; Bobacka, J.; Luostarinen, M.; Rissanen, K.; Lewenstam, A.; Ivaska, A., *J. Solid State Electrochem.* 2005, 9, 312-319.

[152] Pandey, P. C.; Singh, G.; Srivastava, P. K., *Electroanalysis* 2002, 14, 427-432.

[153] Song, F. Y.; Ha, J. H.; Park, B.; Kwak, T. H.; Kim, I. T.; Nam, H.; Cha, G. S., *Talanta* 2002, 57, 263-270.

[154] Sjoberg-Eerola, P.; Bobacka, J.; Sokalski, T.; Mieczkowski, J.; Ivaska, A.; Lewenstam, A., *Electroanalysis* 2004, 16, 379-385.

[155] Shishkanova, T. V.; Sapurina, I.; Stejskal, J.; Kral, V.; Volf, R., *Anal.Chim. Acta* 2005, 553, 160-168.

[156] Yin,X.; Ding,J.; Zhang,S.; Kong,J. *Biosens. Bioelectron.* 2006, 21,2184–2187.

[157] Kocherginsky,N.M.; Wang,Z. *J. Electroanal. Chem* 2007,611, 162–168.

[158] Vinokurov,I.A.; Kankare,J. *J.of Electroanalytical Chemistry* 2003,543, 101-105.

[159] Shoji,E.; Freund,M.S. *J. Am. Chem. Soc.* 2002, 124, 12486.

[160] Alvarez-Romero,G.A.; Morales-Perez,A.; Rojas-Hernandez,A.; Palomar-Pardave,M.; Ramirez-Silva,M.T. *Electroanalysis* 2004, 16, 1236-1243.
[161] Gerard,M.; Chaubey,A.; Malhotra,B.D. *Biosens. Bioelectron.* 2002,17,345–359.
[162] Singh,R.P.; Choi,J.W. *Sensors & Transducers J.* 2009, 104, 1-18
[163] Arshak,.;Velusamy,V.;Korostynska,O.; Oliwa-Stasiak,K.; Adley,C. *IEEE SENSORS J.*2009, 9(12),1942-1951.
[164] Gerard,M.; Chaubey,A.; Malhotra,B.D. *Biosens. Bioelectron.* 2002,17, 345–359.
[165] Yoo,E.-H.; Lee,S.-Y. *Sensors* 2010, 10, 4558-4576.
[166] Wang,*J. Chem. Rev*. 2008, 108, 814-825.
[167] Clark, L.C.; Lyons, C. *Ann. N. Y. Acad. Sci.* 2006, 102, 29–45.
[168] Updike, S.J.; Hicks, G.P. *Nature* 1967, 214, 986-988.
[169] Wang, J., *Chem. Rev.* 2008, *108*, 814–825.
[170] Belanger,D.; Nadreau,J.; Fortier,G. *J. Electroanal. Chem.* 1989,274, 143.
[171]] Ma,M.; Qu,L.; Shi,G. *J. Appl. Polym. Sci. Sep.* 2005, 98(6), 2550–2554.
[172] Mala Ekanayake, E. M. I.; Preethichandra, D. M. G.; Kaneto,*K. IEEE TRANSACTIONS ON INSTRUMENTATION AND MEASUREMENT AUGUST 2008, 57, 8, 1621-1626.*
[173] Parthasarathy,R.V.; Martin,C.R.; *Nature* 1994,369, 298–301.
[174] Baba,A.;Taranekar,P.; Ponnapati,R.R.; Knoll,W.; Advincula,R.C. *ACS Appl. Mater. Interf.*, July 2010, DOI: 10.1021/am100373v
[175] Ozcan,L.; Sahin,Y.; Turk,H. *Biosens. Bioelectron.* 2008,24, 512–517.
[176] Yu, Jian-Guo; Li, Jian-Ping *, Fenxi Huaxue* 2008, 36(9), 1201-1206.
[177] Dhawan,G.; Sumana,G.; Malhotra,B.D. *Biochem. Engin. J.* 2009,44, 42–52.
[178] Jha,S.K.;Kanungo,M.; Nath,A.; D'Souza,S.F. *Biosens. Bioelectron.* 2009,14, 2637.
[179] Castillo-Ortega,M.M; Castillo-Castro,T.D.; Ibarra-Bracamontes,V.J.; Nuño-Donlucas, S.M.; Puig,J.E.; Herrera-Franco,P.J. *Sens. Actuators B* 2007,125, 538.
[180] Adelju,S.B.; Shaw,S.J.; Wallace, G.G. Anal. Chim. Acta,1996,323,107.
[181] Rajesh, Bisht,V.; Takashima,W.; Kaneto,K. *Biomaterials* 2005,26, 3683.
[182] Ahuja,T.; Mir,I.A.; Kumar,D.; Rajesh, *Sens. Actuators B* 2008,134, 140–145.
[183] Korneyeva,L.Ch.; Borisova,A.V.; Yashina,Y.I.; Karyakina,E.E.; Voronin,O.G.; Cosnier,S.; Karyakin,*A.A. Moscow University Chemistry Bulletin, 2010, Vol. 65, No. 1, 49–55.*
[184] Rahman,M.M.; Shiddiky,M.J.A.; Rahman,Md.A.; Shim,Y.-B. *Anal. Biochem.* 2009,384 159–165.
[185] Haccoun,J.; Piro,B.; Noel,V.; Pham,M.C. *Bioelectrochemistry* 2006,68, 218 – 226.
[186] Arya,S.K.; Datta,M.; Malhotra,B.D. *Biosens. Bioelectr.* 2008,23, 1083–1100.
[187] [Dhand C.; Singh SP.; Arya SK.; Datta M.; Malhotra BD. *Anal Chim Acta* 2007,602,244.
[188] Singh, S.; Chaubey, A.; Malhotra, B.D. b. *J. Appl. Polym. Sci.* 2004, 91, 3769–3773.
[189] Singh, S.; Solanki, P.R.; Pandey, M.K.; Malhotra, B.D. *Anal.Chim. Acta* 2006a, 568, 126–132.
[190] Matharu,Z.; Sumana,G.; Arya,S.K.; Singh,S.P.; Gupta,V.; Malhotra,B.D. *Langmuir* 2007, 23, 13188-13192.
[191] Cruz,H.J.; Rosa,C.C.; Oliva,A.G. *Parasitol Res 2002*,88: S4–S7.
[192] Kim,D.M.; Rahman,Md.A.; Do,M.H.; Ban,C.; Shim,Y.-B. *Biosens. Bioelectron.* 2010,25, 1781–1788.

[193] Ramanavicius,A.; Kurilcik,N.; Jursenas,S.; Finkelsteinas,A.; Ramanaviciene,A. *Biosens. Bioelectron.* 23 2007,23, 499–505.

[194] Hafaid,I.; Chebil,S.;Korri-Youssoufi,H.; Bessueille,F.; Errachid,A.; Sassi,Z.; Ali,Z.; Abdelghani,A.; Jaffrezic-Renault,N. *Sens. Actuators B*: 2010,144, 323-331.

[195] Peng,H.; Zhang,L.; Soeller,C.; Travas-Sejdic,*J. Biomaterials* 2009,30, 2132–2148.

[196] Krishnamoorthy, K.; Gokhale, R.S.; Contractor, A.Q.; Kumar, A. *Chem Commun* 2004, 7, 820–1.

[197] Wang,J. *Chem Eur J* 1999,5, 1681–1685.

[198] Lien,T.T.N.; Lam,T.D.; An,V.T.H.; Hoang,T.V.; Quang,D.T.; Khieu,D.Q.; Tsukahara,T.; Lee,Y.H.; Kim,J.S. *Talanta* 2010,80, 1164–1169.

[199] Andreescu.S.; Sadik,O.A. *Pure Appl. Chem.* 2004, Vol. 76, No. 4, 861–878.

[200] Li,W.; Hoa,N.D.; Cho,Y.; Kim,D.; Kim,J.-S. *Sens. Actuators B* 2009,143, 132–138.

[201] Virji,S.; Huang,J.; Kaner,R.B.; Weiller,B.H. Polyaniline nanofiber gas sensors: examination of response mechanisms, *Nano Lett.* 2004,4, 491–496.

[202] Bender,S.; Sadik,O.A. *Environ. Sci. Technol.* 1998, 32, 788–797.

[203] Wang,Y.; Li,Y.; Zhang,J.; Xu,S.; Yang,S. Sun,C. *Anal. Chim. Acta* 2009,646, 78–84.

[204] SINGH,P.R.; CONTRACTOR,A.Q. Intern. *J. Environ. Anal. Chem.* 2005,85, 831–835.

[205] Michalska,A.; Wojciechowski,M.; Jędral,W.; Bulska,E.; Maksymiuk,K. *J .Solid State Electrochem* 2009,13,99–106.

[206] Chowdhury,A.-N.; Ferdousi,S.; Islam,Md.M.; Okajima,T.; Ohsaka,T. *J. Appl. Polym. Sci.* 2007, 104, 1306–1311.

[207] *http://www.pnl.gov/news/release.aspx?id=297*

[208] Tsizh,B.R.; Chokhan,M.I.; Aksimentyeva,O.I.; Konopelnyk,O.I.; Poliovyi,D.O. *Mol. Cryst. Liq. Cryst.* 2008, 497, 254–260.

[209] Maynor,M.S.; Nelson,T.L.; O'Sullivan,C.; Lavigne,J.*J. Org. Lett.* 2007, 9, No. 17, 3217-3220.

[210] Neethirajan ,S.; Freund,M.S.; Jayas,D.S.; Shafai,C.; Thomson,D.J.; White,N.D.G. *Biosystems Engineering* 2010, 106, 395-404.

[211] Stejskal,J.; Sapurina,I.; Trchova,M. *Prog. Polym. Sci.* 2010, doi: 10.1016 /j.prog polym sci.2010.07.006.

[212] Wang,Y.; Tran,D.G.H.; Liao,L.; Duan.X.; Kaner,R.B. *J. Am. Chem. Soc.* 2010, 132 (30), 10365-10373.

[213] Zhang,H.;Wang,J.;Wang,Z.;Zhang,F.;Wang,S. *Synth. Met.* 2009, 159, 277-281.

[214] Xia,L.;Wei,Z.;Wan,M. *J. Colloid Interf. Sci.* 2010, 341, 1-11.

[215] Lange, U.; Roznyatovskaya, N.V.; Mirsky, V.M. *Anal. Chim. Acta* 2008, 614,1-26.

[216] Hatchett, D.W.; Josowicz, M. *Chem.Rev.*2008, 108, 746-769.

[217] Yoon, H.; Jang, J. Adv.Funct.Mater.2009, 19, 1567-1576.

[218] Rajesh, Ahuja, T.; Kumar, D. *Sens Actuators B* 2009, 136, 275-286.

INDEX

A

B

C

D

E

F

G

H

M

N

O

P

T

U

V

W

X

Y

Z